Advanced Quantum Mechanics

An accessible introduction to advanced quantum theory, this graduate-level textbook focuses on its practical applications rather than on mathematical technicalities. It treats real-life examples, from topics ranging from quantum transport to nanotechnology, to equip students with a toolbox of theoretical techniques.

Beginning with second quantization, the authors illustrate its use with different condensed matter physics examples. They then explain how to quantize classical fields, with a focus on the electromagnetic field, taking students from Maxwell's equations to photons, coherent states, and absorption and emission of photons. Following this is a unique master-level presentation on dissipative quantum mechanics, before the textbook concludes with a short introduction to relativistic quantum mechanics, covering the Dirac equation and a relativistic second quantization formalism.

The textbook includes 70 end-of-chapter problems. Solutions to some problems are given at the end of the chapter, and full solutions to all problems are available for instructors at www.cambridge.org/9780521761505.

Yuli V. Nazarov is a Professor in the Quantum Nanoscience Department, the Kavli Institute of Nanoscience, Delft University of Technology. He has worked in quantum transport since the emergence of the field in the late 1980s.

Jeroen Danon is a Researcher at the Dahlem Center for Complex Quantum Systems, Free University of Berlin. He works in the fields of quantum transport and mesoscopic physics.

Advanced Quantum Mechanics

A practical guide

YULI V. NAZAROV

Delft University of Technology

JEROEN DANON

Free University of Berlin

CAMBRIDGE
UNIVERSITY PRESS

University Printing House, Cambridge CB2 8BS, United Kingdom

One Liberty Plaza, 20th Floor, New York, NY 10006, USA

477 Williamstown Road, Port Melbourne, VIC 3207, Australia

314-321, 3rd Floor, Plot 3, Splendor Forum, Jasola District Centre, New Delhi - 110025, India

79 Anson Road, #06-04/06, Singapore 079906

Cambridge University Press is part of the University of Cambridge.

It furthers the University's mission by disseminating knowledge in the pursuit of
education, learning and research at the highest international levels of excellence.

www.cambridge.org
Information on this title: www.cambridge.org/9780521761505

First published 2013

A catalogue record for this publication is available from the British Library

Library of Congress Cataloging in Publication data
Nazarov, Yuli V.
Advanced quantum mechanics : A practical guide / YULI V. NAZAROV AND JEROEN DANON.
pages   cm
ISBN 978-0-521-76150-5 (hardback)
1. Quantum theory – Textbooks.   I. Danon, Jeroen, 1977–   II. Title.
QC174.12.N39   2013
530.12–dc23
2012031315

ISBN 978-0-521-76150-5 Hardback

Additional resources for this publication at www.cambridge.org/9780521761505

Contents

Figure Credits

Courses on advanced quantum mechanics have a long tradition. The tradition is in fact so long that the word "advanced" in this context does not usually mean "new" or "up-to-date." The basic concepts of quantum mechanics were developed in the twenties of the last century, initially to explain experiments in atomic physics. This was then followed by a fast and great *advance* in the thirties and forties, when a quantum theory for large numbers of identical particles was developed. This advance ultimately led to the modern concepts of elementary particles and quantum fields that concern the underlying structure of our Universe. At a less fundamental and more practical level, it has also laid the basis for our present understanding of solid state and condensed matter physics and, at a later stage, for artificially made quantum systems. The basics of this leap forward of quantum theory are what is usually covered by a course on advanced quantum mechanics.

Most courses and textbooks are designed for a fundamentally oriented education: building on basic quantum theory, they provide an introduction for students who wish to learn the advanced quantum theory of elementary particles and quantum fields. In order to do this in a "right" way, there is usually a strong emphasis on technicalities related to relativity and on the underlying mathematics of the theory. Less frequently, a course serves as a brief introduction to advanced topics in advanced solid state or condensed matter.

Such presentation style does not necessarily reflect the taste and interests of the modern student. The last 20 years brought enormous progress in applying quantum mechanics in a very different context. Nanometer-sized quantum devices of different kinds are being manufactured in research centers around the world, aiming at processing quantum information or making elements of nano-electronic circuits. This development resulted in a fascination of the present generation of students with topics like quantum computing and nanotechnology. Many students would like to put this fascination on more solid grounds, and base their understanding of these topics on scientific fundamentals. These are usually people with a practical attitude, who are not immediately interested in brain-teasing concepts of modern string theory or cosmology. They need fundamental knowledge to work with and to apply to "real-life" quantum mechanical problems arising in an unusual context. This book is mainly aimed at this category of students.

The present book is based on the contents of the course Advanced Quantum Mechanics, a part of the master program of the Applied Physics curriculum of the Delft University of Technology. The DUT is a university for practically inclined people, jokingly called "bike-repairmen" by the students of more traditional universities located in nearby cities. While probably meant to be belittling, the joke does capture the essence of the research in Delft. Indeed, the structure of the Universe is not in the center of the physics curriculum in Delft, where both research and education rather concentrate on down-to-earth topics. The DUT is one of the world-leading centers doing research on quantum devices such as semiconductor quantum dots, superconducting qubits, molecular electronics, and many others. The theoretical part of the curriculum is designed to support this research in the most efficient way: after a solid treatment of the basics, the emphasis is quickly shifted to *apply* the theory to understand the essential properties of quantum devices. This book is written with the same philosophy. It presents the fundamentals of advanced quantum theory at an operational level: we have tried to keep the technical and mathematical basis as simple as possible, and as soon as we have enough theoretical tools at hand we move on and give examples how to use them.

The book starts with an introductory chapter on basic quantum mechanics. Since this book is intended for a course on *advanced* quantum mechanics, we assume that the reader is already familiar with all concepts discussed in this chapter. The reason we included it was to make the book more "self-contained," as well as to make sure that we all understand the basics in the same way when we discuss advanced topics. The following two chapters introduce new material: we extend the basic quantum theory to describe many (identical) particles, instead of just one or two, and we show how this description fits conveniently in the framework of second quantization.

We then have all the tools at our disposal to construct simple models for quantum effects in many-particle systems. In the second part of the book (Chapters 4–6), we provide some examples and show how we can understand magnetism, superconductivity, and superfluidity by straightforward use of the theoretical toolbox presented in the previous chapters.

After focusing exclusively on many-*particle* quantum theory in the first parts of the book, we then move on to include *fields* into our theoretical framework. In Chapters 7 and 8, we explain in very general terms how almost any classical field can be "quantized" and how this procedure naturally leads to a very particle-like treatment of the excitations of the fields. We give many examples, but keep an emphasis on the electromagnetic field because of its fundamental importance. In Chapter 9 we then provide the last "missing piece of the puzzle": we explain how to describe the *interaction* between particles and the electromagnetic field. With this knowledge at hand, we construct simple models to describe several phenomena from the field of quantum optics: we discuss the radiative decay of excited atomic states, as well as Cherenkov radiation and Bremsstrahlung, and we give a simplified picture of how a laser works. This third part is concluded with a short introduction on *coherent states*: a very general concept, but in particular very important in the field of quantum optics.

In the fourth part of the book follows a unique master-level introduction to dissipative quantum mechanics. This field developed relatively recently (in the last three decades), and is usually not discussed in textbooks on quantum mechanics. In practice, however, the concept of dissipation is as important in quantum mechanics as it is in classical mechanics. The idea of a quantum system, e.g. a harmonic oscillator, which is brought into a stationary excited eigenstate and will stay there forever, is in reality too idealized: interactions with a (possibly very complicated) environment can dissipate energy from the system and can ultimately bring it to its ground state. Although the problem seems inconceivably hard at first sight (one needs a quantum description of a huge number of degrees of freedom), we show that it can be reduced to a much simpler form, characterizing the environment in terms of its damping coefficient or dynamical susceptibility. After explaining this procedure for the damped oscillator in Chapter 11 and discussing dissipation and fluctuations, in Chapter 12 we extend the picture to a *qubit* (two-level system) in a dissipative environment. We elucidate the role the environment plays in transitions between the two qubit states, and, based on what we find, we provide a very general scheme to classify all possible types of environment.

In the last part (and chapter) of the book, we give a short introduction to relativistic quantum mechanics. We explain how relativity is a fundamental symmetry of our world, and recognize how this leads to the need for a revised "relativistic Schrödinger equation." We follow the search for this equation, which finally leads us to the Dirac equation. Apart from obeying the relativistic symmetry, the Dirac equation predicted revolutionary new concepts, such as the existence of particles and *anti-particles*. Since the existence of anti-particles has been experimentally confirmed, just a few years after Dirac had put forward his theory, we accept their existence and try to include them into our second quantization framework. We then explain how a description of particles, anti-particles, and the electromagnetic field constitutes the basis of *quantum electrodynamics*. We briefly touch on this topic and show how a naive application of perturbation theory in the interaction between radiation and matter leads to divergences of almost all corrections one tries to calculate. The way to handle these divergences is given by the theory of *renormalization*, of which we discuss the basic idea in the last section of the chapter.

The book thus takes examples and applications from many different fields: we discuss the laser, the Cooper pair box, magnetism, positrons, vortices in superfluids, and many more examples. In this way, the book gives a very broad view on advanced quantum theory. It would be very well suited to serve as the principal required text for a master-level course on advanced quantum mechanics which is not exclusively directed toward elementary particle physics. All material in the book could be covered in one or two semesters, depending on the amount of time available per week. The five parts of the book are also relatively self-contained, and could be used separately.

All chapters contain many "control questions," which are meant to slow the pace of the student and make sure that he or she is actively following the thread of the text. These questions could for instance be discussed in class during the lectures. At the end of each chapter there are four to ten larger exercises, some meant to practice technicalities, others presenting more interesting physical problems. We decided to provide in this book the solutions to one or two exercises per chapter, enabling students to independently try to solve a serious problem and check what they may have done wrong. The rest of the solutions are available online for teachers, and the corresponding exercises could be used as homework for the students.

We hope that many students around the world will enjoy this book. We did our absolute best to make sure that no single typo or missing minus sign made it to the printed version, but this is probably an unrealistic endeavor: we apologize beforehand for surviving errors. If you find one, please be so kind to notify us, this would highly improve the quality of a possible next edition of this book.

Finally, we would like to thank our colleagues in the Kavli Institute of Nanoscience at the Delft University of Technology and in the Dahlem Center for Complex Quantum Systems at the Free University of Berlin. Especially in the last few months, our work on this book often interfered severely with our regular tasks, and we very much appreciate the understanding of everyone around us for this. J.D. would like to thank in particular Piet Brouwer and Dganit Meidan: they both were always willing to free some time for very helpful discussions about the content and style of the material in preparation.

Yuli V. Nazarov
Jeroen Danon

PART I

SECOND QUANTIZATION

Elementary quantum mechanics

We assume that the reader is already acquainted with elementary quantum mechanics. An introductory course in quantum mechanics usually addresses most if not all concepts discussed in this chapter. However, there are many ways to teach and learn these subjects. By including this chapter, we can make sure that we understand the basics in the same way. We advise students to read the first six sections (those on classical mechanics, the Schrödinger equation, the Dirac formulation, and perturbation theory) before going on to the advanced subjects of the next chapters, since these concepts will be needed immediately. While the other sections of this chapter address fundamentals of quantum mechanics as well, they do not have to be read right away and are referred to in the corresponding places of the following chapters. The text of this chapter is meant to be concise, so we do not supply rigorous proofs or lengthy explanations. The basics of quantum mechanics should be mastered at an operational level: please also check Table 1.1 and the exercises at the end of the chapter.

1.1 Classical mechanics

Let us start by considering a single particle of mass m, which is moving in a coordinate-dependent potential $V(\mathbf{r})$. In classical physics, the state of this particle at a given moment of time is fully characterized by two vectors, its coordinate $\mathbf{r}(t)$ and its momentum $\mathbf{p}(t)$. Since classical mechanics is a completely deterministic theory, the state of the particle in the future – its position and momentum – can be unambiguously predicted once the initial state of the particle is known. The time evolution of the state is given by Newton's well-known equations

$$\frac{d\mathbf{p}}{dt} = \mathbf{F} = -\frac{\partial V(\mathbf{r})}{\partial \mathbf{r}} \quad \text{and} \quad \frac{d\mathbf{r}}{dt} = \mathbf{v} = \frac{\mathbf{p}}{m}. \tag{1.1}$$

Here the force $\mathbf{F}$ acting on the particle is given by the derivative of the potential $V(\mathbf{r})$, and momentum and velocity are related by $\mathbf{p} = m\mathbf{v}$.

Classical mechanics can be formulated in a variety of equivalent ways. A commonly used alternative to Newton's laws are Hamilton's equations of motion

$$\frac{d\mathbf{p}}{dt} = -\frac{\partial H}{\partial \mathbf{r}} \quad \text{and} \quad \frac{d\mathbf{r}}{dt} = \frac{\partial H}{\partial \mathbf{p}}, \tag{1.2}$$

where the Hamiltonian function $H(\mathbf{r}, \mathbf{p})$ of a particle is defined as the total of its kinetic and potential energy,

$$H = \frac{p^2}{2m} + V(\mathbf{r}). \tag{1.3}$$

One advantage of this formalism is a clear link to the quantum mechanical description of the particle we present below.

An important property of Hamilton's equations of motion is that the Hamiltonian H itself is a constant of motion, i.e. $dH/dt = 0$ in the course of motion.

Control question. Can you prove that $dH/dt = 0$ directly follows from the equations of motion (1.2)?

This is natural since it represents the total energy of the particle, and energy is conserved. This conservation of energy does not hold for Hamiltonians that explicitly depend on time. For instance, an external time-dependent force $\mathbf{F}_{\text{ext}}$ gives an addition $-\mathbf{F}_{\text{ext}} \cdot \mathbf{r}$ to the Hamiltonian. By changing the force, we can manipulate the particle and change its energy.

1.2 Schrödinger equation

The quantum mechanical description of the same particle does not include any new parameters, except for a universal constant $\hbar$. The dynamics of the particle are still determined by its mass m and the external potential $V(\mathbf{r})$. The difference is now that the state of the particle is no longer characterized by just two vectors $\mathbf{r}(t)$ and $\mathbf{p}(t)$, but rather by a continuous function of coordinate $\psi(\mathbf{r}, t)$ which is called the *wave function* of the particle. The interpretation of this wave function is probabilistic: its modulus square $|\psi(\mathbf{r}, t)|^2$ gives the probability density to find the particle at time t at the point $\mathbf{r}$. For this interpretation to make sense, the wave function must obey the normalization condition

$$\int d\mathbf{r} \, |\psi(\mathbf{r}, t)|^2 = 1, \tag{1.4}$$

i.e. the total probability to find the particle anywhere is 1, or in other words, the particle must be *somewhere* at any moment of time.

Since this is a probabilistic description, we never know exactly where the particle is. If the particle is at some time t_0 in a definite state $\psi(\mathbf{r}, t_0)$, then it is generally still impossible to predict at which point in space we will find the particle if we look for it at another time t. However, despite its intrinsically probabilistic nature, quantum mechanics is a deterministic theory. Starting from the state $\psi(\mathbf{r}, t_0)$, the state $\psi(\mathbf{r}, t)$ at any future time t is completely determined by an evolution equation, the time-dependent *Schrödinger equation*

$$i\hbar \frac{\partial \psi}{\partial t} = \hat{H}\psi = \left\{ -\frac{\hbar^2}{2m} \frac{\partial^2}{\partial \mathbf{r}^2} + V(\mathbf{r}) \right\} \psi. \tag{1.5}$$

Erwin Schrödinger (1887–1961)
Shared the Nobel Prize in 1933 with Paul Dirac for "the discovery of new productive forms of atomic theory."

Before 1925, quantum mechanics was an inconsistent collection of results describing several otherwise unexplained observations. In 1925–1926 two different general theories of quantum mechanics were presented: Heisenberg proposed a matrix mechanics of non-commuting observables, explaining discrete energy levels and quantum jumps, while Schrödinger put forward wave mechanics, attributing a function of coordinates to a particle and a wave-like equation governing its evolution. Schrödinger's theory seemed (and seems) less abstract. Besides, it provided a clear mathematical description of the wave–particle duality, a concept actively advocated by Einstein. Therefore, many people favored his theory over Heisenberg's. As we explain in this chapter, both theories are equivalent. However, for many the Schrödinger equation remains the core of quantum mechanics.

Erwin Schrödinger had a life with several restless periods in it. He was born and educated in Vienna, where he stayed till 1920, only interrupted by taking up his duty at the Italian front in the first world war. In the years 1920–1921 he successively accepted positions in Jena, Stuttgart, Breslau, and finally Zurich, where he remained for several years. The intense concentration that allowed him to develop his quantum theory he achieved during a Christmas stay in 1925 in a sanatorium in Arosa. But in 1927 he moved again, this time to Berlin. When the Nazis came to power in 1933 he decided to leave the Reich, and so he became itinerant again: Oxford, Edinburgh, Graz, Rome, Ghent... In 1940 he finally took a position in Dublin where he stayed till his retirement in 1956.

In this notation, $\hat{H}$ is an *operator*: it acts on the function ψ and it returns another function, the time derivative of ψ. This operator $\hat{H}$ is called the Hamiltonian operator, since it represents the total energy of the particle. This is best seen if we rewrite $\hat{H}$ in terms of the momentum operator $\hat{\mathbf{p}} \equiv -i\hbar\partial_{\mathbf{r}}$ and position operator $\hat{\mathbf{r}} \equiv \mathbf{r}$. We then see that $\hat{H}$ is identical to (1.3) but with the vectors $\mathbf{r}$ and $\mathbf{p}$ replaced by their operator analogues.

Not only momentum and position, but any physical property can be expressed in terms of operators. For an arbitrary observable A, we can find the corresponding quantum mechanical operator $\hat{A}$. The average value of this observable, sometimes called expectation value, in a given quantum state reads

$$\langle A \rangle = \int d\mathbf{r}\, \psi^*(\hat{A}\psi), \tag{1.6}$$

ψ^* being the complex conjugate of ψ.

Control question. Can you write the integral in (1.6) for $\langle x^2 \rangle$? And for $\langle p_x^2 \rangle$?

If a state is known to have a definite value of A, let us say A_0, then the wave function of this state must be an eigenfunction of $\hat{A}$, i.e. $\hat{A}\psi = A_0\psi$. We indeed see from (1.6) that this would result in $\langle A \rangle = A_0$. To prove that not only the expectation value of A is A_0, but also that A_0 is actually the *only* value which can be found for A, we use the definition in (1.6) and compute the fluctuations in the observable A,

$$\langle A^2 \rangle - \langle A \rangle^2 = \int d\mathbf{r}\, \psi^*(\hat{A}\hat{A}\psi) - \left(\int d\mathbf{r}\, \psi^*(\hat{A}\psi) \right)^2. \tag{1.7}$$

If ψ in the above formula is an eigenfunction of the operator $\hat{A}$, the two terms cancel each other and the fluctuations in A become zero. This proves the above statement: any eigenstate of the operator $\hat{A}$ has a well-defined value for the observable A.

As we have already mentioned, the Hamiltonian operator represents the total energy of the particle. Therefore, a state with a definite energy E must obey the eigenvalue equation

$$\hat{H}\psi = E\psi. \tag{1.8}$$

We see that the Schrödinger equation predicts a simple time dependence for such a state,

$$\psi(\mathbf{r},t) = \exp\{-iEt/\hbar\}\psi(\mathbf{r}). \tag{1.9}$$

These states are called stationary states, and the eigenvalue equation (1.8) is called the stationary Schrödinger equation.

Control question. Why are these states called stationary, while they retain time dependence? *Hint.* Consider (1.6).

Let us now consider the simple case of a free particle ("free" means here "subject to no forces") in an infinite space. This corresponds to setting the potential $V(\mathbf{r})$ to zero or to a constant value everywhere. In this case the solutions of the Schrödinger equation take the form of plane waves

$$\psi_{\mathbf{p}}(\mathbf{r},t) = \frac{1}{\sqrt{\mathcal{V}}}\exp\{-iEt/\hbar\}\exp\{i\mathbf{p}\cdot\mathbf{r}/\hbar\}, \tag{1.10}$$

where the energy of the particle is purely kinetic, $E = p^2/2m$. The factor $1/\sqrt{\mathcal{V}}$ results from the normalization condition (1.4), where we assumed that the particle dwells in a large but finite volume $\mathcal{V}$. For a free particle, this is of course artificial. To describe an actually infinite space, we have to take the limit $\mathcal{V} \to \infty$. The assumption of a finite $\mathcal{V}$, however, is extremely constructive and we use it throughout the book.

We see that the plane wave states (1.10) are eigenfunctions of the momentum operator, $\hat{\mathbf{p}}\psi_{\mathbf{p}} = \mathbf{p}\psi_{\mathbf{p}}$, and therefore all carry a definite momentum $\mathbf{p}$. In a finite volume, however, only a discrete subset of all momentum states is allowed. To find this set, we have to take into account the boundary conditions at the edges of the volume. Having the volume $\mathcal{V}$ is artificial anyway, so we can choose these conditions at will. An easy and straightforward choice is to take a block volume with dimensions $L_x \times L_y \times L_z$ and assume *periodic boundary conditions* over the dimensions of the block. This gives the conditions

$\psi_{\mathbf{p}}(x + L_x, y, z, t) = \psi_{\mathbf{p}}(x, y + L_y, z, t) = \psi_{\mathbf{p}}(x, y, z + L_z, t) = \psi_{\mathbf{p}}(x, y, z, t)$, resulting in a set of quantized momentum states

$$\mathbf{p} = 2\pi\hbar \left(\frac{n_x}{L_x}, \frac{n_y}{L_y}, \frac{n_z}{L_z} \right), \tag{1.11}$$

where n_x, n_y, and n_z are integers.

Control question. Can you derive (1.11) from the periodic boundary conditions and (1.10) yourself?

As we see from (1.11), there is a single allowed value of $\mathbf{p}$ per volume $(2\pi\hbar)^3/\mathcal{V}$ in momentum space, and therefore the density of momentum states increases with increasing size of the system, $D(\mathbf{p}) = \mathcal{V}/(2\pi\hbar)^3$. Going toward the limit $\mathcal{V} \to \infty$ makes the spacing between the discrete momentum values smaller and smaller, and finally results in a continuous spectrum of $\mathbf{p}$.

The above plane wave function is one of the simplest solutions of the Schrödinger equation and describes a free particle spread over a volume $\mathcal{V}$. In reality, however, wave functions are usually more complex, and they also can have many components. For example, if the plane wave in (1.10) describes a single electron, we have to take the *spin* degree of freedom of the electron into account (see Section 1.7). Since an electron can be in a "spin up" or "spin down" state (or in any superposition of the two), we generally have to use a two-component wave function

$$\psi_{\mathbf{p}}(\mathbf{r}, t) = \left(\begin{array}{c} \psi_{\mathbf{p},\uparrow}(\mathbf{r}, t) \\ \psi_{\mathbf{p},\downarrow}(\mathbf{r}, t) \end{array} \right), \tag{1.12}$$

where the moduli squared of the two components give the relative probabilities to find the electron in the spin up or spin down state as a function of position and time.

1.3 Dirac formulation

In the early days of quantum mechanics, the wave function had been thought of as an actual function of space and time coordinates. In this form, it looks very similar to a *classical field*, i.e. a (multi-component) quantity which is present in every point of coordinate space, such as an electric field $\mathbf{E}(\mathbf{r}, t)$ or a pressure field $p(\mathbf{r}, t)$. However, it appeared to be very restrictive to regard a wave function merely as a function of coordinates, as the Schrödinger formalism implies. In his Ph.D. thesis, Paul Dirac proposed to treat wave functions rather as elements of a multi-dimensional linear vector space, a Hilbert space. Dirac's formulation of quantum mechanics enabled a reconciliation of competing approaches to quantum problems and revolutionized the field.

In Dirac's approach, every wave function is represented by a vector, which can be put as a "ket" $|\psi\rangle$ or "bra" $\langle\psi|$. Operators acting on the wave functions, such as the momentum and position operator, make a vector out of a vector and can therefore be seen as matrices in this vector space. For instance, a Hamiltonian $\hat{H}$ in general produces

$$\hat{H}|\psi\rangle = |\chi\rangle. \tag{1.13}$$

An eigenstate of an observable $\hat{A}$ in this picture can be seen as an eigenvector of the corresponding matrix. For any real physical quantity, this matrix is Hermitian, that is $\hat{A}^\dagger = \hat{A}$. As we know, an eigenstate of the operator $\hat{A}$ has a definite value of A, which now simply is the corresponding eigenvalue of the matrix. By diagonalizing the matrix of an observable one retrieves all possible values which can be found when measuring this observable, and the hermiticity of the matrix guarantees that all these eigenvalues are real.

One of the definitions of a Hilbert space is that the inner product of two vectors in the space, $\langle\psi|\chi\rangle = (\langle\chi|\psi\rangle)^*$, exists. Let us try to make a connection with the Schrödinger approach. In the Hilbert space of functions of three coordinates $\mathbf{r}$, the Dirac notation implies the correspondence $|\psi\rangle \leftrightarrow \psi(\mathbf{r})$ and $\langle\psi| \leftrightarrow \psi^*(\mathbf{r})$, and the inner product of two vectors is defined as

$$\langle\psi|\chi\rangle = \int d\mathbf{r}\, \psi^*(\mathbf{r})\chi(\mathbf{r}). \tag{1.14}$$

We see that the normalization condition (1.4) in this notation reads $|\langle\psi|\psi\rangle|^2 = 1$, and the expectation value of an operator $\hat{A}$ in a state ψ is given by $\langle\psi|\hat{A}|\psi\rangle$.

The dimensionality of the Hilbert space is generally infinite. To give a simple example, let us consider a single spinless particle trapped in a potential well, meaning that $V(\mathbf{r}) \rightarrow \infty$ when $|\mathbf{r}| \rightarrow \infty$. If one solves the Schrödinger equation for this situation, one finds a set of stationary states, or levels, with a discrete energy spectrum

$$\psi_n(\mathbf{r}, t) = \exp\{-iE_n t/\hbar\}\psi_n(\mathbf{r}), \tag{1.15}$$

where n labels the levels. The wave functions $\psi_n(\mathbf{r})$ depend on the potential landscape of the well, and can be quite complicated. This set of eigenstates of the Hamiltonian can be used as a *basis* of the infinite dimensional Hilbert space. This means that, if we denote the basis vectors by $|n\rangle \equiv \psi_n(\mathbf{r}, t)$, we can write any arbitrary vector in this Hilbert space as

$$|\psi\rangle = \sum_n c_n|n\rangle, \tag{1.16}$$

where the sum in principle runs over all integers.

Control question. Can you write the "bra"-version $\langle\psi|$ of (1.16)?

We note that the basis thus introduced possesses a special handy property: it is *orthonormal*, that is, any two basis states satisfy

$$\langle m|n\rangle = \delta_{mn}. \tag{1.17}$$

Conveniently, the normalized eigenfunctions of any Hermitian operator with non-degenerate (all different) eigenvalues form a proper basis. Making use of orthonormality, we can write any linear operator as

$$\hat{A} = \sum_{nm} a_{nm}|n\rangle\langle m|, \tag{1.18}$$

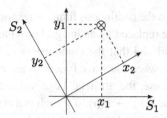

Fig. 1.1 Projections of a vector on the orthogonal axes of a coordinate system. The same point in this two-dimensional space is represented by different vectors in different coordinate systems. The representation (x_1, y_1) in coordinate system S_1 transforms to (x_2, y_2) in coordinate system S_2.

where the complex numbers a_{nm} denote the *matrix elements* of the operator $\hat{A}$, which are defined as $a_{nm} = \langle n|\hat{A}|m\rangle$.

The convenience of the Dirac formulation is the freedom we have in choosing the basis. As is the case with "real" vectors, the basis you choose can be seen as a Cartesian coordinate system, but now in Hilbert space. Let us consider the analogy with usual vectors in two-dimensional space. Any vector then has two components, which are given by the projections of the vector on the orthogonal axes of the coordinate system chosen. When this vector is represented in a different coordinate system, the projections are generally different (see Fig. 1.1). The vector, however, is still the same! Similarly, an arbitrary wave function is defined by a set of its components c_n, which are the projections of the wave function on a certain set of basis vectors. In the case of a discrete set of basis states $|n\rangle$, an arbitrary quantum state $|\psi\rangle$ is written as in (1.16). The projections c_n can then be found from

$$c_n = \langle n|\psi\rangle. \tag{1.19}$$

If we were to choose a different set of basis vectors $|n'\rangle$, the components $c_{n'}$ we would find would be different, but still represent the *same* wave function.

The same picture holds for systems with a continuous spectrum, since, as explained above, a continuous spectrum can be approximated by a discrete spectrum with infinitesimally small spacing. Wave functions in the Schrödinger equation are written in a *coordinate* representation. A wave function $\psi(\mathbf{r})$ can be seen as the projection of the state $|\psi\rangle$ on the continuous set of basis vectors $|\mathbf{r}\rangle$, which are the eigenfunctions of the coordinate operator $\hat{\mathbf{r}}$. The same wave function can of course also be expressed in the basis of plane waves given in (1.10). In the space spanned by these plane waves $|\mathbf{p}\rangle$, we write

$$|\psi\rangle = \sum_{\mathbf{p}} c_{\mathbf{p}}|\mathbf{p}\rangle, \tag{1.20}$$

where the components $c_{\mathbf{p}}$ are given by

$$c_{\mathbf{p}} = \langle \mathbf{p}|\psi\rangle = \int \frac{d\mathbf{r}}{\sqrt{\mathcal{V}}} e^{-\frac{i}{\hbar}\mathbf{p}\cdot\mathbf{r}} \psi(\mathbf{r}). \tag{1.21}$$

These components $c_{\mathbf{p}}$ can be seen as a different representation of the same wave function, in this case the *momentum* representation.

Going over to the continuous limit corresponds to letting $\mathcal{V} \to \infty$. In this limit, the sum in (1.20) can be replaced by an integral provided that the density of states $D(\mathbf{p})$ is included in the integrand and that the components $c_{\mathbf{p}}$ converge to a smooth function $c(\mathbf{p})$ of $\mathbf{p}$. This is the case when the wave functions are concentrated in a finite region of coordinate space. In this case, the artificial large volume $\mathcal{V}$ is redundant and to get rid of it, it is common to rescale $c_{\mathbf{p}} \to c(\mathbf{p})/\sqrt{\mathcal{V}}$ such that $c(\mathbf{p})$ does not depend on $\mathcal{V}$. With this rescaling the relations (1.21) and $\psi(\mathbf{r}) = \langle \mathbf{r}|\psi \rangle$ become the expressions of the standard Fourier transforms, forward and reverse respectively,

$$c(\mathbf{p}) = \int d\mathbf{r}\, e^{-\frac{i}{\hbar}\mathbf{p}\cdot\mathbf{r}} \psi(\mathbf{r}) \quad \text{and} \quad \psi(\mathbf{r}) = \int \frac{d\mathbf{p}}{(2\pi\hbar)^3}\, e^{\frac{i}{\hbar}\mathbf{p}\cdot\mathbf{r}} c(\mathbf{p}). \tag{1.22}$$

Control question. Do you see how the relations (1.22) follow from (1.20) and (1.21), using the rescaling $c_{\mathbf{p}} \to c(\mathbf{p})/\sqrt{\mathcal{V}}$?

So, what are the advantages of the Dirac formulation? First, it brings us from separate quantum particles to the notion of a quantum *system*. A system can be a single particle, or 10^{26} particles interacting with each other, or an electric circuit, or a vacuum – all these examples are considered in this book. In the Dirac formulation they can be described uniformly, since all their quantum states are elements of a Hilbert space. The case of a system consisting of many particles is considered in detail. In the Schrödinger picture, the wave function of such a system becomes a function of all coordinates of all, say, 10^{26}, particles, whereas in the Dirac formulation all allowed quantum states are still simply represented by a vector. Of course, for a many-particle problem more states in the Hilbert space are relevant. However, this does not alter the complexity of the representation: a vector remains a vector irrespective of the number of its components.

Second, the Dirac formulation establishes a "democracy" in quantum mechanics. Given a system, it does not matter which basis is used for a description of a wave function. All bases are equal, and the choice of basis is determined by the personal taste of the descriptor rather than by any characteristic of the system.

Third, the formulation is practically important. It converts any quantum problem for any quantum system to a linear algebra exercise. Since wave functions are now represented by vectors and operators by matrices, finding the eigenstates and eigenvalues of any observable reduces to diagonalizing the corresponding operator matrix.

Let us finish by considering the Dirac formulation for a complex system which is obtained by combining two separate systems. Suppose we have systems A and B which are both described by their own wave functions. We assume that the quantum states of system A are vectors in an M_A-dimensional Hilbert space, spanned by a basis $\{|n_A\rangle\} \equiv \{|1_A\rangle, |2_A\rangle, \ldots, |M_A\rangle\}$, while the states of system B are vectors in an M_B-dimensional space with basis $\{|n_B\rangle\} \equiv \{|1_B\rangle, |2_B\rangle, \ldots, |M_B\rangle\}$. If we combine the two systems and create one large system $A + B$, we find that the states of this combined system can be described in a Hilbert space spanned by the *direct product* of the two separate bases, $\{|n_A\rangle\} \otimes \{|n_B\rangle\} \equiv \{|1_A 1_B\rangle, |1_A 2_B\rangle, |2_A 1_B\rangle, |1_A 3_B\rangle, \ldots\}$, which forms a $(M_A M_B)$-dimensional basis.

1.4 Schrödinger and Heisenberg pictures

Suppose we are interested in the time-dependence of the expectation value of a physical observable A, so we want to know

$$\langle A(t)\rangle = \langle\psi(t)|\hat{A}|\psi(t)\rangle. \tag{1.23}$$

This expression suggests the following approach. First we use the Schrödinger equation to find the *time-dependent* wave function $|\psi(t)\rangle$. With this solution, we can calculate the expectation value of the (time-independent) operator $\hat{A}$ at any moment of time. This approach is commonly called the *Schrödinger picture* since it uses the Schrödinger equation.

We now note that we can write the time-evolution equation for the expectation value as

$$\frac{\partial\langle A\rangle}{\partial t} = \left\langle\frac{\partial\psi}{\partial t}|\hat{A}|\psi\right\rangle + \left\langle\psi|\hat{A}|\frac{\partial\psi}{\partial t}\right\rangle, \tag{1.24}$$

and immediately use the Schrödinger equation to bring it to the form

$$\frac{\partial\langle A\rangle}{\partial t} = \frac{i}{\hbar}\langle\psi|[\hat{H},\hat{A}]|\psi\rangle, \tag{1.25}$$

where the *commutator* $[\hat{A},\hat{B}]$ of two operators $\hat{A}$ and $\hat{B}$ is defined as $[\hat{A},\hat{B}] = \hat{A}\hat{B}-\hat{B}\hat{A}$. This equation in fact hints at another interpretation of time-dependence in quantum mechanics. One can regard the wave function of a system as stationary, and assign the time-dependence to the *operators*. We see that we can reproduce the same time-dependent expectation value $\langle A(t)\rangle$ if we keep the state $|\psi\rangle$ stationary, and let the operator $\hat{A}$ evolve obeying the (Heisenberg) equation of motion

$$\frac{\partial\hat{A}_H}{\partial t} = \frac{i}{\hbar}[\hat{H},\hat{A}_H]. \tag{1.26}$$

Here, the subscript H indicates that we switch to an alternative picture, the *Heisenberg picture*, where

$$\langle A(t)\rangle = \langle\psi|\hat{A}_H(t)|\psi\rangle. \tag{1.27}$$

For many practical problems the Heisenberg picture is the most convenient framework to work in. It allows us to evaluate all dynamics of a system without having to calculate any wave function. It also has a certain physical appeal: it produces quantum mechanical equations of motion for operators which are similar to their classical analogues. For instance, if we use the Hamiltonian in (1.5) to find the Heisenberg equations of motion for the position and momentum operator, we exactly reproduce Newton's equations (1.1) with classical variables replaced by operators. For this, we must postulate the commutation relation between the operators $\hat{\mathbf{p}}$ and $\hat{\mathbf{r}}$,

$$[\hat{p}_\alpha,\hat{r}_\beta] = -i\hbar\delta_{\alpha\beta}, \tag{1.28}$$

α and β labeling the Cartesian components of both vectors. We note that this is consistent with the definition of the operators in the Schrödinger picture, where $\hat{\mathbf{p}} = -i\hbar\partial_{\mathbf{r}}$.

Werner Heisenberg (1901–1976)
Won the Nobel Prize in 1932 for "the creation of quantum mechanics, the application of which has, *inter alia*, led to the discovery of the allotropic forms of hydrogen."

In 1925 Heisenberg was only 23 years old. He was a young Privatdozent in Göttingen and was calculating the energy spectrum of hydrogen, and he wanted to describe the atom in terms of observables only. To escape hay fever, he went to the isolated island of Heligoland, and it was there where he realized that the solution was to make the observables non-commuting, that is, matrices. Within six months he developed, together with more experienced colleagues Max Born and Pascual Jordan, his ideas into the first consistent quantum theory: matrix mechanics.

Heisenberg stayed in Germany during the Nazi period, and was even appointed director of the Kaiser Wilhelm Institute in Berlin. He became one of the principal investigators involved in the "Uranverein," the German nuclear project. This is why at the end of the war Heisenberg was apprehended by the allies and put in an English prison for several months. In 1946, he returned to Göttingen, to become Director of the Max Planck Institute for Physics. When the Institute moved to Munich in 1958, Heisenberg moved along and held his post until his retirement in 1970.

Surprisingly, the Heisenberg picture does not involve stationary states, atomic levels and mysterious wave functions – everything we took for granted from the mere first introductory course of quantum mechanics! All these concepts are not needed here. Nevertheless, solving a problem in the Heisenberg picture produces the same results as in the Schrödinger picture, which *does* involve wave functions, etc. Therefore, quantum mechanics gives a clear and logical example of a physical theory where two individuals can start with very different and seemingly incompatible concepts, and reconcile themselves by predicting identical physical observables. One can draw here an analogy with a modern pluralistic society where persons of opposite views can live together and collaborate successfully (unless they try to convince each other of the validity of their views).

The two opposite pictures – the Heisenberg and Schrödinger pictures – are not isolated, one can interpolate between the two. A resulting "hybrid" picture is called the *interaction picture*. To understand the interpolation, let us arbitrarily split the Hamiltonian into two parts,

$$\hat{H} = \hat{H}_1 + \hat{H}_2. \tag{1.29}$$

We assign to all operators a time-dependence governed by the first part of the Hamiltonian (the subscript I here indicates the interaction picture),

$$\frac{\partial \hat{A}_{\mathrm{I}}}{\partial t} = \frac{i}{\hbar}[\hat{H}_1, \hat{A}_{\mathrm{I}}], \tag{1.30}$$

and the dynamics of the wave functions are determined by the second part,

$$i\hbar \frac{\partial |\psi_I\rangle}{\partial t} = \hat{H}_{2,I}|\psi_I\rangle. \tag{1.31}$$

Note that the Hamiltonian $\hat{H}_{2,I}$ used in (1.31), being an operator, has also acquired the time-dependence governed by (1.30).

Control question. Which choice of $\hat{H}_1$ and $\hat{H}_2$ reproduces the Heisenberg picture? And which reproduces the Schrödinger picture?

This interaction picture is useful for time-dependent perturbation theory outlined in Section 1.6.

1.5 Perturbation theory

The precise analytical solutions to Schrödinger equations are known for very few Hamiltonians. This makes it important to efficiently find approximate solutions, perturbation theory being an indispensable tool for this. Let us consider a "composite" Hamiltonian

$$\hat{H} = \hat{H}_0 + \hat{H}', \tag{1.32}$$

which consists of a Hamiltonian $\hat{H}_0$ with known eigenstates $|n\rangle^{(0)}$ and corresponding energy levels $E_n^{(0)}$, and a perturbation $\hat{H}'$. The perturbation is assumed to be small, we see later the precise meaning of this. This smallness ensures that the energy levels and eigenstates of $\hat{H}$ differ from those of $\hat{H}_0$ only by small corrections, and the goal is thus to find these corrections.

To do so, let us consider an auxiliary Hamiltonian

$$\hat{H} = \hat{H}_0 + \alpha\hat{H}'. \tag{1.33}$$

We expand its eigenstates $|n\rangle$ in the convenient basis of $|n\rangle^{(0)}$,

$$|n\rangle = \sum_m c_{nm}|m\rangle^{(0)}, \tag{1.34}$$

where the coefficients c_{nm} should satisfy the normalization condition

$$\delta_{mn} = \sum_p c_{mp}^* c_{np}. \tag{1.35}$$

Control question. Do you see the equivalence between this normalization condition and (1.17)?

The Schrödinger equation in these notations becomes

$$\left\{ E_n(\alpha) - E_m^{(0)} \right\} c_{nm} = \alpha \sum_p c_{np} M_{mp}, \tag{1.36}$$

where M_{nm} are the matrix elements of the perturbation, $M_{nm} = \langle n|\hat{H}'|m\rangle$.

Control question. Can you derive (1.36) from the Schrödinger equation?

To proceed, we seek for E_n and c_{nm} in the form of a Taylor expansion in α,

$$E_n(\alpha) = E_n^{(0)} + \alpha E_n^{(1)} + \alpha^2 E_n^{(2)} + \ldots,$$
$$c_{nm} = c_{nm}^{(0)} + \alpha c_{nm}^{(1)} + \alpha^2 c_{nm}^{(2)} + \ldots \tag{1.37}$$

The perturbation theory formulated in (1.35) and (1.36) can be solved by subsequent approximations:[1] the corrections of any order N can be expressed in terms of the corrections of lower orders. To make it practical, one restricts to a certain order, and sets α to 1 at the end of the calculation.

The resulting corrections are simple only for the first and second order, and those are widely used. Let us give their explicit form,

$$E_n^{(1)} = M_{nn}, \tag{1.38}$$

$$c_{nm}^{(1)} = \frac{M_{mn}}{E_n^{(0)} - E_m^{(0)}} \quad \text{for } n \neq m, \quad c_{nn}^{(1)} = 0, \tag{1.39}$$

$$E_n^{(2)} = \sum_{p \neq n} \frac{M_{np}M_{pn}}{E_n^{(0)} - E_p^{(0)}}, \tag{1.40}$$

$$c_{nm}^{(2)} = \sum_{p \neq n} \frac{M_{mp}M_{pn}}{(E_n^{(0)} - E_m^{(0)})(E_n^{(0)} - E_p^{(0)})} - \frac{M_{mn}M_{nn}}{(E_n^{(0)} - E_m^{(0)})^2} \quad \text{for } n \neq m,$$

$$c_{nn}^{(2)} = -\frac{1}{2} \sum_{p \neq n} \frac{M_{np}M_{pn}}{(E_n^{(0)} - E_p^{(0)})^2}. \tag{1.41}$$

From these expressions we see that the theory breaks down if there exist *degenerate* eigenstates of $\hat{H}_0$, i.e. $E_n^{(0)} = E_m^{(0)}$ for some $m \neq n$, or in words, there exist multiple eigenstates with the same energy. In this case the above expressions blow up and one has to use a degenerate perturbation theory. This observation also sets a limit on the relative magnitude of $\hat{H}_0$ and $\hat{H}'$. All elements of $\hat{H}'$ must be much smaller than any energy difference $E_n^{(0)} - E_m^{(0)}$ between eigenstates of $\hat{H}_0$.

1.6 Time-dependent perturbation theory

Let us consider again the standard framework of perturbation theory, where the Hamiltonian can be split into a large, uncomplicated, time-independent term, and a small perturbation which we now allow to be time-dependent,

$$\hat{H} = \hat{H}_0 + \hat{H}'(t). \tag{1.42}$$

In this case, we cannot apply the perturbation theory of Section 1.5 since the Hamiltonian $\hat{H}$ has no stationary eigenstates and we cannot write a related stationary eigenvalue problem.

---

[1] Equations (1.35) and (1.36) do not determine c_{nm} unambiguously. The point is that any quantum state is determined up to a phase factor: the state $|\psi'\rangle = e^{i\phi}|\psi\rangle$ is equivalent to the state $|\psi\rangle$ for any phase ϕ. To remove the ambiguity, we require $(\frac{\partial}{\partial \alpha}\langle n|)|n\rangle = 0$.

We have to re-develop the perturbation theory. As a matter of fact, the resulting time-dependent theory appears to be more logical and simpler, facilitating its application for complex systems with many degrees of freedom.

To start, let us note that we could formally solve the Schrödinger equation (1.5) by introducing the *time-evolution operator* $\hat{U}$ defined as

$$|\psi(t)\rangle = \hat{U}(t, t')|\psi(t')\rangle, \tag{1.43}$$

assuming that $t > t'$. That is, $\hat{U}(t, t')$ takes the initial state $|\psi(t')\rangle$ as input and returns the final state $|\psi(t)\rangle$. For a time-independent Hamiltonian, this operator takes the form of an operator exponent,

$$\hat{U}(t, t') = e^{-\frac{i}{\hbar}\hat{H}(t-t')}, \tag{1.44}$$

where the exponent of an operator is defined by the Taylor series of the exponential function

$$e^{\hat{A}} \equiv 1 + \hat{A} + \frac{\hat{A}\hat{A}}{2!} + \frac{\hat{A}\hat{A}\hat{A}}{3!} + \ldots \tag{1.45}$$

What does $\hat{U}(t, t')$ become in the case of a time-dependent $\hat{H}$? It becomes a *time-ordered* exponent,

$$\hat{U}(t, t') = \mathcal{T}\left[\exp\left\{-\frac{i}{\hbar}\int_{t'}^{t} d\tau\, \hat{H}(\tau)\right\}\right], \tag{1.46}$$

$\mathcal{T}$ indicating the time-ordering. To see the meaning of this expression, let us divide the time interval (t, t') into many subintervals of (small) duration Δt, $(t_n, t_n - \Delta t)$, where $t_n = t' + n\Delta t$. Since $\hat{U}(t, t')$ is an evolution operator, it can be written as an operator product of the evolution operators within each subinterval,

$$\hat{U}(t, t') = \hat{U}(t, t - \Delta t) \cdots \hat{U}(t_n, t_n - \Delta t) \cdots \hat{U}(t_1, t'). \tag{1.47}$$

If the subdivision is sufficiently fine, we can disregard the time-dependence of $\hat{H}$ within each subinterval so that we have a long product of corresponding elementary exponents

$$\hat{U}(t, t') = e^{-\frac{i}{\hbar}\hat{H}(t)\Delta t} \cdots e^{-\frac{i}{\hbar}\hat{H}(t_n)\Delta t} \cdots e^{-\frac{i}{\hbar}\hat{H}(t_1)\Delta t}. \tag{1.48}$$

The time-ordered exponent in (1.46) is the limit of this long product for $\Delta t \to 0$. Time-ordering therefore means that the elementary exponents corresponding to later times are placed in a product to the left from those at earlier times.

Control question. Explain the difference between two exponents

$$\mathcal{T}\left[\exp\left\{-\frac{i}{\hbar}\int_{t'}^{t} d\tau\, \hat{H}(\tau)\right\}\right] \quad \text{and} \quad \exp\left\{-\frac{i}{\hbar}\int_{t'}^{t} d\tau\, \hat{H}(\tau)\right\}.$$

Is there a difference if the operators $\hat{H}(t_1)$ and $\hat{H}(t_2)$ commute for all t_1 and t_2?

Since (1.43) presents the formal solution of the Schrödinger equation, and the Heisenberg and Schrödinger pictures are equivalent, the same time-evolution operator helps to solve the Heisenberg equation (1.25) and to determine the evolution of operators in the Heisenberg picture,

$$\hat{A}_H(t) = \hat{U}(t',t)\hat{A}_H(t')\hat{U}(t,t'). \tag{1.49}$$

Note that in the above equation the time indices in the evolution operators appear in a counter-intuitive order (from right to left: evolution from t' to t, then an operator at time t', then evolution from t to t'). This is the consequence of the Heisenberg and Schrödinger pictures being complementary. The "true" time-evolution operator in the Heisenberg picture differs by permutation of the time indices from the operator $U(t',t)$ in the Schrödinger picture. In simple words, where wave functions evolve forward, operators evolve backward. This permutation is actually equivalent to taking the Hermitian conjugate,

$$\hat{U}(t,t') = \hat{U}^\dagger(t',t). \tag{1.50}$$

Indeed, time reversal in the time-ordered exponent (1.46) amounts to switching the sign in front of the $i/\hbar$. Let us also note that the product of two time-evolution operators with permuted indices, $\hat{U}(t',t)\hat{U}(t,t')$, brings the system back to its initial state at time t', and therefore must equal the identity operator, $\hat{U}(t',t)\hat{U}(t,t') = \mathbb{1}$. Combining this with (1.50) proves that *the time-evolution operator is unitary*, $\hat{U}^\dagger(t,t')\hat{U}(t,t') = \mathbb{1}$.

Let us now return to perturbation theory and the Hamiltonian (1.42). The way we split it into two suggests that the interaction picture is a convenient framework to consider the perturbation in. Let us thus work in this picture, and pick $\hat{H}_0$ to govern the time-evolution of the operators. As a result, the perturbation Hamiltonian $\hat{H}'(t)$ acquires an extra time-dependence,[2]

$$\hat{H}'_I(t) = e^{\frac{i}{\hbar}\hat{H}_0 t}\hat{H}'(t)e^{-\frac{i}{\hbar}\hat{H}_0 t}. \tag{1.51}$$

Since the time-dependence of the wave functions is now governed by $\hat{H}'_I(t)$ (see (1.31)), the time-evolution operator for the wave functions in the interaction picture reads

$$\hat{U}_I(t,t') = \mathcal{T}\left[\exp\left\{-\frac{i}{\hbar}\int_{t'}^{t} d\tau\, \hat{H}'_I(\tau)\right\}\right]. \tag{1.52}$$

The time-evolution operator $\hat{U}(t,t')$ in the Schrödinger picture now can be expressed in terms of $\hat{U}_I(t,t')$ as

$$\hat{U}(t,t') = e^{-\frac{i}{\hbar}\hat{H}_0 t}\hat{U}_I(t,t')e^{\frac{i}{\hbar}\hat{H}_0 t'}. \tag{1.53}$$

Control question. Do you see how the relation (1.53) follows from our definition $|\psi_I(t)\rangle = e^{\frac{i}{\hbar}\hat{H}_0 t}|\psi\rangle$?

---

[2] We have made a choice here: at time $t = 0$ the operators in interaction and Schrödinger picture coincide. We could make any other choice of this peculiar moment of time; this would not affect any physical results.

The fact that $\hat{H}'(t)$ is assumed small compared to $\hat{H}_0$ allows us to expand $\hat{U}_I(t,t')$ in orders of $\hat{H}'(t)$. For instance, up to third order this yields

$$\hat{U}_I(t,t') = 1 - \frac{i}{\hbar} \int_{t'}^{t} dt_1 \hat{H}'_I(t_1) - \frac{1}{\hbar^2 2!} \int_{t'}^{t} dt_2 \int_{t'}^{t} dt_1 T\left[\hat{H}'_I(t_2)\hat{H}'_I(t_1)\right]$$
$$+ \frac{i}{\hbar^3 3!} \int_{t'}^{t} dt_3 \int_{t'}^{t} dt_2 \int_{t'}^{t} dt_1 T\left[\hat{H}'_I(t_3)\hat{H}'_I(t_2)\hat{H}'_I(t_1)\right] + \ldots \tag{1.54}$$

We are now dealing with *time-ordered* products of operators $\hat{H}'_I$ corresponding to the time-ordered exponent in (1.52). A time-ordered product is defined as follows: given a sequence of operators $\hat{H}'_I$ at different moments of time, it re-orders the sequence in such a way that terms with an earlier time index are shifted to the right of terms with a later time index. Generally, it requires a permutation of the operators. For example, for two operators

$$T\left[\hat{H}'_I(t_2)\hat{H}'_I(t_1)\right] = \begin{cases} \hat{H}'_I(t_2)\hat{H}'_I(t_1) & \text{if } t_2 > t_1, \\ \hat{H}'_I(t_1)\hat{H}'_I(t_2) & \text{if } t_1 > t_2. \end{cases} \tag{1.55}$$

There exists a simple way to explicitly include this time-ordering in the integrals. Let us illustrate this with the second order term. Using (1.55) we rewrite

$$\hat{U}_I^{(2)}(t,t') = -\frac{1}{\hbar^2 2!} \int_{t'}^{t} dt_2 \int_{t'}^{t} dt_1 \left\{ \hat{H}'_I(t_2)\hat{H}'_I(t_1)\theta(t_2 - t_1) + \hat{H}'_I(t_1)\hat{H}'_I(t_2)\theta(t_1 - t_2) \right\}$$
$$= -\frac{1}{\hbar^2} \int_{t'}^{t} dt_2 \int_{t'}^{t_2} dt_1 \hat{H}'_I(t_2)\hat{H}'_I(t_1), \tag{1.56}$$

where $\theta(t)$ denotes the Heaviside step function,

$$\theta(t) = \begin{cases} 1 & \text{if } t > 0, \\ \frac{1}{2} & \text{if } t = 0, \\ 0 & \text{if } t < 0. \end{cases} \tag{1.57}$$

To arrive at the last equality in (1.56), we exchanged the dummy integral variables $t_1 \leftrightarrow t_2$ in the second term, and combined the two terms. The upper limit of the second integral then effectively becomes t_2, the variable of the first integral. This automatically guarantees that $t_2 > t_1$, which equals the effect of time-ordering. Since we combined two equal terms, the factor 2 in the denominator is canceled. Generally, for the Nth order term, the number of equal terms is $N!$, this always canceling the factor $N!$ in the denominator coming from the expansion of the exponential.

We have now shown how to write a perturbation expansion for the evolution operator $\hat{U}_I$ for an arbitrary perturbation $\hat{H}'(t)$. Evaluating an evolution operator is, however, useless without specifying the initial conditions: the quantum state at some initial moment $t = t_0$. For practical calculations, one would like to take something simple for this initial state, for instance an eigenstate of $\hat{H}_0$. This, however, would deliver a sensible answer only for a rather specific situation: for a perturbation that is switched on *diabatically*. Indeed, for this simple initial state to be relevant, we require that there is no perturbation before the moment t_0, i.e. $\hat{H}'(t) = 0$ for $t < t_0$. Then, at or after the time $t = t_0$ the perturbation rather suddenly jumps to a finite value. This situation can be realized if we have means to change the

Hamiltonian at our will. Typically, however, this is not so, and most perturbations *persist* at all times, thus also before $t = t_0$. In this case, the perturbation inevitably affects the eigenstates at t_0. To choose a reasonable initial condition, we need to know the eigenstates of $\hat{H} + \hat{H}'(t)$, or in other words, to apply the perturbation theory correctly, we need to know the same corrections that we actually hoped to find from this perturbation theory. The latter therefore seems incomplete and void.

The way out is to implement *adiabatic*, that is, very slow switching on of the perturbation $\hat{H}'(t)$. We assume that at $t = -\infty$ the perturbation vanishes and the system is in an eigenstate of $\hat{H}_0$. We also assume that the perturbation evolves further so slowly that the quantum states remain close to the eigenstates of the Hamiltonian, that is, follow the Hamiltonian adiabatically. A handy way to implement adiabatic switching is to include an exponential time-dependence,

$$\hat{H}'(t) \to e^{\eta t}\hat{H}', \tag{1.58}$$

where η is infinitesimally small but positive, and the perturbation is assumed *constant*, $H'(t) = H'$. We see that now the perturbation indeed vanishes at $t = -\infty$ and coincides with the original perturbation at $t = 0$. At sufficiently small η, the system evolves adiabatically, so that if far in the past it was in an eigenstate of $\hat{H}_0$, we will find it always in an eigenstate of the changed operator $\hat{H}_0 + e^{\eta t}\hat{H}'$. As we see in the next section, this requires $\eta \ll \Delta E/\hbar$, with ΔE being the minimal separation of energy levels. In this case, an initial eigenstate of $\hat{H}_0$ at $t = -\infty$ evolves to an eigenstate of $\hat{H}_0 + \hat{H}'$ at $t = 0$. Thus, the modified perturbation (1.58) supplies us the desired corrections to the eigenstates of $\hat{H}_0$ at time $t = 0$. If we use this method to compute the perturbation correction to an eigenstate $|n\rangle^{(0)}$ at $t = -\infty$, we reproduce the results of the time-independent perturbation theory of Section 1.5 for the corresponding state $|n\rangle$.

The time-dependent perturbation theory presented by (1.54) is much simpler in structure and therefore more transparent than the time-independent theory. We see that it is straightforward to write down a term of arbitrarily high order: the expressions do not increase in complexity like the terms of the time-independent theory do. Besides, a term of any order can be computed separately from the other orders. This is why time-dependent perturbation theory is widely applied for systems with infinitely many degrees of freedom and particles, systems of advanced quantum mechanics.

1.6.1 Fermi's golden rule

Quantum mechanics teaches us that a system remains in a given eigenstate forever, while practice shows the opposite: there are *transitions* between different states. For instance, a system in an excited state sooner or later gets to the ground state losing its excess energy. We discuss transitions in detail in later chapters. Here we illustrate the power of time-dependent perturbation theory by deriving Fermi's golden rule, which is an indispensable tool to understand transitions.

Let us assume that the transitions between the eigenstates of an unperturbed Hamiltonian $\hat{H}_0$ are caused by a small perturbation. We use the first-order term of the perturbation theory derived above to calculate transition rates between different eigenstates of $\hat{H}_0$. For the sake of simplicity we concentrate on two states and evaluate the transition rate from the initial state $|i\rangle$ to the final state $|f\rangle$. We assume a time-independent $\hat{H}'$, and implement adiabatic switching as in (1.58). The initial condition we use is that the system is in $|i\rangle$ at time $t = -\infty$. The probability of finding the system in the state $|f\rangle$ at time t can then be expressed in terms of the time-evolution operator $\hat{U}_I(t, -\infty)$ in the interaction picture,[3] $P_f(t) = |\langle f|\hat{U}_I(t, -\infty)|i\rangle|^2$. We need to evaluate this matrix element in the lowest non-vanishing order, which is the first order in $\hat{H}'$. The first order correction to the evolution operator reads

$$\hat{U}_I^{(1)}(t, -\infty) = -\frac{i}{\hbar}\int_{-\infty}^t dt_1 \hat{H}'_I(t_1), \quad \text{with} \quad \hat{H}'_I(t_1) = e^{\frac{i}{\hbar}\hat{H}_0 t_1} e^{\eta t_1} \hat{H}' e^{-\frac{i}{\hbar}\hat{H}_0 t_1}. \tag{1.59}$$

Defining H'_{fi} as the matrix element $\langle f|\hat{H}'|i\rangle$, we can write

$$\langle f|\hat{U}_I^{(1)}(t, -\infty)|i\rangle = -\frac{i}{\hbar}H'_{fi}\int_{-\infty}^t dt_1 e^{\frac{i}{\hbar}(E_f - E_i)t_1 + \eta t_1}, \tag{1.60}$$

where we make use of E_i and E_f, the energy levels of the states $|i\rangle$ and $|f\rangle$ respectively, so that $e^{\frac{i}{\hbar}\hat{H}_0 t}|i,f\rangle = e^{\frac{i}{\hbar}E_{i,f}t}|i,f\rangle$.

Note that this transition is not characterized by the probability $P_f(t)$, but rather by the corresponding *rate*, i.e. the change in the probability per unit of time. Therefore we should evaluate the time derivative of the probability dP_f/dt at $t = 0$, which yields

$$\begin{aligned}\Gamma_{i\to f} = \frac{dP_f}{dt} &= 2\text{Re}\left\{\left(\frac{\partial}{\partial t}\langle f|\hat{U}_I^{(1)}(0, -\infty)|i\rangle\right)\left(\langle f|\hat{U}_I^{(1)}(0, -\infty)|i\rangle\right)^*\right\}\\ &= \frac{2}{\hbar^2}|H'_{fi}|^2\text{Re}\left\{\int_{-\infty}^0 dt_1 e^{-\frac{i}{\hbar}(E_f - E_i)t_1 + \eta t_1}\right\}\\ &= \frac{2}{\hbar}|H'_{fi}|^2\frac{\eta\hbar}{(E_i - E_f)^2 + (\eta\hbar)^2}.\end{aligned} \tag{1.61}$$

We immediately see that the transition rate vanishes in the limit $\eta \to 0$ provided that the difference $E_i - E_f$ remains *finite*. This proves the adiabaticity criterion mentioned above: transition rates disappear as long as $\eta \ll |E_i - E_f|$.

Control question. Do you understand why this disappearing of transition rates guarantees adiabatic evolution of the system?

Note, however, that the rate expression (1.61) can be explicitly evaluated in the limit $\eta \to 0$, using the relation $\lim_{y\to 0^+} y/(x^2+y^2) = \pi\delta(x)$, where $\delta(x)$ is Dirac's delta function. We should remember that the delta function has the strange property that it is equal to zero for all x, except at $x = 0$ where it becomes infinite in such a way that $\int dx\,\delta(x) = 1$.

Control question. Evaluate $\int dx f(x)\delta(x - a)$ for any function $f(x)$.

---

[3] It differs from the evolution operator $\hat{U}$ in the Schrödinger picture only by a phase factor not affecting the probability, see (1.53).

From the way we just wrote it, we can interpret it as a Lorentzian peak centered around $x = 0$ with an infinitely small width but with a fixed area under the peak. Using this delta function, we finally write

$$\Gamma_i = \sum_f \frac{2\pi}{\hbar} |\langle f|\hat{H}'|i\rangle|^2 \delta(E_i - E_f), \tag{1.62}$$

were we sum over all possible final states $|f\rangle$, thereby making Γ_i the total decay rate from the initial state $|i\rangle$. We have reproduced Fermi's famous golden rule.

We see that the rate (1.62) is zero except when $E_f = E_i$, that is, the initial and final state have the same energy. Indeed, since we made sure that all time-dependence in the Hamiltonian is so slow that the system always evolves adiabatically, we expect energy conservation to forbid any transition accompanied by a change in energy. For a discrete energy spectrum, two energies E_i and E_f will never be exactly aligned in practice, and in this case there are no transitions. We do however expect transitions for a continuous spectrum: the sum over final states is to be converted to an integral, and the delta function selects from the continuous distribution of energy levels those providing energy conservation. In Chapter 9 we illustrate this in detail by computing the spontaneous emission rates from excited atoms.

Higher-order terms from the expansion (1.54) can be used to evaluate higher-order transition rates. For instance, the second-order term includes the possibility to make a transition from the initial to the final state *via* another state. This intermediate state is called *virtual* since its energy matches neither the initial nor final energy, this forbidding its true realization. These higher-order rates become particularly important when the perturbation $\hat{H}'$ does not have matrix elements directly between the initial and final states. If the perturbation, however, does have elements between the initial and a virtual state and between this virtual and the final state, transitions can take place, and are second-order with a rate proportional to the second power of $\hat{H}'$. The work-out of this rate by time-dependent perturbation theory (see Exercise 1) proves that the rate is given by Fermi's golden rule with an effective matrix element

$$\langle f|\hat{H}'|i\rangle \rightarrow \sum_v \frac{\langle f|\hat{H}'|v\rangle \langle v|\hat{H}'|i\rangle}{E_i - E_v}. \tag{1.63}$$

Here, the summation is over all possible virtual states. The energy mismatch between the initial and virtual states enters the denominators in this expression.

1.7 Spin and angular momentum

From our introductory courses on quantum mechanics, we know that elementary particles (electrons, neutrons, etc.) all have a special degree of freedom called *spin*. An intuitive picture, which is often given when the concept of spin is introduced, is that the spin of a particle can be regarded as the angular momentum of the particle due to the fact that it is spinning around its own axis. For example, an electron in an atomic orbital could be compared to the earth orbiting the sun: it possesses orbital angular momentum due to its (yearly) revolution around the sun, and, besides that, some angular momentum due to its (daily) revolution around its own axis. Although this picture might help to visualize the concept of spin, in reality it is not accurate. Elementary particles, such as electrons, are structureless point particles, and it is impossible to assign to them a spatial mass distribution revolving around an axis. Let us thus present a more abstract but more adequate way to consider spin.

You might recall from your courses on classical mechanics that it is possible to derive the familiar conservation laws (of momentum, energy, etc.) from the symmetries of space and time. For example, the properties of a system are not expected to depend on where we choose the origin of the coordinate system, in other words, we assume that space is homogeneous. Let us investigate the implications of this statement in more detail. We consider a single free particle, and we assume that its state is given by the wave function $\psi(\mathbf{r})$. A shift of the coordinate system by a vector $\mathbf{a}$ changes the wave function $\psi(\mathbf{r}) \rightarrow \psi'(\mathbf{r}) = \psi(\mathbf{r} + \mathbf{a})$, where the prime indicates the new coordinate system. Since this is nothing but a change of basis, we can look for the "translation operator" $\hat{T}_{\mathbf{a}}$ which shifts the wave function over $\mathbf{a}$, i.e. $\psi'(\mathbf{r}) = \hat{T}_{\mathbf{a}}\psi(\mathbf{r})$. To find $\hat{T}_{\mathbf{a}}$, we expand $\psi(\mathbf{r} + \mathbf{a})$ in a Taylor series,

$$
\begin{aligned}
\psi(\mathbf{r} + \mathbf{a}) &= \psi(\mathbf{r}) + \mathbf{a} \cdot \frac{\partial}{\partial \mathbf{r}} \psi(\mathbf{r}) + \frac{1}{2}\left(\mathbf{a} \cdot \frac{\partial}{\partial \mathbf{r}}\right)^2 \psi(\mathbf{r}) + \dots \\
&= e^{\mathbf{a} \cdot \partial_{\mathbf{r}}} \psi(\mathbf{r}) = e^{-\frac{i}{\hbar}\mathbf{a} \cdot \hat{\mathbf{p}}} \psi(\mathbf{r}),
\end{aligned}
\tag{1.64}
$$

and see that the translation operator is given by $\hat{T}_{\mathbf{a}} = e^{-\frac{i}{\hbar}\mathbf{a} \cdot \hat{\mathbf{p}}}$, where the exponent is defined in terms of its Taylor expansion. An infinitesimally small translation $\delta\mathbf{a}$ is effected by the momentum operator itself, i.e. $\hat{T}_{(\delta\mathbf{a})} = 1 - \frac{i}{\hbar}(\delta\mathbf{a}) \cdot \hat{\mathbf{p}}$.

So what does this have to do with conservation laws? Since we assumed the Hamiltonian of the system to be invariant under translations $\hat{T}_{\mathbf{a}}$, we find that $\hat{T}_{\mathbf{a}}^{\dagger} \hat{H} \hat{T}_{\mathbf{a}} = \hat{H}$ must hold for any $\mathbf{a}$. This implies that $[\hat{H}, \hat{T}_{\mathbf{a}}] = 0$, the translation operator and the Hamiltonian commute. As a result, $\hat{H}$ and $\hat{T}_{\mathbf{a}}$ share a complete set of eigenstates, or, to reason one step further, the eigenstates of the translation operator are stationary states. Now we can finally draw our conclusion. What do we know about the eigenstates of $\hat{T}_{\mathbf{a}}$? The translation operator consists only of products of momentum operators, $\hat{T}_{\mathbf{a}} = 1 - \frac{i}{\hbar}\mathbf{a} \cdot \hat{\mathbf{p}} - \frac{1}{2\hbar^2}(\mathbf{a} \cdot \hat{\mathbf{p}})^2 + \dots$, its eigenstates are therefore the momentum eigenstates. As explained above, these states must be stationary: momentum is thus a conserved quantity.

Another interesting symmetry to investigate is the isotropy of space (space looks the same in all directions). Let us proceed along similar lines as above, and try to find the operators which correspond to conserved quantities for an isotropic Hamiltonian. The (conserved) momentum operator turned out to be the operator for infinitesimal translations, so let us now look for the operator of infinitesimal rotations. In three-dimensional space, any rotation can be decomposed into rotations about the three orthogonal axes. We define the rotation operator $\hat{R}_{\alpha}(\theta)$ as the operator which rotates a wave function over an angle θ about the α-axis (where α is x, y, or z). We are thus interested in the operators $\hat{J}_{\alpha}$ defined by $e^{-\frac{i}{\hbar}\theta \hat{J}_{\alpha}} = \hat{R}_{\alpha}(\theta)$, which yields $\hat{J}_{\alpha} = i\hbar\partial_{\theta}\hat{R}_{\alpha}(\theta)|_{\theta=0}$.

Let us now keep the discussion as general as possible, and look for properties of the $\hat{\mathbf{J}}$ without assuming anything specific about the type of rotations. A useful starting point would be to derive the commutation relations for $\hat{\mathbf{J}}$, i.e. find $[\hat{J}_\alpha, \hat{J}_\beta]$. For this purpose, we investigate the case where $\hat{R}_\alpha(\theta)$ is a simple rotation of the coordinate system about one of the Cartesian axes. You might wonder what other less simple types of rotation there are, but a bit of patience is required, we find this out below. For now, we simply write down the three matrices which rotate a Cartesian coordinate system. For example, for rotations about the z-axis we find

$$
\begin{pmatrix} x' \\ y' \\ z' \end{pmatrix} = \begin{pmatrix} \cos\theta & -\sin\theta & 0 \\ \sin\theta & \cos\theta & 0 \\ 0 & 0 & 1 \end{pmatrix} \begin{pmatrix} x \\ y \\ z \end{pmatrix}, \tag{1.65}
$$

and similar matrices for the other two axes. We calculate $\hat{\mathbf{J}} = i\hbar\partial_\theta\hat{\mathbf{R}}(\theta)|_{\theta=0}$ from these matrices, and find that

$$
[\hat{J}_x, \hat{J}_y] = i\hbar\hat{J}_z, \quad [\hat{J}_y, \hat{J}_z] = i\hbar\hat{J}_x, \quad \text{and} \quad [\hat{J}_z, \hat{J}_x] = i\hbar\hat{J}_y. \tag{1.66}
$$

We use these commutation relations as *definitions* for the operators $\hat{\mathbf{J}}$. The observables associated with these operators, which are conserved quantities, give the angular momentum of the system along the three orthogonal axes.

Control question. Can you verify that the three components of the "quantum analogue" $\hat{\mathbf{L}} = \hat{\mathbf{r}} \times \hat{\mathbf{p}}$ of the classical angular momentum indeed satisfy the commutation relations (1.66)?

It turns out that it is possible to reveal the complete matrix structure of the operators $\hat{\mathbf{J}}$ just based on the commutation relations given above. We start by defining an operator $\hat{J}^2 \equiv \hat{J}_x^2 + \hat{J}_y^2 + \hat{J}_z^2$, corresponding to the square of the total angular momentum. We see that

$$
[\hat{J}^2, \hat{J}_x] = [\hat{J}^2, \hat{J}_y] = [\hat{J}^2, \hat{J}_z] = 0, \tag{1.67}
$$

the total angular momentum operator commutes with all three orthogonal components of $\hat{\mathbf{J}}$. This means that $\hat{J}^2$ and, for example, $\hat{J}_z$ share a common set of eigenstates. It is not difficult to show (see Exercise 5) that any common eigenstate of $\hat{J}^2$ and $\hat{J}_z$ can be written as $|j, m_z\rangle$, such that

$$
\begin{aligned}
\hat{J}^2|j, m_z\rangle &= \hbar^2 j(j+1)|j, m_z\rangle, \\
\hat{J}_z|j, m_z\rangle &= \hbar m_z|j, m_z\rangle,
\end{aligned} \tag{1.68}
$$

where j is either integer or half-integer, and for given j, the number m_z can take one of the $2j + 1$ values $-j, -j+1, \ldots, j-1, j$.

We now derive the matrix structure of $\hat{J}_x$ and $\hat{J}_y$ in this basis of eigenstates of $\hat{J}_z$. Actually, it is easier to work with the operators $\hat{J}_\pm \equiv \hat{J}_x \pm i\hat{J}_y$ (see Exercise 5). The operator $\hat{J}_+$ is a "raising operator," it couples states with quantum number m_z to states with quantum number $m_z + 1$. We can thus write $\hat{J}_+|j, m_z\rangle = \gamma_+(m_z)|j, m_z + 1\rangle$, where we would like to know the matrix element $\gamma_+(m_z)$. To find it, we evaluate

$$
\langle j, m_z|\hat{J}_-\hat{J}_+|j, m_z\rangle = \hbar^2[j(j+1) - m_z(m_z+1)]. \tag{1.69}
$$

Control question. Do you see how to derive (1.69)?

However, per definition we can also write $\langle j, m_z | \hat{J}_- \hat{J}_+ | j, m_z \rangle = \gamma_+(m_z)^* \gamma_+(m_z)$, and comparison with (1.69) gives us the desired $\gamma_+(m_z)$ up to a phase factor. Usually the matrix elements are chosen to be real, so we find

$$\hat{J}_+ | j, m_z \rangle = \hbar \sqrt{j(j+1) - m_z(m_z + 1)} | j, m_z + 1 \rangle, \tag{1.70}$$

and, in a similar way,

$$\hat{J}_- | j, m_z \rangle = \hbar \sqrt{j(j+1) - m_z(m_z - 1)} | j, m_z - 1 \rangle. \tag{1.71}$$

We now return to the issue of the different types of rotation, which we mentioned in the beginning of this section. As you have already checked, the regular orbital angular momentum operators $\hat{\mathbf{L}} = \hat{\mathbf{r}} \times \hat{\mathbf{p}}$ satisfy the commutation relations, and their matrix structure must thus be as derived above. Let us investigate this orbital angular momentum in more detail. If we were to write a wave function in spherical coordinates, $x = r \sin\theta \cos\phi$, $y = r \sin\theta \sin\phi$ and $z = r \cos\theta$, the z-component of $\hat{\mathbf{L}}$ takes the very simple form $\hat{L}_z = -i\hbar\partial_\phi$.

Control question. Do you know how to transform the operator $\hat{\mathbf{r}} \times \hat{\mathbf{p}}$ to spherical coordinates?

The eigenstates of $\hat{L}_z$ can then be easily determined. They are the solutions of $-i\hbar\partial_\phi \psi(r, \theta, \phi) = l_z \psi(r, \theta, \phi)$, which of course are $\psi(r, \theta, \phi) = e^{\frac{i}{\hbar} l_z \phi} f(r, \theta)$. Since the wave function must be single-valued everywhere, we find that the allowed values for l_z are $l_z = \hbar m_z$ where m_z must be an *integer*. This is of course not inconsistent with what we found above for the angular momentum operator $\hat{J}_z$. However, we then concluded that eigenvalues of $\hat{J}_z$ can be integer or *half-integer*. The orbital angular momentum only exploits the first option.

Is there a problem with half-integer values of angular momentum? Suppose we have a state $|\psi\rangle$ which is an eigenstate of $\hat{J}_z$ with eigenvalue $\hbar/2$. Let us apply a full 2π-rotation about the z-axis to this state. We know that the rotation operator reads $\hat{R}_z(\theta) = e^{\frac{i}{\hbar}\theta \hat{J}_z}$. We set θ to 2π and make use of the fact that $|\psi\rangle$ is an eigenstate of $\hat{J}_z$, so that

$$\hat{R}_z(2\pi)|\psi\rangle = e^{\frac{i}{\hbar} 2\pi \frac{\hbar}{2}}|\psi\rangle = -|\psi\rangle. \tag{1.72}$$

The state acquires a factor -1 after a full rotation about the z-axis. This sounds strange, but in quantum mechanics this actually is not a problem! A wave function is in any case only defined up to a phase factor, so $|\psi\rangle$ and $|\psi'\rangle = -|\psi\rangle$ are the same physical state since *any* possible observable is the same for the two states, $\langle\psi|\hat{A}|\psi\rangle = \langle\psi'|\hat{A}|\psi'\rangle$. Therefore, there is no a-priori reason to exclude the possibility of half-integer angular momentum. As explained above, half-integer angular momentum cannot be related to the orbital angular momentum $\hat{\mathbf{L}}$, so there must be another type of "intrinsic" angular momentum. This intrinsic angular momentum is called spin, and is represented by the operators $\hat{\mathbf{S}}$, so that

$$\hat{\mathbf{J}} = \hat{\mathbf{L}} + \hat{\mathbf{S}}. \tag{1.73}$$

The spin angular momentum has nothing to do with actual rotations of the coordinates, so it must be related to rotations of other aspects of the structure of the wave function. What could this structure be? Nothing forbids a wave function to be more complex than just a function of position. It could, for example, be defined in a two-dimensional vector space $\Psi(\mathbf{r}) = (\psi_\uparrow(\mathbf{r}), \psi_\downarrow(\mathbf{r}))^T$, or in an even higher-dimensional vector space. The spin angular momentum would then relate to this vector structure: rotations effected by $e^{-\frac{i}{\hbar}\boldsymbol{\theta}\cdot\hat{\mathbf{S}}}$ can be regarded as rotations *within* this vector space. This means that, if a certain wave function has an N-dimensional vector representation, its spin operators can be written as $N \times N$ matrices. This dimension relates in a simple way to the total spin quantum number s (analogous to j for the total angular momentum): $2s + 1 = N$. A two-dimensional wave function thus describes a particle with spin $s = \frac{1}{2}$, a three-dimensional with spin 1, etc. Wave functions which do not possess a vector structure belong to particles with spin 0. It can be shown that all particles with integer spin must be bosons, and particles with half-integer spin ($\frac{1}{2}, \frac{3}{2}, \dots$) are fermions.

Fortunately, the most common elementary particles, such as electrons, protons and neutrons, are all fermions with spin $\frac{1}{2}$, that is to say the most "simple" fermions allowed by Nature. For this reason, we now focus on spin $-\frac{1}{2}$ particles. We can write down their 2×2 spin operators,

$$\hat{S}_x = \frac{\hbar}{2}\begin{pmatrix} 0 & 1 \\ 1 & 0 \end{pmatrix}, \quad \hat{S}_y = \frac{\hbar}{2}\begin{pmatrix} 0 & -i \\ i & 0 \end{pmatrix}, \quad \text{and} \quad \hat{S}_z = \frac{\hbar}{2}\begin{pmatrix} 1 & 0 \\ 0 & -1 \end{pmatrix}. \tag{1.74}$$

In the basis we have chosen, i.e. the eigenstates of $\hat{S}_z$, a particle with a wave function $(\psi_\uparrow(\mathbf{r}), 0)^T$ is thus a spin $-\frac{1}{2}$ particle with spin quantum number $m_z = \frac{1}{2}$ (or spin up), and a state $(0, \psi_\downarrow(\mathbf{r}))^T$ has a spin quantum number $m_z = -\frac{1}{2}$ (or spin down). These two values, $m_z = \pm\frac{1}{2}$, are the only two possible outcomes if one measures the spin of an electron along a fixed axis. A way to do this would be to make use of the fact that particles with different spin quantum number behave differently in the presence of a magnetic field.

1.7.1 Spin in a magnetic field

As we explained above, a quantum state can be rotated in two different ways. One can simply rotate the coordinates of the wave function (which is achieved by the operator $e^{-\frac{i}{\hbar}\boldsymbol{\theta}\cdot\hat{\mathbf{L}}}$), or one can rotate the spin of the wave function (which is done by $e^{-\frac{i}{\hbar}\boldsymbol{\theta}\cdot\hat{\mathbf{S}}}$). Since these different types of rotation can be performed independently, it is no surprise that there exist Hamiltonians which couple only to the orbital angular momentum of a particle, as well as Hamiltonians which couple only to its spin. An example of the latter is the Hamiltonian describing the coupling between spin and a magnetic field.

Since spin is a form of angular momentum, it adds to the magnetic moment of the particle. This contribution is proportional to the spin, and thus reads

$$\hat{\boldsymbol{\mu}} = \gamma\hat{\mathbf{S}}, \tag{1.75}$$

where the constant γ depends on the type of particle. In the case of free electrons, this constant is $\gamma = |e|/m_e$, where $|e|$ is the absolute value of the electron charge and m_e is the electron mass. The energy of a magnetic moment in an external field $\mathbf{B}$ is given by $-\boldsymbol{\mu} \cdot \mathbf{B}$, so we can write the Hamiltonian for the interaction between the spin of an electron and a magnetic field as

$$\hat{H} = -\mu_B \mathbf{B} \cdot \hat{\boldsymbol{\sigma}}, \tag{1.76}$$

where μ_B denotes the Bohr magneton, $\mu_B = |e|\hbar/2m_e$, and $\hat{\boldsymbol{\sigma}}$ are the three Pauli matrices,

$$\hat{\sigma}_x = \begin{pmatrix} 0 & 1 \\ 1 & 0 \end{pmatrix}, \quad \hat{\sigma}_y = \begin{pmatrix} 0 & -i \\ i & 0 \end{pmatrix}, \quad \text{and} \quad \hat{\sigma}_z = \begin{pmatrix} 1 & 0 \\ 0 & -1 \end{pmatrix}. \tag{1.77}$$

We see that electrons with spin up and down (measured along the axis parallel to the field) acquire an energy difference of $2\mu_B B$. This splitting is usually called the Zeeman splitting of the electron. If the magnetic field and the quantization axis are not aligned, then the Hamiltonian (1.76) causes the spin to precess around the direction of the magnetic field. We investigate this precession in detail in Exercise 7.

1.7.2 Two spins

Let us now consider a system of *two* spin-$\frac{1}{2}$ particles, and investigate the *total* spin of the two-particle system. A natural basis to express the two-particle spin state in is

$$|\uparrow\uparrow\rangle, \quad |\uparrow\downarrow\rangle, \quad |\downarrow\uparrow\rangle, \quad \text{and} \quad |\downarrow\downarrow\rangle. \tag{1.78}$$

We define the operator of total spin as $\hat{\mathbf{S}}_{\text{tot}} = \hat{\mathbf{S}}_{(1)} + \hat{\mathbf{S}}_{(2)}$, the sum of the spin of the two separate particles. The total spin along the z-axis of the four states in (1.78) can now easily be determined. We see that all four states are eigenstates of $\hat{S}_{\text{tot},z}$ with respective eigenvalues m_z of 1, 0, 0, and -1. If we put this result in the context of the discussion in Section 1.7, it seems a bit confusing. The set of eigenvalues we found suggests that the total object (the two electrons) behaves as a spin-1 particle. Its total spin s is one, and the spin along the z-axis can take values ranging from $-s$ to s, i.e. from -1 to 1. The confusing thing is that there obviously exist *two* different states with $m_z = 0$, and this does not fit the picture: a spin-1 particle has only one state with $m_z = 0$.

To find out what is going on, we must examine the four eigenstates more closely and calculate their $\hat{S}_{\text{tot}}^2$ and $\hat{S}_{\text{tot},z}$. The first state, $|\uparrow\uparrow\rangle$, yields the eigenvalues $2\hbar^2$ and $\hbar$ for respectively $\hat{S}_{\text{tot}}^2$ and $\hat{S}_{\text{tot},z}$. This means that the two-particle state $|\uparrow\uparrow\rangle$ indeed behaves as a spin-1 particle with $m_z = 1$, or in other words, using the notation $|s, m_z\rangle$, we find that $|\uparrow\uparrow\rangle = |1, 1\rangle$. We then apply the lowering operator $\hat{S}_{\text{tot},-} = \hat{S}_{(1),-} + \hat{S}_{(2),-}$ to this state, which yields the state $\frac{1}{\sqrt{2}}\{|\uparrow\downarrow\rangle + |\downarrow\uparrow\rangle\}$, where the factor $1/\sqrt{2}$ is included for normalization. We see that we find an $m_z = 0$ state which is a superposition of the two states we discussed above. To identify which $m_z = 0$ state we have found, we evaluate its eigenvalue of $\hat{S}_{\text{tot}}^2$, and find that $\frac{1}{\sqrt{2}}\{|\uparrow\downarrow\rangle + |\downarrow\uparrow\rangle\} = |1, 0\rangle$. Applying the lowering operator once again, we can verify that the state $|\downarrow\downarrow\rangle$ corresponds to $|1, -1\rangle$, as expected.

We thus have found three spin states of a two-electron system which behave like the states of a single spin-1 particle. This means that we still have a fourth state which seems not to fit the picture. This fourth state (it must be orthogonal to the three discussed above) reads $\frac{1}{\sqrt{2}}\{|\uparrow\downarrow\rangle - |\downarrow\uparrow\rangle\}$. To see what kind of state this is, we again apply the operators $\hat{S}_{tot}^2$ and $\hat{S}_{tot,z}$, and find that $\frac{1}{\sqrt{2}}\{|\uparrow\downarrow\rangle - |\downarrow\uparrow\rangle\} = |0,0\rangle$. This completes the picture of the possible two-electron spin states: there are three states with $s = 1$, or in other words the $s = 1$ state forms a spin triplet, and there is a single $s = 0$ state, a spin singlet.

What we have just considered is the simplest non-trivial multi-spin problem there exists. When combining more than two spins, or spins higher than $s = \frac{1}{2}$, the number of possible total spin states grows rapidly. The problem of figuring out the correct total spin states, however, stays of the same complexity and one could treat it in a similar way as we did above. Fortunately, for many combinations of spins this has already been done, and the results can simply be looked up when needed.

In Exercises 2 and 9 we investigate a two-spin system in an external magnetic field, where we also allow for a spin–spin coupling. We use the simplest model for this coupling, sometimes called the Heisenberg model, given by the Hamiltonian $\hat{H}_{spin\text{-}spin} = J\,\hat{\mathbf{S}}_{(1)} \cdot \hat{\mathbf{S}}_{(2)}$. A classical picture which could justify the form of this Hamiltonian is to say that the two spins both carry a magnetic dipole moment (see Section 1.7.1) which then, according to classical electrodynamics, leads to a dipole–dipole coupling. A favored antiparallel spin configuration would correspond to a positive J. We indeed find in Exercise 2 that this interaction is diagonal in the basis of the singlet and triplets, and that it effectively splits the states differing in s by an energy $\hbar^2 J$.

1.8 Two-level system: The qubit

The simplest non-trivial quantum system has only two levels, and therefore its Hilbert space is two-dimensional. There are many realizations of two-level systems that either occur naturally or are artificially made. For instance, a particle confined in a potential well with two separate minima of almost equal depth can be seen as a two-level system if only its two lowermost states are relevant. An electron with spin-$\frac{1}{2}$ confined in a certain orbital state is a two-level system as well, the two possible states corresponding to spin-up and spin-down.

To describe a two-level system, we label the two basis states $|0\rangle$ and $|1\rangle$. By definition of the basis, any arbitrary state of the system can be written with two complex numbers, α and β, as

$$|\psi\rangle = \alpha|0\rangle + \beta|1\rangle, \tag{1.79}$$

where $|\alpha|^2 + |\beta|^2 = 1$.

Control question. Do you see how the constraint $|\alpha|^2 + |\beta|^2 = 1$ follows from the normalization condition (1.4) and (1.17)?

When neither α nor β equals zero, we say that the system is in a *superposition* of $|0\rangle$ and $|1\rangle$ rather than in just one of these basis states. If we should try to determine in which basis state this two-level system is, there is a chance $|\alpha|^2$ of finding $|0\rangle$, and a chance $|\beta|^2$ of finding $|1\rangle$: a system in a superposition is able to give both outcomes.

The notion of superposition forms the basis of the field of *quantum information*, which proposes to use quantum mechanical systems for all kinds of information processing task. The main idea of this field is to use two-level systems as information-carrying units, so-called quantum bits or *qubits*. This is both in contrast and similarity to the "classical" bits of a regular computer. Classical bits can be in either of two possible states, "0" or "1." This suffices to represent any kind of information in a string of bits, and the amount of information stored is proportional to the number of bits used. Similarly, a qubit can also be in the state $|0\rangle$ or $|1\rangle$, but in addition, in contrast to a classical bit, it can also be in any superposition of the two.

A quick way to appreciate the advantage of using quantum information is to discuss the amount of information that can be stored in N qubits. Any state of a qubit can be written as in (1.79) and therefore is characterized by two complex numbers, α and β, or equivalently by four real numbers. These four numbers are not all independent. (i) The normalization condition forms one constraint, fixing a relation between the four numbers. (ii) It turns out that the overall phase factor of a state is irrelevant (the states $|\psi\rangle$ and $e^{i\varphi}|\psi\rangle$ are equivalent), giving the same expectation values for any observable. This leaves just two independent real numbers to characterize any state of the qubit. Let us now try to encode in classical bits the information stored in a single qubit. We assume that for any practical purpose an accuracy of 10^{-19} suffices, which means that any real number can be encoded in $N_{\mathrm{class}} = 64$ bits. For the information stored in our qubit we therefore need $2N_{\mathrm{class}}$ bits. So far there is no dramatic gain: one could even use two "analogue bits" – bits which can take any real value between 0 and 1 – to store the same amount of information. The advantage becomes clear upon increasing the number of qubits. For a set of N qubits, the Hilbert space has dimension 2^N, and an arbitrary superposition of 2^N basis states is thus characterized by 2^N complex or $2 \cdot 2^N$ real numbers.

Control question. What is a natural choice of basis states? Think of the states of N classical bits.

Again, the normalization condition and irrelevance of the overall phase factor reduce this number by two, to $2 \cdot (2^N - 1)$ real numbers. This means we need $2N_{\mathrm{class}} \cdot (2^N - 1)$ classical bits to represent the information stored in N qubits. Thus, the information storage capacity of qubits grows exponentially with N, in contrast to the linear growth for classical information. To give some striking examples: all the music which can be stored in a 80 GB iPod can be written in 33 qubits. A whole life-time of music – 500 times as much – requires just nine extra qubits. Let us assume that all useful information accumulated by humankind is stored in $\sim 10^8$ books. Well, we could encode all this information in just 43 qubits.

In recent years several quantum algorithms have been developed which make use of the advantages of qubits over classical bits. They concern factoring integers, simulating quantum systems and searching databases. Using these algorithms, some of these tasks could be performed exponentially faster on a quantum computer than on a classical one. These discoveries triggered a great interest in the experimental physics of two-level systems, and many physical qubit implementations have been proposed and are being investigated.

Let us now focus on the quantum mechanics of a single qubit. The operator of any physical quantity in the two-dimensional Hilbert space spanned by the basis $\{|1\rangle, |0\rangle\}$ can be represented by a 2×2 matrix. Explicitly, any Hermitian operator can be written as

$$\left(\begin{array}{cc} q+z & x-iy \\ x+iy & q-z \end{array} \right) = q\mathbb{1} + x\hat{\sigma}_x + y\hat{\sigma}_y + z\hat{\sigma}_z, \tag{1.80}$$

where $\mathbb{1}$ is the unit matrix, and the matrices $\hat{\sigma}$ are the Pauli matrices (1.77). The Pauli matrices therefore provide a convenient basis in the space of 2×2 matrices and are frequently used in this context.

Solving the eigenvalue equation $\hat{H}|\psi\rangle = E|\psi\rangle$ for any Hamiltonian now reduces to a 2×2 linear algebra problem in the basis of $|1\rangle$ and $|0\rangle$,

$$\left(\begin{array}{cc} \frac{1}{2}\varepsilon & t \\ t^* & -\frac{1}{2}\varepsilon \end{array} \right) \left(\begin{array}{c} \alpha \\ \beta \end{array} \right) = E \left(\begin{array}{c} \alpha \\ \beta \end{array} \right). \tag{1.81}$$

To simplify, we skip in the Hamiltonian the term proportional to $\mathbb{1}$, which means that we count energy starting from the mean of the two possible energy eigenvalues. Therefore, ε is the energy *difference* $\langle 1|\hat{H}|1\rangle - \langle 0|\hat{H}|0\rangle$, and this defines the diagonal elements of the Hamiltonian in (1.81). Apart from that, there are generally also non-diagonal elements t and t^* present. To illustrate one of the possible physical realizations of this Hamiltonian, let us consider a qubit made from two electron states $|1\rangle$ and $|0\rangle$, which are localized near the minima of a double well potential (see Fig. 1.2). The energy ε is then the difference of the corresponding energy levels, while t accounts for tunneling through the potential barrier separating the states.

Control question. Another way to write the Hamiltonian in (1.81) would be $\hat{H} = \frac{\varepsilon}{2}\{|1\rangle\langle 1| - |0\rangle\langle 0|\} + t|1\rangle\langle 0| + t^*|0\rangle\langle 1|$. Do you see the equivalence?

Control question. A spin-$\frac{1}{2}$ electron in a magnetic field can be regarded as a qubit with Hamiltonian $\hat{H} = \frac{1}{2}g\mu_B\mathbf{B} \cdot \hat{\boldsymbol{\sigma}}$. Let us choose $|0\rangle$ and $|1\rangle$ to be the eigenstates of $\hat{\sigma}_z$ with eigenvalues ± 1. What are ε and t in terms of $\mathbf{B}$ if we write the Hamiltonian in this basis? What are they if $|0\rangle$ and $|1\rangle$ are the eigenstates of $\hat{\sigma}_x$?

If the element t goes to zero, we see that the eigenstates of the Hamiltonian coincide with the basis states $|1\rangle$ and $|0\rangle$. Then, the state $\binom{1}{0} = |1\rangle$ has energy $\varepsilon/2$ and the state $\binom{0}{1} = |0\rangle$ has energy $-\varepsilon/2$. Generally, when $t \neq 0$, this is not the case. With finite t, the

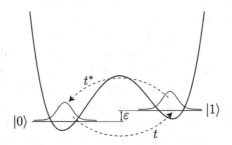

Fig. 1.2   Example of the realization of a qubit. The qubit is formed from two electron states, $|0\rangle$ and $|1\rangle$, close to the minima of a double well potential. The energy difference $\langle 1|\hat{H}|1\rangle - \langle 0|\hat{H}|0\rangle$ is denoted by ε, and tunneling between the two states is enabled by t and t^*.

eigenvalue equation gives the two eigenstates of $\hat{H}$,

$$|\pm\rangle = \frac{1}{\sqrt{2E_{\pm}^2 + E_{\pm}\varepsilon}} \left(\begin{array}{c} \frac{1}{2}\varepsilon + E_{\pm} \\ t^* \end{array} \right), \tag{1.82}$$

with

$$E_{\pm} = \pm E_{\text{qubit}} = \pm\sqrt{\frac{\varepsilon^2}{4} + |t|^2}. \tag{1.83}$$

If we write $t = |t|e^{i\phi}$, the same solutions allow for a compact presentation,

$$|\pm\rangle = \frac{1}{\sqrt{2(1 \pm \cos\theta)}} \left(\begin{array}{c} e^{i\phi/2}(\cos\theta \pm 1) \\ e^{-i\phi/2}\sin\theta \end{array} \right), \tag{1.84}$$

with the angle θ being defined as $\tan\theta = 2|t|/\varepsilon$.

We label the energy eigenstates of a qubit as $|+\rangle$ and $|-\rangle$, to distinguish from the basis states $|0\rangle$ and $|1\rangle$. In the energy basis the Hamiltonian of (1.81) is conveniently diagonal,

$$\hat{H} = \left(\begin{array}{cc} E_{\text{qubit}} & 0 \\ 0 & -E_{\text{qubit}} \end{array} \right), \tag{1.85}$$

making the description of the qubit very simple. The dynamics of a qubit are also important, they are illustrated by Exercise 7.

1.9 Harmonic oscillator

Another simple system which plays a very important role in quantum mechanics is the *harmonic oscillator*: we encounter it many times in the rest of this book, most notably in Chapters 7 and 8. The simplest mechanical version of the harmonic oscillator consists of a single particle confined in a parabolic potential. If we restrict ourselves for simplicity to one dimension, we can write according to (1.5) the Hamiltonian describing the particle as

$$\hat{H} = \frac{\hat{p}^2}{2m} + \frac{m\omega^2}{2}\hat{x}^2, \tag{1.86}$$

where ω characterizes the strength of the confinement and $\hat{p} = -i\hbar\partial_x$.

Since the potential energy becomes infinite for $x \to \pm\infty$, the particle cannot move through the entire space. It thus must be in a bound state, and have a discrete energy spectrum. Let us try to find the eigenenergies of $\hat{H}$ and the corresponding eigenstates. The trick is to introduce two new operators, which at first sight look a bit strange,

$$\hat{a} = \sqrt{\frac{m\omega}{2\hbar}}\left(\hat{x} + \frac{i\hat{p}}{m\omega}\right) \quad \text{and} \quad \hat{a}^\dagger = \sqrt{\frac{m\omega}{2\hbar}}\left(\hat{x} - \frac{i\hat{p}}{m\omega}\right). \tag{1.87}$$

However, if we express the Hamiltonian (1.86) in terms of these new operators, we see that it becomes very compact,

$$\hat{H} = \hbar\omega(\hat{n} + \tfrac{1}{2}), \quad \hat{n} \equiv \hat{a}^\dagger\hat{a}. \tag{1.88}$$

Control question. Can you derive this expression?

Let us note here that the operators $\hat{a}$ and $\hat{a}^\dagger$ obey the commutation relations $[\hat{a}, \hat{a}^\dagger] = 1$, $[\hat{a}, \hat{a}] = 0$, and $[\hat{a}^\dagger, \hat{a}^\dagger] = 0$, which can be easily proven using that $[\hat{x}, \hat{p}] = i\hbar$. In later chapters we recognize that this is a very fundamental set of relations which gives the operators a deeper meaning. However, for now they merely serve to simplify the problem at hand and to allow for a straightforward solution.

Obviously, $\hat{H}$ and $\hat{n}$ can be diagonalized simultaneously, i.e. an eigenstate of $\hat{n}$ is always also an eigenstate of $\hat{H}$. Let us thus focus on the eigenstates and eigenenergies of $\hat{n}$. We label the eigenstates of $\hat{n}$ by their eigenvalues n,

$$\hat{n}|n\rangle = n|n\rangle, \tag{1.89}$$

which then correspond to the eigenenergies $E_n = \hbar\omega(n + \frac{1}{2})$ of $\hat{H}$. We now construct the new state $\hat{a}^\dagger|n\rangle$, and see how it behaves under operation of the Hamiltonian (1.88). We have

$$\hbar\omega(\hat{a}^\dagger\hat{a} + \tfrac{1}{2})\hat{a}^\dagger|n\rangle = \hbar\omega(n + 1 + \tfrac{1}{2})\hat{a}^\dagger|n\rangle, \tag{1.90}$$

where we used the relation $[\hat{a}, \hat{a}^\dagger] = 1$ once. We thus see that $\hat{H}\hat{a}^\dagger|n\rangle = (E_n + \hbar\omega)a^\dagger|n\rangle$, and similarly we find $\hat{H}\hat{a}|n\rangle = (E_n - \hbar\omega)\hat{a}|n\rangle$. The states $\hat{a}^\dagger|n\rangle$ and $\hat{a}|n\rangle$ are thus also eigenstates of $\hat{H}$, with eigenenergies $E_{n\pm1}$.

This implies that with the notation chosen above the states $\hat{a}^\dagger|n\rangle$ and $\hat{a}|n\rangle$ are, up to constant factors, equal to $|n + 1\rangle$ and $|n - 1\rangle$. It is not difficult to find these factors. We have $\hat{a}|n\rangle = c|n - 1\rangle$, and using $\langle n|\hat{a}^\dagger\hat{a}|n\rangle = |c|^2 = n$, we find that $c = \sqrt{n}$ up to an unimportant phase factor. Similarly, we find $\hat{a}^\dagger|n\rangle = \sqrt{n+1}|n + 1\rangle$.

We have thus found a systematic method to derive an infinite "ladder" of eigenstates and eigenenergies of $\hat{H}$ starting from a single state $|n\rangle$. If the state we started with has integer n, we see that the ladder of numbers n is bounded from below: $\hat{a}|0\rangle = 0$, so n cannot be reduced below zero. If, on the other hand, n was non-integer, we could in principle keep lowering n without any bound. How do we decide between integers and non-integers? When n becomes negative we have $\langle n|\hat{a}^\dagger\hat{a}|n\rangle = n < 0$. But $\langle n|\hat{a}^\dagger\hat{a}|n\rangle$ is the norm of the wave function $\hat{a}|n\rangle$! Obviously a norm cannot be negative, so states with negative n are forbidden. The only choice for n which consistently makes sense is therefore integer n, and all states can be constructed from the vacuum as

$$|n\rangle = \frac{(\hat{a}^\dagger)^n}{\sqrt{n!}}|0\rangle. \tag{1.91}$$

We are now ready to find the ground state wave function $\psi_0(x)$ of the harmonic oscillator corresponding to $|0\rangle$. We make use of the property $\hat{a}|0\rangle = 0$, which in terms of $\hat{p}$ and $\hat{x}$ reads

$$\left(\hat{x} + \frac{i\hat{p}}{m\omega}\right)\psi_0(x) = 0. \tag{1.92}$$

This is a simple first-order differential equation which we can easily solve. Its normalized solution reads

$$\psi_0(x) = \left(\frac{m\omega}{\pi\hbar}\right)^{\frac{1}{4}} \exp\left\{-\frac{m\omega}{2\hbar}x^2\right\}. \tag{1.93}$$

Now that we have an explicit expression for $\psi_0(x)$ we can derive all wave functions of excited states $\psi_n(x)$ by applying the operator $\hat{a}^\dagger$ consecutively n times to $\psi_0(x)$, of course dividing out the factors $\sqrt{n+1}$. This procedure only involves taking derivatives, and is thus relatively straightforward although the expressions become cumbersome for larger n. In fact, the solutions are proportional to the Hermite polynomials $H_n(x\sqrt{m\omega/\hbar})$.

1.10 The density matrix

A wave function is the most complete way to describe the state of any system. Knowing the wave function of the whole Universe, we could know everything about every detail of it. Unfortunately, we do not know this wave function. And even if we did, we would probably not be interested in really every detail ... This is a typical situation in quantum mechanics: we either do not know the exact wave function of a complicated system or do not want to know.

Traditionally this is illustrated with a *statistical ensemble*. To make an exemplary statistical ensemble, we take many relatively small independent quantum systems, such as atoms. We bring each atom in one of several quantum states labeled $|i\rangle$. We then know the wave function of the whole ensemble: it is a (long) direct product of all individual atomic wave functions. Let us now close our eyes and randomly pick an atom from the ensemble. Since we do not know which atom we took, we do not know its quantum state. What do we know? If the atoms were in classical states labeled i, we would know the probability distribution p_i: we would know the chance of picking an atom which is in the state i. The quantum generalization of this classical concept is given by the *density matrix*. For the ensemble in question, the density matrix reads

$$\hat{\rho} = \sum_i p_i |i\rangle \langle i|, \tag{1.94}$$

it is an operator (matrix) in the Hilbert space of atomic states $|i\rangle$. This density matrix can be used to evaluate the expectation value of any observable of the randomly picked atom,

$$\langle \hat{A} \rangle = \sum_i p_i \langle i|\hat{A}|i\rangle = \text{Tr}[\hat{A}\hat{\rho}]. \tag{1.95}$$

This expression now contains the probabilistic nature of the quantum states involved as well as the statistical distribution of the particles in the ensemble. We note that from the definition given above it follows that $\text{Tr}[\hat{\rho}] = 1$: the trace of a density matrix equals one.

A traditional application of statistical ensembles is in thermodynamics. In this case, the states $|i\rangle$ are eigenstates of the atomic Hamiltonian with energies E_i. The probabilities $p_i = \exp(-E_i/k_B T)/Z$ are the Boltzmann factors corresponding to the temperature T, where k_B is Boltzmann's constant and the factor Z is to be determined from the condition $\sum_i p_i = 1$. This implies that the density matrix of a thermodynamic ensemble can be written compactly as

$$\hat{\rho} = \frac{\exp(-\hat{H}/k_B T)}{\text{Tr}\left[\exp(-\hat{H}/k_B T)\right]}. \tag{1.96}$$

Another situation concerns a small system, say a qubit, that interacts weakly with many other degrees of freedom – in principle with the rest of the Universe. Here we have to deal with incomplete knowledge: we care about the small system and not about the rest of the Universe. We are solely interested in the dynamics of the qubit, so that figuring out the wave function for the combined qubit–Universe system is practically impossible. To get rid of all redundant variables characterizing the Universe, we can perform a *partial trace* over these variables to obtain the reduced density matrix in the Hilbert space of the qubit.

To illustrate this, let us suppose that the state of the total system is described by a wave function of many variables, $|\psi\rangle = \sum_{i,j} \psi_{ij} |i\rangle |j\rangle$, where the index i labels the states of the qubit and j those of the rest of the Universe. To compare this with the definition of the statistical ensemble in (1.94), we recognize that we now have the simplest ensemble possible: it consists of one single member, $|\psi\rangle$, which is realized with probability 1. The density matrix for the whole system thus reads

$$\hat{R} = |\psi\rangle\langle\psi| = \sum_{i,i',j,j'} \psi_{i'j'}^* \psi_{ij} |i\rangle |j\rangle \langle i'| \langle j'|. \tag{1.97}$$

Since we care only about the state of the qubit, we trace out all other degrees of freedom to find the reduced density matrix of the qubit,

$$\hat{\rho} = \text{Tr}_j[\hat{R}] = \sum_j \langle j|\psi\rangle \langle\psi|j\rangle = \sum_{i,i',j} \psi_{i'j}^* \psi_{ij} |i\rangle \langle i'| = \sum_{i,i'} \rho_{ii'} |i\rangle \langle i'|. \tag{1.98}$$

The relevant information about the state of the Universe is now encoded in the coefficients $\rho_{ii'}$.

Control question. Suppose that the wave function of a system depends only on two coordinates, x and y. If the system is in the state $\psi(x, y)$, its full density matrix reads $\hat{\rho}(x, x', y, y') = \psi(x, y)\psi^*(x', y')$. Do you know how to calculate the reduced density matrix $\hat{\rho}(x, x') = \text{Tr}_y[\hat{\rho}]$?

We can now deduce some useful properties of the density matrix. From the definition of $\hat{\rho}$, we can derive a time-evolution equation for the density matrix,

$$\frac{\partial \hat{\rho}}{\partial t} = -\frac{i}{\hbar}[\hat{H}, \hat{\rho}]. \tag{1.99}$$

Note that this equation looks very similar to the Heisenberg equation of motion for operators, the sign however being opposite.

Control question. How does this relate to the definition of the expectation value of an operator $\hat{A}$? (See (1.95).)

If a system (in contact with the Universe) can still be described by a single wave function $|i\rangle$, its density matrix reads simply $\hat{\rho} = |i\rangle\langle i|$, and therefore obeys the relation

$$\hat{\rho}^2 = \hat{\rho}. \tag{1.100}$$

Indeed, if we diagonalize such a density matrix, we see that all eigenvalues are zero except for a single one, which equals one and corresponds to the eigenvector $|i\rangle$. Such a density matrix is called *pure*. The opposite case of a completely random mixture of states is described by a density matrix with a set of equal eigenvalues. This implies that it is diagonal, $\hat{\rho} = (1/N)\mathbb{1}$, with N being the dimension of the Hilbert space. For any density matrix given, one can thus make a "purity test" by accessing $\text{Tr}[\hat{\rho}^2]$. The value obtained is always between zero and one, $1/N \leq \text{Tr}[\hat{\rho}^2] \leq 1$. The maximum corresponds to a pure quantum state, while the minimum corresponds to a random mixture.

The most complete characterization of a density matrix is given by the complete set of its eigenvalues ρ_i. All specific properties of a given density matrix which can be characterized by a single number therefore can be calculated as a function of the eigenvalues, $\sum_i f(\rho_i)$. As shown above, the number $\text{Tr}[\hat{\rho}^2] = \sum_i \rho_i^2$ expresses the purity of a density matrix.

Another interesting value, given by

$$S = -\text{Tr}[\hat{\rho} \ln \hat{\rho}] = -\sum_i \rho_i \ln \rho_i, \tag{1.101}$$

is called the *entropy* of the density matrix. We see that in the case of a pure state $S = 0$ and that for a completely random mixture $S = \ln N$. This entropy thus coincides with the definition of entropy in statistical physics and characterizes the number of possible states the system can be in. The thermodynamical density matrix (1.96) maximizes the entropy for a fixed average value of total energy of the system.

1.11 Entanglement

It is traditional to put forward *entanglement* as the most spectacular, fascinating, and even mysterious feature that distinguishes between quantum and classical physics. If a two-part system is in an entangled state, this means that it is impossible to adequately describe the state of one part of the system without considering the other part. In simple words, if we cut the system into two and determine the state of each part, we are going to miss something.

To understand this in formal terms, let us consider a larger system W which we divide into two subsystems, A and B. The procedure of such division is called *bipartition*. As explained in Section 1.3, the Hilbert space $\mathcal{W}$ of the larger system is spanned by the direct product of the Hilbert spaces of the subsystems, $\mathcal{W} = \mathcal{A} \otimes \mathcal{B}$. If the states $|n_A\rangle$ denote the basis states in $\mathcal{A}$ and $|m_B\rangle$ those in $\mathcal{B}$, then any state in $\mathcal{W}$ can be written as $|\psi_W\rangle = \sum_{n,m} c_{nm}|n_A\rangle \otimes |m_B\rangle$, or shorter, $|\psi_W\rangle = \sum_{n,m} c_{nm}|n_A m_B\rangle$. Entanglement is then defined through the following negation: the two subsystems are in an entangled state if it is *not* possible to assign a separate wave function to each of them, i.e. if it is not possible to write $|\psi_W\rangle = |\psi_A\rangle \otimes |\psi_B\rangle$.

Let us illustrate this with the simplest example of a system consisting of two qubits, A and B. The total Hilbert space of this system is spanned by the basis $\{|00\rangle, |01\rangle, |10\rangle, |11\rangle\}$. We construct a superposition of two of the basis states

$$|\psi_+\rangle = \frac{1}{\sqrt{2}}(|01\rangle + |10\rangle),\tag{1.102}$$

which is a valid quantum state for the two-qubit system. It is, however, impossible to write down two separate state vectors for the two qubits such that $|\psi_+\rangle = |\psi_A\rangle \otimes |\psi_B\rangle$. The state $|\psi_+\rangle$ is therefore an entangled state. To give a contrasting example of a superposition state which looks complex but is *not* entangled, we consider

$$\frac{1}{2}(|00\rangle + |01\rangle + |10\rangle + |11\rangle) = \left\{\frac{1}{\sqrt{2}}(|0\rangle + |1\rangle)\right\} \otimes \left\{\frac{1}{\sqrt{2}}(|0\rangle + |1\rangle)\right\}.\tag{1.103}$$

This two-qubit state can be written as a direct product of two superpositions.

From these examples we see that it is not always straightforward to determine whether a given state is entangled or not. Unfortunately there is no entanglement operator $\hat{A}_{\text{ent}}$ with which the degree of entanglement in a state $|\psi\rangle$ can be found simply as the expectation value $\langle\psi|\hat{A}_{\text{ent}}|\psi\rangle$. This means that entanglement is not a physical observable. To characterize entanglement, we can use the entropy as defined in (1.101). The degree of entanglement between the subsystems A and B of system W, in a pure quantum state $|\psi_W\rangle$, is given by the entropy of the reduced density matrix of one of the subsystems, for instance $\hat{\rho}^{(A)}$. To understand this, let us first consider the case of a non-entangled state, $|\psi_W\rangle = |\psi_A\rangle \otimes |\psi_B\rangle$. In this case, the reduced density matrix $\hat{\rho}^{(A)} = |\psi_A\rangle\langle\psi_A|$ is pure, and its entropy is therefore zero. If we now introduce entanglement between the subsystems A and B, their reduced density matrices have an increased entropy, indicating a statistical uncertainty in each part. In that case the entropies $S_A, S_B > 0$, while the density matrix of the whole system is pure, that is, $S_{A+B} = 0$! There exists a set of special states of W that are called *maximally* entangled. For those states, both reduced density matrices $\hat{\rho}^{(A)}$ and $\hat{\rho}^{(B)}$ are proportional to the unity matrix, and therefore the entropies S_A and S_B reach their maximum.

Since entanglement cannot be associated with any observable, it is in strict terms not a physical quantity but rather a mathematical property of the wave function of a composite system. A quantum system evolves according to its Hamiltonian and does not care about entanglement. In fact, the mere notion of bipartition is unnatural since it breaks the symmetry with respect to the choice of basis in the Hilbert space of W. The bases in $\mathcal{W}$ that are direct products of the bases in $\mathcal{A}$ and $\mathcal{B}$ acquire, in a way, a higher status, and this violates the "democracy" mentioned in Section 1.3. While entanglement is not significant for the natural evolution of a quantum system, it is important for we humans that try to observe, manipulate, and make use of this quantum system. Entanglement becomes significant in the framework of quantum information, both theoretical and practical. To illustrate this significance, let us give an example of an application in quantum information: a scheme for *quantum teleportation*, i.e. the transfer of a quantum state from one qubit to another.

We must first outline an important difference between quantum and classical information. A carrier of classical information (think of a book) can be read many times without degradation of information. This is not the case for information contained in a quantum carrier, such as a qubit. As we know from Section 1.8, the state of a qubit is most generally given by $|\psi\rangle = \alpha|0\rangle + \beta|1\rangle$. Suppose that this state is the outcome of a quantum computation. We want to know the result of our calculation, so we are interested in the values α and β, and we perform a measurement on the qubit. Unfortunately, there is no measurement device that can give us α and β directly. The wave function is an amplitude of a probability distribution, and a probability distribution cannot be characterized by a single measurement. The best we can do is to design a measurement apparatus that can distinguish whether the qubit is in the state $|0\rangle$ or $|1\rangle$. The outcome of the measurement will then be "0" with probability $|\alpha|^2$ and "1" with probability $|\beta|^2$. To characterize α and β, we should repeat the measurement many times on the same state. Unfortunately this is not simply possible: the mere process of measuring destroys the original state $|\psi\rangle$ of the qubit, and it becomes either $|0\rangle$ or $|1\rangle$ according to the outcome of the measurement. One says that the quantum information is destroyed by reading it, this being a fundamental difference compared to classical information.

A way out would be to make many copies of the unknown qubit state before the measurement. Measuring all the copies would allow us to accumulate enough information to evaluate α and β with any accuracy desired. But again, the laws of quantum mechanics forbid us to proceed like this. The *no-cloning theorem* states that it is impossible to copy a qubit without destroying the original. This is a simple consequence of the fact that the Schrödinger equation is linear. If we were able to successfully clone a qubit, the resulting state would read

$$(\alpha|0\rangle + \beta|1\rangle)(\alpha|0\rangle + \beta|1\rangle) = \alpha^2|00\rangle + \alpha\beta(|10\rangle + |01\rangle) + \beta^2|11\rangle, \qquad (1.104)$$

a state quadratic in α and β. Since both α and β only appear linearly in the input state $\alpha|0\rangle + \beta|1\rangle$, there is no way that the state (1.104) can be the result of any evolution obeying the Schrödinger equation.

The good news, however, is that any unknown quantum state can be transferred from one qubit to another *without any loss* of information. In the context of the above discussion this sounds quite improbable, so let us outline the scheme of this quantum teleportation. Since quantum information does not make sense without humans involved, let us consider two individuals, Angelo and Bertha, who live far apart. Angelo possesses a piece of quantum information (a qubit labeled "1," in the state $|\psi_1\rangle = \alpha|0_1\rangle + \beta|1_1\rangle$) which he would like to transfer to Bertha. Suppose that it is for some reason impossible to put the qubit in a box and just send it by mail. It then seems that there is no way to get the full quantum information of the qubit to Bertha.

The way out is to use an entangled state. Angelo and Bertha prepare in advance a pair of qubits in a maximally entangled state

$$|\psi_{23}\rangle = \frac{1}{\sqrt{2}}(|0_2 1_3\rangle - |1_2 0_3\rangle), \qquad (1.105)$$

where we have labeled the qubits "2" and "3." Angelo takes qubit 2 and Bertha qubit 3, and they separate. After Angelo has acquired qubit number 1 – the one he wants to transfer to Bertha – the total three-qubit state reads

$$|\psi_{123}\rangle = |\psi_1\rangle \otimes |\psi_{23}\rangle = \frac{1}{\sqrt{2}}(\alpha|0_1\rangle + \beta|1_1\rangle) \otimes (|0_2 1_3\rangle - |1_2 0_3\rangle)$$

$$= \frac{1}{\sqrt{2}}(\alpha|0_1 0_2 1_3\rangle - \alpha|0_1 1_2 0_3\rangle + \beta|1_1 0_2 1_3\rangle - \beta|1_1 1_2 0_3\rangle).$$

(1.106)

The states of any two qubits can be presented in the so-called Bell basis of maximally entangled states

$$|\Psi^{(1)}\rangle = \frac{1}{\sqrt{2}}(|00\rangle + |11\rangle), \qquad |\Psi^{(3)}\rangle = \frac{1}{\sqrt{2}}(|01\rangle + |10\rangle),$$

$$|\Psi^{(2)}\rangle = \frac{1}{\sqrt{2}}(|00\rangle - |11\rangle), \qquad |\Psi^{(4)}\rangle = \frac{1}{\sqrt{2}}(|01\rangle - |10\rangle).$$

(1.107)

Control question. Prove that the four Bell states are maximally entangled.

Angelo has a device that can measure two qubits in this basis. With this device, he performs a measurement on the two qubits he possesses, numbers 1 and 2. Since the total wave function of the three qubits can be written as

$$|\psi_{123}\rangle = \frac{1}{2}\Big\{|\Psi_{12}^{(1)}\rangle(-\beta|0_3\rangle + \alpha|1_3\rangle) + |\Psi_{12}^{(2)}\rangle(\beta|0_3\rangle + \alpha|1_3\rangle)$$

$$+ |\Psi_{12}^{(3)}\rangle(-\alpha|0_3\rangle + \beta|1_3\rangle) + |\Psi_{12}^{(4)}\rangle(-\alpha|0_3\rangle - \beta|1_3\rangle)\Big\},$$

(1.108)

the state of the third qubit, the one kept by Bertha, after the measurement will be either $\binom{\beta}{\alpha}$, $\binom{-\beta}{\alpha}$, $\binom{-\alpha}{-\beta}$, or $\binom{-\alpha}{\beta}$, depending on the outcome of the measurement performed by Angelo. All these four states can be brought to the state $\alpha|0_3\rangle + \beta|1_3\rangle$ by a simple unitary operation. If Angelo makes a phone call to Bertha and tells her the outcome of his measurement, Bertha knows which operation to apply. So finally qubit number 3 is brought to exactly the same quantum state as qubit 1 was in before the measurement. Note that the whole procedure destroys the state of the first qubit, and therefore does not violate the no-cloning theorem. This illustrates the significance of entanglement. Entanglement is frequently called a *resource*, something useful for humans.

Table 1.1 Summary: Elementary quantum mechanics

Basics of quantum mechanics

state of a particle: wave function $\psi(\mathbf{r}, t)$ or wave vector $|\psi\rangle$

probabilistic interpretation: $|\psi(\mathbf{r}, t)|^2$ or $|\langle \mathbf{r}|\psi\rangle|^2$ gives probability of finding the particle at $(\mathbf{r}, t)$

normalization condition: $\int d\mathbf{r} \, |\psi(\mathbf{r}, t)|^2 = 1$ or $|\langle\psi|\psi\rangle|^2 = 1$

physical observables: operators $\hat{A}$, their expectation value $\langle A \rangle = \int d\mathbf{r} \, \psi^* \hat{A} \psi = \langle\psi|\hat{A}|\psi\rangle$

Dynamics of a quantum state

Schrödinger picture: $i\hbar \dfrac{\partial \psi}{\partial t} = \hat{H}\psi$

Heisenberg picture: $\dfrac{\partial \hat{A}}{\partial t} = \dfrac{i}{\hbar}[\hat{H}, \hat{A}]$

interaction picture: $\hat{H} = \hat{H}_0 + \hat{H}_1$, $\quad \dfrac{\partial \hat{A}_I}{\partial t} = \dfrac{i}{\hbar}[\hat{H}_0, \hat{A}_I]$, $\quad i\hbar \dfrac{\partial \psi_I}{\partial t} = \hat{H}_{1,I}\psi_I$

Perturbation theory, split $\hat{H} = \hat{H}_0 + \hat{H}'$ into large "simple" term and small perturbation

time-independent: expand eigenstates $|n\rangle = \sum_m c_{nm}|m^{(0)}\rangle$ and energies E_n in powers of H'/H_0

$$E_n^{(1)} = \langle n^{(0)}|\hat{H}'|n^{(0)}\rangle, \quad c_{nm}^{(1)} = \frac{\langle m^{(0)}|\hat{H}'|n^{(0)}\rangle}{E_n^{(0)} - E_m^{(0)}}, \quad E_n^{(2)} = \sum_{p \neq n} \frac{|\langle m^{(0)}|\hat{H}'|n^{(0)}\rangle|^2}{E_n^{(0)} - E_m^{(0)}}$$

time-dependent: expand time-evolution operator $\hat{U}_I(t, t')$ in powers of H'/H_0

$$\hat{U}_I(t, t') = 1 - \frac{i}{\hbar} \int_{t'}^{t} dt_1 \hat{H}'_I(t_1) - \frac{1}{\hbar^2} \int_{t'}^{t} dt_2 \int_{t'}^{t_2} dt_1 \hat{H}'_I(t_2)\hat{H}'_I(t_1)$$
$$+ \frac{i}{\hbar^3} \int_{t'}^{t} dt_3 \int_{t'}^{t_3} dt_2 \int_{t'}^{t_2} dt_1 \hat{H}'_I(t_3)\hat{H}'_I(t_2)\hat{H}'_I(t_1) + \dots$$

Fermi's golden rule: $\Gamma_{i \to f} = \frac{2\pi}{\hbar} |\langle f|\hat{H}'|i\rangle|^2 \delta(E_i - E_f)$

Spin

isotropy of space $\to$ conserved quantity $\hat{\mathbf{J}}$ with $[\hat{J}_x, \hat{J}_y] = i\hbar \hat{J}_z$, $[\hat{J}_y, \hat{J}_z] = i\hbar \hat{J}_x$, and $[\hat{J}_z, \hat{J}_x] = i\hbar \hat{J}_y$

$\hat{\mathbf{J}} = \hat{\mathbf{L}} + \hat{\mathbf{S}}$, with $\hat{\mathbf{L}} = \hat{\mathbf{r}} \times \hat{\mathbf{p}}$ orbital angular momentum and $\hat{\mathbf{S}}$ "intrinsic" angular momentum

spin-$\frac{1}{2}$ operators: $\hat{\mathbf{S}} = \frac{\hbar}{2}\hat{\boldsymbol{\sigma}}$, with Pauli matrices $\hat{\sigma}_x = \left(\begin{smallmatrix} 0 & 1 \\ 1 & 0 \end{smallmatrix}\right)$, $\hat{\sigma}_y = \left(\begin{smallmatrix} 0 & -i \\ i & 0 \end{smallmatrix}\right)$,

                 and $\hat{\sigma}_z = \left(\begin{smallmatrix} 1 & 0 \\ 0 & -1 \end{smallmatrix}\right)$

spin-$\frac{1}{2}$ in magnetic field: $\hat{H} = -\mu_B \mathbf{B} \cdot \hat{\boldsymbol{\sigma}}$.

Two-level system

Hamiltonian: $\hat{H} = H_x \hat{\sigma}_x + H_y \hat{\sigma}_y + H_z \hat{\sigma}_z$, with $|\psi\rangle = \alpha|0\rangle + \beta|1\rangle$

             $= E_{\text{qubit}}\left\{ |+\rangle\langle+| - |-\rangle\langle-| \right\}$, with $E_{\text{qubit}} = \sqrt{H_x^2 + H_y^2 + H_z^2}$, in "qubit" basis

---

Harmonic oscillator

Hamiltonian (1D): $\hat{H} = \dfrac{\hat{p}^2}{2m} + \dfrac{m\omega^2}{2}\hat{x}^2 = \hbar\omega(\hat{a}^\dagger \hat{a} + \frac{1}{2})$, with $\hat{a} = \sqrt{\dfrac{m\omega}{2\hbar}}\left(\hat{x} + \dfrac{i\hat{p}}{m\omega}\right)$

eigenenergies: $E_n = \hbar\omega(n + \frac{1}{2})$, with $n = 0, 1, 2, \dots$

Density matrix, allows for *statistical* uncertainty in quantum description

density matrix: $\hat{\rho} = \sum_i p_i |i\rangle\langle i|$, where p_i is the *statistical* probability of finding $|i\rangle$

physical observables: $\langle A \rangle = \text{Tr}[\hat{A}\hat{\rho}]$

in thermodynamic equilibrium: $\hat{\rho} = \exp(-\hat{H}/k_B T)/\text{Tr}\left[\exp(-\hat{H}/k_B T)\right]$

dynamics: $\dfrac{\partial \hat{\rho}}{\partial t} = -\dfrac{i}{\hbar}[\hat{H}, \hat{\rho}]$

Entanglement of two parts of a system if it is *not* possible to write the system's total wave function as a direct product of two wave functions for the two parts

Exercises

1. *Second-order Fermi's golden rule* (solution included). We consider a Hamiltonian $\hat{H}_0$ and we would like to calculate transition rates between the eigenstates of $\hat{H}_0$ due to a small perturbation $\hat{H}'$. Suppose that we are interested in transitions between levels that are not directly coupled by the perturbation, i.e. $\langle f|\hat{H}'|i\rangle = 0$. This means that the first-order transition amplitude, $\langle f|\hat{U}_I^{(1)}(t, -\infty)|i\rangle$, vanishes and Fermi's golden rule from Section 1.6.1 predicts $\Gamma_{i\to f} = 0$. In that case we should investigate the second-order correction $\hat{U}_I^{(2)}(t, -\infty)$, which can lead to a "second-order" transition rate similar to Fermi's (first-order) golden rule.

   a. Prove that the summation $\sum_v |v\rangle\langle v|$ performed over a complete set of orthonormal basis states $|v\rangle$ equals the identity operator.

   b. Take expression (1.56) for the second-order correction to the amplitude. Making use of the relation proven at (a), cast it into the form

   $$\langle f|\hat{U}_I^{(2)}(t, -\infty)|i\rangle =$$
   $$-\frac{1}{\hbar^2}\sum_v \int_{-\infty}^t dt_2 \int_{-\infty}^{t_2} dt_1 e^{\frac{i}{\hbar}\hat{E}_f t_2 + \eta t_2}\langle f|\hat{H}'|v\rangle e^{-\frac{i}{\hbar}\hat{E}_v t_2} e^{\frac{i}{\hbar}\hat{E}_v t_1 + \eta t_1}\langle v|\hat{H}'|i\rangle e^{-\frac{i}{\hbar}\hat{E}_i t_1}.$$

   c. Compute the transition rate from $|i\rangle$ to $|f\rangle$,

   $$\Gamma_{i\to f}^{(2)} = \frac{dP_f^{(2)}}{dt} = 2\,\mathrm{Re}\left\{\left(\frac{\partial}{\partial t}\langle f|\hat{U}_I^{(2)}(0, -\infty)|i\rangle\right)\left(\langle f|\hat{U}_I^{(2)}(0, -\infty)|i\rangle\right)^*\right\},$$

   and show that in the limit $\eta \to 0$

   $$\Gamma_{i\to f}^{(2)} = \frac{2\pi}{\hbar}\left|\sum_v \frac{H'_{fv}H'_{vi}}{E_v - E_i}\right|^2 \delta(E_i - E_f),$$

   assuming that $E_v \neq E_i$ for all v.

2. *Two coupled spins in a magnetic field* (solution included). Consider two spins, L and R, in a magnetic field along the z-axis, i.e. $\mathbf{B} = (0, 0, B)$. The magnetic moments of the two spins are coupled to each other so that the total Hamiltonian reads

   $$\hat{H} = g\mu_B \mathbf{B}\cdot(\hat{\mathbf{S}}_L + \hat{\mathbf{S}}_R) + J\hat{\mathbf{S}}_L\cdot\hat{\mathbf{S}}_R.$$

   a. Write this Hamiltonian in the basis $\{|\uparrow\uparrow\rangle, |\uparrow\downarrow\rangle, |\downarrow\uparrow\rangle, |\downarrow\downarrow\rangle\}$.

   b. Diagonalize this Hamiltonian. What are its eigenstates?

   c. Express the density matrix $\hat{\rho}(T)$ as a function of temperature, assuming that the system is in thermodynamic equilibrium. Use expression (1.96).

   d. What is in thermodynamic equilibrium the average value of the total spin in z direction, $\langle \hat{S}_R^z + \hat{S}_L^z\rangle$? Discuss this expectation value in the limits $T \to 0$ and $T \to \infty$.

3. *Hermitian matrices.*

   a. Prove that the eigenvalues of any Hermitian matrix are real.

   b. Prove that the normalized eigenfunctions of an arbitrary Hermitian operator $\hat{A}$ with non-degenerate eigenvalues form an orthonormal basis.

Hint. Compute $\langle n|\hat{A}|m\rangle$ and use the fact that $|n\rangle$ and $|m\rangle$ are eigenfunctions of $\hat{A}$, $\hat{A}|n,m\rangle = A_{n,m}|n,m\rangle$.

c. Now assume that one of the eigenvalues of $\hat{A}$ is twofold degenerate, i.e. $\hat{A}|n_1\rangle = A_n|n_1\rangle$ and $\hat{A}|n_2\rangle = A_n|n_2\rangle$. Using the simple linear transformation $|n_1'\rangle = |n_1\rangle$ and $|n_2'\rangle = c|n_1\rangle + |n_2\rangle$ we can construct two orthogonal eigenvectors. Given that $|n_1\rangle$ and $|n_2\rangle$ are normalized, find c. The same procedure can of course also be followed for higher-order degeneracies.

d. Prove that if two Hermitian matrices commute, $[\hat{A}, \hat{B}] = 0$, they have a common complete set of eigenfunctions. First consider the case where both $\hat{A}$ and $\hat{B}$ have only non-degenerate eigenvalues.

Hint. Consider vectors like $\hat{A}\hat{B}|v\rangle$ and $\hat{B}\hat{A}|v\rangle$. Then reason that the statement also holds for operators with degenerate eigenvalues.

4. *Two-level system.* The states $|\psi_1\rangle$ and $|\psi_2\rangle$ form an orthonormal basis of the Hilbert space of a two-level system. We define a new basis $\{|\phi_1\rangle, |\phi_2\rangle\}$ by

$$|\phi_1\rangle = \frac{1}{\sqrt{2}}(|\psi_1\rangle + |\psi_2\rangle) \quad \text{and} \quad |\phi_2\rangle = \frac{1}{\sqrt{2}}(|\psi_1\rangle - |\psi_2\rangle).$$

The operator $\hat{A}$ in the original basis reads as

$$\hat{A} = \begin{pmatrix} a & t \\ t & b \end{pmatrix}.$$

Find the matrix representation of this operator in the new basis.

5. *Operators of angular momentum.* The commutation relations for the operators of angular momentum read

$$[\hat{J}_x, \hat{J}_y] = i\hbar\hat{J}_z, \quad [\hat{J}_y, \hat{J}_z] = i\hbar\hat{J}_x, \quad \text{and} \quad [\hat{J}_z, \hat{J}_x] = i\hbar\hat{J}_y.$$

We know that $\hat{J}^2$ and $\hat{J}_z$ share a complete set of eigenstates, so let us denote the states in this set as $|\alpha, \beta\rangle$, where $\hat{J}^2|\alpha, \beta\rangle = \alpha|\alpha, \beta\rangle$ and $\hat{J}_z|\alpha, \beta\rangle = \beta|\alpha, \beta\rangle$.

a. We define the two operators $\hat{J}_\pm \equiv \hat{J}_x \pm i\hat{J}_y$. Show that

$$\hat{J}_z\hat{J}_\pm|\alpha, \beta\rangle = (\beta \pm \hbar)\hat{J}_\pm|\alpha, \beta\rangle.$$

b. The operators $\hat{J}_\pm$ thus effectively raise or lower the eigenvalue of $\hat{J}_z$ by one unit of $\hbar$. Show that for any eigenstate $|\alpha, \beta\rangle$, it must be that $\langle\hat{J}_z^2\rangle \leq \langle\hat{J}^2\rangle$, or in other words, $-\sqrt{\alpha} \leq \beta \leq \sqrt{\alpha}$: there exist a maximal and minimal $\beta_{\max}$ and $\beta_{\min}$ for any given α.

c. Show that $\hat{J}_-\hat{J}_+ = \hat{J}^2 - \hat{J}_z^2 - \hbar\hat{J}_z$.

d. Consider $\hat{J}_-\hat{J}_+|\alpha, \beta_{\max}\rangle$ and $\hat{J}_+\hat{J}_-|\alpha, \beta_{\min}\rangle$, and show that $\beta_{\min} = -\beta_{\max}$.

e. We define $j \equiv \beta_{\max}/\hbar$. Show that (i) all allowed eigenvalues of $\hat{J}^2$ can be written as $\hbar^2 j(j + 1)$ with j integer or half-integer, and (ii) for a given j, the eigenvalue of $\hat{J}_z$ can be $-j\hbar, (-j + 1)\hbar, \ldots, (j - 1)\hbar, j\hbar$.

6. *Harmonic oscillator.* We consider a particle in a one-dimensional parabolic potential, the same as in Section 1.9. The Hamiltonian thus reads

$$\hat{H} = \frac{\hat{p}^2}{2m} + \frac{m\omega^2}{2}\hat{x}^2,$$

and has eigenstates $|n\rangle$ with energies $E_n = \hbar\omega(n + \frac{1}{2})$. Calculate the expectation value for both the kinetic and potential energy of a particle in the state $|n\rangle$.

7. *Spin in a magnetic field.* The Hamiltonian describing the effect of a magnetic field on a single spin reads

$$\hat{H} = g\mu_B \mathbf{B} \cdot \hat{\mathbf{S}},$$

where $\mathbf{B}$ is the magnetic field and $\hat{\mathbf{S}} \equiv \frac{1}{2}\hat{\sigma}$ the vector spin operator. The g-factor is assumed to be negative, $g < 0$.

Initially, the system is in the ground state, and the spin is thus aligned with the magnetic field. At time $t = 0$ the direction of the magnetic field suddenly changes: it is rotated over an angle θ. Since the change is sudden, the wave function does not adjust at $t = 0$, and after the change, it is no longer one of the two eigenstates of the new Hamiltonian. We thus expect a non-trivial time evolution. Let us choose the z-axis along the new direction of the field, so that the new Hamiltonian reads

$$\hat{H} = g\mu_B B \hat{S}_z \equiv \frac{E_z}{2} \begin{pmatrix} -1 & 0 \\ 0 & 1 \end{pmatrix},$$

$E_z = |g|\mu_B B$ being the Zeeman energy, i.e. the energy splitting between the spin up and down states.

a. Find the wave function in this basis assuming that the rotation was about the y-axis.
b. Write down the Schrödinger equation in this basis.
c. Solve the equation with the initial condition derived in (a).
d. Use the result from (c) to find the time-dependent expectation values $\langle \hat{S}_x(t) \rangle$ and $\langle \hat{S}_y(t) \rangle$.
e. We would like to derive the Heisenberg equations of motion for the operators $\hat{S}_x$ and $\hat{S}_y$. To this end, first write down the Heisenberg equation of motion for a general operator represented by a Hermitian 2×2 matrix. The time-dependence of the specific operators $\hat{S}_x$ and $\hat{S}_y$ then follows from implementing the right initial conditions, i.e. $\hat{S}_{x,y}(0) = \frac{1}{2}\hat{\sigma}_{x,y}$. Write down these solutions.
f. To find the time-dependent *expectation values* $\langle \hat{S}_x(t) \rangle$ and $\langle \hat{S}_y(t) \rangle$, use the fact that the wave function does not change in the Heisenberg picture and is thus given by the expression found in (a). Check if you reproduce the result from (d).

8. *Time-dependent and time-independent perturbation theory.* Show that, in the case of a time-independent perturbation, the matrix elements of the first-order correction to the time-evolution operator

$$\langle m | \hat{U}_I^{(1)}(0, -\infty) | n \rangle,$$

are identical to the correction factors $c_{nm}^{(1)}$ to the basis states as presented in Section 1.5. Do not forget to switch on the perturbation adiabatically.

9. *Two coupled spins in two different magnetic fields.* We consider two spins, L and R, in two *different* magnetic fields $\mathbf{B}_L$ and $\mathbf{B}_R$. The magnetic moments of the spins are coupled to each other so that the full Hamiltonian reads

$$\hat{H} = g\mu_B(\mathbf{B}_L \cdot \hat{\mathbf{S}}_L + \mathbf{B}_R \cdot \hat{\mathbf{S}}_R) + J\hat{\mathbf{S}}_L \cdot \hat{\mathbf{S}}_R.$$

Let us for simplicity assume equally large, *opposite* fields, $\mathbf{B}_R = -\mathbf{B}_L \equiv \mathbf{B}$, and choose the quantization axes along their direction.

a. Work in the basis $\{|\uparrow\uparrow\rangle, |\uparrow\downarrow\rangle, |\downarrow\uparrow\rangle, |\downarrow\downarrow\rangle\}$. Express the Hamiltonian in this basis, and find its eigenvalues. Express the two eigenstates of the Hamiltonian involving $|\uparrow\downarrow\rangle$ and $|\downarrow\uparrow\rangle$ in terms of the angle α defined by $\tan\alpha = J/2|\mathbf{B}|$. If the g-factor is positive, $g > 0$, what is the ground state of the system?

b. What is the natural bipartition in this system?

c. Assuming that the system is in the ground state found at (a), find the reduced density matrices for the left and right spin, $\hat{\rho}_L$ and $\hat{\rho}_R$.

d. Now we are ready to characterize the entanglement between the spins. Compute the entropy for both matrices $\hat{\rho}_L$ and $\hat{\rho}_R$ and show that it is a function of α. What is the degree of entanglement in the limits $g\mu_B|\mathbf{B}| \gg |J|$ and $\mathbf{B} \to 0$?

10. *Density matrix for a spin in a magnetic field.* Consider a spin in a magnetic field, described by the Hamiltonian

$$\hat{H} = E_z \begin{pmatrix} 1 & 0 \\ 0 & -1 \end{pmatrix},$$

in the basis $\{|\uparrow\rangle, |\downarrow\rangle\}$.

a. Assume thermodynamic equilibrium. Write down the density matrix of this system as a function of the temperature T.

b. At time $t = 0$ the direction of the field is suddenly changed to the x-direction. The Hamiltonian therefore reads for $t > 0$

$$\hat{H} = E_z \begin{pmatrix} 0 & 1 \\ 1 & 0 \end{pmatrix}.$$

We will describe the system after $t = 0$ in the basis of eigenstates of the new Hamiltonian, $|+\rangle = \frac{1}{\sqrt{2}}(|\uparrow\rangle + |\downarrow\rangle)$ and $|-\rangle = \frac{1}{\sqrt{2}}(|\uparrow\rangle - |\downarrow\rangle)$. Express the density matrix found at (a) in this basis.

c. After the field has changed, the density matrix is no longer in equilibrium. Use the time-evolution equation (1.99) and the initial condition found at (b) to find the time-dependent density matrix at $t > 0$.

d. Using the density matrix found, determine the three components of the average spin for $t > 0$. How do they depend on T?

Solutions

1. *Second-order Fermi's golden rule.*

a. Since the states $|v\rangle$ form a complete basis, we can write any arbitrary state as $|\psi\rangle = \sum_i a_i|v_i\rangle$. Then it is clear that $\sum_j |v_j\rangle \langle v_j|\psi\rangle = |\psi\rangle$.

b. This is straightforward. Do not forget to implement $\hat{H}'(t) = \hat{H}'e^{\eta t}$.

c. Also this derivation is pretty straightforward. In the end you again need the limit

$$\lim_{\eta\to 0^+} \frac{2\eta\hbar}{(E_i - E_f)^2 + (2\eta\hbar)^2} = \pi\,\delta(E_i - E_f).$$

2. *Two coupled spins in a magnetic field.*

  a. The Hamiltonian reads

$$\hat{H} = \begin{pmatrix} B + \frac{1}{4}J & 0 & 0 & 0 \\ 0 & -\frac{1}{4}J & \frac{1}{2}J & 0 \\ 0 & \frac{1}{2}J & -\frac{1}{4}J & 0 \\ 0 & 0 & 0 & -B + \frac{1}{4}J \end{pmatrix},$$

    where we, for simplicity of notation, set $g\mu_B = 1$.

  b. Diagonalization of this Hamiltonian yields

$$\hat{H} = \begin{pmatrix} B & 0 & 0 & 0 \\ 0 & 0 & 0 & 0 \\ 0 & 0 & -J & 0 \\ 0 & 0 & 0 & -B \end{pmatrix},$$

    with the set of eigenstates $\{|\uparrow\uparrow\rangle, \frac{1}{\sqrt{2}}(|\uparrow\downarrow\rangle + |\downarrow\uparrow\rangle), \frac{1}{\sqrt{2}}(|\uparrow\downarrow\rangle - |\downarrow\uparrow\rangle), |\downarrow\downarrow\rangle\}$. Note that we subtracted a constant energy offset of $\frac{1}{4}J$.

  c. Using the Hamiltonian found in (b), we write

$$\hat{\rho} = \frac{e^{-\beta\hat{H}}}{\mathrm{Tr}\{e^{-\beta\hat{H}}\}} = \frac{1}{1 + e^{\beta J} + 2\cosh\beta B} \begin{pmatrix} e^{-\beta B} & 0 & 0 & 0 \\ 0 & 1 & 0 & 0 \\ 0 & 0 & e^{\beta J} & 0 \\ 0 & 0 & 0 & e^{\beta B} \end{pmatrix},$$

    where $\beta \equiv 1/k_B T$ is the inverse temperature.

  d. Calculating $\mathrm{Tr}\{(\hat{S}_R^z + \hat{S}_L^z)\hat{\rho}\}$ yields

$$\langle \hat{S}_R^z + \hat{S}_L^z \rangle = \frac{-2\sinh\beta B}{1 + e^{\beta J} + 2\cosh\beta B}.$$

In the limit $T \to \infty$, or $\beta \to 0$, the result is 0, as expected. When the temperature becomes so large that all energy differences of the levels become negligible ($k_B T \gg B, J$), there is no energetic preference for any of the four states. The system is in an equal mixture of the four states, yielding an average spin of zero.

In the opposite limit, $T \to 0$ or $\beta \to \infty$, we find that the average spin can be either -1 or 0, depending on whether $B > J$ or $J > B$. This also was to be expected: depending on the magnitude of B and J, either $|\downarrow\downarrow\rangle$ or $\frac{1}{\sqrt{2}}(|\uparrow\downarrow\rangle - |\downarrow\uparrow\rangle)$ is the ground state of the system.

2

Identical particles

We are now ready for new material. In this chapter we consider systems which contain not just one or two particles, but any (large) number of particles. After presenting the Schrödinger equation for a general N-particle system, we focus on the case where all N particles are *identical*. For a classical system this would have no special implications, it would just mean that all particles have equal properties like charge, mass, etc. For a quantum system, however, we show that the situation is completely different: most solutions of the Schrödinger equation for a system of identical particles are redundant – they cannot occur in nature. The physical solutions of the Schrödinger equation are either completely symmetric with respect to particle exchange (for bosons) or completely antisymmetric (for fermions). Finally, we explain how this allows us to represent any quantum state of many identical particles in a very elegant and intuitive way in so-called Fock space – the space spanned by all physical solutions of the N-particle Schrödinger equation. Table 2.1 summarizes the topics covered.

2.1 Schrödinger equation for identical particles

Let us start with some very general considerations about the quantum mechanics of many-particle systems. The wave function of a many-particle state can be written as a function of the coordinates of all the particles,

$$\psi(\mathbf{r}_1, \mathbf{r}_2, \ldots, \mathbf{r}_N) \equiv \psi(\{\mathbf{r}_i\}), \tag{2.1}$$

where N denotes the number of particles. We use the shorthand notation $\{\mathbf{r}_i\}$ to denote the set of coordinates $\mathbf{r}_1, \mathbf{r}_2, \ldots, \mathbf{r}_N$. The Schrödinger equation describing the evolution of this wave function reads generally

$$\hat{H} = \sum_{i=1}^{N} \left\{ -\frac{\hbar^2}{2m_i} \frac{\partial^2}{\partial \mathbf{r}_i^2} + V_i(\mathbf{r}_i) \right\} + \frac{1}{2} \sum_{i \neq j} V_{ij}^{(2)}(\mathbf{r}_i, \mathbf{r}_j) + \frac{1}{6} \sum_{i \neq j \neq k} V_{ijk}^{(3)}(\mathbf{r}_i, \mathbf{r}_j, \mathbf{r}_k) + \ldots \tag{2.2}$$

The first term in this equation is just the sum of the one-particle Hamiltonians of all particles. The second and third terms add to this the interactions between the particles: the second presents the pairwise interaction between all particles, the third gives the three-particle interactions, and there could be many more exotic interactions up to $V^{(N)}$ giving the N-particle interactions.

In the special case of a system of N *identical* particles, this Hamiltonian acquires an important symmetry: it becomes invariant under any permutation of particles. Let us explain this in more detail. If all particles are identical, they are assumed to all have the same mass and feel the same potential. More importantly, however, the interactions between the particles also reflect their identity. Quite generally, the interaction energy of two particles depends on their coordinates, $E_{int} = V_{12}^{(2)}(\mathbf{r}_1, \mathbf{r}_2)$. If the particles are truly identical, one can permute them – put particle 1 in the place of particle 2, and particle 2 in the place of particle 1 – without changing the interaction energy. Thus, for identical particles $V_{12}^{(2)}(\mathbf{r}_1, \mathbf{r}_2) = V_{21}^{(2)}(\mathbf{r}_2, \mathbf{r}_1)$ should hold. Also the more exotic interactions, that involve three and more particles, should be symmetric with respect to permutations. For instance, the three-particle interaction satisfies

$$V_{123}^{(3)}(\mathbf{r}_1, \mathbf{r}_2, \mathbf{r}_3) = V_{132}^{(3)}(\mathbf{r}_1, \mathbf{r}_3, \mathbf{r}_2) = V_{231}^{(3)}(\mathbf{r}_2, \mathbf{r}_3, \mathbf{r}_1)$$
$$= V_{213}^{(3)}(\mathbf{r}_2, \mathbf{r}_1, \mathbf{r}_3) = V_{312}^{(3)}(\mathbf{r}_3, \mathbf{r}_1, \mathbf{r}_2) = V_{321}^{(3)}(\mathbf{r}_3, \mathbf{r}_2, \mathbf{r}_1),$$

where we list all six possible permutations of three particles.

Control question. How many permutations are possible for N particles? How many *pair* permutations are possible?

So, however complicated the structure of a Hamiltonian for identical particles, the most important feature of it is its symmetry. The Hamiltonian does not change upon permutation of the coordinates of any pair of particles. More formally, we can state that for any pair of particles i and j

$$\hat{H}(\dots, \mathbf{r}_i, \mathbf{r}_i', \dots, \mathbf{r}_j, \mathbf{r}_j', \dots) = \hat{H}(\dots, \mathbf{r}_j, \mathbf{r}_j', \dots, \mathbf{r}_i, \mathbf{r}_i', \dots). \tag{2.3}$$

Here the coordinates $\mathbf{r}'$ refer to the coordinates of the wave function on which the Hamiltonian acts, and $\mathbf{r}$ to the coordinates of the resulting wave function.

Control question. Do you understand this notation? Can you explain that for a single free particle one can write $\hat{H}(\mathbf{r}, \mathbf{r}') = -\frac{\hbar^2}{2m} \nabla_{\mathbf{r}} \delta(\mathbf{r} - \mathbf{r}') \nabla_{\mathbf{r}'}$?

The many-particle Hamiltonian as presented in (2.2) for the case of identical particles thus reads

$$\hat{H} = \sum_{i=1}^{N} \left\{ -\frac{\hbar^2}{2m} \frac{\partial^2}{\partial \mathbf{r}_i^2} + V_i(\mathbf{r}_i) \right\} + \sum_{i=2}^{N} \sum_{j=1}^{i-1} V^{(2)}(\mathbf{r}_i, \mathbf{r}_j) + \sum_{i=3}^{N} \sum_{j=2}^{i-1} \sum_{k=1}^{j-1} V^{(3)}(\mathbf{r}_i, \mathbf{r}_j, \mathbf{r}_k) + \dots$$
$$\tag{2.4}$$

Control question. Do you understand how permutation symmetry is incorporated into this Hamiltonian?

One can also state the permutation symmetry of the Hamiltonian by saying that $\hat{H}$ *commutes* with all pair permutation operators, i.e. $\hat{H}\hat{P}_{ab} = \hat{P}_{ab}\hat{H}$, where the permutation operator $\hat{P}_{ab}$ effects pair permutation of the particles a and b. This implies that $\hat{H}$ and $\hat{P}_{ab}$ must possess a complete common set of eigenfunctions (see Exercise 3 of Chapter 1). So, by just investigating the symmetries of the permutation operators we thus reveal properties of the eigenfunctions of any many-particle Hamiltonian!

Unfortunately, the eigenfunctions of a symmetric Hamiltonian do not have to be of the same symmetry as the Hamiltonian itself, which you probably already learned in your elementary course on quantum mechanics. An illustrative example of this fact is given by the atomic orbitals that are eigenstates of a spherically-symmetric Hamiltonian. While the s-orbitals are also spherically symmetric, the p- and d-orbitals – "hourglasses" and "clover leafs" – are not. A simpler example concerns a particle confined in a one-dimensional potential well, with $V(x) = 0$ for $|x| < L/2$ and $V(x) = \infty$ for $x \geq L/2$. The Hamiltonian describing this particle is symmetric with respect to reflection, $x \to -x$. Its eigenfunctions, however, found from solving the Schrödinger equation, are either symmetric,

$$\psi_n^s(x) = \sqrt{\frac{2}{L}} \cos\left([2n+1]\pi \frac{x}{L}\right), \quad \text{with } n = 0, 1, 2, \ldots, \tag{2.5}$$

or antisymmetric

$$\psi_n^a(x) = \sqrt{\frac{2}{L}} \sin\left(2n\pi \frac{x}{L}\right), \quad \text{with } n = 0, 1, 2, \ldots \tag{2.6}$$

So indeed, although $\hat{H}(x) = \hat{H}(-x)$ in this case, its eigenfunctions can also be antisymmetric, $\psi_n^a(x) = -\psi_n^a(-x)$.

The same holds for the permutation symmetric many-particle Hamiltonian. The case of two identical particles is relatively simple: we know that the pair permutation operator obeys $(\hat{P}_{12})^2 = \mathbb{1}$, so the only possible eigenvalues of $\hat{P}_{12}$ are 1 and -1. The two eigenfunctions therefore are either symmetric, $\hat{P}_{12}\psi_s = \psi_s$, or antisymmetric, $\hat{P}_{12}\psi_a = -\psi_a$, with respect to permutation of the particles. For three and more particles, the symmetry classification gets more complicated. Different permutation operators do not always commute with each other (e.g. $\hat{P}_{12}\hat{P}_{23} \neq \hat{P}_{23}\hat{P}_{12}$), and therefore there does not exist a complete set of common eigenfunctions for all permutation operators. However, one can still form a completely symmetric and a completely antisymmetric eigenfunction, i.e. one that remains unchanged, $\hat{P}_{ab}\psi_s = \psi_s$, and one that flips sign, $\hat{P}_{ab}\psi_a = -\psi_a$, under permutation of any pair of particles. Further there may also exist states which form *partially symmetric* subspaces: any pair permutation performed on a state in a partially symmetric subspace produces a linear combination of states in that same subspace. Let us rephrase this in the formal language introduced above. Suppose the states ψ_1 and ψ_2 form a partially symmetric subspace of an N-particle system. Any pair permutation $\hat{P}_{ab}$ performed on the state ψ_1 or ψ_2 then produces a linear combination of the same two states: $\hat{P}_{ab}\psi_1 = \alpha_{11}\psi_1 + \alpha_{21}\psi_2$ and $\hat{P}_{ab}\psi_2 = \alpha_{12}\psi_1 + \alpha_{22}\psi_2$. Thus, in this partially symmetric subspace all permutation operations are represented by two-dimensional matrices. This is in contrast with their one-dimensional representations in the fully symmetric and fully antisymmetric case (there they are represented by a simple number, 1 or -1).

To illustrate this, we now treat the case of three particles explicitly. Using the shorthand notation $|ijk\rangle$, where i labels the state of the first particle, j of the second, and k of the third,

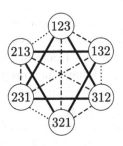

Fig. 2.1 All permutations of three particles. The six states obtained by permutation of the particles' coordinates can be represented by the vertices of a hexagon. The states connected by a triangle (thick lines) can be obtained from each other by cyclic permutation. Pair permutations connect states from different triangles and are represented by thin lines: dotted, dashed, and dotted-dashed.

we can make a fully symmetric wave function,

$$|(123)_s\rangle = \frac{1}{\sqrt{6}}\{|123\rangle + |231\rangle + |312\rangle + |132\rangle + |321\rangle + |213\rangle\}, \qquad (2.7)$$

which is symmetric under any pair permutation of two particles, meaning that $\hat{P}_{ab}|(123)_s\rangle = |(123)_s\rangle$, and a fully antisymmetric wave function,

$$|(123)_a\rangle = \frac{1}{\sqrt{6}}\{|123\rangle + |231\rangle + |312\rangle - |132\rangle - |321\rangle - |213\rangle\}, \qquad (2.8)$$

which is antisymmetric under any pair permutation, $\hat{P}_{ab}|(123)_a\rangle = -|(123)_a\rangle$. This is easy to see from Fig. 2.1. There are two triangles of states connected by thick lines. The components of the wave function $|(123)_a\rangle$ belonging to different thick triangles come with different sign. Any pair permutation (thin lines) swaps the thick triangles, thus changing the sign of the wave function. The functions $|(123)_s\rangle$ and $|(123)_a\rangle$ are two linearly independent combinations of all six permutations of $|123\rangle$. We included a normalization factor of $1/\sqrt{6}$ so that, if $|123\rangle$ is normalized, then also $|(123)_{s,a}\rangle$ are normalized.

There exist, however, other linearly independent combinations of the permutations of $|123\rangle$ which exhibit more complicated symmetry. Let us consider the set of two functions

$$|(123)_{+A}\rangle = \frac{1}{\sqrt{3}}\{|123\rangle + \varepsilon|231\rangle + \varepsilon^*|312\rangle\},$$
$$\qquad (2.9)$$
$$|(123)_{-A}\rangle = \frac{1}{\sqrt{3}}\{|132\rangle + \varepsilon^*|321\rangle + \varepsilon|213\rangle\}.$$

Here we defined $\varepsilon \equiv e^{i2\pi/3}$ so that $\varepsilon^3 = 1$ and $\varepsilon^2 = \varepsilon^{-1} = \varepsilon^*$. Such functions are transformed to themselves by a cyclic permutation of particles, $|123\rangle \to |231\rangle$ and so on. This cyclic permutation gives $\hat{P}_{cyc}|(123)_{+A}\rangle = \varepsilon|(123)_{+A}\rangle$ and $\hat{P}_{cyc}|(123)_{-A}\rangle = \varepsilon^*|(123)_{-A}\rangle$.[1] A *pair* permutation switches between the two functions $|(123)_{\pm A}\rangle$. Therefore, any permutation of particles in the two-dimensional subspace spanned by these functions produces again a state within the same subspace. In other words, all permutation operators can be

---

[1] For this reason, these functions are also used in solid state theory to represent the trigonal symmetry of $2\pi/3$ rotations. This symmetry is also evident in Fig. 2.1.

represented by two-dimensional matrices in the subspace. The functions are said to form a two-dimensional irreducible presentation of the permutation group.

There is an alternative set of two functions

$$|(123)_{-B}\rangle = \frac{1}{\sqrt{3}}\{|123\rangle + \varepsilon^*|231\rangle + \varepsilon|312\rangle\},$$

$$|(123)_{+B}\rangle = \frac{1}{\sqrt{3}}\{|132\rangle + \varepsilon|321\rangle + \varepsilon^*|213\rangle\},$$

(2.10)

which have identical transformation properties with respect to all permutations. They are said to belong to the same two-dimensional irreducible representation.

Control question. Give the matrix representation of the permutation operator $\hat{P}_{13}$ in the partially symmetric subspaces spanned by (2.9) and (2.10).

In principle these partial symmetries could have been exploited by Nature as well. There is no a-priori reason why wave functions must belong to a one-dimensional vector space. One could also create partially symmetric multi-dimensional representations of quantum states which are eigenstates of a permutation symmetric Hamiltonian. Fortunately, however, it seems that Nature has chosen the most simple option: only fully symmetric and fully antisymmetric states are to be found.

2.2 The symmetry postulate

The general observation that the wave function of a set of identical particles is either fully symmetric or fully antisymmetric can now be expressed in the form of an exact definition. We put it here as a postulate, which distinguishes all particles as either *bosons* or *fermions*.

---

symmetry postulate

The wave function of N identical bosons is completely symmetric with respect to particle pair permutations.

The wave function of N identical fermions is completely antisymmetric with respect to particle pair permutations.

One of the implications of this postulate is that quantum particles are more than just identical, they are truly *indistinguishable*. Let us illustrate this statement with a comparison with classical objects, such as ping-pong balls. We can imagine having two identical ping-pong balls, which we distinguish by putting one ball in our left pocket and the other in our right pocket (see Fig. 2.2). Then there are two well distinguishable objects: a ball "sitting in our right pocket" and a ball "sitting in our left pocket". Let us try to follow the same reasoning for two quantum particles, where the two pockets are now two quantum states, $|L\rangle$ and $|R\rangle$. Suppose that the two particles are bosons (but the argument also holds for

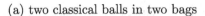

(a) two classical balls in two bags

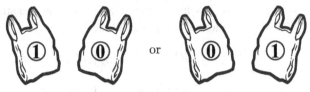

(b) two quantum balls in two quantum bags

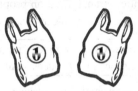

Fig. 2.2 The possible ways to distribute two balls, labeled "0" and "1", over two bags, so that there is one ball in each bag. (a) If the balls and bags are classical, then there are obviously two possibilities. (b) For quantum balls in "quantum bags" there is only one possibility due to the symmetry postulate.

fermions). We would again like to put each particle in a separate "pocket," i.e. we would like to create the state $|L\rangle|R\rangle$ or $|R\rangle|L\rangle$, thereby distinguishing the particles. We then realize that this pocket trick cannot be implemented. Since the wave function must be symmetric with respect to permutation, the only permitted state with one ball in each pocket is the superposition

$$\psi_s = \frac{1}{\sqrt{2}}(|L\rangle|R\rangle + |R\rangle|L\rangle). \tag{2.11}$$

Neither $|L\rangle|R\rangle$ nor $|R\rangle|L\rangle$ can be realized. It is clear that, when the particles are in the state ψ_s, it is impossible to tell which of them is in state $|L\rangle$ and which in state $|R\rangle$: they both are in both states!

Control question. Why is there a square root in the above formula?

2.2.1 Quantum fields

If one contemplates the implications of the symmetry postulate, it soon gets strange, and one is forced to draw some counterintuitive conclusions. For instance, the postulate seems to contradict the most important principle of physics, the *locality principle*. We would expect that objects and events that are separated by large distances in space or time must be independent of each other and should not correlate. Suppose Dutch scientists pick up an electron and confine it in a quantum dot, while a group of Japanese colleagues performs a similar experiment (Fig. 2.3). The two electrons picked up do not know about each other, have never been in a close proximity, nor do the two research teams communicate with each other. However, due to the magic of the symmetry postulate, the wave function of these two identical electrons must be antisymmetric. Therefore, the electrons correlate! Why?

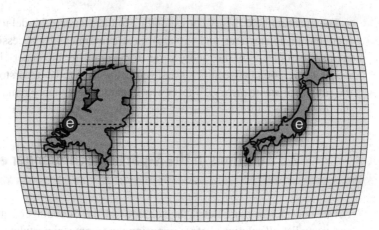

Fig. 2.3   Electrons separated by large distances in space and time all originate from the same quantum field. This explains their correlations, imposed by the symmetry postulate.

The attempts to solve this paradox have led to the most important theoretical finding of the 20th century: there are no particles! At least, particles are not small separate actors in the otherwise empty scene of space and time, as they have been envisaged by generations of pre-quantum scientists starting from Epicurus. What we see as particles are in fact quantized excitations of corresponding *fields*. There is a single entity – the electron field – which is responsible for all electrons in the world. This field is the same in Japan and in the Netherlands, thus explaining the correlations. The field persists even if no electrons are present: the physical vacuum, which was believed to be empty, is in fact not.

The electron field is similar to a classical field – a physical quantity defined in space and time, such as an electric field in a vacuum or a pressure field in a liquid. Starting from the next chapter, most of the material in this book illustrates and explains the relation between particles and fields: a relation which is a dichotomy in classical science, and a unity in quantum mechanics. We see fields emerging from indistinguishable particles, and, conversely, also see particles emerging when a classical field is quantized. The consequences of these relations are spelled out in detail.

The discovery of the concept just outlined was revolutionary. It has radically changed the way we understand and perceive the Universe. Theory able to treat particles and fields in a unified framework – quantum field theory – has become a universal language of science. Any imaginable physical problem can be formulated as a quantum field theory (though frequently it is not required, and often it does not help to solve the problem).

Although condensed matter provides interesting and stimulating examples of quantum field theories (we present some in later chapters), the most important applications of the framework are aimed at explaining fundamental properties of the Universe and concern elementary particles and cosmology. As such, the theory must be relativistic to conform with the underlying symmetries of the space–time continuum. The proper relativistic theory of quantized fields provides a more refined understanding of the symmetry postulate. It gives the relation between spin and statistics: particles with half-integer spin are bound to be fermions while particles with integer spin are bosons.

The quantum field theory corresponding to the Standard Model unifies three of the four fundamental interactions and explains most experimentally accessible properties of elementary particles. Still the theory is far from complete: the underlying symmetries of this world, the relation between fields and geometry, have to be understood better (those being the subject of *string theory*). This is why quantum field theory remains, and will remain in the foreseeable future, a very active and challenging research area.

2.3 Solutions of the *N*-particle Schrödinger equation

Let us consider in more detail the solutions of the Schödinger equation for N particles, and try to list *all* solutions before imposing the symmetry postulate. To simplify the work, we skip the interactions between the particles. The Hamiltonian is then given by the first line of (2.4), and is just a sum of one-particle Hamiltonians. The Schrödinger equation becomes

$$\hat{H}\psi = \sum_{i=1}^{N} \left\{ -\frac{\hbar^2}{2m}\frac{\partial^2}{\partial \mathbf{r}_i^2} + V(\mathbf{r}_i) \right\} \psi = E\psi(\{\mathbf{r}_i\}). \tag{2.12}$$

We can find the general solution of this equation in terms of one-particle eigenfunctions $\varphi_n(\mathbf{r})$ that satisfy

$$\left\{ -\frac{\hbar^2}{2m}\frac{\partial^2}{\partial \mathbf{r}^2} + V(\mathbf{r}) \right\} \varphi_n(\mathbf{r}) = E_n\varphi_n(\mathbf{r}). \tag{2.13}$$

Here n labels the one-particle quantum states, and to distinguish them from the many-particle states, we call them *levels*.

Control question. What are the levels φ_n and their energies E_n for identical particles in a one-dimensional infinite square well?

These levels φ_n can then be used to construct N-particle solutions. The general solution to (2.12) can be written as an N-term product of one-particle solutions φ_n,

$$\psi_{\{\ell_i\}}(\{\mathbf{r}_i\}) = \varphi_{\ell_1}(\mathbf{r}_1)\varphi_{\ell_2}(\mathbf{r}_2)\ldots\varphi_{\ell_N}(\mathbf{r}_N), \tag{2.14}$$

where the integer index ℓ_i gives the level in which the ith particle is situated. The set of indices $\{\ell_i\} = \{\ell_1, \ell_2, \ell_3, \ldots, \ell_N\}$ can be any set of (positive) integers, and to each set of numbers corresponds a single wave function (2.14). The set of all $\psi_{\{\ell_i\}}$ we can construct, forms a complete basis of the space of all wave functions of N identical particles.

Control question. Prove that this basis is orthonormal.

To illustrate, with three particles we could create a three-particle state $\varphi_1(\mathbf{r}_1)\varphi_1(\mathbf{r}_2)\varphi_1(\mathbf{r}_3)$, describing the situation where all three particles occupy the first level. Another allowed three-particle state, $\varphi_8(\mathbf{r}_1)\varphi_7(\mathbf{r}_2)\varphi_6(\mathbf{r}_3)$, has the first particle in the eighth level, the second in the seventh level and the third in the sixth level. We could go on like this forever: every state $\varphi_{\ell_1}(\mathbf{r}_1)\varphi_{\ell_2}(\mathbf{r}_2)\varphi_{\ell_3}(\mathbf{r}_3)$ is formally a solution of the Schrödinger equation.

The energy corresponding to the many-particle state (2.14) is given by the sum of the one-particle energies,

$$E = \sum_{i=1}^{N} E_{n_i}. \tag{2.15}$$

It is instructive to re-sum this expression in the following way. We introduce the (integer) occupation numbers n_k, which tell us how many times the level k appears in the index $\{\ell_i\}$, or, in other words, how many particles occupy level k. Obviously, we can rewrite

$$E = \sum_{k=1}^{\infty} E_k n_k. \tag{2.16}$$

The point we would like to concentrate on now is that there are plenty of many-particle states which have the same total energy E. We know that in quantum mechanics this is called *degeneracy*. Let us find the degeneracy of an energy eigenvalue that corresponds to a certain choice of occupation numbers $\{n_1, n_2, \dots\} \equiv \{n_k\}$. For this, we have to calculate the number of ways in which the particles can be distributed over the levels, given the numbers $\{n_k\}$ of particles in each level. Let us start with the first level: it has to contain n_1 particles. The first particle which we put in the first level is to be chosen from N particles, the second particle from $N - 1$ remaining particles, etc. Therefore, the number of ways to fill the first level is

$$W_1 = \frac{1}{n_1!} N(N-1) \cdots (N - n_k + 1) = \frac{N!}{n_1!(N - n_1)!}. \tag{2.17}$$

Control question. Do you understand why the factor $n_1!$ appears in the denominator of (2.17)?

Let us now fill the second level. Since we start with $N - n_1$ remaining particles,

$$W_2 = \frac{(N - n_1)!}{n_2!(N - n_1 - n_2)!}. \tag{2.18}$$

The total number of ways to realize a given set of occupation numbers $\{n_k\}$ is

$$W = \prod_{i=1}^{\infty} W_i = \frac{N!}{n_1! \, n_2! \cdots n_\infty!}, \tag{2.19}$$

as illustrated in Fig. 2.4. An infinite number of levels to sum and multiply over should not confuse us, since for finite N most of the levels are empty, and do not contribute to the sums and products.

These considerations on degeneracy provide a way to connect the seemingly artificial and abstract symmetry postulate with experimental facts. If there were no symmetry postulate dictating that a wave function be either symmetric or antisymmetric, then the number W would indeed give the degeneracy of a state with a certain set of occupation numbers $\{n_k\}$, i.e. it would give the number of possible wave functions which realize a specific energy $E = \sum_n E_k n_k$. Since the degeneracy of E enhances the probability of finding the system at this energy, this degeneracy would manifest itself strongly in statistical mechanics. In this case, as shown by Gibbs, the entropy of an ideal gas would not

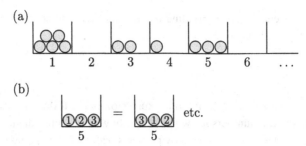

Fig. 2.4 The number W can be understood directly from a simple statistical consideration. Suppose we have a state with occupation numbers $\{5, 0, 2, 1, 3, 0, \dots\}$, as depicted in (a). There are $N!$ ways to fill the levels like this: the first particle we put in level 1 can be chosen out of N particles, the second out of $N-1$ particles, etc. However, since the particles are identical, the order in each level does not matter. Indeed, as shown in (b), the factors $\varphi_5(\mathbf{r}_1)\varphi_5(\mathbf{r}_2)\varphi_5(\mathbf{r}_3)$ and $\varphi_5(\mathbf{r}_3)\varphi_5(\mathbf{r}_1)\varphi_5(\mathbf{r}_2)$ in the many-particle wave function (2.14) are equal. Therefore $N!$ has to be divided by the number of permutations possible in each level, $n_1! \, n_2! \cdots$

be proportional to its volume. This proportionality, however, is an experimental fact, and suggests that the degeneracy of every energy level is one. Already, before the invention of quantum theory, Gibbs showed that if one assumes the gas particles to be *indistinguishable*, the degeneracy of all energy levels reduces to one, resulting in the right expression for the entropy. The symmetry postulate provided the explanation for this absence of degeneracy and for the indistinguishability of identical particles. From the postulate it follows that there is only one *single* quantum state corresponding to each set of occupation numbers. We show this in the two following sections.

2.3.1 Symmetric wave function: Bosons

Let us now explicitly construct wave functions that *do* satisfy the symmetry postulate. We start with bosons, so we have to build completely symmetric wave functions. Suppose we have a set of occupation numbers $\{n_k\}$, and we pick *any* of the wave functions (2.14) that correspond to this set. All other basis states satisfying the same $\{n_k\}$ can be obtained from this state by permutation of particles. We construct a linear combination of all these permuted states, taking them with equal weight C,

$$\Psi^{(S)}_{\{n_k\}} = C \sum_P P \left\{ \varphi_{\ell_1}(\mathbf{r}_1)\varphi_{\ell_2}(\mathbf{r}_2) \dots \varphi_{\ell_N}(\mathbf{r}_N) \right\}, \tag{2.20}$$

where the summation is over all different permutations. The result obtained does not change upon any pair permutation, that is, it presents the completely symmetric wave function sought. We are only left to determine the unknown C from the following normalization condition,

$$1 = \langle \Psi^{(S)} | \Psi^{(S)} \rangle = |C|^2 \sum_{P,P'} \langle P | P' \rangle = |C|^2 \sum_P \langle P | P \rangle$$

$$= |C|^2 \times (\text{number of permutations}). \tag{2.21}$$

Albert Einstein (1879–1955)
Won the Nobel Prize in 1921 for "his services to theoretical physics, and especially for his discovery of the law of the photoelectric effect."

Albert Einstein is probably the most famous physicist of the last three centuries, if not ever. He is most widely known for his theory of relativity, but his contributions to theoretical physics are numerous and cover almost all fields. The work for which he was awarded the Nobel prize was his explanation of the photoelectric effect, introducing the idea that light consists of individual discrete quanta – photons. This idea of quantization became a major inspiration for the development of quantum mechanics in the 1920s.

In 1924 Einstein received a letter from the Indian scientist Satyendra Nath Bose about the statistical properties of photons. Bose had discovered that the inconsistencies between experiments and the theory of radiation could be solved by assuming that photons were actually *indistinguishable*. Several journals had rejected Bose's manuscript on this topic, so that he had decided to write personally to Einstein. Einstein immediately agreed with Bose's idea. He translated the manuscript to German, and sent it together with a strong recommendation to the famous *Zeitschrift für Physik*, where it was then accepted immediately. The special statistics resulting from Bose's idea became known as *Bose–Einstein* statistics, and the particles obeying it were called *bosons*.

After Einstein graduated in 1900 from the Eidgenössische Polytechnische Schule in Zurich, he spent almost two years in vain looking for a post as lecturer. Finally, he accepted a job as assistant examiner at the Swiss patent office in Bern. While being employed at the patent office, he worked on his dissertation and started publishing papers on theoretical physics. The year 1905 became known as his *annus mirabilis* in which he published four groundbreaking works which would later revolutionize physics: he explained the photoelectric effect and Brownian motion, he presented his theory of special relativity, and he derived the famous relation between mass and energy $E = mc^2$.

These papers quickly earned him recognition, and in 1908 he was appointed Privatdozent in Bern. In the following years he moved to Zurich, Prague, and back to Zurich, and in 1914 he finally became Director of the Kaiser Wilhelm Institute for Physics in Berlin. When Hitler came to power in January 1933, Einstein was visiting the US to give a series of lectures. He immediately decided not to return to Germany, and accepted a post at Princeton University. Einstein became an American citizen in 1940, and stayed there till his death in 1955.

But this number of permutations we know already: it is W, as calculated in (2.19). Therefore,

$$C = \sqrt{\frac{n_1! \, n_2! \cdots n_\infty!}{N!}}. \tag{2.22}$$

Let us illustrate this procedure for the case of three particles. Suppose we have a state which is characterized by the occupation numbers $\{2, 1, 0, 0, \dots\}$: two of the particles are in the first level and the third is in the second level. We pick a wave function satisfying these occupations, $\varphi_1(\mathbf{r}_1)\varphi_1(\mathbf{r}_2)\varphi_2(\mathbf{r}_3)$. From this state we construct an equal superposition of all possible permutations, as described above, and include the prefactor $\sqrt{2 \cdot 1}/\sqrt{6}$,

$$\Psi^{(S)} = \frac{1}{\sqrt{3}}\left\{\varphi_1(\mathbf{r}_1)\varphi_1(\mathbf{r}_2)\varphi_2(\mathbf{r}_3) + \varphi_1(\mathbf{r}_1)\varphi_2(\mathbf{r}_2)\varphi_1(\mathbf{r}_3) + \varphi_2(\mathbf{r}_1)\varphi_1(\mathbf{r}_2)\varphi_1(\mathbf{r}_3)\right\}, \tag{2.23}$$

which is the desired fully symmetric wave function.

Let us emphasize here that the symmetrization procedure creates one *unique* symmetric wave function for a given set of occupation numbers. So, indeed, imposing the symmetry postulate removes all degeneracies in this case.

Control question. Can you construct a three-particle fully symmetric wave function corresponding to the occupation numbers $\{1, 1, 1, 0, \dots\}$?

2.3.2 Antisymmetric wave function: Fermions

For the case of identical fermions, the symmetry postulate requires a fully antisymmetric wave function. To build such a wave function, we proceed in a way similar to that followed for bosons.

We first fix the set of occupation numbers $\{n_k\}$. We must note here, however, that if any of the $n_k \geq 2$, there is no chance that we can build a completely antisymmetric function. This follows from the fact that any basis state which satisfies this choice is automatically symmetric with respect to permutation of the particles located in the multiply occupied level k. This gives us the celebrated *Pauli principle*: all occupation numbers for fermions are either 0 or 1, there is no way that two fermions can share the same level.

We now pick one of the basis states fitting the set of occupation numbers. Again, all other basis states that satisfy the same set are obtained by permutations from this state. To proceed, we have to introduce the concept of *parity* of a permutation. We define that a permutation is:

- *odd* if obtained by an odd number of pair permutations, and
- *even* if obtained by an even number of pair permutations,

both starting from the same initial state. We then construct a linear combination of all basis states, taking those which are even with a weight C and those which are odd with $-C$. The resulting linear combination then changes sign upon permutation of any pair of particles, and is therefore the completely antisymmetric state we were looking for,

$$\Psi^{(A)}_{\{n_k\}} = C \sum_P (-1)^{\text{par}(P)} P \left\{\varphi_{\ell_1}(\mathbf{r}_1)\varphi_{\ell_2}(\mathbf{r}_2) \dots \varphi_{\ell_N}(\mathbf{r}_N)\right\}, \tag{2.24}$$

Enrico Fermi (1901–1954)

Won the Nobel Prize in 1938 for "his demonstrations of the existence of new radioactive elements produced by neutron irradiation, and for his related discovery of nuclear reactions brought about by slow neutrons."

In 1926, Enrico Fermi was a young lecturer at the University of Florence in Italy, and he was studying the statistical properties of particles obeying Pauli's exclusion principle. He showed that assuming a comparable exclusion principle for the identical particles in an ideal gas, one could derive a theory for the quantization of the gas, which was perfectly consistent with thermodynamics. In the same year, Paul Dirac was investigating in a more general context the statistical properties of identical quantum particles. He concluded that there are two types of quantum particle: those obeying Bose–Einstein statistics, and those obeying the statistics Fermi had developed for quantizing the ideal gas. The latter became known as *Fermi–Dirac* statistics, and the particles obeying it as *fermions*.

In 1927, Fermi gained a professorial position at the University of Rome. He directed his group there mainly toward the field of nuclear physics, excelling in both theoretical and experimental research. During this time, he developed his famous theory of beta decay, and made also some important experimental steps toward nuclear fission. In 1938, after receiving the Nobel prize, Fermi moved with his family to New York, mainly because of the threat the fascist regime of Mussolini posed for his Jewish wife. He first accepted a position at Columbia University, and moved in 1946 to the University of Chicago, where he stayed until his death in 1954. During the second world war, he was heavily involved in the Manhattan project, the development of the American nuclear bomb.

where par(P) denotes the parity of the permutation P, so that par(P) = 0 for P even, and par(P) = 1 for P odd. The above superposition can be written in a very compact way using the definition of the determinant of a matrix. If we define the elements of the $N \times N$ matrix $\mathbf{M}$ to be $M_{ij} = \varphi_{\ell_i}(\mathbf{r}_j)$, then we see that the linear combination (2.24) can be written as

$$\Psi_{\{n_k\}}^{(A)} = C \det(\mathbf{M}). \tag{2.25}$$

This determinant is called the *Slater determinant*.

Control question. In the above equation, i labels the levels and j labels the particles. Why are both dimensions of the matrix N?

We still have to determine C. As in the bosonic case, it follows from the normalization condition $\langle \Psi^{(A)} | \Psi^{(A)} \rangle = 1$. We find the same result for C, but since every n_k is either 0 or 1, we can simplify

$$C = \sqrt{\frac{n_1! \, n_2! \cdots n_\infty!}{N!}} = \frac{1}{\sqrt{N!}}. \tag{2.26}$$

Control question. Can you write down the antisymmetrized three-particle wave function corresponding to the occupation numbers $\{1, 1, 1, 0, \ldots\}$?

Following this antisymmetrization procedure, we end up with one unique antisymmetric wave function for a given set of occupation numbers. Therefore, we see that also in the case of fermions, the symmetrization postulate removes all degeneracies from the many-particle states.

2.3.3 Fock space

To proceed with the quantum description of identical particles, we need the Hilbert space of all possible wave functions of all possible particle numbers. This space is called Fock space, and in this section we show how to construct it.

In the previous sections we explained that for a system of many identical particles most solutions of the Schrödinger equation are actually redundant: only completely symmetric and antisymmetric states occur in Nature. In fact, it turns out that for both bosons and fermions any set of occupation numbers corresponds to exactly one unique many-particle wave function that obeys the symmetry postulate. This suggests that all basis states of the many-particle Hilbert space can be characterized by just a set of occupation numbers. On this idea the construction of Fock space is based.

Let us now show formally how to construct Fock space in four steps:

1. We start with the space of all possible solutions of the Schrödinger equation for N particles. This space is represented by the basis states in (2.14).
2. We combine all these basis states for all possible N, and introduce a special basis state corresponding to $N = 0$. This state is called the *vacuum*.
3. We construct (anti)symmetric linear combinations of these basis states to obtain states that satisfy the symmetry postulate.
4. We throw away all other states since they do not satisfy the postulate.

The resulting set of basis states spans Fock space, and any of these basis states can be uniquely characterized by a set of occupation numbers. Making use of this, we introduce the "ket" notation for the basis states in Fock space,

$$|\{n_k\}\rangle \equiv |n_1, n_2, \ldots\rangle, \tag{2.27}$$

where the number n_k is the occupation number of the level k. Of course, in order to translate a vector in Fock space to an actual many-particle wave function, one needs to know all wave functions $\psi_k(\mathbf{r})$ corresponding to the levels.

The special state defined in step 2 above, the vacuum, reads

$$|0, 0, \ldots\rangle \equiv |0\rangle. \tag{2.28}$$

With this vacuum state included, we have constructed a complete basis spanning the Hilbert space of all possible wave functions for any N-particle system, where N can range from zero to infinity. The basis is orthonormal, so that

$$\langle\{n_k\}|\{n'_k\}\rangle = \prod_k \delta_{n_k, n'_k}. \tag{2.29}$$

Control question. Why is there a product over k in this expression?

We hope that it is clear that a vector in Fock space represents in a very intuitive and physical way the actual many-particle wave function of a system. From the vector we can immediately see how many particles are present in the system, how they are distributed over the levels, and thus what the total energy of the system is. As we see in the next chapters, working in Fock space allows for the development of a very convenient tool for solving many-body problems.

Table 2.1 Summary: Identical particles

Schrödinger equation for many identical particles

$$\hat{H} = \sum_{i=1}^{N} \left\{ -\frac{\hbar^2}{2m} \frac{\partial^2}{\partial \mathbf{r}_i^2} + V_i(\mathbf{r}_i) \right\} + \sum_{i=2}^{N} \sum_{j=1}^{i-1} V^{(2)}(\mathbf{r}_i, \mathbf{r}_j) + \sum_{i=3}^{N} \sum_{j=2}^{i-1} \sum_{k=1}^{j-1} V^{(3)}(\mathbf{r}_i, \mathbf{r}_j, \mathbf{r}_k) + \ldots$$

the permutation symmetry of this Hamiltonian imposes restrictions on the allowed symmetries of the many-particle wave function

Symmetry postulate

there exist only *bosons* and *fermions*

bosons: the N-particle wave function is fully symmetric under particle pair permutation

fermions: the N-particle wave function is fully antisymmetric under particle pair permutation

Solutions of the N-particle Schrödinger equation, ignoring interactions $V^{(2)}$, $V^{(3)}$, ...

general: N-term product of single-particle solutions, $\psi_{\{\ell_i\}}(\{\mathbf{r}_i\}) = \varphi_{\ell_1}(\mathbf{r}_1)\varphi_{\ell_2}(\mathbf{r}_2)\ldots\varphi_{\ell_N}(\mathbf{r}_N)$,

where $\{\ell_i\}$ is the set of levels occupied (N numbers)

bosons: symmetric combination $\Psi_{\{n_k\}}^{(S)} = \sqrt{\dfrac{n_1!\, n_2!\cdots n_\infty!}{N!}} \sum_P P\left\{\varphi_{\ell_1}(\mathbf{r}_1)\varphi_{\ell_2}(\mathbf{r}_2)\ldots\varphi_{\ell_N}(\mathbf{r}_N)\right\}$,

where $\{n_k\}$ is the set of occupation numbers (maximally N numbers)

fermions: antisymmetric combination $\Psi_{\{n_k\}}^{(A)} = \dfrac{1}{\sqrt{N!}} \sum_P (-1)^{\mathrm{par}(P)} P\left\{\varphi_{\ell_1}(\mathbf{r}_1)\varphi_{\ell_2}(\mathbf{r}_2)\ldots\varphi_{\ell_N}(\mathbf{r}_N)\right\}$,

where n_k is either 0 or 1

Fock space

each set of $\{n_k\}$ corresponds to exactly *one* allowed many-particle wave function

the basis of $|\{n_k\}\rangle \equiv |n_1, n_2, \ldots\rangle$ is thus *complete*

Exercises

1. *Pairwise interaction for two bosons* (solution included). Consider particles in the system of levels labeled by n, corresponding to the states $|n\rangle$, with *real* wave functions $\psi_n(\mathbf{r})$ and energies E_n. There is a weak pairwise interaction between the particles,

$$\hat{H}_{\text{int}} = U(\mathbf{r}_1 - \mathbf{r}_2) = U\delta(\mathbf{r}_1 - \mathbf{r}_2).$$

This interaction is called *contact* interaction since it is present only if two particles are at the same place.

a. Assume that there are two bosonic particles in the system. Write down the wave functions of all possible stationary states in terms of two-particle basis states $|nm\rangle \equiv \psi_n(\mathbf{r}_1)\psi_m(\mathbf{r}_2)$.

b. Express the matrix element $\langle kl|\hat{H}_{\text{int}}|nm\rangle$ of the contact interaction in terms of the integral

$$A_{klmn} = \int d\mathbf{r}\, \psi_k(\mathbf{r})\psi_l(\mathbf{r})\psi_m(\mathbf{r})\psi_n(\mathbf{r}).$$

Note that this integral is fully symmetric in its indices k, l, m, and n.

c. Compute the correction to the energy levels to first order in U. Express the answer in terms of A_{klmn}.

d. Now assume that there are N bosonic particles in one level, say $|1\rangle$. Compute the same correction.

2. *Particle on a ring* (solution included). Consider a system of levels in a ring of radius R. The coordinates of a particle with mass M are conveniently described by just the angle θ in the polar system (Fig. 2.5a). The Hamiltonian in this coordinate reads

$$\hat{H} = -\frac{\hbar^2}{2MR^2}\left(\frac{\partial}{\partial\theta}\right)^2,$$

if there is no magnetic flux penetrating the ring. Otherwise, the Hamiltonian is given by

$$\hat{H} = \frac{\hbar^2}{2MR^2}\left(-i\frac{\partial}{\partial\theta} - \nu\right)^2,$$

(a)   (b)   (c)

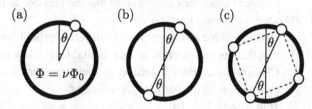

Fig. 2.5   (a) Particle on a ring penetrated by a flux ν. The polar angle θ parameterizes the particle's coordinate. (b) Two strongly repelling particles on a ring go to opposite points on the ring, forming a rigid molecule. (c) A molecule formed by $N = 4$ repelling particles.

ν expressing the flux in the ring in dimensionless units: $\nu = \Phi/\Phi_0$. Here, $\Phi_0 = 2\pi\hbar/e$ is the flux quantum, where e is the electric charge of the particle. The effect of this flux is an extra phase $\phi = \nu(\theta_1 - \theta_2)$ acquired by the particle when it travels from the point θ_1 to θ_2. Indeed, the transformation $\Psi(\theta) \to \Psi(\theta)\exp(-i\nu\theta)$ eliminates the flux from the last Hamiltonian.

Owing to rotational symmetry, the wave functions are eigenfunctions of angular momentum, $|m\rangle = (2\pi)^{-1/2}e^{im\theta}$. An increase of θ by a multiple of 2π brings the particle to the same point. Therefore, the wave functions must be 2π-periodic. This restricts m to integer values.

a. Let us first carry out an exercise with the levels. Compute their energies. Show that at zero flux the levels are double-degenerate, meaning that $|m\rangle$ and $|-m\rangle$ have the same energy. Show that the energies are periodic functions of ν.

b. We assume $\nu = 0$ and concentrate on two levels with $m = \pm 1$. Let us denote the levels as $|+\rangle$ and $|-\rangle$. There are four degenerate states for two particles in these levels, $|++\rangle$, $|-+\rangle$, $|+-\rangle$, and $|--\rangle$. How many states for boson particles are there? How many for fermion particles? Give the corresponding wave functions.

c. Compute the contact interaction corrections for the energy levels of two bosons. Does the degeneracy remain? Explain why.

d. Explain why the results of the previous exercise cannot be immediately applied for point (b) although the wave functions of the levels can be made real by transforming the basis to $|a\rangle = \{|+\rangle - |-\rangle\}/i\sqrt{2}$ and $|b\rangle = \{|+\rangle + |-\rangle\}/\sqrt{2}$.

3. *Pairwise interaction for two fermions.* Let us take the setup of Exercise 1 and consider electrons (fermions with spin $\frac{1}{2}$) rather than bosons. The levels are thus labeled $|n\sigma\rangle$, with $\sigma = \{\uparrow, \downarrow\}$ labeling the spin direction and n indicating which "orbital" the electron occupies.

a. Put two electrons into the same orbital level n. Explain why only one state is permitted by the symmetry postulate and give its wave function. Compute the correction to its energy due to contact interaction.

b. How does the wave function change upon permutation of the electrons' coordinates?

c. Now put two electrons into two different orbital levels n and m. How many states are possible? Separate the states according to the total electron spin into singlet and triplet states and give the corresponding wave functions.

d. Compute again the corrections to the energies due to the contact interaction.

4. *Pairwise interaction for three fermions.*

a. Consider three electrons in three different orbital levels n, m, and p. We assume that there is one electron in each orbital level. How many states are then possible? Classify these states according to their transformation properties with respect to permutation of the particles' coordinates.
   Hint. Use (2.7–2.10).

b. Classify the states found in (a) according to total spin. Compute the matrix elements of the contact interaction (assuming real orbital wave functions) and find the interaction-induced splitting of these otherwise degenerate states.

c. Consider now the case where two electrons are in the same orbital level n while the third one is in m. Classify the states again. Find the interaction-induced splitting.

5. *Molecule on a ring.* Let us return to the ring setup described in Exercise 2. We now put two particles in the ring and assume a strong long-range repulsion between the particles. In this case, at low energy the particles are situated in opposite points of the ring: they form a sort of rigid molecule. The resulting two-particle wave function is again a function of a single coordinate θ (see Fig. 2.5(b)).

a. Assume that the particles are distinguishable while having the same mass. Write down the wave functions corresponding to discrete values of momentum, and compute the corresponding energies.

b. What changes if the particles are indistinguishable bosons?
    Hint. A rotation of the molecule by an angle π permutes the particles.

c. Characterize the energy spectrum assuming the particles are fermions.

d. Consider a molecule made of N strongly repelling particles ($N = 4$ in Fig. 2.5(c)). Give the energy spectrum for the cases where the particles are bosons and fermions.

6. *Anyons.* In 1977, Jon Magne Leinaas and Jan Myrheim figured out that the symmetry postulate would not apply in two-dimensional space. In 1982, Frank Wilczek proposed a realization of this by constructing hypothetical particles that he called *anyons*. Those are obtained by attributing a magnetic flux to the usual fermions or bosons. When one anyon makes a loop around another anyon, it does not necessarily end up in the same state it started from, but it acquires an extra phase shift due to the "flux" present in the loop. It is believed that elementary excitations of the two-dimensional electron gas in the Fractional Quantum Hall regime are in fact anyons.

a. Consider the two-particle molecule from Exercise 5. Suppose the molecule is rotated over an angle α. By what angle is particle 1 then rotated with respect to particle 2?

b. Attach a flux ν to each particle. What is the phase change acquired by the molecule after rotation over an angle α? What is the phase change acquired in the course of a π-rotation that permutes the particles?

c. Show that the effect of this flux can be ascribed to particle *statistics*. The same effect is achieved if one considers fluxless particles of which the wave function acquires a phase factor $e^{i\beta}$ upon permutation. Express β in terms of ν.

d. Find the energy spectrum of the molecule.

Solutions

1. *Pairwise interaction for two bosons.* This is a relatively simple problem.

a. The symmetrized wave functions read

$$\text{if } n \neq m \qquad \frac{|nm\rangle + |mn\rangle}{\sqrt{2}}, \qquad \text{with } E = E_n + E_m,$$

$$\text{if } n = m \qquad |nn\rangle, \qquad \text{with } E = 2E_n.$$

b. The matrix element between two non-symmetrized states reads

$$\langle kl|\hat{H}_{\text{int}}|nm\rangle = U \int d\mathbf{r}_1 d\mathbf{r}_2 \, \psi_k(\mathbf{r}_1)\psi_l(\mathbf{r}_2)\psi_m(\mathbf{r}_1)\psi_n(\mathbf{r}_2)\delta(\mathbf{r}_1 - \mathbf{r}_2)$$

$$= UA_{klmn}.$$

c. Substituting the symmetrized states found at (a), we get $\delta E = U2A_{nnmm}$ if $n \neq m$. One would guess that $\delta E = U2A_{nnnn}$ for $n = m$, yet this is incorrect, in fact $\delta E = UA_{nnnn}$.

d. One has to recall that the interaction is between each pair of particles,

$$\hat{U} = \sum_{1 \leq i < j \leq N} U(\mathbf{r}_i - \mathbf{r}_j).$$

Each term in the above sum equally contributes to the correction. Therefore, $\delta E(N) = \frac{1}{2}N(N-1)UA_{1111}$.

2. *Particle on a ring.* This is a more complicated problem.

a. Just by substituting the wave function $\propto \exp(im\theta)$ into the Schrödinger equation one obtains

$$E_m = \frac{\hbar^2}{2MR^2}(m - v)^2.$$

Plotting this versus v yields an infinite set of parabolas which is periodic in v with period 1.

b. Three states are symmetric with respect to particle exchange,

$$|++\rangle, \quad \frac{|-+\rangle + |+-\rangle}{\sqrt{2}}, \quad \text{and} \quad |--\rangle,$$

and therefore suitable for bosonic particles. For fermions only one single antisymmetric state $\frac{1}{\sqrt{2}}\{|-+\rangle - |+-\rangle\}$ is suitable.

c. The corrections read

$$\delta E_{++} = \frac{U}{(2\pi)^2} \int d\phi_1 d\phi_2 \delta(\phi_1 - \phi_2) = U/2\pi,$$

$$\delta E_{--} = U/2\pi,$$

$$\delta E_{+-} = U/\pi,$$

and the degeneracy is lifted. Indeed, $|++\rangle$ and $|--\rangle$ have angular momentum ± 2, so they have the same energy, while the momentum of the third state is 0.

d. In the new basis, the interaction has non-diagonal matrix elements. While those can be disregarded in the case of non-degenerate levels, they become important for degenerate ones.

3

Second quantization

In this chapter we describe a technique to deal with identical particles that is called *second quantization*. Despite being a technique, second quantization helps a lot in understanding physics. One can learn and endlessly repeat newspaper-style statements *particles are fields*, *fields are particles* without grasping their meaning. Second quantization brings these concepts to an operational level.

Since the procedure of second quantization is slightly different for fermions and bosons, we have to treat the two cases separately. We start with considering a system of identical bosons, and introduce *creation* and *annihilation* operators (CAPs) that change the number of particles in a given many-particle state, thereby connecting different states in Fock space. We show how to express occupation numbers in terms of CAPs, and then get to the heart of the second quantization: the *commutation relations* for the bosonic CAPs. As a next step, we construct *field* operators out of the CAPs and spend quite some time (and energy) presenting the physics of the particles in terms of these field operators. Only afterward, do we explain why all this activity has been called second quantization. In Section 3.4 we then present the same procedure for fermions, and we mainly focus on what is different as compared to bosons. The formalism of second quantization is summarized in Table 3.1.

3.1 Second quantization for bosons

We now use the framework of identical particles we built up earlier. We begin with a basis of single-particle states – the levels. As explained in the previous chapter, these levels are the allowed one-particle states in the system considered. For instance, for a square well with infinitely high potential walls they are the single-particle eigenfunctions we know from our introductory course on quantum mechanics, and for particles in a large volume the levels are the plane wave solutions we discussed in Section 1.3.

In any case, we can label all available levels with an index k, and then use state vectors in Fock space to represent all many-particle states. In this section we focus on bosons, so the many-particle basis states in Fock space read $|\{n_k\}\rangle$, where the numbers $\{n_k\} \equiv \{n_1, n_2, \dots\}$ give the occupation numbers of the levels k. Since we are dealing with bosons, the numbers $n_1, n_2, \dots$ can all be any positive integer (or zero of course).

We now introduce an *annihilation* operator in this space. Why introduce? Usually, an operator in quantum mechanics can be immediately associated with a physical quantity, the average value of this quantity in a certain state to be obtained with the operator acting on the wave function. Would it not be simpler to first discuss the corresponding physical quantity, instead of just "introducing" an operator? Well, you must believe the authors that the annihilation operator is no exception: there indeed is a physical quantity associated with it. However, let us wait till the end of the section for this discussion.

Instead, we define the bosonic annihilation operator $\hat{b}_k$ by what it *does*: it removes a particle from the level k. Therefore, when it acts on an arbitrary basis state $|\{n_k\}\rangle$, it transforms it into another basis state that is characterized by a different set of occupation numbers. As expected from the definition of $\hat{b}_k$, most occupation numbers eventually remain the same, and the only change is in the occupation number of the level k: it reduces by 1, from n_k to $n_k - 1$. We thus write formally

$$\hat{b}_k|\ldots, n_k, \ldots\rangle = \sqrt{n_k}|\ldots, n_k - 1, \ldots\rangle. \tag{3.1}$$

As we see, it does not simply flip between the states, it also multiplies the state with an ugly-looking factor $\sqrt{n_k}$. We will soon appreciate the use of this factor, but for now just regard it as part of the definition of $\hat{b}_k$.

Let us first consider the Hermitian conjugated operator, which is obtained by transposition and complex conjugation, $\hat{b}_{nm}^\dagger = (\hat{b}_{mn})^*$. It has a separate meaning: it is a *creation* operator,[1] and its action on a basis state is defined as

$$\hat{b}_k^\dagger|\ldots, n_k, \ldots\rangle = \sqrt{n_k + 1}|\ldots, n_k + 1, \ldots\rangle, \tag{3.2}$$

that is it adds a particle into the level k.

Control question. Can you show how the factor $\sqrt{n_k + 1}$ in (3.2) directly follows from the definition of the annihilation operator (3.1)?

3.1.1 Commutation relations

Let us now come to the use of the ugly square-roots in (3.2) and (3.1). For this, we consider products of CAPs, $\hat{b}_k \hat{b}_k^\dagger$ and $\hat{b}_k^\dagger \hat{b}_k$. We now make use of the definitions given above to write

$$\hat{b}_k^\dagger \hat{b}_k|\ldots, n_k, \ldots\rangle = \hat{b}_k^\dagger \sqrt{n_k}|\ldots, n_k - 1, \ldots\rangle = \sqrt{(n_k - 1) + 1}\sqrt{n_k}|\ldots, n_k, \ldots\rangle$$
$$= n_k|\ldots, n_k, \ldots\rangle. \tag{3.3}$$

We see that the basis states in Fock space are eigenstates of the operator $\hat{b}_k^\dagger \hat{b}_k$ with eigenvalues n_k. In other words, when this operator acts on some state $|g\rangle$, it gives the number of particles in the level k,

$$\langle g|\hat{b}_k^\dagger \hat{b}_k|g\rangle = n_k. \tag{3.4}$$

Therefore, the operator $\hat{b}_k^\dagger \hat{b}_k$ is also called the *number operator* for the level k, and it is sometimes denoted $\hat{n}_k \equiv \hat{b}_k^\dagger \hat{b}_k$. It is now clear why the square-root prefactors in (3.2) and

---

[1] Some people have a difficulty remembering that $\hat{b}$ annihilates and $\hat{b}^\dagger$ creates. A confusion might come from the association of † with the sign of the cross, which may symbolize death in the Christianity-based cultures. The following mnemonics could help: *death is a new birth* or just + *adds a particle*.

(3.1) are a very convenient choice: they provide us with a very simple operator which we can use to retrieve the number of particles in any of the levels!

> **Control question.** Can you interpret the above equation for the case where the state $|g\rangle$ is not one of the basis functions $|\{n_k\}\rangle$?

Using this number operator we can construct another useful operator: the operator of the total number of particles in the system,

$$\hat{N} = \sum_k \hat{n}_k = \sum_k \hat{b}_k^\dagger \hat{b}_k. \tag{3.5}$$

Let us now look at another product of CAPs, $\hat{b}_k \hat{b}_k^\dagger$. Doing the same algebra as above, we obtain

$$\hat{b}_k \hat{b}_k^\dagger |\ldots, n_k, \ldots\rangle = \hat{b}_k \sqrt{n_k + 1} |\ldots, n_k + 1, \ldots\rangle = \sqrt{n_k + 1}\sqrt{n_k + 1} |\ldots, n_k, \ldots\rangle$$
$$= (n_k + 1)|\ldots, n_k, \ldots\rangle. \tag{3.6}$$

We see that this product is also a diagonal operator in the basis of $|\{n_k\}\rangle$, and it has eigenvalues $n_k + 1$. It is important to note that the difference of the products $\hat{b}_k \hat{b}_k^\dagger$ and $\hat{b}_k^\dagger \hat{b}_k$, i.e. the commutator of $\hat{b}_k$ and $\hat{b}_k^\dagger$, is just 1,

$$\hat{b}_k \hat{b}_k^\dagger - \hat{b}_k^\dagger \hat{b}_k = [\hat{b}_k, \hat{b}_k^\dagger] = 1. \tag{3.7}$$

This property does not depend on a concrete basis, it always holds for CAPs in the same level. Since CAPs that act on different levels always commute, we can now formulate the general *commutation rules* for bosonic CAPs.

---

commutation relations for boson CAPs

$$[\hat{b}_k, \hat{b}_{k'}^\dagger] = \delta_{kk'}$$
$$[\hat{b}_k, \hat{b}_{k'}] = [\hat{b}_k^\dagger, \hat{b}_{k'}^\dagger] = 0$$

These commutation rules are not only simple and universal, but they could serve as an alternative basis of the whole theory of identical particles. One could start with the commutation rules and derive everything else from there. Besides, they are very convenient for practical work: most calculations with CAPs are done using the commutation rules rather than the ugly definitions (3.1) and (3.2).

3.1.2 The structure of Fock space

We see that the CAPs provide a kind of navigation through Fock space. One can use them to go step by step from any basis state to any other basis state. Let us illustrate this by showing how we can create many-particle states starting from the the vacuum $|0\rangle$. First we note that we cannot annihilate any particle from the vacuum, since there are no particles in it. We see from the definition (3.1) that for any level k

$$\hat{b}_k |0\rangle = 0. \tag{3.8}$$

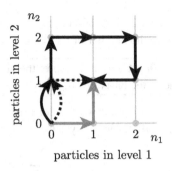

Fig. 3.1 There are many different ways to create the state $|1, 1\rangle$ from the state $|0, 0\rangle$ using creation and annihilation operators: $\hat{b}_2^\dagger \hat{b}_1^\dagger$ (gray arrows), $\hat{b}_1^\dagger \hat{b}_2^\dagger$ (dotted arrows), or why not $\hat{b}_1 \hat{b}_2 \hat{b}_1^\dagger \hat{b}_1^\dagger \hat{b}_2^\dagger \hat{b}_2^\dagger$ (black arrows)?

However, we are free to *create* particles in the vacuum. We can build any basis state out of the vacuum by applying the creation operators $\hat{b}_k^\dagger$ as many times as needed to fill all levels with the desired number of particles. Therefore,

$$|\{n_k\}\rangle = \frac{1}{\sqrt{\prod_k n_k!}} \prod_k \left(\hat{b}_k^\dagger\right)^{n_k} |0\rangle. \tag{3.9}$$

The order of the operators in the above formula is not important since all $\hat{b}_k^\dagger$ commute with each other. The prefactor ensures the correct normalization given the square-root in the definition (3.1).

Control question. Prove the above formula.

In this way the CAPs give us insight into the structure of Fock space. One could say that the space looks like a *tree*: a basis state is a node, and from this node one can make steps in all directions corresponding in all possible $\hat{b}_k^{(\dagger)}$. But actually the space looks more like a *grate*. In distinction from an ideal tree, there are several different paths – sequences of $\hat{b}_k$s and $\hat{b}_k^\dagger$s – connecting a pair of given nodes. Let us illustrate this with an example in the simple Fock space spanned by only two levels, 1 and 2. The basis states can then be presented as the nodes of a two-dimensional square grate labeled by n_1 and n_2, such as depicted in Fig. 3.1. We see that there are multiple paths from any basis state to another. For instance, to get from the vacuum $|0, 0\rangle$ to the state $|1, 1\rangle$, one could apply either of the operators $\hat{b}_2^\dagger \hat{b}_1^\dagger$ (as indicated with gray arrows) and $\hat{b}_1^\dagger \hat{b}_2^\dagger$ (dotted arrows), or even $\hat{b}_1 \hat{b}_2 \hat{b}_1^\dagger \hat{b}_1^\dagger \hat{b}_2^\dagger \hat{b}_2^\dagger$ (black arrows).

3.2 Field operators for bosons

We now introduce another type of operator, the so-called *field operators*. To construct field operators from CAPs, we first note that the CAPs as presented in (3.1) and (3.2) are defined in the convenient framework of Fock space. The basis spanned by the eigenstates of the single-particle Hamiltonian, i.e. levels with a definite energy. By defining the CAPs in this way, we have made a very specific choice for the basis. However, as we have already emphasized in Chapter 1, the choice of a basis to work in is always completely democratic: we can formulate the theory in any arbitrary basis. Let us now try to transform the CAPs, as defined in Fock space, to a representation in the basis of eigenstates of the coordinate operator $\hat{\mathbf{r}}$.

To start, we recall how to get a wave function in a certain representation if it is known in another representation. Any one-particle wave function in the level representation is fully characterized by the set of its components $\{c_k\}$. This means that the wave function can be written as the superposition

$$\psi(\mathbf{r}) = \sum_k c_k \psi_k(\mathbf{r}), \qquad (3.10)$$

$\psi_k(\mathbf{r})$ being the wave functions of the level states. This wave function can be regarded in the coordinate representation as a function $\psi(\mathbf{r})$ of coordinate, or, equivalently, in the level representation as a (possibly discrete) function c_k of the level index k. In Section 1.3 we illustrated this correspondence of different representations in more detail by comparing the coordinate representation with the plane wave representation of a free particle.

The idea now is that we transform the CAPs to another basis, in a similar way to the wave functions discussed above. We replace $c_k \rightarrow \hat{b}_k$ in (3.10), which brings us to the definition of the *field operator*,

$$\hat{\psi}(\mathbf{r}) = \sum_k \hat{b}_k \psi_k(\mathbf{r}). \qquad (3.11)$$

Since $\hat{b}_k^{(\dagger)}$ annihilates (creates) a particle in the level k, the field operator $\hat{\psi}^{(\dagger)}(\mathbf{r})$ annihilates (creates) a particle at the point $\mathbf{r}$.

Let us look at some simple properties of field operators. We note that the commutation relations take the following elegant form:

---

commutation relations for boson field operators

$$[\hat{\psi}(\mathbf{r}), \hat{\psi}^\dagger(\mathbf{r}')] = \delta(\mathbf{r} - \mathbf{r}')$$
$$[\hat{\psi}(\mathbf{r}), \hat{\psi}(\mathbf{r}')] = [\hat{\psi}^\dagger(\mathbf{r}), \hat{\psi}^\dagger(\mathbf{r}')] = 0$$

So, bosonic field operators commute unless they are at the same point ($\mathbf{r} = \mathbf{r}'$). These commutation relations can be derived from the definition (3.11) if one notes the *completeness* of the basis, that is expressed by

$$\sum_k \psi_k^*(\mathbf{r})\psi_k(\mathbf{r}') = \delta(\mathbf{r} - \mathbf{r}'). \qquad (3.12)$$

Control question. Are you able to present this derivation explicitly?

3.2.1 Operators in terms of field operators

We now show how we can generally express the one- and two-particle operators for a many-particle system in terms of field operators. To this end, we first consider a simple

Vladimir Fock (1898–1974)

Vladimir Fock received his education from the University of Petrograd in difficult times. In 1916, shortly after enrolling, he interrupted his studies to voluntarily join the army and fight in World War I. He got injured (causing a progressive deafness which he suffered from the rest of his life), but survived and returned to Petrograd in 1918. By then, a civil war had broke out in Russia resulting in nationwide political and economical chaos. Fortunately, to ensure, despite the situation, a good education for the brightest students, a new institute was opened in Petrograd, where Fock could continue his studies.

After receiving his degree in 1922, Fock stayed in Petrograd, closely following the development of quantum theory. When Schrödinger published his wave theory in 1926, Fock immediately recognized the value of it and started to develop Schrödinger's ideas further. When he tried to generalize Schrödinger's equation to include magnetic fields, he independently derived the Klein–Gordon equation (see Chapter 13), even before Klein and Gordon did. In the subsequent years, Fock introduced the Fock states and Fock space, which later became fundamental concepts in many-particle theory. In 1930 he also improved Hartree's method for finding solutions of the many-body Schrödinger equation by including the correct quantum statistics: this method is now know as the Hartree–Fock method (see e.g. Chapter 4).

In 1932 Fock was finally appointed professor at the University of Leningrad (as the city was called since 1924). He stayed there until his retirement, making important contributions to a wide range of fields, including general relativity, optics, theory of gravitation, and geophysics. He was brave enough to confront Soviet "philosophers" who stamped the theory of relativity as being bourgeois and idealistic: a stand that could have cost him his freedom and life.

system consisting of only a single particle. In the basis of the levels – the allowed single particle states ψ_k – we can write any operator acting on a this system as

$$\hat{f} = \sum_{k,k'} f_{k'k} |\psi_{k'}(\mathbf{r})\rangle\langle\psi_k(\mathbf{r})|, \tag{3.13}$$

where the coefficients $f_{k'k}$ are the matrix elements $f_{k'k} = \langle\psi_{k'}(\mathbf{r})|\hat{f}|\psi_k(\mathbf{r})\rangle$. For a many-particle system, this operator becomes a sum over all particles,

$$\hat{F} = \sum_{i=1}^{N} \sum_{k,k'} f_{k'k} |\psi_{k'}(\mathbf{r}_i)\rangle\langle\psi_k(\mathbf{r}_i)|, \tag{3.14}$$

where $\mathbf{r}_i$ now is the coordinate of the ith particle. Since the particles are identical, the coefficients $f_{k'k}$ are equal for all particles.

We see that, if in a many-particle state the level k is occupied by n_k particles, the term with $f_{k'k}$ in this sum will result in a state with $n_k - 1$ particles in level k and $n_{k'} + 1$ particles in level k'. In Fock space we can write for the term $f_{k'k}$

$$\sum_{i=1}^{N} f_{k'k} |\psi_{k'}(\mathbf{r}_i)\rangle \langle \psi_k(\mathbf{r}_i)| \ldots, n_{k'}, \ldots, n_k, \ldots\rangle = C f_{k'k} |\ldots, n_{k'} + 1, \ldots, n_k - 1, \ldots\rangle,$$

(3.15)

where the constant C still has to be determined. It follows from the following reasoning. The original state as we write it in Fock space is a normalized symmetrized state: it consists of a sum over all possible permutations of products of single particle wave functions (2.20) with a normalization constant $C_{\text{old}} = \sqrt{n_1! \cdots n_{k'}! \cdots n_k! \cdots /N!}$. The operator in (3.15) then produces a sum over all permutations obeying the *new* occupation numbers. Since in each term of this sum, any of the $n_{k'} + 1$ wave functions $\psi_{k'}$ can be the one which was created by the operator $|\psi_{k'}\rangle\langle\psi_k|$, there are formed $n_{k'} + 1$ copies of the sum over all permutations. The state vector $|\ldots, n_{k'} + 1, \ldots, n_k - 1, \ldots\rangle$ with which we describe this new state comes by definition with a constant $C_{\text{new}} = \sqrt{n_1! \cdots (n_{k'} + 1)! \cdots (n_k - 1)! \cdots /N!}$. Combining all this, we see that

$$C = (n_{k'} + 1) \frac{C_{\text{old}}}{C_{\text{new}}} = \sqrt{n_{k'} + 1} \sqrt{n_k}.$$

(3.16)

This allows for the very compact notation

$$\hat{F} = \sum_{k,k'} f_{k'k} \hat{b}_{k'}^{\dagger} \hat{b}_k,$$

(3.17)

where we again see the use of the ugly square root factors in the definition of the CAPs. Of course, we can write equivalently

$$\hat{F} = \int d\mathbf{r}_1 d\mathbf{r}_2 f_{\mathbf{r}_1 \mathbf{r}_2} \hat{\psi}^{\dagger}(\mathbf{r}_1) \hat{\psi}(\mathbf{r}_2),$$

(3.18)

where we have switched to the representation in field operators. The coefficient $f_{\mathbf{r}_1 \mathbf{r}_2}$ is now given by the matrix element $f_{\mathbf{r}_1 \mathbf{r}_2} = \langle \mathbf{r}_1 | \hat{f} | \mathbf{r}_2 \rangle$.

Let us illustrate how one can use this to present very compactly single-particle operators in a many-particle space. Suppose we would like to find the many-particle density operator $\hat{\rho}$. We know that for a single particle this operator is diagonal in the coordinate basis, $\hat{\rho} = |\mathbf{r}\rangle\langle\mathbf{r}|$. We now simply use (3.18) to find

$$\hat{\rho} = \int d\mathbf{r}_1 d\mathbf{r}_2 \, \rho_{\mathbf{r}_1 \mathbf{r}_2} \hat{\psi}^{\dagger}(\mathbf{r}_1) \hat{\psi}(\mathbf{r}_2) = \hat{\psi}^{\dagger}(\mathbf{r}) \hat{\psi}(\mathbf{r}).$$

(3.19)

We can check this result by calculating the operator of the total number of particles

$$\hat{N} = \int d\mathbf{r} \, \hat{\rho}(\mathbf{r}) = \int d\mathbf{r} \, \hat{\psi}^{\dagger}(\mathbf{r}) \hat{\psi}(\mathbf{r}) = \sum_{k,k'} \int d\mathbf{r} \, \psi_{k'}^{*}(\mathbf{r}) \psi_k(\mathbf{r}) \hat{b}_{k'}^{\dagger} \hat{b}_k$$

$$= \sum_{k,k'} \delta_{kk'} \hat{b}_{k'}^{\dagger} \hat{b}_k = \sum_k \hat{b}_k^{\dagger} \hat{b}_k.$$

(3.20)

Comparing this with (3.5) we see that we have indeed found the right expression.

Control question. Which property of the level wave functions has been used to get from the first to the second line in the above derivation?

3.2.2 Hamiltonian in terms of field operators

The true power of field operators is that they can provide a complete and closed description of a dynamical system of identical particles without invoking any wave functions or the Schrödinger equation. Since the dynamics of a quantum system is determined by its Hamiltonian, our next step is to get the Hamiltonian in terms of field operators.

Let us start with a system of non-interacting particles. Since in this case the many-particle Hamiltonian is just the sum over all one-particle Hamiltonians, we can write it as in (3.18),

$$
\begin{aligned}
\hat{H}^{(1)} &= \int d\mathbf{r}_1 d\mathbf{r}_2 \left\langle \mathbf{r}_1 \left| \frac{\hat{p}^2}{2m} + V(\hat{\mathbf{r}}) \right| \mathbf{r}_2 \right\rangle \hat{\psi}^\dagger(\mathbf{r}_1)\hat{\psi}(\mathbf{r}_2) \\
&= \int d\mathbf{r}\, \hat{\psi}^\dagger(\mathbf{r}) \left\{ -\frac{\hbar^2 \nabla^2}{2m} + V(\mathbf{r}) \right\} \hat{\psi}(\mathbf{r}),
\end{aligned}
\tag{3.21}
$$

where we have used the fact that $\langle \mathbf{r}_1 | \frac{1}{2m}\hat{p}^2 + V(\hat{\mathbf{r}})|\mathbf{r}_2\rangle = \left\{ -\frac{\hbar^2}{2m}\nabla^2_{\mathbf{r}_1} + V(\mathbf{r}_1) \right\} \delta(\mathbf{r}_1 - \mathbf{r}_2)$.

The same Hamiltonian can also be expressed in terms of CAPs that create and annihilate particles in the levels ψ_k. Since the levels are per definition the eigenstates of the single particle Hamiltonian $\frac{1}{2m}\hat{p}^2 + V(\hat{\mathbf{r}})$, we can substitute $\hat{\psi}(\mathbf{r}) = \sum_{\mathbf{k}} \hat{b}_{\mathbf{k}}\psi_k(\mathbf{r})$ in (3.21) and write

$$
\begin{aligned}
\hat{H}^{(1)} &= \sum_{k,k'} \int d\mathbf{r}\, \psi_k^*(\mathbf{r}) \left\{ -\frac{\hbar^2}{2m}\nabla^2 + V(\mathbf{r}) \right\} \psi_{k'}(\mathbf{r})\hat{b}_k^\dagger \hat{b}_{k'} \\
&= \sum_{k,k'} \delta_{kk'}\varepsilon_{k'}\hat{b}_k^\dagger \hat{b}_{k'} = \sum_k \varepsilon_k \hat{n}_k,
\end{aligned}
\tag{3.22}
$$

where ε_k is the energy of the level ψ_k. We see that the representation in CAPs reduces to a very intuitive form: the operator $\hat{H}^{(1)}$ counts the number of particles in each level and multiplies it with the energy of the level. It is easy to see that any basis state in Fock space is an eigenstate of this operator, its eigenvalue being the total energy of the state.

The following step is to include also interaction between particles into the Hamiltonian. We consider here pairwise interaction only. We have seen that in the Schrödinger equation for N identical particles the pairwise interaction is given by

$$
\hat{H}^{(2)} = \frac{1}{2} \sum_{i \neq j} V^{(2)}(\mathbf{r}_i - \mathbf{r}_j),
\tag{3.23}
$$

where the summation runs over all particles. We assume here that the interaction between two particles depends only on the distance between the particles, not on the precise location of the pair in space. Let us rewrite this Hamiltonian in a form where the number of particles is of no importance. We do so by making use of the particle density defined as

$$\rho(\mathbf{r}) = \sum_i \delta(\mathbf{r} - \mathbf{r}_i). \tag{3.24}$$

This provides us a handy way to deal with the sums over particle coordinates. For any function $A(\mathbf{r})$ of a coordinate, or function $B(\mathbf{r}, \mathbf{r}')$ of two coordinates, we can now write

$$\sum_i A(\mathbf{r}_i) = \int d\mathbf{r}\, \rho(\mathbf{r}) A(\mathbf{r}), \tag{3.25}$$

$$\sum_{i,j} B(\mathbf{r}_i, \mathbf{r}_j) = \int d\mathbf{r}\, d\mathbf{r}'\, \rho(\mathbf{r}) \rho(\mathbf{r}') B(\mathbf{r}, \mathbf{r}'). \tag{3.26}$$

So we rewrite the Hamiltonian as

$$\hat{H}^{(2)} = \frac{1}{2} \int d\mathbf{r}_1 d\mathbf{r}_2 V^{(2)}(\mathbf{r}_1 - \mathbf{r}_2) \hat{\rho}(\mathbf{r}_1)[\hat{\rho}(\mathbf{r}_2) - \delta(\mathbf{r}_1 - \mathbf{r}_2)]. \tag{3.27}$$

The δ-function in the last term is responsible for the "missing" term $i \neq j$ in the double sum in (3.23): a particle does not interact with itself! Now we use $\hat{\rho}(\mathbf{r}) = \hat{\psi}^\dagger(\mathbf{r})\hat{\psi}(\mathbf{r})$ to obtain

$$\hat{H}^{(2)} = \frac{1}{2} \int d\mathbf{r}_1 d\mathbf{r}_2 V^{(2)}(\mathbf{r}_1 - \mathbf{r}_2) \hat{\psi}^\dagger(\mathbf{r}_1)\hat{\psi}^\dagger(\mathbf{r}_2)\hat{\psi}(\mathbf{r}_2)\hat{\psi}(\mathbf{r}_1). \tag{3.28}$$

Control question. How does the order of the field operators in (3.28) incorporate the fact that a particle does not interact with itself?
Hint. Use the commutation relations for field operators.

Another instructive way to represent the pairwise interaction in a many-particle system is in terms of the CAPs, which create and annihilate particles in the states with certain wave vectors (or momenta). This will reveal the relation between the pairwise interaction and scattering of two particles. We rewrite the Hamiltonian (3.28) in the wave vector representation, substituting $\hat{\psi}(\mathbf{r}) = \frac{1}{\sqrt{\mathcal{V}}}\sum_{\mathbf{k}} \hat{b}_{\mathbf{k}} e^{i\mathbf{k}\cdot\mathbf{r}}$, to obtain

$$\hat{H}^{(2)} = \frac{1}{2\mathcal{V}} \sum_{\mathbf{k},\mathbf{k}',\mathbf{q}} V^{(2)}(\mathbf{q})\, \hat{b}^\dagger_{\mathbf{k}+\mathbf{q}} \hat{b}^\dagger_{\mathbf{k}'-\mathbf{q}} \hat{b}_{\mathbf{k}'} \hat{b}_{\mathbf{k}}. \tag{3.29}$$

Control question. Do you see how to derive (3.29) from (3.28)?
Hint. Switch to the coordinates $\mathbf{R} = \mathbf{r}_1 - \mathbf{r}_2$ and $\mathbf{r}_2$, and then perform the integrations in (3.28) explicitly. Remember that $\int d\mathbf{r} = \mathcal{V}$.

Let us now focus on the physical meaning of (3.29). We concentrate on a single term in the sum, characterized by the wave vectors $\mathbf{k}$, $\mathbf{k}'$, and $\mathbf{q}$. Suppose that this term acts on an initial state which contains two particles with wave vectors $\mathbf{k}$ and $\mathbf{k}'$. The term annihilates both these particles and creates two new particles, with wave vectors $\mathbf{k}+\mathbf{q}$ and $\mathbf{k}'-\mathbf{q}$. If we now compare the initial and final states, we see that effectively the particles have exchanged the momentum $\mathbf{q}$ with each other. In other words, this term describes a scattering event in which the total momentum is conserved. We can depict this interaction in a cartoon style as in Fig. 3.2.

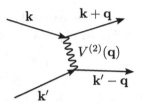

A term in a pairwise interaction Hamiltonian pictured as a scattering event. Two particles with wave vectors **k** and **k′** can interact and thereby exchange momentum **q**. After this interaction the particles have wave vectors **k** + **q** and **k′** − **q**. The amplitude of the process is proportional to the Fourier component $V^{(2)}(\mathbf{q})$ of the interaction potential.

3.2.3 Field operators in the Heisenberg picture

Let us now investigate the time dependence of the CAPs and field operators when considering them in the Heisenberg picture. In this picture the wave function of a system of identical particles is "frozen," and all operators move (see Section 1.4). Assuming non-interacting particles and making use of the Hamiltonian (3.22), we see that the Heisenberg equation for CAPs becomes

$$ i\hbar\frac{\partial \hat{b}_k(t)}{\partial t} = [\hat{b}_k, \hat{H}^{(1)}] = \varepsilon_k \hat{b}_k(t), \tag{3.30} $$

its solution being $\hat{b}_k(t) = e^{-\frac{i}{\hbar}\varepsilon_k t}\hat{b}_k(0)$. If we rewrite this in coordinate representation, we get the evolution equation for field operators

$$ i\hbar\frac{\partial \hat{\psi}(\mathbf{r}, t)}{\partial t} = \left\{ -\frac{\hbar^2\nabla^2}{2m} + V(\mathbf{r}) \right\} \hat{\psi}(\mathbf{r}, t). \tag{3.31} $$

This equation is linear and therefore still easy to solve. If we also include interactions, the equations for the field operators are still easy to write down. Importantly, the equations are *closed*: to solve for the dynamics of field operators, one does not need any variables other than the field operators. This is why a set of time-evolution equations for field operators together with the commutation relations is said to define a quantum *field theory*. In the presence of interactions, however, the equations are no longer linear. The general solutions are very difficult to find and in most cases they are not known.

3.3 Why second quantization?

Let us now add some explanations which we have postponed for a while. We have learned a certain technique called second quantization. Why? From a practical point of view, the resulting equations for systems with many particles are much simpler and more compact than the original many-particle Schrödinger equation. Indeed, the original scheme operates with a wave function of very many variables (the coordinates of all particles), which obeys

Pascual Jordan (1902–1980)
Never won the Nobel Prize, supposedly because of his strong support of the Nazi party.

Shortly after receiving his Ph.D. degree at the University of Göttingen in 1924, Pascual Jordan was asked by the 20 years older Max Born to help him understand Heisenberg's manuscript on the formulation of quantum mechanics in terms of non-commuting observables. Born had immediately suspected a relation with matrix algebra, but he lacked the mathematical background to work out a consistent picture. Within a week Jordan showed that Born's intuition was correct, and that it provided a rigid mathematical basis for Heisenberg's ideas. In September 1925, Jordan was the first to derive the famous commutation relation $[\hat{x}, \hat{p}] = i\hbar$. Later that year, he wrote a paper on "Pauli statistics" in which he derived the Fermi–Dirac statistics before either Fermi or Dirac did. The only reason why fermions are nowadays not called jordanions is that Born delayed publication by forgetting about the manuscript while lecturing in the US for half a year.

In the years 1926–27, Jordan published a series of papers in which he introduced the canonical anti-commutation relations and a way to quantize wave fields. These works are probably his most important, and are commonly regarded as the birth of quantum field theory. After moving to Rostock in 1928, Jordan became more interested in the mathematical and conceptual side of quantum mechanics, and he lost track of most of the rapid developments in the field he started. In 1933 he joined the Nazi party (he even became a storm trooper), which damaged his relations with many colleagues. Despite his strong verbal support of the party, the Nazis also did not trust him fully: quantum mechanics was still regarded as a mainly Jewish business. Jordan therefore ended up practically in scientific isolation during the war.

After the war, his radical support of the Nazis caused him severe trouble. Only in 1953 did he regain his full rights as professor at the University of Hamburg, where he had been since 1951 and stayed till his retirement in 1970.

a differential equation involving all these variables. In the framework of second quantization, we reduced this to an equation for an operator of one single variable, (3.31). Apart from the operator "hats," it looks very much like the Schrödinger equation for a single particle,

$$i\hbar \frac{\partial \psi(\mathbf{r}, t)}{\partial t} = \hat{H}\psi = \left\{ -\frac{\hbar^2 \nabla^2}{2m} + V(\mathbf{r}) \right\} \psi(\mathbf{r}, t), \qquad (3.32)$$

and is of the same level of complexity.

From an educational point of view, the advantage is enormous. To reveal it finally, let us take the expectation value of (3.31) and introduce the (deliberately confusing) notation $\psi(\mathbf{r}, t) = \langle \hat{\psi}(\mathbf{r}, t) \rangle$. Owing to the linearity of (3.31), the equation for the expectation value formally coincides with (3.32). However, let us put this straight: this equation is not for a wave function anymore. It is an equation for a *classical field*.[2] Without noticing, we have thus got all the way from particles to fields and have understood that these concepts are just two facets of a single underlying concept: that of the quantum field.

Another frequently asked question is why the technique is *called* second quantization. Well, the name is historic rather than rational. A quasi-explanation is that the first quantization is the replacement of numbers – momentum and coordinate – by non-commuting operators, whereas the wave function still is a number,

$$[\hat{p}_x, \hat{x}] \neq 0, \qquad [\psi^*, \psi] = 0. \tag{3.33}$$

As a "second" step, one "quantizes" the wave function as well, so that finally

$$[\hat{\psi}^\dagger, \hat{\psi}] \neq 0. \tag{3.34}$$

Once again, this is an explanation for the name and not for the physics involved. The wave function is not really quantized, it is not replaced with an operator! The field operator is the replacement of the field rather than that of a wave function.

3.4 Second quantization for fermions

As we see below, the second quantization technique for fermions is very similar to the one for bosons. We proceed by repeating the steps we made for bosons and concentrate on the differences.

We again start from a basis in the single-particle space – the basis of levels – and label the levels with an index k. The first difference we note is that fermions are bound to have half-integer spin,[3] whereas bosons have integer spin. For bosons the simplest case is given by spinless ($s = 0$) bosons. We did not mention spin in the section on bosons, since we tacitly assumed the bosons to be spinless. For fermions, however, the simplest case is given by particles with two possible spin directions ($s = \frac{1}{2}$). A complete set of single particle basis states now includes an index for spin. For example, the set of free particle plane waves now reads

$$\psi_{\mathbf{k}, \uparrow} = \frac{1}{\sqrt{\mathcal{V}}} \exp(i\mathbf{k} \cdot \mathbf{r}) \begin{pmatrix} 1 \\ 0 \end{pmatrix},$$

$$\psi_{\mathbf{k}, \downarrow} = \frac{1}{\sqrt{\mathcal{V}}} \exp(i\mathbf{k} \cdot \mathbf{r}) \begin{pmatrix} 0 \\ 1 \end{pmatrix}, \tag{3.35}$$

---

[2] The field at hand is a complex scalar field. By this it differs from, say, an electromagnetic field which is a vector field. However, we will quantize the electromagnetic field soon and see that the difference is not essential.

[3] Provided they are *elementary* particles. There are examples of excitations in condensed matter that are naturally regarded as spinless fermion particles.

where the two components of the wave functions correspond to the two different directions of spin. One thus has to remember always that all levels are labeled by a wave vector $\mathbf{k}$ *and* a spin index $\sigma = \{\uparrow, \downarrow\}$. In the rest of the section we just use the labels k, which are shorthand for the full index, $k \equiv \{\mathbf{k}, \sigma\}$.

Now we can construct a basis in Fock space as we did before. We again label the basis states in Fock space with sets of occupation numbers $\{n_k\} \equiv \{n_1, n_2, \ldots\}$, and express the orthonormality of the basis as

$$\langle \{n_k\} | \{n'_k\} \rangle = \prod_k \delta_{n_k, n'_k}. \tag{3.36}$$

We note that for fermions these occupation numbers can take only two values, 0 or 1. A set of occupation numbers therefore reduces to just a long sequence of zeros and ones.

This defines the basis in Fock space. Let us note, however, that at this moment still *nothing* in the basis reflects the *most important* difference between bosons and fermions: the fact that a bosonic wave function is symmetric, whereas a fermionic one is antisymmetric. We could use the same basis presented above to describe some fancy bosons that for some reason strongly dislike sharing the same level.[4] So the question is: how do we incorporate the information about the antisymmetry of the wave function into the second quantization scheme?

3.4.1 Creation and annihilation operators for fermions

The only place to put the information that a fermionic many-particle wave function is antisymmetric, is in the structure of the CAPs. To understand this, let us do a step back and consider identical *distinguishable* particles that prefer not to share the same level. In this case, the number of possible (basis) states is no longer restricted by the symmetry postulate. So we must consider a space of all possible products of basis states, not only the (anti-)symmetrized ones. For instance, a state where a first particle is put in level 5, a second particle in level 3, and a third one in level 2, is different from a state with the same occupation numbers but a different filling order. Using elementary *creation* operators $\hat{c}_k^\dagger$ we can rephrase this statement as

$$\hat{c}_2^\dagger \hat{c}_3^\dagger \hat{c}_5^\dagger |0\rangle \neq \hat{c}_3^\dagger \hat{c}_2^\dagger \hat{c}_5^\dagger |0\rangle, \tag{3.37}$$

where $|0\rangle$ represents the vacuum. We require here that the order of application of the operators corresponds to the order of the filling of the levels. We see that if the particles were distinguishable, the order of the operators determines which particle sits in which level. The two states created in (3.37) are thus in fact the very same state, but with the particles in levels 2 and 3 *permuted*. For the *indistinguishable* fermions, such a permutation should result in a minus sign in front of the many-particle wave function.

We can now build the Fock space for fermions. To conform to the antisymmetry requirement, we impose that two states with the same occupation numbers but a different filling

---

[4] Such models of *impenetrable* boson particles have been actually put forward.

order must differ by a minus sign if the number of permutations needed to go from the one state to the other is odd. For example,

$$\hat{c}_2^\dagger \hat{c}_3^\dagger \hat{c}_5^\dagger |0\rangle = -\hat{c}_2^\dagger \hat{c}_5^\dagger \hat{c}_3^\dagger |0\rangle \quad \text{and} \quad \hat{c}_2^\dagger \hat{c}_3^\dagger \hat{c}_5^\dagger |0\rangle = \hat{c}_3^\dagger \hat{c}_5^\dagger \hat{c}_2^\dagger |0\rangle. \tag{3.38}$$

This imposes a specific requirement on the commutation relations of the fermionic CAPs: they must *anticommute*. Indeed, for two particles we can write analogously to (3.38)

$$\hat{c}_n^\dagger \hat{c}_m^\dagger |0\rangle = -c_m^\dagger c_n^\dagger |0\rangle, \tag{3.39}$$

which is satisfied provided

$$\hat{c}_n^\dagger \hat{c}_m^\dagger + \hat{c}_m^\dagger \hat{c}_n^\dagger \equiv \{c_n^\dagger, c_m^\dagger\} = 0. \tag{3.40}$$

The antisymmetry is crucial here. If we run these procedures for the fancy impenetrable bosons mentioned above, we have to require $\hat{c}_n^\dagger \hat{c}_m^\dagger |0\rangle = c_m^\dagger c_n^\dagger |0\rangle$ and again end up with *commuting* creation operators.

It is convenient to require that a product of a creation and annihilation operator again presents the occupation number operator, as it does for bosons. This leads us to a unique definition of CAPs for fermions. A word of warning is appropriate here: the fermionic CAPs are in fact simple and nice to work with, but this does not apply to their definition. So brace yourself!

The annihilation operator $\hat{c}_k$ for a fermion in the level k is defined as

$$\hat{c}_k | \ldots \ldots, 1_k, \ldots \rangle = (-1)^{\sum_{i=1}^{k-1} n_i} | \ldots \ldots, 0_k, \ldots \rangle, \tag{3.41}$$

$$\hat{c}_k | \ldots \ldots, 0_k, \ldots \rangle = 0. \tag{3.42}$$

Here the zeros and ones refer to the occupation number of the kth level, as indicated with the index k. The annihilation operator thus does what we expect: it removes a particle from the level k if it is present there. Besides, it multiplies the function with a factor ± 1. Inconveniently enough, this factor depends on the occupation of all levels *below* the level k. It becomes $+1(-1)$ if the total number of particles in these levels is even (odd). The conjugated operator to $\hat{c}_k$ creates a particle in the level k as expected,

$$\hat{c}_k^\dagger | \ldots \ldots, 0_k, \ldots \rangle = (-1)^{\sum_{i=1}^{k-1} n_i} | \ldots \ldots, 1_k, \ldots \rangle. \tag{3.43}$$

Control question. Can you find the values of the matrix elements $\langle 1, 1, 1 | \hat{c}_1 | 1, 1, 1 \rangle$, $\langle 1, 0, 1 | \hat{c}_2 | 1, 1, 1 \rangle$, and $\langle 1, 1, 1 | \hat{c}_3^\dagger | 1, 1, 0 \rangle$?

What is really strange about these definitions is that they are subjective. They explicitly depend on the way *we* have ordered the levels. We usually find it convenient to organize the levels in order of increasing energy, but we could also choose decreasing energy or any other order. The point is that the signs of the matrix elements of fermionic CAPs are subjective indeed. Therefore, the CAPs do not purport to correspond to any physical quantity. On the contrary, the matrix elements of their products do not depend on the choice of the level ordering. Therefore, the products correspond to physical quantities.

Let us show this explicitly by calculating products of CAPs. We start with $\hat{c}_k^\dagger \hat{c}_{k'}^\dagger$ for $k \neq k'$. We apply this product to a state $|\ldots, 0_k, \ldots, 0_{k'}, \ldots\rangle$ since any other state produces zero. From the definition (3.43) we find

$$
\begin{aligned}
\hat{c}_k^\dagger \hat{c}_{k'}^\dagger |\ldots, 0_k, \ldots, 0_{k'}, \ldots\rangle &= \hat{c}_k^\dagger (-1)^{\sum_{i=1}^{k'-1} n_i} |\ldots, 0_k, \ldots, 1_{k'}, \ldots\rangle \\
&= (-1)^{\sum_{i=1}^{k-1} n_i} (-1)^{\sum_{i=1}^{k'-1} n_i} |\ldots, 1_k, \ldots, 1_{k'}, \ldots\rangle.
\end{aligned}
\tag{3.44}
$$

Changing the order of the two operators, we get

$$
\begin{aligned}
\hat{c}_{k'}^\dagger \hat{c}_k^\dagger |\ldots, 0_k, \ldots, 0_{k'}, \ldots\rangle &= \hat{c}_{k'}^\dagger (-1)^{\sum_{i=1}^{k-1} n_i} |\ldots, 1_k, \ldots, 0_{k'}, \ldots\rangle \\
&= (-1)^{\sum_{i=1}^{k-1} n_i} (-1)^{1+\sum_{i=1}^{k'-1} n_i} |\ldots, 1_k, \ldots, 1_{k'}, \ldots\rangle = -\hat{c}_k^\dagger \hat{c}_{k'}^\dagger |\ldots, 0_k, \ldots, 0_{k'}, \ldots\rangle.
\end{aligned}
\tag{3.45}
$$

So, the creation operators anticommute as expected.

For the product $\hat{c}_k^\dagger \hat{c}_{k'}$ we can proceed in the same way to find the expected result

$$
\hat{c}_k^\dagger \hat{c}_{k'} |\ldots, 0_k, \ldots, 1_{k'}, \ldots\rangle = -\hat{c}_{k'} \hat{c}_k^\dagger |\ldots, 0_k, \ldots, 1_{k'}, \ldots\rangle.
\tag{3.46}
$$

In a similar way one can find for the case $k = k'$ the relations

$$
\hat{c}_k^\dagger \hat{c}_k |\{n\}\rangle = n_k |\{n\}\rangle,
\tag{3.47}
$$

$$
\hat{c}_k \hat{c}_k^\dagger |\{n\}\rangle = (1 - n_k)|\{n\}\rangle,
\tag{3.48}
$$

so $\hat{c}_k^\dagger \hat{c}_k + \hat{c}_k \hat{c}_k^\dagger = 1$. Conveniently, the representation of the occupation numbers in terms of CAPs is the same as for bosons, $\hat{n}_k = \hat{c}_k^\dagger \hat{c}_k$. The operator of the total number of particles reads

$$
N = \sum_k \hat{n}_k = \sum_k \hat{c}_k^\dagger \hat{c}_k.
\tag{3.49}
$$

We summarize the commutation relations for the fermionic CAPs as follows:

---

commutation relations for fermion CAPs

$$
\{\hat{c}_k, \hat{c}_{k'}^\dagger\} = \delta_{kk'}
$$

$$
\{\hat{c}_k, \hat{c}_{k'}\} = \{\hat{c}_k^\dagger, \hat{c}_{k'}^\dagger\} = 0
$$

Let us conclude this section with a few comments on the structure of the Fock space for fermions. We have again one special state, the vacuum $|0\rangle$. By definition, $\hat{c}_k |0\rangle = 0$ for all k since there are no particles to be annihilated in the vacuum. We have already noted that all states can be obtained from the vacuum state by creating particles one by one in it, that is, by applying the proper sequence of creation operators,

$$
|\{n_k\}\rangle = \prod_{k \in \text{occupied}} \hat{c}_k^\dagger |0\rangle.
\tag{3.50}
$$

The product runs over the occupied levels only, those for which $n_k = 1$. A distinction from the similar formula for bosons is that the order in the sequence is now in principle important: fermionic creation operators do not commute. However, while they anticommute, the

difference between different sequences of the same operators is minimal: it is just a factor
± 1. Another important distinction from bosons is that the fermionic Fock space spanned
by a finite number of levels M is of a finite dimension 2^M (where M is taken to include the
spin degeneracy).

Control question. Can you explain why the dimension is such? If not, count the
number of basis states in the Fock space for $M = 1, 2, \ldots$

3.4.2 Field operators

The field operators, the Hamiltonian in terms of field operators, and the dynamic equations
generated by this Hamiltonian almost literally coincide with those for bosons. One of the
differences is spin: there are (at least) two sorts of field operator corresponding to particles
of different spin directions,

$$\hat{\psi}_\sigma(\mathbf{r}) = \sum_{\mathbf{k}} \hat{c}_{\mathbf{k}\sigma} \psi_{\mathbf{k}}(\mathbf{r}). \tag{3.51}$$

The most important difference, however, is in the commutation relations. Since the field
operators are linear combinations of CAPs, they anticommute as well:

---

commutation relations for fermion field operators

$$\{\hat{\psi}_\sigma(\mathbf{r}), \hat{\psi}_{\sigma'}^\dagger(\mathbf{r}')\} = \delta(\mathbf{r} - \mathbf{r}')\delta_{\sigma\sigma'}$$
$$\{\hat{\psi}_\sigma(\mathbf{r}), \hat{\psi}_{\sigma'}(\mathbf{r}')\} = \{\hat{\psi}_\sigma^\dagger(\mathbf{r}), \hat{\psi}_{\sigma'}^\dagger(\mathbf{r}')\} = 0$$

Remarkably (and quite conveniently) this change in commutation relations hardly
affects the form of the Hamiltonians. Quite similarly to their boson counterparts, the
non-interacting Hamiltonian and the pairwise interaction term read

$$\hat{H}^{(1)} = \sum_\sigma \int d\mathbf{r}\, \hat{\psi}_\sigma^\dagger(\mathbf{r}) \left\{ -\frac{\hbar^2 \nabla^2}{2m} + V(\mathbf{r}) \right\} \hat{\psi}_\sigma(\mathbf{r}), \tag{3.52}$$

$$\hat{H}^{(2)} = \frac{1}{2} \sum_{\sigma,\sigma'} \int d\mathbf{r}_1 d\mathbf{r}_2\, V^{(2)}(\mathbf{r}_1 - \mathbf{r}_2) \hat{\psi}_\sigma^\dagger(\mathbf{r}_1) \hat{\psi}_{\sigma'}^\dagger(\mathbf{r}_2) \hat{\psi}_{\sigma'}(\mathbf{r}_2) \hat{\psi}_\sigma(\mathbf{r}_1). \tag{3.53}$$

In the wave vector representation, the interaction term now reads

$$\hat{H}^{(2)} = \frac{1}{2\mathcal{V}} \sum_{\mathbf{k},\mathbf{k}',\mathbf{q}} \sum_{\sigma,\sigma'} V^{(2)}(\mathbf{q})\, \hat{c}_{\mathbf{k}+\mathbf{q},\sigma}^\dagger \hat{c}_{\mathbf{k}'-\mathbf{q},\sigma'}^\dagger \hat{c}_{\mathbf{k}',\sigma'} \hat{c}_{\mathbf{k},\sigma}. \tag{3.54}$$

This can still be interpreted in terms of two-particle scattering, such as depicted in Fig. 3.2.
A minor difference arises from the presence of spin indices, since the total spin of both
particles is conserved in the course of the scattering.

The Heisenberg equations for the fermionic CAPs, although derived from different commutation rules, are also almost identical to those for the boson CAPs,

$$i\hbar\frac{\partial \hat{c}_k(t)}{\partial t} = [\hat{c}_k, \hat{H}^{(1)}] = \varepsilon_k \hat{c}_k(t), \tag{3.55}$$

so that $\hat{c}_k(t) = e^{-\frac{i}{\hbar}\varepsilon_k t}\hat{c}_k(0)$. Transforming this time-evolution equation to the coordinate representation, we obtain

$$i\hbar\frac{\partial \hat{\psi}_\sigma(\mathbf{r}, t)}{\partial t} = \left\{-\frac{\hbar^2\nabla^2}{2m} + V(\mathbf{r})\right\}\hat{\psi}_\sigma(\mathbf{r}, t), \tag{3.56}$$

indeed the same equation as for bosons. However, there is a physical difference. As discussed, for bosons we can take the expectation value of the above equation and come up with an equation that describes the dynamics of a classical field. It is a *classical limit* of the above equation and is relevant if the occupation numbers of the levels involved are large, $n_k \gg 1$. Formally, we could do the same with (3.56). However, the resulting equation is for a quantity $\langle\hat{\psi}(\mathbf{r}, t)\rangle$, the expectation value of a fermionic field. This expectation value, however, is zero for all physical systems known. One can infer this from the fact that the matrix elements of single fermionic CAPs are in fact subjective quantities (see the discussion above). The resulting equation does therefore not make any sense. This is usually formulated as follows: *the fermionic field does not have a classical limit*. Another way of seeing this is to acknowledge that for fermions, occupation numbers can never be large, n_k is either 0 or 1, and therefore there is no classical limit.

3.5 Summary of second quantization

The most important facts about second quantization are summarized in Table 3.1 below. Some of the relations turn out to be identical for bosons and fermions: in those cases we use the generic notation $\hat{a}$ for the annihilation operator. Otherwise we stick to the convention $\hat{b}$ for bosons and $\hat{c}$ for fermions.

Let us end this chapter with a few practical tips concerning CAPs. In all applications of the second quantization technique, you sooner or later run into expressions consisting of a long product of CAPs stacked between two basis vectors of the Fock space. In order to derive any useful result, you must be able to manipulate those products of CAPs, i.e. you must master the definition and commutation relations of the CAPs at an operational level. Any expression consisting of a long string of CAPs can be handled with the following simple rules:

1. Do not panic. You can do this, it is a routine calculation.
2. Try to reduce the number of operators by converting them to occupation number operators $\hat{a}_l^\dagger \hat{a}_l |\{n_k\}\rangle = n_l|\{n_k\}\rangle$.
3. In order to achieve this, permute the operators using the commutation rules.

4. Do not calculate parts which reduce obviously to zero. Use common sense and the definition of the CAPs to guess whether an expression is zero before evaluating it.

Let us illustrate this by evaluating the following expression involving fermionic operators:

$$\langle \{n_k\} | \hat{c}^\dagger_{l_1} \hat{c}^\dagger_{l_2} \hat{c}_{l_3} \hat{c}_{l_4} | \{n_k\} \rangle = \ldots ?$$

The last rule comes first. The bra and ket states are identical, and this allows us to establish relations between the level indices l_1, l_2, l_3, and l_4. The two annihilation operators in the expression kill particles in the ket state in the levels l_3 and l_4. In order to end up in $|\{n_k\}\rangle$ again and thus have a non-zero result, the particles in these levels have to be re-created by the creation operators. So we have to concentrate on two possibilities only,[5]

$$l_1 = l_3, \quad l_2 = l_4, \quad l_1 \neq l_2,$$
$$\text{or} \quad l_1 = l_4, \quad l_2 = l_3, \quad l_1 \neq l_2.$$

We now focus on the first possibility, and try to reduce the CAPs to occupation number operators,

$$\langle \{n_k\} | \hat{c}^\dagger_{l_1} \hat{c}^\dagger_{l_2} \hat{c}_{l_1} \hat{c}_{l_2} | \{n_k\} \rangle \qquad \text{(permute the second and the third)}$$
$$= - \langle \{n_k\} | \hat{c}^\dagger_{l_1} \hat{c}_{l_1} \hat{c}^\dagger_{l_2} \hat{c}_{l_2} | \{n_k\} \rangle \qquad \text{(use reduction rule)}$$
$$= - \langle \{n_k\} | \hat{c}^\dagger_{l_1} \hat{c}_{l_1} | \{n_k\} \rangle n_{l_2} \qquad \text{(once again)}$$
$$= - n_{l_1} n_{l_2}.$$

Treating the second possibility in the same way, we find

$$\langle \{n_k\} | \hat{c}^\dagger_{l_1} \hat{c}^\dagger_{l_2} \hat{c}_{l_2} \hat{c}_{l_1} | \{n_k\} \rangle \qquad \text{(permute the third and the fourth)}$$
$$= - \langle \{n_k\} | \hat{c}^\dagger_{l_1} \hat{c}^\dagger_{l_2} \hat{c}_{l_1} \hat{c}_{l_2} | \{n_k\} \rangle \qquad \text{(permute the second and the third)}$$
$$= + \langle \{n_k\} | \hat{c}^\dagger_{l_1} \hat{c}_{l_1} \hat{c}^\dagger_{l_2} \hat{c}_{l_2} | \{n_k\} \rangle \qquad \text{(use reduction rule twice)}$$
$$= + n_{l_1} n_{l_2}.$$

---

[5] The case $l_1 = l_2 = l_3 = l_4$ gives zero since all levels can only be occupied once. The term $\hat{c}_{l_3} \hat{c}_{l_4}$ then always produces zero.

Table 3.1 Summary: Second quantization

	Bosons	**Fermions**
Many-particle wave function:	$\psi(\{\mathbf{r}_i\}, t)$	
	fully symmetric	fully antisymmetric
Fock space, where the basis states are labeled by the sets $\{n_k\}$ of occupation numbers		
occupation numbers can be:	non-negative integers	0 or 1
Creation and annihilation operators: commute		anticommute
	$[\hat{b}_k, \hat{b}_{k'}] = 0$	$\{\hat{c}_k, \hat{c}_{k'}\} = 0$
	$[\hat{b}_k^\dagger, \hat{b}_{k'}^\dagger] = 0$	$\{\hat{c}_k^\dagger, \hat{c}_{k'}^\dagger\} = 0$
	$[\hat{b}_k, \hat{b}_{k'}^\dagger] = \delta_{kk'}$	$\{\hat{c}_k, \hat{c}_{k'}^\dagger\} = \delta_{kk'}$

Occupation number operator

number of particles in the level k:
$$\hat{n}_k \equiv \hat{a}_k^\dagger \hat{a}_k$$

total number of particles:
$$\hat{N} = \sum_k \hat{n}_k = \sum_k \hat{a}_k^\dagger \hat{a}_k$$

Hamiltonian

with particle–particle interactions: the same, with spin

$$\hat{H} = \sum_{\mathbf{k}} \varepsilon_{\mathbf{k}} \hat{b}_{\mathbf{k}}^\dagger \hat{b}_{\mathbf{k}}$$

$$+ \frac{1}{2\mathcal{V}} \sum_{\mathbf{k},\mathbf{k}',\mathbf{q}} V(\mathbf{q})\, \hat{b}_{\mathbf{k}+\mathbf{q}}^\dagger \hat{b}_{\mathbf{k}'-\mathbf{q}}^\dagger \hat{b}_{\mathbf{k}'} \hat{b}_{\mathbf{k}}$$

Field operators:
$$\hat{\psi}(\mathbf{r}) = \sum_k \hat{a}_k \psi_k(\mathbf{r})$$

dynamics are (without interactions) the same as the Schrödinger equation for the wave function

Heisenberg equation:
$$i\hbar \frac{\partial}{\partial t} \hat{\psi}(\mathbf{r}, t) = \left\{ -\frac{\hbar^2 \nabla^2}{2m} + V(\mathbf{r}) \right\} \hat{\psi}(\mathbf{r}, t)$$

Exercises

In all Exercises, we use the following notation for the CAPs: the operators $\hat{b}$ and $\hat{c}$ are respectively bosonic and fermionic operators, while $\hat{a}$ is an operator that can be either bosonic or fermionic. Matrices M_{kl} with k, l labeling the levels are denoted as $\check{M}$.

1. *Applications of commutation rules* (solution included).
   a. Use the commutation relations to show that

   $$[\hat{a}_k^\dagger \hat{a}_l, \hat{a}_m] = -\delta_{mk}\hat{a}_l,$$

   for both bosonic and fermionic CAPs.
   b. Given the Hamiltonian of non-interacting particles

   $$\hat{H} = \sum_k \varepsilon_k \hat{n}_k = \sum_k \varepsilon_k \hat{a}_k^\dagger \hat{a}_k,$$

   use the Heisenberg equations of motion to show that

   $$\frac{d}{dt}\hat{a}_k = -\frac{i}{\hbar}\varepsilon_k \hat{a}_k.$$

   c. Compute for an ideal Bose-gas of spinless particles the matrix element

   $$\langle g|\hat{b}_{i1}\hat{b}_{i2}\hat{b}_{i3}^\dagger \hat{b}_{i4}^\dagger|g\rangle,$$

   where $|g\rangle$ is a number state in Fock space, $|g\rangle \equiv |\{n_k\}\rangle$. The level indices $i_1 \dots i_4$ can be different or coinciding. Compute the matrix element considering all possible cases.
   d. Do the same for $\langle g|\hat{b}_{i1}^\dagger \hat{b}_{i2}\hat{b}_{i3}^\dagger \hat{b}_{i4}|g\rangle$.
   e. Compute a similar matrix element, $\langle g|\hat{c}_{i1}^\dagger \hat{c}_{i2}\hat{c}_{i3}^\dagger \hat{c}_{i4}|g\rangle$ for the fermion case.

2. *Fluctuations.* Any one-particle operator $\hat{A}$ can be written using creation and annihilation operators as

   $$\hat{A} = \sum_{i,j} A_{ij}\hat{a}_i^\dagger \hat{a}_j.$$

   a. Calculate the expectation value $\langle g|\hat{A}|g\rangle \equiv \langle A\rangle$, $|g\rangle$ being a number state, $|g\rangle \equiv |\{n_k\}\rangle$. Is there a difference for fermions and bosons?
   b. Calculate the variance $\langle \hat{A}^2\rangle - \langle \hat{A}\rangle^2$ for fermions and bosons, making use of the results found in 1(d) and 1(e).

3. *Linear transformations of CAPs.* Let us consider a set of bosonic CAPs, $\hat{b}_k$ and $\hat{b}_k^\dagger$. A linear transformation A, represented by the matrix $\check{L}$, transforms the CAPs to another set of operators,

   $$\hat{b}_m' = \sum_k L_{mk}\hat{b}_k.$$

   a. Find the conditions on the matrix $\check{L}$ representing the transformation A under which the resulting operators are also bosonic CAPs, that is, they satisfy the bosonic commutation rules.

A more general linear transformation B can also mix the two types of operator,

$$\hat{b}'_m = \sum_k L_{mk}\hat{b}_k + \sum_k M_{mk}\hat{b}_k^\dagger.$$

  b. Find the conditions on the matrices $\check{L}$ and $\check{M}$ of the more general transform B under which the resulting operators are bosonic CAPs.

  c. Consider the transformations A and B in application to fermionic CAPs and find the conditions under which the transformed set remains a set of fermionic CAPs.

4. *Exponents of boson operators.*

  a. Find

$$[\hat{b}^n, \hat{b}^\dagger].$$

  b. Employ the result of (a) to prove that for an arbitrary sufficiently smooth function F it holds that

$$[F(\hat{b}), \hat{b}^\dagger] = F'(\hat{b}).$$

  c. Use the result of (b) to prove that

$$F(\hat{b})\exp\left(\alpha\hat{b}^\dagger\right) = \exp\left(\alpha\hat{b}^\dagger\right)F(\hat{b} + \alpha).$$

  d. Show that

$$\exp\left(\alpha\hat{b}^\dagger + \beta\hat{b}\right) = \exp\left(\alpha\hat{b}^\dagger\right)\exp\left(\beta\hat{b}\right)e^{\alpha\beta/2}.$$

  Hint. Consider the operator

$$\hat{O}(s) = \exp\left(-s\alpha\hat{b}^\dagger\right)\exp\left(-s\beta\hat{b}\right)\exp\left(s(\alpha\hat{b}^\dagger + \beta\hat{b})\right),$$

  and derive the differential equation it satisfies as a function of s.

5. *Statistics in disguise.*

  a. Consider the operator $\hat{Z} \equiv \exp\left(-\sum_{k,l}\hat{c}_k^\dagger M_{kl}\hat{c}_l\right)$, where $\check{M}$ is a Hermitian matrix. Show that

$$\mathrm{Tr}\{\hat{Z}\} = \det[1 + \exp(-\check{M})].$$

  Hint. Diagonalize $\check{M}$.

  b. Compute $\mathrm{Tr}\left\{\exp\left(-\sum_{k,l}\hat{b}_k^\dagger M_{kl}\hat{b}_l\right)\right\}$.

Solutions

1. *Applications of commutation rules.*

  a. For bosons,

$$[\hat{b}_k^\dagger\hat{b}_l, \hat{b}_m] = \hat{b}_k^\dagger\hat{b}_l\hat{b}_m - \hat{b}_m\hat{b}_k^\dagger\hat{b}_l = \hat{b}_k^\dagger\hat{b}_m\hat{b}_l - (\delta_{km} + \hat{b}_k^\dagger\hat{b}_m)\hat{b}_l = -\delta_{km}\hat{b}_l,$$

where after the second equality sign we make use of commutation relations for two annihilation operators to transform the first term, and that for a creation and an annihilation operator to transform the second term. For fermions,

$$[\hat{c}_k^\dagger \hat{c}_l, \hat{c}_m] = \hat{c}_k^\dagger \hat{c}_l \hat{c}_m - \hat{c}_m \hat{c}_k^\dagger \hat{c}_l = -\hat{c}_k^\dagger \hat{c}_m \hat{c}_l - (\delta_{km} - \hat{c}_k^\dagger \hat{c}_m) \hat{c}_l = -\delta_{km} \hat{c}_l,$$

where we make use of anticommutation rather than commutation relations.

b. The Heisenberg equation of motion for the operator $\hat{a}_k$ reads

$$\frac{d}{dt} \hat{a}_k = \frac{i}{\hbar} [\hat{H}, \hat{a}_k].$$

We compute the commutator using the relation found in (a),

$$\left[\sum_l \varepsilon_l \hat{a}_l^\dagger \hat{a}_l, \hat{a}_k \right] = -\varepsilon_k \hat{a}_k,$$

which proves the equation of motion given in the problem.

c. Since the numbers of particles are the same in $\langle g|$ and $|g\rangle$, the matrix element is non-zero only if each creation of a particle in a certain level is compensated by an annihilation in the same level. This gives us three distinct cases:

I. $i_1 = i_3$, $i_2 = i_4$, and $i_1 \neq i_2$. In this case, we start by commuting the second and third operator,

$$\langle g | \hat{b}_{i1} \hat{b}_{i2} \hat{b}_{i1}^\dagger \hat{b}_{i2}^\dagger | g \rangle = \langle g | \hat{b}_{i1} \hat{b}_{i1}^\dagger \hat{b}_{i2} \hat{b}_{i2}^\dagger | g \rangle = \langle g | \hat{b}_{i1} \hat{b}_{i1}^\dagger | g \rangle (1 + n_{i2}) = (1 + n_{i2})(1 + n_{i1}),$$

where we have made use of $\hat{b}_k \hat{b}_k^\dagger | g \rangle = (1 + n_k) | g \rangle$.

II. $i_1 = i_4$, $i_2 = i_3$, and $i_1 \neq i_2$. In this case, the first and fourth operators have to be brought together. We commute the fourth operator with the middle two,

$$\langle g | \hat{b}_{i1} \hat{b}_{i2} \hat{b}_{i2}^\dagger \hat{b}_{i1}^\dagger | g \rangle = \langle g | \hat{b}_{i1} \hat{b}_{i1}^\dagger \hat{b}_{i2} \hat{b}_{i2}^\dagger | g \rangle = \langle g | \hat{b}_{i1} \hat{b}_{i1}^\dagger | g \rangle (1 + n_{i2}) = (1 + n_{i2})(1 + n_{i1}),$$

thus getting the same as in the previous case.

III. All indices are the same, $i_1 = i_2 = i_3 = i_4$. The matrix element can then be computed directly giving

$$\langle g | \hat{b}_{i1} \hat{b}_{i1} \hat{b}_{i1}^\dagger \hat{b}_{i1}^\dagger | g \rangle = (2 + n_{i1})(1 + n_{i1}),$$

where we use the definitions (3.1) and (3.2).

Collecting all three cases, we obtain

$$\langle g | \hat{b}_{i1} \hat{b}_{i2} \hat{b}_{i3}^\dagger \hat{b}_{i4}^\dagger | g \rangle = (\delta_{i1,i3} \delta_{i2,i4} + \delta_{i1,i3} \delta_{i2,i4})$$
$$\times \left\{ (1 + n_{i1})(1 + n_{i2}) - \tfrac{1}{2} \delta_{i1,i2} n_{i1} (1 + n_{i2}) \right\}.$$

d. We proceed in the same way considering the cases

I. $i_1 = i_2$, $i_3 = i_4$, and $i_1 \neq i_3$ gives $n_{i1} n_{i3}$.

II. $i_1 = i_4$, $i_3 = i_2$, and $i_1 \neq i_3$ gives $n_{i1} n_{i3}$.

III. $i_1 = i_2 = i_3 = i_4$ also gives $n_{i1} n_{i3}$.

Collecting all cases again, we now obtain

$$\langle g | \hat{b}_{i1}^\dagger \hat{b}_{i2} \hat{b}_{i3}^\dagger \hat{b}_{i4} | g \rangle = (\delta_{i1,i2} \delta_{i3,i4} + \delta_{i1,i4} \delta_{i2,i3})(1 - \tfrac{1}{2} \delta_{i1,i3}) n_{i1} n_{i3}.$$

e. Here we have the same cases as in (d). In the cases I and III, we do not have to commute any operators and the matrix element is readily reduced to $n_{i1}n_{i3}$. For case II we derive

$$\langle g|\hat{c}_{i1}^{\dagger}\hat{c}_{i3}\hat{c}_{i3}^{\dagger}\hat{c}_{i1}|g\rangle = \langle g|\hat{c}_{i1}^{\dagger}\hat{c}_{i1}\hat{c}_{i3}\hat{c}_{i3}^{\dagger}|g\rangle = n_{i1}(1 - n_{i3}).$$

Collecting the three cases thus yields

$$\langle g|\hat{c}_{i1}^{\dagger}\hat{c}_{i2}\hat{c}_{i3}^{\dagger}\hat{c}_{i4}|g\rangle = \delta_{i1,i2}\delta_{i3,i4}n_{i1}n_{i3} + \delta_{i1,i4}\delta_{i2,i3}n_{i1}(1 - n_{i3}).$$

It might seem that this expression does not account for case III, but in fact it does. If we set all indices equal on the right-hand side, we get $n_i n_i + n_i(1 - n_i)$. But in any number state, $n_i(1 - n_i) = 0$ for all i, dictated by fermionic statistics.

PART II

EXAMPLES

In the next three chapters we give examples of applications of the second quantization technique. All our examples come from the field of *condensed matter* physics. It would be possible and probably more "fundamental" to take examples from elementary particle physics. However, viewed ironically, the physics of elementary particles is essentially physics of the vacuum, that is, of an "empty space." All observables – the elementary particles – are excitations of the vacuum, an interesting event is that of a collision between two particles, and observers have to run an experiment for years to get a sufficient number of interesting collisions. This is in sharp contrast with condensed matter systems, where really many identical particles are present. These particles are close to each other and are interacting (colliding) all the time.

Fortunately, the technique of second quantization does not care about the actual number of particles in a system. It deals as easily with excitations in a condensed matter system containing $\sim 10^{23}$ interacting particles, as with elementary particles flying through the vacuum. The technique can thus be applied to real macroscopic systems, and, as we will see, be used to develop some qualitative understanding of phenomena such as superconductivity, superfluidity and magnetism. We think that this connection with reality (we all know what a magnet is) makes the examples from condensed matter physics more appealing.

Still, the usual questions asked about condensed matter systems sound very similar to those of "fundamental" importance. The most important question to be asked is always: what are the properties of the *ground state* of a system. The ground state has the lowest energy possible, it is achieved when the system relaxes at a sufficiently low temperature. In elementary particle physics, the ground state is simply the vacuum. Also in many condensed matter systems the ground state turns out to be rather boring. Surprisingly enough, however, the notion of being "interesting" can be defined with an almost mathematical accuracy. In many cases, an interesting ground state possesses a *broken symmetry* of a kind, which is the case in all examples given in this part of the book. Another property of the ground state, which can make it interesting, is the (elementary) *excitations* which may appear with the ground state as background. In the case of a vacuum, the excitations are the elementary particles themselves. In condensed matter systems, the excitations can be of many types such as particle–hole excitations, phonons, magnons, plasmons, and many more.

The three examples we consider – magnetism, superconductivity, and superfluidity: you should have heard about those – are traditionally presented in detail in textbooks on solid state and condensed matter physics. The aims of the present book are different, and the level of detail in treating these phenomena is much lower. Our main goal is to exemplify the technique of second quantization. We show that, while keeping the models as simple as possible, we can use this technique to understand many interesting features of the systems considered.

4

Magnetism

In this chapter (summarized in Table 4.1) we present a simple model to describe the phenomenon of magnetism in metals. Magnetism is explained by spin ordering: the spins and thus magnetic moments of particles in the material tend to line up. This means that one of the spin projections is favored over another, which would be natural if an external magnetic field is applied. Without the field, both spin projections have the same Zeeman energy, and we would expect the same numbers of particles with opposite spin, and thus a zero total magnetic moment. Magnetism must thus have a slightly more complicated origin than just an energy difference between spin states.

As we have already seen in Chapter 1, interaction between the spins of particles can lead to an energy splitting between parallel and anti-parallel spin states. A reasonable guess is thus that the *interactions* between spin-carrying particles are responsible for magnetism. The conduction electrons in a metal can move relatively freely and thus collide and interact with each other frequently. Let us thus suppose that magnetism originates from the interaction between these conduction electrons.

We first introduce a model describing the conduction electrons as non-interacting free particles. Within this framework we then propose a simple wave function which can describe a magnetic state in the metal. We then include the Coulomb interaction between the electrons in the model, and investigate what this does with the energy of our trial wave function. We see that under certain conditions the magnetic state acquires a lower energy than the non-magnetic one, and the ground state becomes magnetic! Of course, our model is a simple "toy model," but it nevertheless captures the basic features of magnetism and gives us a qualitative physical picture of the phenomenon.

After we have established a magnetic ground state due to the electron–electron interactions, we investigate the elementary excitations in a magnet. We look at single particle excitations and electron–hole pairs, and we finally discuss magnons: spin-carrying low-energy excitations which exist in magnets but not in normal metals.

4.1 Non-interacting Fermi gas

Let us first try to find a magnetic ground state. To this end, we start by treating the magnet as a normal (non-magnetic) metal, then add electron–electron interactions, and investigate under what circumstances the ground state could become magnetic.

We start by introducing the famous *non-interacting* Fermi gas (NIF). As a model, the NIF is widely used to describe delocalized electrons in metals. These electrons move in a periodic potential set up by the atomic structure of the metal, and their eigenfunctions are in principle Bloch waves. Bloch waves, however, are frequently approximated by conventional plane waves, and we will do the same. We label all electronic levels by an effective momentum vector and a spin state, $\mathbf{k}$ and σ. The energy of each level is then given by $E(\mathbf{k}) = \hbar^2 k^2/2m$, with m being the effective electron mass.

What is the ground state of the NIF for a given number of electrons? Per definition it is the state in which each electron has the lowest energy possible. However, since electrons are fermions, they cannot all just occupy the lowest level $\mathbf{k} = 0$. The ground state is thus the state where all levels starting from $E = 0$ to a certain energy $E = E_F$ are occupied. Of course, E_F (the Fermi energy) depends on the total number of electrons. In terms of wave vectors, this means that all states up to $|\mathbf{k}| = k_F$ are filled, where k_F is defined through $E_F = \hbar^2 k_F^2/2m$. This corresponds in reciprocal space to a filled sphere with radius k_F, the Fermi sphere. In the notation of second quantization, we can write this ground state as

$$|g\rangle = \prod_{|\mathbf{k}|<k_F} \hat{c}^\dagger_{\mathbf{k},\uparrow} \hat{c}^\dagger_{\mathbf{k},\downarrow} |0\rangle. \tag{4.1}$$

Control question. Can you explain why (4.1) is the ground state?

With this simple picture in mind, we can evaluate various useful quantities that characterize the NIF. Let us for example calculate the electron density and the energy density. It is easy to see that the total number of electrons N and their total energy E are

$$N = 2_s \sum_{|\mathbf{k}|<k_F} 1, \tag{4.2}$$

$$E = 2_s \sum_{|\mathbf{k}|<k_F} E(\mathbf{k}), \tag{4.3}$$

where we use the symbol 2_s to account for the fact that there are actually two levels for each $\mathbf{k}$ corresponding to the two possible spin projections $\sigma \in \{\uparrow, \downarrow\}$.

We now convert the discrete sums over $\mathbf{k}$ into integrals. The density of allowed $\mathbf{k}$ states is $D(\mathbf{k}) = \mathcal{V}/(2\pi)^3$, so we can replace

$$\sum_{\mathbf{k}} \rightarrow \mathcal{V} \int \frac{d\mathbf{k}}{(2\pi)^3}. \tag{4.4}$$

If we use this conversion for the expressions in (4.2) and (4.3), we can evaluate explicitly

$$n = \frac{N}{\mathcal{V}} = 2_s \int_{|\mathbf{k}|<k_F} \frac{d\mathbf{k}}{(2\pi)^3} \quad = 2_s \int_0^{k_F} \frac{4\pi k^2 dk}{(2\pi)^3} = 2_s \frac{k_F^3}{6\pi^2}, \tag{4.5}$$

$$\frac{E}{\mathcal{V}} = 2_s \int_0^{k_F} E(k) \frac{4\pi k^2 dk}{(2\pi)^3} = 2_s \frac{\hbar^2 k_F^2}{2m} \frac{k_F^3}{10\pi^2} \quad = \frac{3}{5} E_F n. \tag{4.6}$$

The last relation shows that the average energy per particle is $\frac{3}{5} E_F$. From the lower line we see that we can express the energy density also just in terms of k_F (apart from the constants

$\hbar$ and m). The density n is also a function of k_F only. We thus use $k_F = (6\pi^2 n/2_s)^{1/3}$, and express the energy density in terms of the electron density,

$$\frac{E}{\mathcal{V}} = \left(\frac{6\pi^2}{2_s}\right)^{\frac{2}{3}} \frac{3\hbar^2}{10m} n^{\frac{5}{3}}. \tag{4.7}$$

Control question. With semiconductor heterostructures, one can realize a *two-dimensional* electron gas, which can be modeled as an NIF as well. In this case, the density and energy density are respectively the number of electrons and the energy per unit area. What is in this case the relation between $E/\mathcal{A}$ (where $\mathcal{A}$ denotes the area of the system) and n?

4.2 Magnetic ground state

4.2.1 Trial wave function

As noted in the introduction, we assume that the magnetic field produced by ferromagnets originates from an alignment of the magnetic moments, i.e. spins, of the conduction electrons. Magnetism can thus be seen as an excess of spin. We choose the spin quantization axis parallel to the direction of magnetization, and then say that the magnetic state is just a state where the numbers of electrons with opposite spin projections, $N_\uparrow$ and $N_\downarrow$ are different. Let us characterize such a magnetic state with polarization P defined as

$$P = \frac{N_\uparrow - N_\downarrow}{N}, \quad \text{so that} \quad N_{\uparrow,\downarrow} = \frac{N}{2}(1 \pm P), \tag{4.8}$$

P ranging from -1 to 1. The states $P = \pm 1$ correspond to the fully polarized states of the two possible magnetization directions, and $P = 0$ is the non-magnetic state.

Per definition, the ground state of the NIF (4.1) is always non-magnetic. We expect that a magnetic ground state can occur only if the interaction between electrons is taken into account. Therefore, the magnetic ground state has to be an eigenfunction of a full Hamiltonian including interactions,

$$\hat{H} = \hat{H}^{(1)} + \hat{H}^{(2)}, \tag{4.9}$$

where

$$\hat{H}^{(1)} = \sum_{\mathbf{k},\sigma} E(\mathbf{k}) \hat{c}^\dagger_{\mathbf{k},\sigma} \hat{c}_{\mathbf{k},\sigma}, \tag{4.10}$$

$$\hat{H}^{(2)} = \frac{1}{2\mathcal{V}} \sum_{\mathbf{k},\mathbf{k}',\mathbf{q}} \sum_{\sigma,\sigma'} U(\mathbf{q}) \hat{c}^\dagger_{\mathbf{k}+\mathbf{q},\sigma} \hat{c}^\dagger_{\mathbf{k}'-\mathbf{q},\sigma'} \hat{c}_{\mathbf{k}',\sigma'} \hat{c}_{\mathbf{k},\sigma}. \tag{4.11}$$

In $\hat{H}^{(2)}$, the coefficient $U(\mathbf{q})$ stands for the Fourier transform of the pairwise electron–electron interaction potential $U(\mathbf{r})$.

A point of concern is that the eigenfunctions of this Hamiltonian are not precisely known, this situation being very typical for systems of interacting identical particles. One might think that this is not a big issue in our times because of the ready availability of powerful computers which can handle very large matrices. We also do not know precisely the roots of a fifth power polynomial, but a computer can give it to us with any desired accuracy. But although computational approaches in the field of interacting particles can be useful and informative, they have to be based on heuristic and often uncontrollable assumptions, and they are far from giving us answers with any accuracy required. The fundamental problem is the high dimension of Fock space. Owing to this, an exponentially big number of basis vectors is required to represent the wave functions. One could not fit such information into the memories of conventional computers nor process it. A quantum computer might be helpful for these problems...

Luckily, we can learn a lot about the physics of many particles with approximate methods. In this section, we apply the *trial* wave function method. This method is best suited for searching a ground state. The idea is to replace a true eigenfunction of the Hamiltonian with a simple wave function that has some adjustable parameters. One calculates the expectation value of the energy as a function of these parameters and looks for the values of these parameters that correspond to the energy minimum of the system. This provides a "best fit" of the true ground state with a simple wave function.

Control question. Can you give an example of an analogous method from a different field of physics?

Let us introduce the simplest wave function that corresponds to a magnetic state. The idea is to regard electrons of opposite spin as two different kinds of particle and allow for different densities of particles of each kind. The total number of particles is fixed, however, so the wave function has a single adjustable parameter: the polarization P. Instead of two different densities, it is convenient to use two different Fermi vectors, $k_{F,\uparrow}$ and $k_{F,\downarrow}$. So the wave function reads

$$|g_P\rangle = \left(\prod_{|\mathbf{k}| < k_{F,\uparrow}} \hat{c}^{\dagger}_{\mathbf{k},\uparrow} \right) \left(\prod_{|\mathbf{k}| < k_{F,\downarrow}} \hat{c}^{\dagger}_{\mathbf{k},\downarrow} \right) |0\rangle. \tag{4.12}$$

4.3 Energy

The aim for now is to calculate the energy $E = \langle g_P | \hat{H} | g_P \rangle$ as a function of the polarization P. This calculation consists of two parts. The single-particle term in the Hamiltonian $\hat{H}^{(1)}$ represents the total kinetic energy of the system, and the interaction term $\hat{H}^{(2)}$ gives the potential energy due to the interaction potential $U(\mathbf{r})$. We treat these terms separately.

4.3.1 Kinetic energy

Let us evaluate the kinetic energy corresponding to the trial wave function $|g_P\rangle$. Formally speaking, we should evaluate $\langle g_P | \hat{H}^{(1)} | g_P \rangle$ using the Hamiltonian as defined in (4.10) and

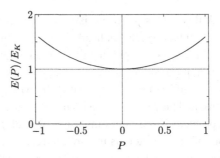

Fig. 4.1 Kinetic energy of the polarized state versus the polarization.

(4.11). We understand, however, that this gives the same as just summing up the energies of the levels filled,

$$\langle gP|\hat{H}^{(1)}|gP\rangle = \sum_{\mathbf{k},\sigma} E(k)n_{\mathbf{k},\sigma}, \quad \text{with} \quad n_{\mathbf{k},\sigma} = \theta(k_{\mathrm{F},\sigma} - |\mathbf{k}|). \tag{4.13}$$

The calculation is very similar to the one done in (4.6). The only difference is that we have to sum up the contributions of two spin directions, $E \propto k_{\mathrm{F},\uparrow}^5 + k_{\mathrm{F},\downarrow}^5$. In addition, we should express each $k_{\mathrm{F},\sigma}$ in terms of N_σ. Actually, we can use the fact that energy density of each sort is proportional to the particle density to the power 5/3 to find that

$$E(P) = \frac{E_K}{2}\left\{\left(\frac{2N_\downarrow}{N}\right)^{\frac{5}{3}} + \left(\frac{2N_\uparrow}{N}\right)^{\frac{5}{3}}\right\} = \frac{E_K}{2}\left\{(1+P)^{\frac{5}{3}} + (1-P)^{\frac{5}{3}}\right\}, \tag{4.14}$$

where $E_K \equiv E(0)$, the energy of the non-magnetic state.

Control question. Can you justify the above formula?

A plot of the energy $E(P)$ versus polarization is given in Fig. 4.1. We see that creating a non-zero polarization costs kinetic energy. Indeed, if we compare the completely polarized state with the unpolarized one having the same total number of particles, we find a larger k_F in the polarized case. This increases the Fermi energy and therefore the energy per particle. It is also easy to see from (4.7). A fully polarized state has the same density of particles n, but lacks the spin degeneracy, so does not have the factor 2_s.

4.3.2 Potential energy

Before we start with the actual evaluation of potential energy, let us make some general remarks about the model of the electron–electron interaction commonly used in solid state physics. It is not directly related to our problem, but it is a nice illustration of a way of thinking that allows us to simplify otherwise complicated and unsolvable issues.

Electrons are charged particles, so the pairwise interaction must satisfy Coulomb's law. The corresponding interaction potential and its Fourier transform read

$$U(r) = \frac{e^2}{4\pi\varepsilon_0 r} \quad \text{and} \quad U(q) = \frac{e^2}{\varepsilon_0 q^2}. \tag{4.15}$$

Importantly, the Coulomb interaction is *long range*: it falls off very slowly upon increasing distance, with a dependence $\propto 1/r$. In terms of the Fourier transform, this implies a divergence of $U(\mathbf{q})$ at small $\mathbf{q}$.

However, if we pick a pair of electrons in a metal and look at their energy as a function of their coordinates, we do not see this long-range potential. The interaction energy dies out as soon as their separation exceeds the average distance between the electrons in the metal. What happens is that all other electrons adjust their positions constantly to the position of the pair picked and *screen* the potential produced by the pair. One can thus say that, due to the long range of the potential, an act of interaction – a collision – always involves many electrons. Indeed, attempts to treat electrons in metals starting from the long-range interaction potential quickly run into enormous technical difficulties.

The way out is to use in the second-quantization Hamiltonian an *effective* pairwise interaction that corresponds to a *screened* short-range potential,

$$U(r) = \frac{e^2}{4\pi\varepsilon_0 r}e^{-r/r_C} \quad \text{and} \quad U(q) = \frac{e^2}{\varepsilon_0(q^2 + r_C^{-2})}, \tag{4.16}$$

r_C being the screening radius, for typical metals of the order of the inter-electron distance. This adjustment has the status of a model. The true Coulomb potential is replaced by an effective one and this is supposed to mimic some complicated many-body effects. The model proves to be very successful when applied, and is partially justified by theoretical reasoning.

We actually use an even simpler model. Since we believe that the effective interaction potential is of a short range, let us take zero range. We thus assume a so-called contact potential,

$$U(r) = U\delta(r) \quad \text{and} \quad U(q) = U, \tag{4.17}$$

meaning that the electrons interact only if they are at the very same place.

We are now ready to evaluate the potential energy. We apply the above results to the actual interaction Hamiltonian,

$$\langle gP|\hat{H}^{(2)}|gP\rangle = \frac{U}{2\mathcal{V}}\sum_{\mathbf{k},\mathbf{k}',\mathbf{q}}\sum_{\sigma,\sigma'}\langle gP|\hat{c}^\dagger_{\mathbf{k}+\mathbf{q},\sigma}\hat{c}^\dagger_{\mathbf{k}'-\mathbf{q},\sigma'}\hat{c}_{\mathbf{k}',\sigma'}\hat{c}_{\mathbf{k},\sigma}|gP\rangle. \tag{4.18}$$

We see that what we have to calculate has exactly the form of the example we considered in Section 3.5: the expectation value of a product of two creation operators and two annihilation operators. We can thus use the results of our previous analysis. The first observation we made was that there are two possible combinations of the indices of the operators which give a non-zero result. In terms of our $\mathbf{k}$s and $\mathbf{q}$s, the two options are (i) $\mathbf{k} + \mathbf{q} = \mathbf{k}$, $\mathbf{k}' - \mathbf{q} = \mathbf{k}'$, and (ii) $\mathbf{k} + \mathbf{q} = \mathbf{k}'$, $\sigma = \sigma'$, $\mathbf{k}' - \mathbf{q} = \mathbf{k}$. These two possibilities correspond to two physically different contributions to the energy.

One can picture an interaction event as described by (4.18) by the diagrams shown in Fig. 4.2. We start with the same diagram as shown in Chapter 3 representing a term in $\hat{H}^{(2)}$, as shown in Fig. 4.2(a). The incoming (outgoing) arrows correspond to annihilation (creation) operators in certain levels. The interaction (the wiggly line) annihilates two electrons (in $\mathbf{k}, \sigma$ and $\mathbf{k}', \sigma'$) and creates two electrons in two other states ($\mathbf{k} + \mathbf{q}, \sigma$ and $\mathbf{k}' - \mathbf{q}, \sigma'$).

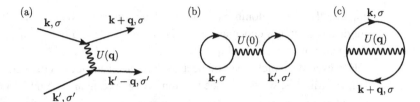

(a) A term from the Hamiltonian $\hat{H}^{(2)}$. (b) Representing the Hartree energy. (c) Representing the Fock energy.

The two different possibilities for the **k**s and **q**s found above are depicted in Fig. 4.2(b,c). The first possibility, $\mathbf{k}+\mathbf{q} = \mathbf{k}$ and $\mathbf{k}'-\mathbf{q} = \mathbf{k}'$, gives rise to the *Hartree* energy of the system. If we incorporate this in our diagram, and connect the arrows representing the same levels with each other, we get the diagram of Fig. 4.2(b). Each line represents a *pairing* of the corresponding CAPs, $\langle \hat{c}_k^\dagger \hat{c}_k \rangle$. We see that it demands that $\mathbf{q} = 0$.

The Hartree contribution to the energy becomes

$$E_\mathrm{H} = \frac{U}{2\mathcal{V}} \sum_{\mathbf{k},\sigma} \sum_{\mathbf{k}',\sigma'} n_{\mathbf{k},\sigma} n_{\mathbf{k}',\sigma'}. \tag{4.19}$$

The double sum is actually just a square of a single sum, which is

$$\sum_{\mathbf{k},\sigma} n_{\mathbf{k},\sigma} = N, \tag{4.20}$$

so that

$$E_\mathrm{H} = \frac{U}{2\mathcal{V}} N^2 = \frac{U\mathcal{V}n^2}{2}. \tag{4.21}$$

We see that the Hartree energy does not depend on the polarization P, and is just a function of the particle density. This Hartree energy has a simple physical meaning. It is the result of looking for combinations of creation and annihilation operators in a Hamiltonian that describe electron density, and replacing those by their expectation values, or in other words, making the electron density a classical variable. The potential energy of this classical density profile is the Hartree energy.

Control question. Suppose we had taken the long-range potential (4.15) for our model. What kind of trouble would we have run into?

The other possibility we mentioned above, $\mathbf{k}+\mathbf{q} = \mathbf{k}'$, $\sigma = \sigma'$, and $\mathbf{k}'-\mathbf{q} = \mathbf{k}$, gives rise to the *Fock* energy, also called *exchange* energy. We see that in this case both occupation numbers involved come with the same spin projection. In terms of diagrams, the Fock energy corresponds to connecting the lines as shown in Fig. 4.2(c). It reads

$$E_\mathrm{Fock} = -\frac{U}{2\mathcal{V}} \sum_{\mathbf{k},\mathbf{q}} \sum_\sigma n_{\mathbf{k},\sigma} n_{\mathbf{k}+\mathbf{q},\sigma} = -\frac{U}{2\mathcal{V}} \sum_{\mathbf{k},\mathbf{k}'} \sum_\sigma n_{\mathbf{k},\sigma} n_{\mathbf{k}',\sigma}. \tag{4.22}$$

Note that if U had still been a function of $\mathbf{q}$, the last simplification could not have been made, and the expression would have been very hard to evaluate. The double sum above can again be regarded as the square of a single sum. However, we now have two squares corresponding to two different spin projections. Since

$$\sum_{\mathbf{k}} n_{\mathbf{k},\sigma} = N_\sigma, \tag{4.23}$$

we find that

$$E_{\text{Fock}} = -\frac{U}{2\mathcal{V}}\left(N_\uparrow^2 + N_\downarrow^2\right) = -E_H \frac{1+P^2}{2}. \tag{4.24}$$

We have found a simple but rather surprising answer. First of all, the Fock energy *does* depend on the polarization, while the pairwise interaction does not. Secondly, the Fock energy is negative, in contrast to the Hartree energy.

A crucial observation is that the sum of the interaction energies is lowest for a fully polarized state $P = \pm 1$. Why should this be so? Our interaction model is that of the simplified contact potential. The interaction energy comes only from electrons that happen to be at the same point. In a fully polarized state, however, all electrons have the same spin projection and therefore *never* come to the same point. This is forbidden by their fermionic nature, by the Pauli exclusion principle. This observation explains everything. If electrons are polarized, they cancel the positive energy due to their coulomb interaction, since they just *cannot come together*. So it is understandable that, given the fact that there is contact repulsion between the electrons, the system energetically favors a non-zero polarization.

4.3.3 Energy balance and phases

Now we can determine whether a polarized state has a lower total energy than a non-polarized. The full energy consists of the kinetic energy and the Hartree and Fock terms, and reads

$$E(P) = \frac{E_K}{2}\left\{(1+P)^{\frac{5}{3}} + (1-P)^{\frac{5}{3}}\right\} + \frac{E_H}{2}(1-P^2). \tag{4.25}$$

It is convenient to characterize the strength of the interaction by the dimensionless parameter E_H/E_K. For small values of this parameter the interaction is weak, and we expect the kinetic energy to dominate. The non-magnetic state is then energetically favorable. In the opposite case, for large E_H/E_K, the polarization-dependent Fock energy wins and the ground state is polarized.

Control question. Can you express E_H/E_K in terms of U, n, and m?

As an example, we plot the total energy and its separate contributions as a function of P in Fig. 4.3(a). For this plot we chose the value $E_H/E_K = 1.8$.

Control question. What can you say about the ground state for $E_H/E_K = 1.8$, as used in the plot?

For general E_H/E_K, the optimal polarization can be determined as being the state with minimal total energy, that is, $\partial E(P)/\partial P = 0$. The corresponding relation reads

$$\frac{\partial E(P)}{\partial P} = 0 \quad \Rightarrow \quad \frac{5}{6}\frac{(1+P)^{\frac{2}{3}} - (1-P)^{\frac{2}{3}}}{P} = \frac{E_H}{E_K}. \tag{4.26}$$

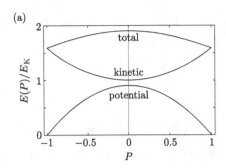
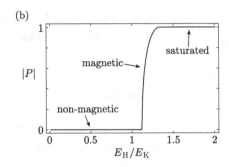

Fig. 4.3 (a) Total, kinetic, and potential energies versus polarization. For this plot we chose $E_H/E_K = 1.8$. (b) The polarization versus the interaction strength E_H/E_K and the resulting phases.

Although it is impossible to solve this equation analytically with respect to P, it suffices to plot the optimal polarization versus E_H/E_K, as in Fig. 4.3(b). From the plot, we understand the following:

1. If the interaction is weak, $E_H/E_K < 10/9$, the ground state is non-magnetic.
2. A transition to a magnetic state occurs at $E_H/E_K = 10/9$. The polarization is still vanishing at the transition point and gradually increases above it.
3. At $E_H/E_K > \frac{5}{6} \cdot 2^{2/3}$ the ground state is completely polarized.

Of course our model is too simplistic to account for all details of ferromagnetism in metals. However, we manage to capture several qualitative features of the phenomenon.

4.4 Broken symmetry

As mentioned, interesting ground states of condensed matter are frequently associated with a certain symmetry breaking. How does this relate to our results? The ground state of the NIF has several symmetries. One can, for instance, rotate the gas without any change of its state. The polarization P was used above like a scalar, but this was only possible because we chose the spin quantization axis in a certain direction when we wrote (4.12), and we assumed the magnetization to be parallel to this axis. Since the choice of this axis is arbitrary and could have been in any direction, the polarization is actually a vector $\mathbf{P}$.[1] If we rotate a vector, it *does* change. A magnetic (polarized) state does not stay the same if we rotate the system, and therefore the rotational symmetry is broken. Let us stress that the Hamiltonian still remains symmetric: it is its ground state that looses a symmetry. The mechanism we considered results in a certain polarization and fixes its amplitude, but it does not fix its direction.

This implies that ground states with a broken symmetry are highly degenerate. Different polarization directions correspond to distinct physical states with the same energy. Since it

---

[1] As a matter of fact, it is a pseudovector. Pseudovectors do not change sign upon inversion of all coordinates.

requires an infinitesimally small energy to change from one of those states to another, there must be excitations possible with arbitrarily small energy. The lowest-lying excitations are thus slow spatial (long wave length) variations of the order parameter. In magnets, those excitations are called *magnons*, and we address them in the next section. Yoichiro Nambu discovered this relation between the breaking of a continuous symmetry and the existence of low-energy bosonic excitations for the case of a metal with a superconducting ground state (see Chapter 5). Later, it was recognized and proven by Jeffrey Goldstone that the statement is general and holds for any broken continuous symmetry. Therefore, such excitations are generally called Goldstone or Nambu–Goldstone bosons.

It is less obvious from our consideration that another important symmetry is broken as well in a magnetic state: the symmetry with respect to time-reversal. Indeed, a finite polarization produces a magnetic field. This field can be seen as the result of microscopic currents flowing along the surface of the magnet. Upon time reversal, the sign of all those currents changes, so also does the sign of the magnetic field produced. Therefore, the time-reversal symmetry is broken in a magnetic state.

Our consideration has shown that magnetism arises by means of a *second-order* phase transition which is characterized by a continuous change of the state involved. Indeed, the polarization along with the energy is continuous in the transition point. Such phase transitions are typical for spontaneously broken symmetries.

4.5 Excitations in ferromagnetic metals

We have thus found the ground state of a ferromagnet, or at least a reasonable approximation to it. Let us now consider the possible excited states. We restrict ourselves to a ground state that is not polarized completely, $|P| \neq 1$, so that electrons of both polarizations are present in the solid. This is the case in most ferromagnetic metals. For definiteness, let us assume an excess of electrons with spin "up" along the axis we have chosen, i.e. $P > 0$.

4.5.1 Single-particle excitations

We start with *single-particle* excitations. These are obtained from the ground state by either adding an electron to an empty level or extracting an electron from a filled level. These excited states are commonly called *electrons*[2] and *holes*, and are defined as follows:

$$|e_{\mathbf{k},\sigma}\rangle = \hat{c}_{\mathbf{k},\sigma}^{\dagger}|g\rangle, \quad \text{with } |\mathbf{k}| > k_{F,\sigma}, \tag{4.27}$$

$$|h_{-\mathbf{k},-\sigma}\rangle = \hat{c}_{\mathbf{k},\sigma}|g\rangle, \quad \text{with } |\mathbf{k}| < k_{F,\sigma}. \tag{4.28}$$

---

[2] Fastidiously, a proper name would be a quasi-electron, or, more generally, a quasi-particle state. "Quasi" is here meant to distinguish the electrons (or particles) existing on the background of the physical vacuum from similar excitations on the background of the ground state in a solid. However, to avoid too many "quasies", we mostly omit them.

Yoichiro Nambu (b. 1921)
Won the Nobel Prize in 2008 for "the discovery of the mechanism of spontaneous broken symmetry in subatomic physics."

Yoichiro Nambu was educated in tough times. His physics studies at the Imperial University of Tokyo were shortened to 2.5 years, so that the students could be drafted for the war. In the army, Nambu spent most of his time in an army radar laboratory, performing unsuccessful attempts to construct a short-wavelength radar system. After the war, he gained a research position at the University of Tokyo, but in fact, due to the poor living conditions in post-war Japan, he had to spend most of his time looking for food. Since housing was scarce in Tokyo, he even literally lived in his office for several years: his colleagues regularly found him in the morning sleeping on his desk.

An invitation to spend some time at Princeton University in 1952 was thus happily accepted. Although life in the US seemed paradisaical compared to Japan, Nambu had serious problems adapting to the scientific climate. He suffered a depression, and published only a single paper during his two years in Princeton. He nevertheless made a good impression and was offered a position in Chicago in 1954, where his productivity revived again. He stayed in Chicago till his retirement in 1991.

Nambu's main field of research is elementary particle physics, but besides he always had an interest in many-body physics. In 1957 he attended a talk by Schrieffer explaining the new BCS theory for superconductivity. This triggered Nambu to think about the symmetries in a superconductor. He realized that the broken gauge symmetry in the BCS theory was not a problem, but actually led to an explanation of superconductivity in terms of spinless bosonic particles. He then found that, in the field of elementary particle physics also, the existence of pions could be explained as resulting from a spontaneously broken symmetry. This made Nambu suspect that the connection between broken symmetries and the emergence of new types of particle might be more general. Soon this was indeed confirmed with a general proof by Jeffrey Goldstone.

Note that we define the creation of a hole with momentum $\hbar\mathbf{k}$ and spin σ to be the removal of an electron with *opposite* momentum and spin. In this way one can regard holes as being genuine particles: creation of the particle $|h_{\mathbf{k},\sigma}\rangle$ adds $\hbar\mathbf{k}$ and σ to the total momentum and spin of the system.

These electron and hole states, as well as the ground state, are number states having a well-defined number of particles in each level. We thus can use the relations for the kinetic (4.13), Hartree (4.21), and Fock (4.22) energies to compute the total energy of the single-particle excitations. We are interested in excitation energies, that is, in energies relative to the ground state energy, or in other words, the energy the system *gains*

by adding an electron or a hole to the ground state. A straightforward calculation yields
($n_\sigma \equiv N_\sigma / \mathcal{V}$),

$$E_e(\mathbf{k}, \sigma) = \ \ E(k) + Un - Un_\sigma, \tag{4.29}$$

$$E_h(\mathbf{k}, \sigma) = -E(k) - Un + Un_{-\sigma}. \tag{4.30}$$

The first, second and third terms in these expressions come respectively from the kinetic,
Hartree, and Fock energies.

There are two ways to visualize this result. In Fig. 4.4(a) we plot the electron energies
(solid lines), and hole energies taken with a minus sign (dotted lines). We see two parabolas
that resemble the parabolic spectrum of free electrons, but in this case both parabolas have a
vertical shift which depends on the spin direction. Qualitatively it thus looks as if a Zeeman
magnetic field acts on the particles, providing a spin-splitting of $2B$, where $B \equiv UnP/2$.
However, there is no external magnetic field applied: the splitting observed is due to the
exchange interaction between electrons. For this reason, B is called the *exchange field* of a
ferromagnet.

To understand the restrictions imposed on the values of the electron and hole energies
by stability of the ground state, we should consider our magnet in contact with an electron reservoir with Fermi energy E_F. Stability then implies that a charge transfer between
the reservoir and the magnet should not be energetically favorable, otherwise there would
be a charge flow to or from the magnet. This constraint can be written as $E_e > E_F$ and
$E_h > -E_F$ for any k, σ, implying that the minimum electron energy should be the same for
both spin directions, that is,

$$\frac{\hbar^2 k_{F,\uparrow}^2}{2m} - \frac{\hbar^2 k_{F,\downarrow}^2}{2m} = 2B. \tag{4.31}$$

Control question. Equation 4.31 *must* be (4.26) in disguise. Why is it so? Can you
prove the equivalence of the two equations?

Control question. What is the difference between the Fermi energy of non-interacting
and interacting electrons as follows from (4.29)?

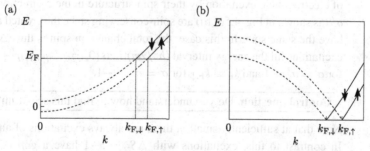

Fig. 4.4   Single-particle excitations in a metallic ferromagnet (hole energies are plotted with dotted lines). (a) The electron and
negated hole energies plotted together resemble the parabolic spectrum of free electrons subject to an exchange
splitting. (b) All electron and hole energies are positive if counted from the Fermi energy E_F.

Another way of visualization (see Fig. 4.4(b)) is to plot the electron and hole energies counted from E_F and $-E_F$, respectively. In this way all energies are positive.

4.5.2 Electron–hole pairs

More complicated excited states can be formed by applying more CAPs to the ground state. Let us consider excitations that do not change the number of particles in the system, unlike the single-particle excitations. The simplest way to produce such an excitation is to move an electron from a filled level to an empty one. This results in a hole in a previously filled level and an electron in a previously empty level. We call this the creation of an *electron–hole pair*. In terms of CAPs, the resulting state reads

$$|eh_{\mathbf{k},\mathbf{q};\sigma,\sigma'}\rangle = \Theta(\mathbf{k},\mathbf{q},\sigma,\sigma')\hat{c}^\dagger_{\mathbf{k}+\mathbf{q}/2,\sigma}\hat{c}_{\mathbf{k}-\mathbf{q}/2,\sigma'}|g\rangle. \tag{4.32}$$

We see that electron–hole pair creation is not possible for arbitrary $\mathbf{k}$ and $\mathbf{q}$. Indeed, to be able to create the state in (4.32), the level with $(\mathbf{k}+\frac{1}{2}\mathbf{q},\sigma)$ must be empty in the ground state, while the level $(\mathbf{k}-\frac{1}{2}\mathbf{q},\sigma')$ must be filled. Therefore, an electron–hole pair can only be created if $|\mathbf{k}+\frac{1}{2}\mathbf{q}| > k_{F,\sigma}$ and $|\mathbf{k}-\frac{1}{2}\mathbf{q}| < k_{F,\sigma'}$. These restrictions are automatically imposed by the electronic CAPs (a creation operator acting on an already filled state returns zero, the same for an annihilation operator acting on an empty state), but to keep track of this when calculating the energy spectrum of the excitations, we include a Θ-function which is 1 when $|\mathbf{k}+\frac{1}{2}\mathbf{q}| > k_{F,\sigma}$ and $|\mathbf{k}-\frac{1}{2}\mathbf{q}| < k_{F,\sigma'}$, and zero otherwise.

The excitation in (4.32) bears a total momentum $\hbar\mathbf{q}$. It is also a number state, so its energy is easy to evaluate,

$$E_{eh}(\mathbf{k},\mathbf{q};\sigma,\sigma') = E_e(\mathbf{k}+\tfrac{1}{2}\mathbf{q},\sigma) + E_h(\mathbf{k}-\tfrac{1}{2}\mathbf{q},-\sigma') = \frac{\hbar^2(\mathbf{k}\cdot\mathbf{q})}{m} + B(\sigma'-\sigma), \quad (4.33)$$

where in the last expression σ and σ' are ± 1 for spin up and down. Logically enough, the energy of an electron–hole pair is the sum of the energies of its constituents.

We see from (4.33) that for a given wave vector $\mathbf{q}$, the spectrum of the electron–hole excitations is continuous: the energies corresponding to different $\mathbf{k}$ fill a finite interval. In Fig. 4.5 we plot this continuous spectrum, and we see that we can distinguish three classes of electron–hole excitation by their spin structure in the z-direction. Excitations with $\sigma = \sigma'$ (as shown in Fig. 4.5(a,b)) are spin-conserving since the created and annihilated electron have the same spin. In this case, the total change in spin is thus $\Delta S_z = 0$. These spinless excitations fill the energy interval $(\hbar^2/2m)[\max\{0, q^2 - 2qk_F\}; q^2 + 2qk_F]$, where $k_F \equiv k_{F,\uparrow}$ for $\sigma = \sigma' = 1$ and $k_F \equiv k_{F,\downarrow}$ for $\sigma = \sigma' = -1$.

Control question. Do you understand how to find this spectrum?

Note that at sufficiently small q, there are always excitations of an arbitrary small energy. In contrast to this, excitations with $\Delta S_z = -1$ have a *gap* of energy $2B$ at $q = 0$, as shown in Fig. 4.5(c). Indeed, such an excitation is equivalent to an electron transfer from the $\uparrow$-band to the $\downarrow$-band and has to cost the energy splitting $2B$. The gap reduces with increasing q and finally vanishes, the energies being distributed in the interval

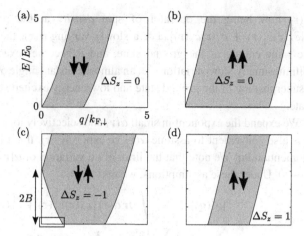

Fig. 4.5 Electron–hole excitations in a ferromagnet. The electron–hole continuum is plotted for different spin projections σ, σ'. For all plots $k_{F,\uparrow} = 2k_{F,\downarrow}$ is assumed.

$(\hbar^2/2m)[\max\{0, (q-k_{F,\uparrow})^2 - k_{F,\downarrow}^2\}; (q+k_{F,\uparrow})^2 - k_{F,\downarrow}^2]$. The last scenario concerns excitations with $\Delta S_z = 1$ (see Fig. 4.5(d)): they do not exist at all for q below a critical value. Above that value, they fill the energy interval $(\hbar^2/2m)[\max\{0, (q-k_{F,\downarrow})^2 - k_{F,\uparrow}^2\}; (q+k_{F,\downarrow})^2 - k_{F,\uparrow}^2]$.

Control question. What is the critical q for excitations with $\Delta S_z = 1$? At which q does the gap for $\Delta S_z = -1$ vanish?

All this gives a straightforward picture of single-particle and electron–hole excitations in ferromagnets. It is an important step in the understanding of advanced quantum mechanics to figure out that this picture is incomplete, and to learn how to complete it.

4.5.3 Magnons

To understand what we have omitted, let us consider an alternative way to produce excited states of low energy, the Goldstone modes mentioned earlier. We will use a trick commonly implemented in *gauge theories*, to be discussed in more detail in later chapters. Let us recall that the polarized ground state is not unique. Another ground state, with the same energy, can be obtained by a rotation, say, about the x-axis over an angle α,

$$|g'\rangle = \exp\{-i\alpha\hat{S}_x\}|g\rangle, \tag{4.34}$$

$\hat{S}_x$ being the x-component of the spin operator. This only changes the direction of the polarization, nothing more.

The trick is to make this rotation angle a *slowly-varying* function of the coordinates, $\alpha(\mathbf{r})$. We then can write the rotated state as

$$|g''\rangle = \exp\left\{-i \int d\mathbf{r}\,\alpha(\mathbf{r})\hat{s}_x(\mathbf{r})\right\} |g\rangle, \tag{4.35}$$

$\hat{s}_x(\mathbf{r})$ now being the operator of spin *density* at the point $\mathbf{r}$, defined as $\hat{s}_x(\mathbf{r}) = \frac{1}{2}\{\hat{c}_\uparrow^\dagger(\mathbf{r})\hat{c}_\downarrow(\mathbf{r}) + \hat{c}_\downarrow^\dagger(\mathbf{r})\hat{c}_\uparrow(\mathbf{r})\}$. For a slowly-varying $\alpha(\mathbf{r})$, the state $|g'\rangle$ has approximately the same energy as the ground state, and it becomes exactly the same in the limit of inifinitesimally slow variation, i.e. an almost constant angle α. The state $|g'\rangle$ should thus be a superposition of the ground state and low-energy excited states orthogonal to the ground state.

We expand the exponent in small $\alpha(\mathbf{r})$ and selectively look at terms of first order in $\alpha(\mathbf{r})$. It is also convenient to assume $\alpha(\mathbf{r}) \propto \exp(-i\mathbf{q} \cdot \mathbf{r})$, this filters out an excited state with momentum $\hbar\mathbf{q}$. We note that the limit of no variation of $\alpha(\mathbf{r})$ thus corresponds to the limit $q \to 0$. Under these assumptions, we find

$$
\begin{aligned}
|g''(\mathbf{q})\rangle &= -\frac{i}{2}\int d\mathbf{r}\,\alpha(\mathbf{r})\Big\{\hat{c}_\uparrow^\dagger(\mathbf{r})c_\downarrow(\mathbf{r}) + c_\downarrow^\dagger(\mathbf{r})c_\uparrow(\mathbf{r})\Big\}|g\rangle \\
&\propto \sum_{\mathbf{k}}\Big\{\hat{c}_{\mathbf{k}+\mathbf{q}/2,\uparrow}^\dagger \hat{c}_{\mathbf{k}-\mathbf{q}/2,\downarrow} + \hat{c}_{\mathbf{k}+\mathbf{q}/2,\downarrow}^\dagger \hat{c}_{\mathbf{k}-\mathbf{q}/2,\uparrow}\Big\}|g\rangle \\
&\propto \sum_{\mathbf{k}}\Theta(\mathbf{k},\mathbf{q},\downarrow,\uparrow)\,\hat{c}_{\mathbf{k}+\mathbf{q}/2,\downarrow}^\dagger \hat{c}_{\mathbf{k}-\mathbf{q}/2,\uparrow}|g\rangle.
\end{aligned}
\tag{4.36}
$$

To arrive at the last line, we use the property of excitations with $\Delta S_z = 1$ mentioned above: they do not exist at sufficiently small $\mathbf{q}$. We can thus skip the second term of the middle line. We now identify the resulting state as a *superposition* of the $\Delta S_z = -1$ electron–hole states with all $\mathbf{k}$ allowed. In the limit $q \to 0$, the allowed excitations fill the shell $k_{F,\downarrow} < k < k_{F,\uparrow}$. By construction, the energy of the excitation created in (4.36) must vanish when $q \to 0$. Therefore, this superposition has a lower energy than any of its constituents, since the energies of the electron–hole excitations with $\Delta S_z = -1$ are at least $2B$ (see again Fig. 4.5(c)).

We have just built a *magnon* state. Let us normalize it and associate with it an operator $\hat{m}_\mathbf{q}^\dagger$ that creates a magnon with wave vector $\mathbf{q}$,

$$
|g''(\mathbf{q})\rangle = \hat{m}_\mathbf{q}^\dagger|g\rangle, \quad \text{where} \quad \hat{m}_\mathbf{q}^\dagger \equiv \frac{1}{\sqrt{N_\uparrow - N_\downarrow}}\sum_{\mathbf{k}}\Theta(\mathbf{k},\mathbf{q},\downarrow,\uparrow)\,\hat{c}_{\mathbf{k}+\mathbf{q}/2,\downarrow}^\dagger \hat{c}_{\mathbf{k}-\mathbf{q}/2,\uparrow}.
\tag{4.37}
$$

Direct computation gives the commutator

$$
[\hat{m}_\mathbf{q}^\dagger, \hat{m}_{\mathbf{q}'}] = \frac{\delta_{\mathbf{q}\mathbf{q}'}}{N_\uparrow - N_\downarrow}\sum_{\mathbf{k}}\Big\{\hat{c}_{\mathbf{k},\uparrow}^\dagger \hat{c}_{\mathbf{k},\uparrow} - \hat{c}_{\mathbf{k},\downarrow}^\dagger \hat{c}_{\mathbf{k},\downarrow}\Big\}.
\tag{4.38}
$$

We now make one further approximation. We replace the electronic CAPs in (4.38) by their expectation values in the ground state. The justification is as follows: the electronic CAPs are equally contributed to by many independent levels, eventually all levels in the shell, so we expect its relative fluctuations to be small.

The replacement yields *boson* commutation relations for the magnon CAPs,

$$
[\hat{m}_\mathbf{q}^\dagger, \hat{m}_{\mathbf{q}'}] = \delta_{\mathbf{q}\mathbf{q}'}.
\tag{4.39}
$$

We thus recognize that we can regard magnons as particles obeying boson statistics. Magnons are excitations on the background of a "new vacuum," the ground state of the ferromagnet, and they cannot exist without this background.

Control question. Can you give the magnon creation operator for the ground state with opposite polarization? And for a ground state polarized in the x-direction?

A magnon can be regarded as a *bound* state of an electron and hole, a kind of a molecule where the electron and hole remain separated by a distance of the order of k_F^{-1}.

We thus achieve a complete description of all low-energy excited states of a metallic ferromagnet by constructing a bosonic Fock space with the aid of magnon CAPs and the new vacuum $|g\rangle$. An arbitrary state is then given by the magnon occupation numbers of each level $\mathbf{q}$, the set $|\{n_\mathbf{q}\}\rangle$, and the energy of this state is simply $E = \sum_\mathbf{q} E_{\text{mag}}(\mathbf{q}) n_\mathbf{q}$.

4.5.4 Magnon spectrum

The gauge trick has helped us to find the magnon states, but not their energies. To find those, we have to resort to the original Hamiltonian (4.9) and find its matrix elements in the basis of the electron–hole states $|eh_{\mathbf{k},\mathbf{q}}\rangle$ with $\Delta S_z = 1$ and certain $\mathbf{q}$,

$$H_{\mathbf{k},\mathbf{k}'} = \langle eh_{\mathbf{k}',\mathbf{q}}|\hat{H}|eh_{\mathbf{k},\mathbf{q}}\rangle. \tag{4.40}$$

The elements of $\hat{H}^{(1)}$ in (4.10), corresponding to the kinetic energy term, are diagonal and trivial. The elements corresponding to the interaction term, $\hat{H}^{(2)}$ as given in (4.11), are more involved. They require averaging of eight CAPs over the ground state – four in the two basis states $\langle eh_{\mathbf{k}',\mathbf{q}}|$ and $|eh_{\mathbf{k},\mathbf{q}}\rangle$, and four in the Hamiltonian $\hat{H}^{(2)}$ – and are evaluated with the rules we derived in Chapter 3. The structure of possible pairings of CAPs is illustrated by the diagrams in Fig. 4.6. Open line ends correspond to CAPs coming with electron and hole states: on the left side of the diagrams an electron in the state $|\mathbf{k} + \frac{1}{2}\mathbf{q}, \downarrow\rangle$ is created and one in the state $|\mathbf{k} - \frac{1}{2}\mathbf{q}, \uparrow\rangle$ is annihilated (or equivalently, a hole is created), and on the right side an electron in $|\mathbf{k}' + \frac{1}{2}\mathbf{q}, \downarrow\rangle$ is annihilated and one in $|\mathbf{k}' - \frac{1}{2}\mathbf{q}, \uparrow\rangle$ is created (or a hole is annihilated).

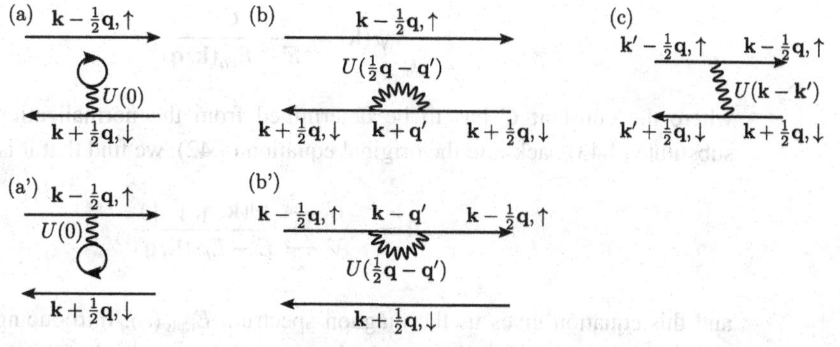

Fig. 4.6 Matrix elements of interaction between electron–hole states. Those in (a) and (b) correspond to respectively Hartree and Fock corrections to single-particle energies, while the diagram in (c) provides an important non-diagonal contribution.

All possible effects of the interaction Hamiltonian $\hat{H}^{(2)}$ are indicated in the diagrams. Diagrams (a) and (b) represent the Hartree and Fock corrections to the electron energy, while (a') and (b') give them for the hole energy. An important element is the non-diagonal term given by diagram (c). Assuming again a contact potential (4.17), we see that this matrix element reads

$$H_{\mathbf{k},\mathbf{k}'} = -\frac{U}{\mathcal{V}}. \tag{4.41}$$

We then write down and solve the Schrödinger equation in this basis. This *is* an approximation: in this case the basis is incomplete. There are matrix elements of the Hamiltonian between, say, electron–hole states and states involving two electrons and holes, and those we disregard. Nevertheless, this approximation is well-established, universally applied, and comes in a variety of equivalent forms. It is called the *random-phase* approximation. The motivation is to keep the structure of excited states as simple as possible, and the hope is that this assumption provides qualitatively correct results. There is actually a similarity with the trial wave function method: we seek for a wave function of an excited state assuming some sort of restricted basis of electron–hole states.

Explicitly, in this case the allowed wave functions for the excitations can take the form $|\psi\rangle = \sum_{\mathbf{k}} \psi(\mathbf{k})|eh_{\mathbf{kq}}\rangle \Theta(\mathbf{k}, \mathbf{q}, \downarrow, \uparrow)$. We thus arrive at a Schödinger equation for this wave function,

$$E\psi(\mathbf{k}) = E_{eh}(\mathbf{k}, \mathbf{q})\psi(\mathbf{k}) - \frac{U}{\mathcal{V}} \sum_{k'} \psi(\mathbf{k}')\Theta(\mathbf{k}', \mathbf{q}, \downarrow, \uparrow). \tag{4.42}$$

This eigenvalue equation has a continuous set of solutions at $E > E_g(q) \equiv \min_{\mathbf{k}}\{E_{eh}(\mathbf{k}, \mathbf{q})\}$. They are the unbound electron–hole pairs discussed in Section 4.5.2. There exists also a solution at $E < E_g(q)$, corresponding to a bound electron–hole pair, i.e. a magnon. To solve the equation in this interval, we make use of the fact that the last term does not depend on $\mathbf{k}$. We can denote this last term, including the whole sum over $\mathbf{k}'$, by C, and then solve for the wave function,

$$\psi(\mathbf{k}) = \frac{C}{E - E_{eh}(\mathbf{k}, \mathbf{q})}, \tag{4.43}$$

where the constant C has to be determined from the normalization condition. If we substitute (4.43) back into the original equation (4.42), we find that it is only consistent if

$$1 = -\frac{U}{\mathcal{V}} \sum_{k} \frac{\Theta(\mathbf{k}, \mathbf{q}, \downarrow, \uparrow)}{E - E_{eh}(\mathbf{k}, \mathbf{q})}, \tag{4.44}$$

and this equation gives us the magnon spectrum $E_{\text{mag}}(q)$. Introducing the magnons, we have been assuming that $E_{\text{mag}} \to 0$ when $q \to 0$. Let us check if this holds, and substitute $E = 0$ and $q = 0$. This gives

$$1 = \frac{U}{\mathcal{V}} \frac{N_\uparrow - N_\downarrow}{2B}, \tag{4.45}$$

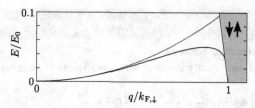

Fig. 4.7   Zoom-in on the gap of the $\Delta S_z = -1$ electron–hole continuum, indicated with a rectangle in Fig. 4.5(c). We plotted the magnon spectrum given by Eq. 4.44 (solid line) and a parabolic approximation (dotted line).

yet another equivalent form of the self-consistency equation given above. Expanding (4.44) in small E and q yields a parabolic magnon spectrum at $q \ll k_F$,

$$E_{\mathrm{mag}}(q) = \frac{2B}{5} \frac{k_{F,\uparrow}^2 + 3k_{F,\uparrow}k_{F,\downarrow} + k_{F,\downarrow}^2}{k_{F,\uparrow}^2 + k_{F,\uparrow}k_{F,\downarrow} + k_{F,\downarrow}^2} \frac{q^2}{\left(k_{F,\uparrow} + k_{F,\downarrow}\right)^2} \simeq B(q/k_F)^2. \qquad (4.46)$$

At $q \simeq k_F$, the spectrum is obtained by solving Eq. 4.44 numerically. An example is shown in Fig. 4.7. Upon increasing q, the magnon energy deviates from the parabolic law and eventually merges with the electron–hole continuum.

Table 4.1 Summary: Magnetism

Non-interacting Fermi gas

ground state: all states up to the Fermi level are filled, $|g_{\text{NIF}}\rangle = \displaystyle\prod_{|\mathbf{k}|<k_{\text{F}}} \hat{c}^{\dagger}_{\mathbf{k},\uparrow} \hat{c}^{\dagger}_{\mathbf{k},\downarrow} |0\rangle$

energy density: $\dfrac{E}{\mathcal{V}} = \dfrac{\hbar^2 k_{\text{F}}^5}{10\pi^2 m} = \dfrac{3}{5} E_{\text{F}} n \equiv \dfrac{E_0}{\mathcal{V}}$

Add electron–electron interactions

Coulomb interaction: $U(r) = \dfrac{e^2}{4\pi\varepsilon_0 r} e^{-r/r_C}$, screened with radius r_C

model interaction: $U(r) = U\delta(r)$, contact interaction

Hamiltonian: $\hat{H}^{(2)} = \dfrac{U}{2\mathcal{V}} \displaystyle\sum_{\mathbf{k},\mathbf{k}',\mathbf{q}} \sum_{\sigma,\sigma'} \hat{c}^{\dagger}_{\mathbf{k}+\mathbf{q},\sigma} \hat{c}^{\dagger}_{\mathbf{k}'-\mathbf{q},\sigma'} \hat{c}_{\mathbf{k}',\sigma'} \hat{c}_{\mathbf{k},\sigma}$

Trial wave function

method: guess ground state wave function with an adjustable parameter, minimize energy

try: finite spin polarization, $|g_P\rangle = \left(\displaystyle\prod_{|\mathbf{k}|<k_{\text{F},\uparrow}} \hat{c}^{\dagger}_{\mathbf{k},\uparrow} \right) \left(\displaystyle\prod_{|\mathbf{k}|<k_{\text{F},\downarrow}} \hat{c}^{\dagger}_{\mathbf{k},\downarrow} \right) |0\rangle$, parameter $P = \dfrac{N_{\uparrow} - N_{\downarrow}}{N}$

energy: single-particle: $\langle g_P | \hat{H}^{(1)} | g_P \rangle = \displaystyle\sum_{\mathbf{k},\sigma} E(k) n_{\mathbf{k},\sigma} = \dfrac{E_0}{2} \left\{ (1+P)^{\frac{5}{3}} + (1-P)^{\frac{5}{3}} \right\}$, favors $P = 0$

Hartree term: $\langle g_P | \hat{H}^{(2)} | g_P \rangle_{\text{H}} = \dfrac{U}{2\mathcal{V}} \left(\displaystyle\sum_{\mathbf{k},\sigma} n_{\mathbf{k},\sigma} \right)^2 = \dfrac{UN^2}{2\mathcal{V}}$, is P-independent

Fock term: $\langle g_P | \hat{H}^{(2)} | g_P \rangle_X = -\dfrac{U}{2\mathcal{V}} \displaystyle\sum_{\mathbf{k},\mathbf{k}'} \sum_{\sigma} n_{\mathbf{k},\sigma} n_{\mathbf{k}',\sigma} = -\dfrac{UN^2}{2\mathcal{V}} \dfrac{1+P^2}{2}$, favors large P

the ground state becomes magnetic when $(UN/E_{\text{F}}) > \frac{4}{3}$

Excitations

electron: $|e_{\mathbf{k},\sigma}\rangle = \hat{c}^{\dagger}_{\mathbf{k},\sigma} |g_P\rangle$, with energy $E_e(\mathbf{k},\sigma) = E(k) + Un - Un_{\sigma}$

hole: $|h_{-\mathbf{k},-\sigma}\rangle = \hat{c}_{\mathbf{k},\sigma} |g_P\rangle$, with energy $E_h(\mathbf{k},\sigma) = -E(k) - Un + Un_{-\sigma}$

e–h pair: $|eh_{\mathbf{k},\mathbf{q};\sigma,\sigma'}\rangle = \hat{c}^{\dagger}_{\mathbf{k}+\mathbf{q}/2,\sigma} \hat{c}_{\mathbf{k}-\mathbf{q}/2,\sigma'} |g_P\rangle$, with energy $E_{eh}(\mathbf{k},\mathbf{q};\sigma,\sigma') = \dfrac{\hbar^2}{m}(\mathbf{k}\cdot\mathbf{q}) + B(\sigma' - \sigma)$

magnon: $|m_{\mathbf{q}}\rangle = \dfrac{1}{\sqrt{NP}} \displaystyle\sum_{\mathbf{k}} \hat{c}^{\dagger}_{\mathbf{k}+\mathbf{q}/2,\downarrow} \hat{c}_{\mathbf{k}-\mathbf{q}/2,\uparrow} |g_P\rangle$, assuming majority spin up

excitation energy $E_m(\mathbf{q}) \simeq B(q/k_{\text{F}})^2$ for small q

---

Broken symmetry

the polarization points in a specific direction: rotational symmetry is broken

breaking of a continuous symmetry guarantees existence of Goldstone bosons → magnons

Exercises

1. *Anderson model of a magnetic impurity* (solution included). P. W. Anderson has suggested a model for a magnetic impurity embedded in a non-ferromagnetic metal. It is described by the Hamiltonian

$$\hat{H} = \hat{H}_{\text{metal}} + \hat{H}_{\text{imp}} + \hat{H}_c,$$

$$\hat{H}_{\text{metal}} = \sum_{\mathbf{k},\sigma} E(k)\,\hat{c}_{\mathbf{k},\sigma}^{\dagger}\hat{c}_{\mathbf{k},\sigma},$$

$$\hat{H}_{\text{imp}} = \epsilon_0 \hat{n} + \tfrac{1}{2} U \hat{n}(\hat{n} - 1), \quad \text{with} \quad \hat{n} = \sum_{\sigma} \hat{a}_{\sigma}^{\dagger} \hat{a}_{\sigma},$$

$$\hat{H}_c = \sum_{\mathbf{k},\sigma} t_{\mathbf{k}} \left(\hat{a}_{\sigma}^{\dagger} \hat{c}_{\mathbf{k},\sigma} + \hat{c}_{\mathbf{k},\sigma}^{\dagger} \hat{a}_{\sigma} \right).$$

Here, electron energies are counted from the Fermi energy E_{F} of the metal, $\hat{c}_{\mathbf{k},\sigma}^{(\dagger)}$ are the CAPs of the electron levels in the metal. The impurity houses a single localized level, $\hat{a}_{\sigma}^{(\dagger)}$ being the corresponding CAPs, and $\hat{n}$ being the operator of the electron number at the impurity. Interactions between the localized electrons (the term proportional to U) make double occupation of the localized level energetically unfavorable. To simplify, we assume that all energy parameters characterizing the problem, that is, ϵ_0, U and

$$\Gamma(E) \equiv 2\pi \sum_{\mathbf{k}} t_{\mathbf{k}}^2 \delta\big(E(k) - E\big),$$

are much smaller than E_{F}. Under this assumption, one may disregard the energy dependence of Γ. The Anderson model can be solved exactly. However, the solution is too complex to be outlined here, and we proceed with approximate methods.

a. Disregard $\hat{H}_c$ first. Find the ground state of the whole system. Give the interval of the parameter ϵ_0 where in the ground state the impurity is "magnetic," that is, houses a single electron with either spin up or spin down.

b. Perturbations. Compute the second-order corrections in $\hat{H}_c$ for all ground states. Show that the wave function of the magnetic ground state with an accuracy up to first order in $\hat{H}_c$ reads

$$|\uparrow, g\rangle^{(1)} = \left\{ \hat{a}_{\uparrow}^{\dagger} + \sum_{\mathbf{k}} \alpha_{\mathbf{k}} \hat{c}_{\mathbf{k},\uparrow}^{\dagger} + \sum_{\mathbf{k}} \beta_{\mathbf{k}} \hat{a}_{\uparrow}^{\dagger} \hat{a}_{\downarrow}^{\dagger} \hat{c}_{\mathbf{k},\downarrow} \right\} |0\rangle |g\rangle,$$

$|0\rangle$ being the vacuum of the impurity level, and $|g\rangle$ the ground state of the metal. Give explicit expressions for $\alpha_{\mathbf{k}}$ and $\beta_{\mathbf{k}}$ in first order.

c. Trial wave function. Treat the above expression as a trial wave function. Compute the energy corresponding to the wave function and derive expressions for $\alpha_{\mathbf{k}}$ and $\beta_{\mathbf{k}}$ for which this energy is minimal. Derive a self-consistency equation.

2. *Interaction as a perturbation.* Here we start with a non-magnetic non-interacting Fermi gas and treat the interaction between electrons ($\hat{H}^{(2)}$ in (4.11)) as a perturbation.

    a. Compute the first-order interaction correction to the energy of a number state $|\{n_\mathbf{k}\}\rangle$ in terms of the interaction potential $U(q)$.

    b. Make use of the resulting expression to give the correction to the ground state energy and to the energies of single-particle excitations, i.e. of single electrons added to the system.

    c. Derive a simplified form of this correction valid near the Fermi surface. Determine the corrections to the Fermi energy and Fermi velocity.

3. *Magnon entanglement.* Let us define a bipartition of the electron Fock space as follows: the space U is made of all states where electrons are created with spin up and D of all states with electron spin down. The original vacuum state can be seen as a product of the two vacua for spin up and spin down electrons, $|0\rangle = |0_U\rangle|0_D\rangle$.

    a. Demonstrate that the ground state given by (4.12) is a product state in this bipartition.

    b. A single-magnon state given by (4.37) is entangled in this bipartition. Compute the reduced density matrix of this state in U and determine the entanglement entropy associated with the single magnon.

    c. Consider a ground state obtained from (4.12) by a rotation about the x-axis by an angle α. Compute its entanglement entropy.

4. *Spin waves in a non-ferromagnetic metal.* It turns out that electron–hole excitations with wave functions similar to those of magnons may also exist in non-ferromagnetic metals. They are called spin waves. In distinction from magnons, the spectrum of spin waves is linear in q at $q \to 0$ and is characterized by a velocity v_S.

    a. Rewrite (4.42) for a non-ferromagnetic metal.

    b. Assume $q \ll k_F$ and simplify the resulting equation. Show that the wave function is concentrated in a narrow region of $\mathbf{k}$-space close to the Fermi surface, and describe the shape of this region.

    c. Under which condition does the equation have a spin wave solution? What is v_S in units of the Fermi velocity?

Solutions

1. *Anderson model of a magnetic impurity.*

    a. The possible ground states differ by number of electrons at the impurity. The corresponding impurity energies for $n = 0, 1, 2$ read $0, \epsilon_0, 2\epsilon_0 + U$, while the energy of the metal remains the same. We conclude that the "magnetic" state is the ground state provided that $-U < \epsilon_0 < 0$.

    b. Let us start with the second-order corrections to the "magnetic" ground state, $n = 1$. For definiteness, we assume spin up. Two types of virtual state contribute to the correction: (i) $n = 0$ with an extra electron in the level $\mathbf{k}, \uparrow$ in the metal, and (ii) $n = 2$ with an electron extracted from the level $\mathbf{k}, \downarrow$. The energy differences with the ground state are for (i) $E(k) - \epsilon_0$, and for (ii) $U + \epsilon_0 - E(k)$. The correction thus reads

$$E^{(2)} = -\sum_{\mathbf{k}} t_{\mathbf{k}}^2 \left(\frac{\Theta(E(k))}{-\epsilon_0 + E(k)} + \frac{\Theta(-E(k))}{U + \epsilon_0 - E(k)} \right).$$

We note that both denominators are positive, provided the "magnetic" state is the ground state, that is, $-U < \epsilon_0 < 0$. We then use the definition of Γ to change the sum over $\mathbf{k}$ into an integration over energy,

$$E^{(2)} = -\frac{\Gamma}{2\pi} \int dE \left(\frac{\Theta(E)}{-\epsilon_0 + E} + \frac{\Theta(-E)}{U + \epsilon_0 - E} \right).$$

These integrals diverge at large energies and must be cut-off at some energy E_{cut}. This gives

$$E^{(2)} = -\frac{\Gamma}{2\pi} \left\{ \ln \left(\frac{E_{\text{cut}}}{-\epsilon_0} \right) + \ln \left(\frac{E_{\text{cut}}}{U + \epsilon_0} \right) \right\},$$

where we used the fact that $E_{\text{cut}} \gg U, \epsilon_0$. The correction formally diverges near the ends of the "magnetic" interval, indicating a problem with perturbation theory.

The corrections to the other ground states are computed in a similar fashion. The difference is that the virtual states contributing to these corrections are only of one type: those with $n = 1$. We obtain

$$n = 0 \quad \text{with} \quad E^{(2)} = -\frac{\Gamma}{\pi} \ln \left(\frac{E_{\text{cut}}}{\epsilon_0} \right),$$

$$n = 2 \quad \text{with} \quad E^{(2)} = -\frac{\Gamma}{\pi} \ln \left(\frac{E_{\text{cut}}}{-U - \epsilon_0} \right).$$

The corrections to the wave function of the "magnetic" ground state are indeed proportional to the wave functions of the virtual states with

$$\alpha_{\mathbf{k}} = \frac{t_{\mathbf{k}}}{\epsilon_0 - E(k)} \quad \text{and} \quad \beta_{\mathbf{k}} = \frac{t_{\mathbf{k}}}{-U - \epsilon_0 + E(k)}.$$

c. The trial wave function must be normalized first, so it reads

$$|\uparrow, g\rangle = \left\{ \sqrt{C} \hat{a}_{\uparrow}^{\dagger} + \sum_{\mathbf{k}} \alpha_{\mathbf{k}} \hat{c}_{\mathbf{k},\uparrow}^{\dagger} + \sum_{\mathbf{k}} \beta_{\mathbf{k}} \hat{a}_{\uparrow}^{\dagger} \hat{a}_{\downarrow}^{\dagger} \hat{c}_{\mathbf{k},\downarrow} \right\} |0\rangle |g\rangle,$$

where we define $A = \sum_{\mathbf{k}} \alpha_{\mathbf{k}}^2 \Theta(E(k))$, $B = \sum_{\mathbf{k}} \beta_{\mathbf{k}}^2 \Theta(-E(k))$, and $C = 1 - A - B$. We assume real $\alpha_{\mathbf{k}}$ and $\beta_{\mathbf{k}}$. The terms $\hat{H}_{\text{metal}}$ and $\hat{H}_{\text{imp}}$ are diagonal in these wave functions and are readily evaluated,

$$\langle \hat{H}_{\text{metal}} \rangle = \sum_{\mathbf{k}} |E(k)| \left\{ \Theta(E(k)) \alpha_{\mathbf{k}}^2 + \Theta(-E(k)) \beta_{\mathbf{k}}^2 \right\},$$

$$\langle \hat{H}_{\text{imp}} \rangle = C\epsilon_0 + B(2\epsilon_0 + U).$$

The coupling term requires more concentration. It has contributions from cross-terms that differ in n, and gives

$$\langle \hat{H}_c \rangle = 2\sqrt{C} \sum_{\mathbf{k}} t_{\mathbf{k}} \left\{ \Theta(E(k))\alpha_{\mathbf{k}} + \Theta(-E(k))\beta_{\mathbf{k}} \right\}.$$

To minimize the total energy, we differentiate it with respect to $\alpha_{\mathbf{k}}$ and $\beta_{\mathbf{k}}$, and set this derivative to zero. This yields

$$(|E(k)| - \epsilon_0 + S)\alpha_{\mathbf{k}} = -t_{\mathbf{k}}\sqrt{C},$$
$$(|E(k)| + \epsilon_0 + U + S)\beta_{\mathbf{k}} = -t_{\mathbf{k}}\sqrt{C},$$

where

$$S \equiv -\frac{1}{\sqrt{C}} \sum_{\mathbf{k}} t_{\mathbf{k}} \left\{ \alpha_{\mathbf{k}}\Theta(E(k)) + \beta_{\mathbf{k}}\Theta(-E(k)) \right\}.$$

Next we express $\alpha_{\mathbf{k}}$ and $\beta_{\mathbf{k}}$ in terms of S, and substitute them back into the definition of S. This yields a self-consistency equation, where we replace the summation by an integration over energies,

$$S = \sum_{\mathbf{k}} t_{\mathbf{k}}^2 \left(\frac{\Theta(E(k))}{|E(k)| - \epsilon_0 + S} + \frac{\Theta(-E(k))}{|E(k)| + \epsilon_0 + U + S} \right)$$
$$= \frac{\Gamma}{2\pi} \int dE \left(\frac{\Theta(E)}{|E| - \epsilon_0 + S} + \frac{\Theta(-E)}{|E| + \epsilon_0 + U + S} \right).$$

Inspecting the resulting equation, we recognize that S plays the role of an energy shift of the impurity level: strangely enough, this shift is opposite for electrons and holes. The value of the shift is determined self-consistently for any ϵ_0, U, and Γ. This shift stays finite at the ends of the "magnetic" interval, and thus in principle removes the problem with perturbation theory.

5

Superconductivity

The next example of second quantization techniques we consider concerns *superconductivity*. Superconductivity was discovered in 1911 in Leiden by Kamerlingh Onnes, in the course of his quest to achieve ultra-low temperatures. He was applying freshly invented refrigeration techniques to study the low-temperature behavior of materials using every method available at that time. The most striking discovery he made, was that the electrical resistance of mercury samples abruptly disappeared below a critical temperature of 4.2 K – the metal underwent a transition into a novel phase.

After this first discovery, superconductivity has since been observed in a large number of metals and metallic compounds. Despite long-lasting efforts to find a material that is superconducting at room temperature, today superconductivity still remains a low-temperature phenomenon, the critical temperature varying from compound to compound. A huge and expensive quest for so-called high-temperature superconductivity took place rather recently, in the 1980s and 1990s. Although the record temperature has been increased by a factor of 5 during this time, it still remains as low as 138 K.

The most important manifestation of superconductivity is of course the absence of electrical resistance. In a superconductor, electric current can flow without any dissipation. Such a *supercurrent*, once excited in a superconducting ring, would in principle persist forever, at least as long as the superconductivity of the ring is preserved. Another remarkable property of superconductors is the Meissner effect: superconductors "expel" magnetic fields. If a superconductor is placed in an external magnetic field, supercurrents are induced in a narrow layer beneath its surface, and these currents exactly cancel the magnetic field inside the superconductor. A popular public demonstration of this effect shows a piece of superconductor levitating above a magnet.

These fascinating manifestations reveal very little about the *origin* of superconductivity, and it took almost 50 years to pinpoint it. The first satisfactory microscopic theory of superconductivity was put forward by Bardeen, Cooper, and Schrieffer in 1957. Below we present a slightly simplified formulation of this theory. It became clear that the peculiarities of the superconducting state are due to a spontaneous breaking of a very fundamental symmetry, the gauge symmetry. This means that a piece of superconductor is in a *macroscopic coherent state*: its state is characterized by a phase that is very similar to the phase of a wave function. This property is utilized in various nano-devices made to realize elementary quantum-mechanical systems.

It is said that superconductivity could never have been predicted theoretically. Theorists would never have dared to think about such an exotic state without a firm knowledge that it exists. However, the explanatory theory of superconductivity was a very successful application of second quantization ideas, and it has changed our ways of understanding condensed matter. Besides, it had a large influence in the field of nuclear physics and in theories of elementary particles.

5.1 Attractive interaction and Cooper pairs

Soon we will be able to prove that superconductivity requires *attractive* interaction between electrons. This already sounds ill-conceived, since electrostatics dictates that electrons always repel each other. How can the interaction then become attractive? First of all, we have seen already that the effective interaction between a pair of electrons in a metal is in fact weakened due to screening. Secondly, electrons also interact with the surrounding ions which set up the periodic lattice of the solid. This lattice becomes locally deformed by the presence of an extra electron. If two such electrons come close to each other, they deform the lattice more strongly than each separate electron would do. This leads to an energy gain if two electrons are brought together, and the interaction can become effectively attractive. In a way, this is similar to what happens if one puts two steel balls on a suspended rubber sheet: the sheet will deform and the balls will be pushed together. In a superconductor, the deformation of the lattice consists of phonons, so it is said that phonons *mediate* the attractive interaction between electrons. The attractive interaction competes with the weakened electrostatic repulsion and indeed wins in approximately one third of the metallic compounds. A more quantitative discussion of this competition is left as Exercise 5 at the end of this chapter.

The next step is to understand how attractive interaction can lead to superconductivity. There is a popular cartoon used to mislead students. It says that the interaction binds electrons together so that they form pairs, so-called *Cooper pairs*. The lowest energy state of such a Cooper pair has zero total momentum, and this can be realized when two electrons with opposite momenta pair up. Besides, the lowest energy is achieved when the two paired electrons have opposite spin, so that the total spin of a Cooper pair is zero, meaning that Cooper pairs effectively are bosons. At sufficiently low temperatures, all bosons occupy the lowest energy state of zero momentum. They thus undergo Bose condensation[1] and form a condensate, a charged superfluid liquid. The supercurrents mentioned above are eventually flows of this liquid.

The drawback of this simplified picture is the "natural" assumption that Cooper pairs are particle-like objects and do not have other degrees of freedom except for position. It is easy to see that this is not so. Let us try to make a Cooper pair using two electronic creation operators. Since we only care about pairs with zero total momentum and spin, we take creation operators with opposite momenta and spins and construct a creation operator for a Cooper pair,

$$\hat{Z}_{\mathbf{k}}^{\dagger} = \hat{c}_{\mathbf{k},\uparrow}^{\dagger} \hat{c}_{-\mathbf{k},\downarrow}^{\dagger}. \tag{5.1}$$

---

[1] If this does not ring a bell, do not worry. We discuss Bose condensation in the next chapter.

Heike Kamerlingh Onnes (1853–1926)
Won the Nobel Prize in 1913 for "his investigations on the properties of matter at low temperatures which led, *inter alia*, to the production of liquid helium."

Heike Kamerlingh Onnes was educated at the universities of Groningen (The Netherlands) and Heidelberg (Germany), where he studied physics and chemistry. After meeting Van der Waals in the mid 1870s, Kamerlingh Onnes developed a strong interest in the behavior of gases. When he was appointed professor in physics at Leiden University, he started a quest to test Van der Waals' law using gases in extreme conditions, more specifically at very low temperatures. In 1908, in the course of this research, Kamerlingh Onnes became the first to liquefy helium, reaching temperatures down to 1.7 K.

Driven to test a gas law, Kamerlingh Onnes always produced relatively large quantities of liquefied gases, and this allowed him to investigate the low-temperature behavior of other materials as well. At the time, there was a lively debate about zero-temperature electrical resistivity: it was predicted to grow infinitely, as well as to decrease gradually to zero.

In 1911, Kamerlingh Onnes placed purified samples of mercury in his newly developed cryostats and measured their resistance at the lowest temperatures, and the resistance seemed to disappear completely! This was remarkable, since the temperature was still significantly above 0 K. The team thus spent a lot of effort looking for a short circuit or some other problem in the setup. The story goes that the answer was due to a student of the instrument-makers school who was assisting in the experiments. His job was to watch the helium vapor pressure, making sure that it stayed low enough. During one measurement however, the student fell asleep, causing the sample to warm up gradually. The researchers measuring the resistance suddenly observed a step-like increase during this measurement. After they found out what had happened, it became clear that the sudden complete vanishing of electrical resistance of mercury below a temperature of about 4.2 K was real. Kamerlingh Onnes published the results, and named the newly found state "superconducting."

This still leaves a great number of possible Cooper pairs to be created, one for each value of the electron wave vector $\mathbf{k}$. We have to take care of this.

Let us first specify the model in use. We start with the usual single-particle Hamiltonian $\hat{H}^{(1)}$ and add to this the interaction term

$$\hat{H}^{(2)} = \frac{1}{2\mathcal{V}} \sum_{\mathbf{k},\mathbf{k}',\mathbf{q}} \sum_{\sigma,\sigma'} U(\mathbf{q})\, \hat{c}^{\dagger}_{\mathbf{k}+\mathbf{q},\sigma} \hat{c}^{\dagger}_{\mathbf{k}'-\mathbf{q},\sigma'} \hat{c}_{\mathbf{k}',\sigma'} \hat{c}_{\mathbf{k},\sigma}. \tag{5.2}$$

As in the previous chapter, we make use of a contact interaction potential $U(\mathbf{q}) = U$, and since the interaction now is attractive, we take $U < 0$. We make yet another simplification: since our goal is to describe the system purely in terms of pairs of electrons, we interpret the terms in $\hat{H}^{(2)}$ as a product of an operator that creates an electron pair, $\hat{c}^{\dagger}_{\mathbf{k}+\mathbf{q},\sigma}\hat{c}^{\dagger}_{\mathbf{k}'-\mathbf{q},\sigma'}$, and an operator that annihilates a pair, $\hat{c}_{\mathbf{k}',\sigma'}\hat{c}_{\mathbf{k},\sigma}$. The simplified picture presented above suggests that we should then concentrate on pairs with opposite momenta and spins, so we keep only terms with $\sigma = -\sigma'$ and $\mathbf{k} = -\mathbf{k}'$, and throw away all others. The resulting model Hamiltonian then contains only the operators $\hat{Z}_{\mathbf{k}}$ and $\hat{Z}^{\dagger}_{\mathbf{k}}$,

$$\hat{H}^{(2)} = \frac{U}{2\mathcal{V}} \sum_{\mathbf{k},\mathbf{q}} \sum_{\sigma} \hat{c}^{\dagger}_{\mathbf{k}+\mathbf{q},\sigma} \hat{c}^{\dagger}_{-\mathbf{k}-\mathbf{q},-\sigma} \hat{c}_{-\mathbf{k},-\sigma} \hat{c}_{\mathbf{k},\sigma} = \frac{U}{\mathcal{V}} \Big(\sum_{\mathbf{k}} \hat{Z}^{\dagger}_{\mathbf{k}} \Big) \Big(\sum_{\mathbf{k}} \hat{Z}_{\mathbf{k}} \Big). \tag{5.3}$$

This we can rewrite in an even more compact form

$$\hat{H}^{(2)} = \frac{\mathcal{V}}{U} \hat{\Delta}^{\dagger} \hat{\Delta}, \tag{5.4}$$

where we have introduced the operator

$$\hat{\Delta} \equiv \frac{U}{\mathcal{V}} \sum_{\mathbf{k}} \hat{Z}_{\mathbf{k}}. \tag{5.5}$$

This operator has the dimension of energy and, as we will see, plays a key role in the theory.

5.1.1 Trial wave function

We make use of a trial wave function method, as we did when we described ferromagnetism. In contrast to the ferromagnetic trial wave function, the superconducting one does not look self-explanatory. To understand it better, let us first start with the familiar ground state of a NIF and rewrite it in terms of the pair creation operators,

$$|g\rangle = \prod_{|\mathbf{k}| < k_{\mathrm{F}}} \hat{c}^{\dagger}_{\mathbf{k},\uparrow} \hat{c}^{\dagger}_{-\mathbf{k},\downarrow} |0\rangle = \prod_{|\mathbf{k}| < k_{\mathrm{F}}} \hat{Z}^{\dagger}_{\mathbf{k}} |0\rangle. \tag{5.6}$$

We thus certainly create a pair at all $|\mathbf{k}| < k_{\mathrm{F}}$ and certainly do not create any pairs at $|\mathbf{k}| > k_{\mathrm{F}}$. Sounds dogmatic, does it not? We know that quantum mechanics allows for some flexibility in this question. We thus allow for some more generality, and think in terms of *superpositions* of quantum states. Let us take two states, $|0_{\mathbf{k}}\rangle$ (no Cooper pair present), and $|2_{\mathbf{k}}\rangle \equiv \hat{Z}^{\dagger}_{\mathbf{k}}|0\rangle$ (a Cooper pair present). Their superposition

$$u|0_{\mathbf{k}}\rangle + v|2_{\mathbf{k}}\rangle, \tag{5.7}$$

is also an eligible quantum state. It is, however, all but usual: it has zero particles with probability $|u|^2$ and two particles with probability $|v|^2$.

We can create the above state from the vacuum by applying the operator $u + v\hat{Z}^{\dagger}_{\mathbf{k}}$. The trial wave function of the superconducting state which we use involves such superpositions for all $\mathbf{k}$, i.e. also for $k > k_{\mathrm{F}}$,

(*left to right*)**John Bardeen** (1908–1991), **Leon Cooper** (b. 1930),
and **Robert Schrieffer** (b. 1931)
Won the Nobel Prize in 1972 for "their jointly developed theory of superconductivity, usually called the BCS-theory."

After superconductivity was discovered in 1911, its origin remained a mystery for almost 50 years. Almost all great physicists of the first half of the 20th century (Einstein, Heisenberg, Bohr, and many others) tried to devise a theoretical explanation for it, but failed.

In the 1950s, John Bardeen was already a famous physicist. He was one of the inventors of the transistor, for which he would be awarded the Nobel Prize in 1956 (making him the only person who received the Nobel Prize in physics twice). In 1955, Bardeen met Leon Cooper, and told him about his latest endeavor: he wanted to solve the mystery of superconductivity. Cooper, who had then just received his Ph.D. in a different field, was not aware of the impressive collection of great minds that already had got stuck in the problem, and he agreed to join Bardeen.

In the same year, David Pines had shown that interaction with phonons could create an attractive force between electrons. Cooper took this as starting point, but soon got stuck in complicated calculations. When thinking about it (i.e. not calculating) on a long train ride, he realized that he should focus on electrons near the Fermi surface. This led him to the conclusion that an attractive force also leads to formation of pairs of electrons, but this still did not explain superconductivity.

Meanwhile, Robert Schrieffer, a graduate student of Bardeen, had joined the project, and he started thinking about the implications of the idea of Cooper pairs. In the subway in New York, he suddenly realized that if all electrons have formed pairs, one needs to describe the behavior of all Cooper pairs at the same time instead of concentrating on each individual pair. He showed his ideas first to Cooper, and then (after Bardeen had returned from Stockholm to collect his first Nobel Prize) to Bardeen. Both were convinced, and they published their results in 1957.

$$|gs\rangle = \prod_{\mathbf{k}} (u_{\mathbf{k}} + v_{\mathbf{k}} \hat{Z}_{\mathbf{k}}^{\dagger})|0\rangle. \qquad (5.8)$$

The amplitudes $u_{\mathbf{k}}$ and $v_{\mathbf{k}}$ are adjustable parameters which we use to minimize the energy corresponding to the wave function. The work to do is to compute this energy and then actually perform the minimization.

5.1.2 Nambu boxes

To work with this trial wave function, we need a convenient partition of the Fock space. Let us pick two electronic levels, $|\mathbf{k}, \uparrow\rangle$ and $|-\mathbf{k}, \downarrow\rangle$. We write down the basis vectors in the two-particle space by filling these levels with electrons. There are of course four possibilities, which we label as follows:

$$|0\rangle \equiv |0_{\mathbf{k}\uparrow} 0_{-\mathbf{k}\downarrow}\rangle, \ |2\rangle \equiv |1_{\mathbf{k}\uparrow} 1_{-\mathbf{k}\downarrow}\rangle, \ |+\rangle \equiv |1_{\mathbf{k}\uparrow} 0_{-\mathbf{k}\downarrow}\rangle, \ \text{and} \ |-\rangle \equiv |0_{\mathbf{k}\uparrow} 1_{-\mathbf{k}\downarrow}\rangle. \qquad (5.9)$$

So we have a tiny four-dimensional Fock subspace for this pair of levels. We call this subspace a Nambu box (Fig. 5.1), with index $\mathbf{k}$. The total many-particle Fock space is a direct product of all Nambu boxes. The advantage of such a partition is that the operators that constitute our Hamiltonian, the occupation numbers $\hat{n}_{\mathbf{k}}$ and the pair creation operators $\hat{Z}_{\mathbf{k}}^{\dagger}$, act only within the corresponding boxes.

We further note that the trial wave function (5.8) does not involve states with only one electron in the box, i.e. $|+\rangle$ or $|-\rangle$. So let us forget for the moment about one-electron states and concentrate on the zero-electron state $|0\rangle$ and the two-electron state $|2\rangle$. In the remaining two-dimensional basis, the relevant operators can be presented as 2×2 matrices,

$$\hat{Z}_{\mathbf{k}}^{\dagger} = \begin{pmatrix} 0 & 1 \\ 0 & 0 \end{pmatrix}, \quad \hat{Z}_{\mathbf{k}} = \begin{pmatrix} 0 & 0 \\ 1 & 0 \end{pmatrix}, \quad \text{and} \quad \hat{n}_{\mathbf{k}\uparrow} + \hat{n}_{-\mathbf{k}\downarrow} = \begin{pmatrix} 2 & 0 \\ 0 & 0 \end{pmatrix}. \qquad (5.10)$$

Or, to put it in words, $\hat{Z}^{\dagger}$ makes $|2\rangle$ from $|0\rangle$, $\hat{Z}$ makes $|0\rangle$ from $|2\rangle$, and there are two particles in the state $|2\rangle$.

Let us now turn our attention again to the trial wave function $|gs\rangle$ presented above. It is clear that we can express this wave function as a direct product of wave functions $|g_{\mathbf{k}}\rangle$ defined in different Nambu boxes,

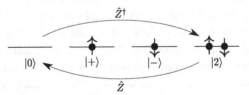

Fig. 5.1   A Nambu box can be in four states: in $|0\rangle$ there are no electrons present, in $|+\rangle$ there is one electron in the state $|\mathbf{k}, \uparrow\rangle$ present, in $|-\rangle$ one electron in $|-\mathbf{k}, \downarrow\rangle$ is present, and in $|2\rangle$ both these electrons are present. The operators $\hat{Z}_{\mathbf{k}}^{(\dagger)}$ couple the states $|0\rangle$ and $|2\rangle$.

$$|gs\rangle = \prod_{\mathbf{k}} |g_{\mathbf{k}}\rangle, \tag{5.11}$$

where the terms $|g_{\mathbf{k}}\rangle$ are nothing but superpositions of the one-electron and the two-electron state with amplitudes $u_{\mathbf{k}}$ and $v_{\mathbf{k}}$,

$$|g_{\mathbf{k}}\rangle = (u_{\mathbf{k}} + v_{\mathbf{k}} \hat{Z}_{\mathbf{k}}^{\dagger})|0\rangle = u_{\mathbf{k}}|0\rangle + v_{\mathbf{k}}|2\rangle. \tag{5.12}$$

The normalization condition $\langle g_{\mathbf{k}}|g_{\mathbf{k}}\rangle = 1$ imposes a constraint on $u_{\mathbf{k}}$ and $v_{\mathbf{k}}$,

$$u_{\mathbf{k}}^{*} u_{\mathbf{k}} + v_{\mathbf{k}}^{*} v_{\mathbf{k}} = 1. \tag{5.13}$$

5.2 Energy

We are now ready to evaluate the energy of our trial wave function. We start with the usual kinetic energy, but now add a term that comes with the chemical potential $\mu = E_{\mathrm{F}}$,

$$\hat{H}^{(1)} = \sum_{\mathbf{k},\sigma} E(\mathbf{k}) \hat{c}_{\mathbf{k},\sigma}^{\dagger} \hat{c}_{\mathbf{k},\sigma} - \mu \hat{N} \equiv \sum_{\mathbf{k},\sigma} \varepsilon_{\mathbf{k}} \hat{n}_{\mathbf{k},\sigma}. \tag{5.14}$$

The reason for this addition is that we do not want to minimize the trial wave function at a given number of particles. Instead, we want to keep the $u_{\mathbf{k}}$ and $v_{\mathbf{k}}$ (which determine the expected number of particles in the Nambu boxes) as free parameters. What we are doing is actually a method known from variational calculus. We introduce a Lagrangian multiplier corresponding to the number of particles, the multiplier in this case being the chemical potential. We see from the above equation that the effect of the inclusion of the chemical potential in $\hat{H}^{(1)}$ is to change the reference point for the single-particle energies. The energies $\varepsilon_{\mathbf{k}}$ are now counted starting from the Fermi energy.

The kinetic energy of the trial ground state $\langle gs|\hat{H}^{(1)}|gs\rangle$ is evaluated straightforwardly. We replace the operators of the occupation numbers with their expectation values for the trial wave function, and using (5.10) we obtain

$$\langle gs|\hat{n}_{\mathbf{k}\uparrow} + \hat{n}_{-\mathbf{k}\downarrow}|gs\rangle = \langle g_{\mathbf{k}}|\hat{n}_{\mathbf{k}\uparrow} + \hat{n}_{-\mathbf{k}\downarrow}|g_{\mathbf{k}}\rangle = \begin{pmatrix} v_{\mathbf{k}}^{*} \\ u_{\mathbf{k}}^{*} \end{pmatrix} \begin{pmatrix} 2 & 0 \\ 0 & 0 \end{pmatrix} \begin{pmatrix} v_{\mathbf{k}} \\ u_{\mathbf{k}} \end{pmatrix} = 2|v_{\mathbf{k}}|^{2}. \tag{5.15}$$

The total kinetic energy thus reads

$$E_{\mathrm{K}} = 2 \sum_{\mathbf{k}} \varepsilon_{\mathbf{k}} |v_{\mathbf{k}}|^{2}. \tag{5.16}$$

It is instructive to minimize this contribution to the energy before we start thinking about interactions. We can minimize each term in the sum (5.16) separately. If $\varepsilon_{\mathbf{k}} > 0$, the optimal $v_{\mathbf{k}}$ is obviously 0. If $\varepsilon_{\mathbf{k}} < 0$, the minimal contribution corresponds to the maximal $|v_{\mathbf{k}}|$. Owing to the constraint (5.13) this maximum is $|v_{\mathbf{k}}| = 1$, in which case $u_{\mathbf{k}} = 0$.

Control question. What is the resulting state?

Let us next try to evaluate the interaction contribution $\langle g_S | \hat{H}^{(2)} | g_S \rangle$ to the energy, where $\hat{H}^{(2)}$ is given by (5.4). We thus want to find the potential energy

$$E_P = \langle \hat{H}^{(2)} \rangle = \frac{\mathcal{V}}{U} \langle \hat{\Delta}^\dagger \hat{\Delta} \rangle, \tag{5.17}$$

i.e. the expectation value of $\hat{H}^{(2)}$ in the state $|g_S\rangle$. We note that both $\hat{\Delta}^\dagger$ and $\hat{\Delta}$ contain a sum over all momenta, in fact $\langle \hat{\Delta}^\dagger \hat{\Delta} \rangle \propto \sum_{\mathbf{k},\mathbf{k}'} \langle \hat{Z}_{\mathbf{k}}^\dagger \hat{Z}_{\mathbf{k}'} \rangle$. The dominant contribution to (5.17) comes from the terms with $\mathbf{k} \neq \mathbf{k}'$, so that we can safely use $\langle \hat{\Delta}^\dagger \hat{\Delta} \rangle \approx \langle \hat{\Delta}^\dagger \rangle \langle \hat{\Delta} \rangle$. The potential energy then becomes

$$\langle \hat{H}^{(2)} \rangle = \frac{\mathcal{V}}{U} |\langle \hat{\Delta} \rangle|^2 = \frac{\mathcal{V}}{U} |\Delta|^2. \tag{5.18}$$

The last equation defines a number $\Delta \equiv \langle \hat{\Delta} \rangle$, and we have to evaluate it. To do so, let us first evaluate the expectation value of $\hat{Z}_{\mathbf{k}}$,

$$\langle \hat{Z}_{\mathbf{k}} \rangle = \langle g_{\mathbf{k}} | \hat{Z}_{\mathbf{k}} | g_{\mathbf{k}} \rangle = \begin{pmatrix} v_{\mathbf{k}}^* \\ u_{\mathbf{k}}^* \end{pmatrix} \begin{pmatrix} 0 & 0 \\ 1 & 0 \end{pmatrix} \begin{pmatrix} v_{\mathbf{k}} \\ u_{\mathbf{k}} \end{pmatrix} = u_{\mathbf{k}}^* v_{\mathbf{k}}. \tag{5.19}$$

Using (5.4), we then obtain

$$\Delta = \frac{U}{\mathcal{V}} \sum_{\mathbf{k}} \langle \hat{Z}_{\mathbf{k}} \rangle = \frac{U}{\mathcal{V}} \sum_{\mathbf{k}} u_{\mathbf{k}}^* v_{\mathbf{k}}, \tag{5.20}$$

which now allows us to write for the potential energy

$$E_P = \frac{\mathcal{V}}{U} \left| \sum_{\mathbf{k}} u_{\mathbf{k}}^* v_{\mathbf{k}} \right|^2. \tag{5.21}$$

Control question. What is Δ for the normal (non-superconducting) NIF?

In the superconducting state, $\Delta \neq 0$, and it characterizes the state of the superconductor in a way very comparable to how the magnetization P characterizes the state of a ferromagnet (see the previous chapter). Since $U < 0$, the potential energy E_P is negative and we thus clearly see how, in our simple model, an attractive electron–electron interaction can make the superconducting state favorable.

5.2.1 Energy minimization

The total energy, which we still have to minimize with respect to $u_{\mathbf{k}}, v_{\mathbf{k}}$, thus reads

$$E = 2 \sum_{\mathbf{k}} \varepsilon_{\mathbf{k}} |v_{\mathbf{k}}|^2 + \frac{\mathcal{V}}{U} \left| \sum_{\mathbf{k}} u_{\mathbf{k}}^* v_{\mathbf{k}} \right|^2. \tag{5.22}$$

The minimization is relatively easy, and can be done analytically. As a first step, we compute the variation of the total energy when we introduce small changes to a particular $v_{\mathbf{k}}$ and $u_{\mathbf{k}}$ ($v_{\mathbf{k}}^*$ and $u_{\mathbf{k}}^*$ do *not* change),

$$\delta E = 2\varepsilon_{\mathbf{k}} v_{\mathbf{k}}^* \delta v_{\mathbf{k}} + \Delta^* u_{\mathbf{k}}^* \delta v_{\mathbf{k}} + \Delta v_{\mathbf{k}}^* \delta u_{\mathbf{k}}. \tag{5.23}$$

Then we make use of the fact that the variations of $v_{\mathbf{k}}$ and $u_{\mathbf{k}}$ are not independent, but they are related by the constraint (5.13), which implies that

$$u_{\mathbf{k}}^* \delta u_{\mathbf{k}} + v_{\mathbf{k}}^* \delta v_{\mathbf{k}} = 0. \tag{5.24}$$

This gives the relation $\delta v_{\mathbf{k}} = -(u_{\mathbf{k}}^*/v_{\mathbf{k}}^*)\delta u_{\mathbf{k}}$, so that we obtain

$$\delta E = (-2u_{\mathbf{k}}^* \varepsilon_{\mathbf{k}} - \Delta^* u_{\mathbf{k}}^{*2}/v_{\mathbf{k}}^* + \Delta v_{\mathbf{k}}^*)\,\delta u_{\mathbf{k}}. \tag{5.25}$$

We know that at the minimum $\delta E/\delta u_{\mathbf{k}} = 0$ must hold, which finally yields the two conjugated equations

$$\begin{aligned}
-2\varepsilon_{\mathbf{k}} u_{\mathbf{k}}^* - \Delta^* u_{\mathbf{k}}^{*2}/v_{\mathbf{k}}^* + \Delta v_{\mathbf{k}}^* &= 0, \\
-2\varepsilon_{\mathbf{k}} u_{\mathbf{k}} - \Delta u_{\mathbf{k}}^2/v_{\mathbf{k}} + \Delta^* v_{\mathbf{k}} &= 0.
\end{aligned} \tag{5.26}$$

We now need to exclude Δ^* from the two expressions above. To do so, we multiply the first equation with $v_{\mathbf{k}}$, the second with $(u_{\mathbf{k}}^*)^2/v_{\mathbf{k}}^*$, and then sum them up and multiply with $|v_{\mathbf{k}}|^2$. This finally yields

$$2\varepsilon_{\mathbf{k}} u_{\mathbf{k}}^* v_{\mathbf{k}} = \Delta\left(|v_{\mathbf{k}}|^2 - |u_{\mathbf{k}}|^2\right), \tag{5.27}$$

where we have once again used the constraint (5.13).

To solve (5.27), we take the modulus square of both sides and use the normalization constraint again to obtain an expression for the product $|u_{\mathbf{k}}|^2|v_{\mathbf{k}}|^2$,

$$4|u_{\mathbf{k}}|^2|v_{\mathbf{k}}|^2 = \frac{|\Delta|^2}{\varepsilon_{\mathbf{k}}^2 + |\Delta|^2}. \tag{5.28}$$

And using the constraint once more, we figure out that

$$|v_{\mathbf{k}}|^2, |u_{\mathbf{k}}|^2 = \frac{1}{2}\left(1 \mp \frac{\varepsilon_{\mathbf{k}}}{\sqrt{\varepsilon_{\mathbf{k}}^2 + |\Delta|^2}}\right), \tag{5.29}$$

so we have finally found the magnitude of $u_{\mathbf{k}}$ and $v_{\mathbf{k}}$. However, if you did the above calculation yourself, you will have found that you still cannot determine which of the two terms ($|v_{\mathbf{k}}|^2$ or $|u_{\mathbf{k}}|^2$) has the plus sign and which the minus sign. For this, we use the fact that for the non-superconducting NIF (with $\Delta = 0$) we must find $|v_{\mathbf{k}}|^2 = 1$ and $|u_{\mathbf{k}}|^2 = 0$ whenever $\varepsilon_{\mathbf{k}} < 0$ and $|v_{\mathbf{k}}|^2 = 0$ and $|u_{\mathbf{k}}|^2 = 1$ otherwise.[2] This fixes for $|v_{\mathbf{k}}|^2$ the minus, and allows us to find

$$u_{\mathbf{k}}^* v_{\mathbf{k}} = -\frac{1}{2}\frac{\Delta}{\sqrt{\varepsilon_{\mathbf{k}}^2 + |\Delta|^2}}. \tag{5.30}$$

---

[2] Remember that we included the chemical potential into the definition of the $\varepsilon_{\mathbf{k}}$. This makes $\varepsilon_{\mathbf{k}}$ negative for levels below the Fermi level.

By now, we have found $u_{\mathbf{k}}$ and $v_{\mathbf{k}}$, but we still need to evaluate Δ. Before doing this, it is instructive to first compute the contribution of a given box to the total energy with the $u_{\mathbf{k}}$ and $v_{\mathbf{k}}$ found. We obtain for a single box

$$
\begin{aligned}
E_{\mathbf{k}} &= 2\varepsilon_{\mathbf{k}}|v_{\mathbf{k}}|^2 + u_{\mathbf{k}}^* v_{\mathbf{k}} \Delta^* + u_{\mathbf{k}} v_{\mathbf{k}}^* \Delta \\
&= \varepsilon_{\mathbf{k}} - \sqrt{\varepsilon_{\mathbf{k}}^2 + |\Delta|^2}.
\end{aligned}
\tag{5.31}
$$

This energy is always negative. It goes to $2\varepsilon_{\mathbf{k}}$ for $\varepsilon_{\mathbf{k}} \to -\infty$ and to zero for $\varepsilon_{\mathbf{k}} \to +\infty$, similar to the corresponding quantity for a NIF, but being always smaller than that.

Let us now determine Δ. We combine (5.20) and (5.30), which yields an equation giving Δ in terms of itself,

$$
\Delta = \frac{|U|}{2\mathcal{V}} \sum_{\mathbf{k}} \frac{\Delta}{\sqrt{\varepsilon_{\mathbf{k}}^2 + |\Delta|^2}} = \frac{|U|}{2} \int \frac{d\mathbf{k}}{(2\pi)^3} \frac{\Delta}{\sqrt{\varepsilon_{\mathbf{k}}^2 + |\Delta|^2}}.
\tag{5.32}
$$

An equation of this type is called a *self-consistency* equation. Since the integrand depends on the energies $\varepsilon_{\mathbf{k}}$ only, it is instructive to change the variables of integration,

$$
\frac{d\mathbf{k}}{(2\pi)^3} \;\to\; \nu(\varepsilon)d\varepsilon,
\tag{5.33}
$$

where $\nu(\varepsilon)$ is the density of states at a given energy ε.

Control question. Can you derive an explicit expression for $\nu(\varepsilon)$?

We expect the main contribution to the integral to come from energies close to the Fermi energy. With this accuracy we may set $\nu(\varepsilon) \approx \nu(0)$. The equation then becomes

$$
1 = \frac{\nu(0)|U|}{2} \int d\varepsilon \frac{1}{\sqrt{\varepsilon^2 + |\Delta|^2}}.
\tag{5.34}
$$

This integral logarithmically diverges at energies much larger than Δ. We take care of this by introducing some cut-off energy E_{cut}[3] and integrate ε from $-E_{\text{cut}}$ to E_{cut}. We evaluate the integral in the leading logarithmic approximation to obtain

$$
1 = \nu(0)|U| \ln\left(\frac{2E_{\text{cut}}}{|\Delta|}\right), \quad \text{so that} \quad |\Delta| = 2E_{\text{cut}} \exp\left\{-\frac{1}{\nu(0)|U|}\right\}.
\tag{5.35}
$$

Typically, the attractive interaction is rather small, $\nu(0)|U| \ll 1$. Under these conditions, superconductivity creates its own energy scale Δ, which is not present in a NIF. Although this energy scale is very small – exponentially small – in comparison with E_{cut}, superconductivity can persist at an arbitrarily small attraction. This explains both the fact that superconductivity is wide-spread and the fact that it exists at low temperatures only.

---

[3] It was shown that this cut-off energy is of the order of the upper phonon energy. This is because phonons are the source of the attractive interaction. This is why $E_{\text{cut}} \ll E_{\text{F}}$ and not of the same order, as a naive calculation would suggest.

5.3 Particles and quasiparticles

At this point, we have minimized our trial wave function and obtained the superconducting ground state. Interestingly enough, we do not have to stop here. Using our simple model, we can learn not only about the ground state, but we can also understand the excitations of the superconductor.

Let us put an extra electron into a superconductor which is in the ground state we just found. The electron will have a certain wave vector $\mathbf{k}$ and spin, and thus it will be put into a certain Nambu box. So now it is time to recall the two states $|+\rangle$ and $|-\rangle$, which we had temporarily omitted from our consideration. If the extra electron comes with spin up, it will bring the box labeled $\mathbf{k}$ to the state $|+\rangle$. If its spin is down, the box with label $-\mathbf{k}$ ends up in the state $|-\rangle$. This gives us a simple picture of the excited states in a superconductor: some Nambu boxes are no longer in superpositions of $|0\rangle$ and $|2\rangle$, but in the state $|+\rangle$ or $|-\rangle$.

What is the energy of such an elementary excitation? The states $|+\rangle$ and $|-\rangle$ do not take part in the superconductivity and their energy is the same as in the NIF: just the kinetic term $\varepsilon_{\mathbf{k}}$. However, we must also take into account that the corresponding box is not any longer in the state $|g_{\mathbf{k}}\rangle$ and therefore does not contribute to the ground state energy as calculated above. We have already evaluated a contribution of $|g_{\mathbf{k}}\rangle$ to the total energy, see (5.31) above. Therefore, the net energy of the excitation reads

$$\epsilon_{\mathbf{k}}^{\mathrm{ex}} = \varepsilon_{\mathbf{k}} - E_{\mathbf{k}} = \sqrt{\varepsilon_{\mathbf{k}}^2 + |\Delta|^2}. \tag{5.36}$$

We see that the excitation energy is always positive, as it should be. In Fig. 5.2 is plotted the spectrum of these excitations. The solid line gives the excitation energy as a function of the pure kinetic energy of the added electron, whereas the dotted line shows the original particle energy $\varepsilon_{\mathbf{k}}$ as it would be in the NIF. The spectrum is said to have a *gap*, meaning that one cannot create excitations of arbitrarily small energy. We see indeed that there is a minimum energy for an excitation, and the magnitude of this gap happens to be $|\Delta|$.

The energy of a state in which several boxes are excited is just the sum of the individual excitation energies. So, in a way, the excitations behave as non-interacting identical particles. This brings us to an interesting point. In the first chapters, we managed to build a

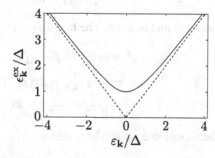

Fig. 5.2   The spectrum of elementary excitations in a superconductor. The dotted line gives the asymptotes $|\varepsilon_{\mathbf{k}}|$.

scheme of second quantization for "real" identical particles. By working with this scheme, we have now found the electronic excitations of a superconductor. These excitations are *quasiparticles*, existing on the background of the superconducting ground state. It would be neat to treat them in a second quantization scheme as well. Could we devise such a scheme? Or, to put it more technically, can we construct CAPs for these quasiparticles?

Since the excited states are Nambu boxes filled with single electrons, it seems natural to try just the same CAPs as for particles, e.g. $\hat{c}^\dagger_{\mathbf{k},\uparrow}$ and $\hat{c}_{\mathbf{k},\uparrow}$ for the excited state $|+\rangle$. However, this does not work. To see this, let us try to extract a quasiparticle, i.e. remove an excitation, from the superconducting ground state. This should give zero, since the ground state does not contain any excitations. Let us see what happens. We apply the proposed quasiparticle annihilation operator $\hat{c}_{\mathbf{k},\uparrow}$ to the ground state wave function,

$$\hat{c}_{\mathbf{k},\uparrow}|g_\mathbf{k}\rangle = \hat{c}_{\mathbf{k},\uparrow}(u_\mathbf{k} + v_\mathbf{k}\hat{c}^\dagger_{\mathbf{k},\uparrow}\hat{c}^\dagger_{-\mathbf{k},\downarrow})|0\rangle = v_\mathbf{k}c^\dagger_{-\mathbf{k},\downarrow}|0\rangle \propto |-\rangle. \qquad (5.37)$$

Surprisingly, we do not get zero, but instead we bring the box into another excited state! This of course happens because the ground state contains an admixture of the state $|2\rangle$, so that both $\hat{c}^\dagger_{\mathbf{k},\uparrow}$ and $\hat{c}_{\mathbf{k},\uparrow}$ can create and annihilate quasiparticles.

Obviously we need another set of CAPs, let us call them $\hat{\gamma}^\dagger_{\mathbf{k},\uparrow}$ and $\hat{\gamma}_{\mathbf{k},\uparrow}$. To make them good CAPs, we require that they satisfy the same anti-commutation relations as the original CAPs (the excitations are single electrons, and thus fermions). In order to do this, let us take a linear combination of the particle CAPs available in the box,

$$\begin{aligned} \hat{\gamma}^\dagger_{\mathbf{k},\uparrow} &= \alpha\hat{c}^\dagger_{\mathbf{k},\uparrow} + \beta\hat{c}_{-\mathbf{k},\downarrow}, \\ \hat{\gamma}_{\mathbf{k},\uparrow} &= \alpha^*\hat{c}_{\mathbf{k},\uparrow} + \beta^*\hat{c}^\dagger_{-\mathbf{k},\downarrow}, \end{aligned} \qquad (5.38)$$

with coefficients α and β. If we then impose the fermionic anti-commutation relation $\hat{\gamma}^\dagger_{\mathbf{k},\sigma}\hat{\gamma}_{\mathbf{k}',\sigma'} + \hat{\gamma}_{\mathbf{k}',\sigma'}\hat{\gamma}^\dagger_{\mathbf{k},\sigma} = \delta_{\mathbf{k},\mathbf{k}'}\delta_{\sigma,\sigma'}$, we find the requirement (see Exercise 3 of Chapter 3)

$$|\alpha|^2 + |\beta|^2 = 1. \qquad (5.39)$$

Let us try to fix the coefficients α and β explicitly. Being the quasiparticle creation and annihilation operators, they should obey

$$\begin{aligned} \hat{\gamma}^\dagger_{\mathbf{k},\uparrow}|g_\mathbf{k}\rangle &= |+\rangle = \hat{c}^\dagger_{\mathbf{k},\uparrow}|0\rangle, \\ \hat{\gamma}_{\mathbf{k},\uparrow}|g_\mathbf{k}\rangle &= 0, \end{aligned} \qquad (5.40)$$

which allows us to find α and β. The first line of (5.40) yields

$$\begin{aligned} \hat{\gamma}^\dagger_{\mathbf{k},\uparrow}|g_\mathbf{k}\rangle &= (\alpha\hat{c}^\dagger_{\mathbf{k},\uparrow} + \beta\hat{c}_{-\mathbf{k},\downarrow})(u_\mathbf{k} + v_\mathbf{k}\hat{c}^\dagger_{\mathbf{k},\uparrow}\hat{c}^\dagger_{-\mathbf{k},\downarrow})|0\rangle \\ &= (\alpha u_\mathbf{k}\hat{c}^\dagger_{\mathbf{k},\uparrow} - \beta v_\mathbf{k}\hat{c}^\dagger_{\mathbf{k},\uparrow}[1 - \hat{n}_{-\mathbf{k},\downarrow}])|0\rangle \\ &= (\alpha u_\mathbf{k} - \beta v_\mathbf{k})\hat{c}^\dagger_{\mathbf{k},\uparrow}|0\rangle, \end{aligned} \qquad (5.41)$$

and from the second line we similarly derive

$$\hat{\gamma}_{\mathbf{k},\uparrow}|g_\mathbf{k}\rangle = (\alpha^*v_\mathbf{k} + \beta^*u_\mathbf{k})\hat{c}^\dagger_{-\mathbf{k},\downarrow}|0\rangle. \qquad (5.42)$$

Comparing this with (5.40), we find the two constraints

$$\alpha u_{\mathbf{k}} - \beta v_{\mathbf{k}} = 1,$$
$$\alpha^* v_{\mathbf{k}} + \beta^* u_{\mathbf{k}} = 0. \tag{5.43}$$

We see that

$$\alpha = u_{\mathbf{k}}^* \quad \text{and} \quad \beta = -v_{\mathbf{k}}^*, \tag{5.44}$$

satisfies these two constraints, as well as the normalization condition (5.39).

We have thus completed our task, and found the quasiparticle CAPs to be

$$\hat{\gamma}_{\mathbf{k},\uparrow}^{\dagger} = u_{\mathbf{k}}^* \hat{c}_{\mathbf{k},\uparrow}^{\dagger} - v_{\mathbf{k}}^* \hat{c}_{-\mathbf{k},\downarrow},$$
$$\hat{\gamma}_{\mathbf{k},\uparrow} = u_{\mathbf{k}} \hat{c}_{\mathbf{k},\uparrow} - v_{\mathbf{k}} \hat{c}_{-\mathbf{k},\downarrow}^{\dagger}. \tag{5.45}$$

Since the quasiparticles are independent, the Hamiltonian describing the excitations of the superconductor reads

$$\hat{H}_{\text{qp}} = \sum_{\mathbf{k},\sigma} \epsilon_{\mathbf{k}}^{\text{ex}} \hat{\gamma}_{\mathbf{k},\sigma}^{\dagger} \hat{\gamma}_{\mathbf{k},\sigma}. \tag{5.46}$$

5.4 Broken symmetry

The self-consistency equation (5.32) defines the modulus of Δ but not its phase. Different choices of phase correspond to different superconducting ground states of the same energy. This is similar to what we found for the ferromagnetic state in the previous chapter: energy considerations then determined the magnitude of the magnetization P, but not the direction of P. Such a situation is typical if a symmetry is broken in a ground state. For the case of the ferromagnet, we argued that it is the rotational symmetry that is broken. Now we of course wonder which symmetry is actually broken in a superconductor.

This appears to be the *gauge* symmetry, a symmetry intrinsically related to the "ambiguous" nature of quantum mechanics. Given a single-particle wave function $|\psi\rangle$, one can multiply it with a phase factor (or run a *gauge transform*),

$$|\psi\rangle \rightarrow e^{i\chi}|\psi\rangle, \tag{5.47}$$

resulting in a wave function which describes precisely the same physical reality. This freedom is called *gauge symmetry*. Since many-particle wave functions are made of products of single-particle wave functions, a gauge transform in Fock space works as

$$|\{n_k\}\rangle \rightarrow e^{i\chi N}|\{n_k\}\rangle, \tag{5.48}$$

N being the total number of particles in the state $|\{n_k\}\rangle$. Combining this with the definitions of the CAPs, we observe that they do not stay invariant with respect to a gauge transform but change according to

$$\hat{c}_k^{\dagger} \rightarrow e^{i\chi} \hat{c}_k^{\dagger}. \tag{5.49}$$

This is nothing to worry about, since all usual physical observables involve products of CAPs with the same number of creation and annihilation operators (e.g. the particle number operator $\hat{n}_k = \hat{c}_k^\dagger \hat{c}_k$). So we see that the vector $e^{i\chi N}|\{n_k\}\rangle$ indeed describes the same physical state as $|\{n_k\}\rangle$ does.

However, this argument does not hold in a superconductor. Since the pair creation operators are made of two single-particle creation operators, and the superconducting order parameter Δ consists of expectation values of these pair creation operators, we see that

$$\hat{Z}_k^\dagger \to e^{i2\chi}\hat{Z}_k^\dagger \quad \text{implies} \quad \Delta \to e^{-i2\chi}\Delta. \tag{5.50}$$

Therefore, Δ *does* change upon a gauge transform, and we conclude that the gauge symmetry is broken in the superconducting state. It is also said that the ground state of a superconductor is a macroscopic quantum-coherent state. It is characterized by a phase in the very same manner as a bare wave function is.

Since one cannot measure the phase of a wave function, one is not able to measure the *absolute* phase of a superconductor. One can, however, measure the relative phase, i.e. a phase difference between two superconductors. We should probably stress that quantities which one cannot measure directly are not necessarily unphysical. Take an electrostatic potential as an example. One can never measure the absolute value of the potential although one can measure the potential difference between two conducting bodies.

Since electrons are charged, the gauge symmetry breaking in a superconductor is immediately related to electricity. The gradients of the superconducting phase are electric supercurrents, and the time derivative of the superconducting phase is proportional to the voltage of a superconductor. We illustrate the consequences of gauge symmetry breaking further in the context of coherent states (see Chapter 10).

Table 5.1 Summary: Superconductivity

Non-interacting Fermi gas

ground state: all states up to the Fermi level are filled, $|g_{\text{NIF}}\rangle = \displaystyle\prod_{|\mathbf{k}|<k_F} \hat{c}_{\mathbf{k},\uparrow}^{\dagger} \hat{c}_{\mathbf{k},\downarrow}^{\dagger} |0\rangle$

Add electron–electron interactions, Cooper pairs

idea: assume *attractive* electron–electron interaction

focus on pairs of electrons with opposite spin and momentum $\rightarrow$ Cooper pairs

model: attractive contact interaction: $U(r) = U\delta(r)$, with $U < 0$

Hamiltonian: $\hat{H}^{(2)} = \dfrac{U}{2\mathcal{V}} \displaystyle\sum_{\mathbf{k},\mathbf{k}',\mathbf{q}} \sum_{\sigma,\sigma'} \hat{c}_{\mathbf{k}+\mathbf{q},\sigma}^{\dagger} \hat{c}_{\mathbf{k}'-\mathbf{q},\sigma'}^{\dagger} \hat{c}_{\mathbf{k}',\sigma'} \hat{c}_{\mathbf{k},\sigma}$

in terms of Cooper pairs, $\hat{H}^{(2)} \approx \dfrac{\mathcal{V}}{U} \hat{\Delta}^{\dagger} \hat{\Delta}$, with $\hat{\Delta} \equiv \dfrac{U}{\mathcal{V}} \displaystyle\sum_{\mathbf{k}} \hat{c}_{\mathbf{k},\uparrow}^{\dagger} \hat{c}_{-\mathbf{k},\downarrow}^{\dagger}$

Trial wave function

try: uncertainty in particle number,

$|g_S\rangle = \displaystyle\prod_{\mathbf{k}} \left(u_{\mathbf{k}} + v_{\mathbf{k}} \hat{c}_{\mathbf{k},\uparrow}^{\dagger} \hat{c}_{-\mathbf{k},\downarrow}^{\dagger} \right) |0\rangle$, parameters $u_{\mathbf{k}}$ and $v_{\mathbf{k}}$

energy: $\langle g_S | \hat{H}^{(1)} | g_S \rangle = \displaystyle\sum_{\mathbf{k}} \varepsilon_{\mathbf{k}} |v_{\mathbf{k}}|^2$, with $\varepsilon_{\mathbf{k}} \equiv E(k) - E_F$, favors NIF ground state

$\langle g_S | \hat{H}^{(2)} | g_S \rangle = \dfrac{\mathcal{V}}{U} \left| \displaystyle\sum_{\mathbf{k}} u_{\mathbf{k}}^* v_{\mathbf{k}} \right|^2 = \dfrac{\mathcal{V}}{U} |\Delta|^2$

minimize: $|v_{\mathbf{k}}|^2 = \dfrac{1}{2} - \dfrac{\varepsilon_{\mathbf{k}}}{2\sqrt{\varepsilon_{\mathbf{k}}^2 + |\Delta|^2}}$, $|u_{\mathbf{k}}|^2 = \dfrac{1}{2} + \dfrac{\varepsilon_{\mathbf{k}}}{2\sqrt{\varepsilon_{\mathbf{k}}^2 + |\Delta|^2}}$,   and $u_{\mathbf{k}}^* v_{\mathbf{k}} = -\dfrac{1}{2} \dfrac{\Delta}{\sqrt{\varepsilon_{\mathbf{k}}^2 + |\Delta|^2}}$

self-consistency equation for $\Delta = \dfrac{|U|}{2\mathcal{V}} \displaystyle\sum_{\mathbf{k}} \dfrac{\Delta}{\sqrt{\varepsilon_{\mathbf{k}}^2 + |\Delta|^2}}$

Excitations

extra holes or electrons, but since $|g_S\rangle$ is not a number state, the $\hat{c}^{(\dagger)}$ are the *wrong* operators

instead quasiparticle: $|q_{\mathbf{k},\sigma}\rangle = \left(u_{\mathbf{k}}^* \hat{c}_{\mathbf{k},\sigma}^{\dagger} - v_{\mathbf{k}}^* \hat{c}_{-\mathbf{k},-\sigma} \right) |g_S\rangle$, with energy $\epsilon_{\mathbf{k}}^{\text{ex}} = \sqrt{\varepsilon_{\mathbf{k}}^2 + |\Delta|^2}$

Broken symmetry

the superconducting ground state has a definite *phase*: gauge symmetry is broken

Exercises

1. *Bogoliubov–de Gennes equation* (solution included). Let us derive an equivalent alternative description of a superconducting metal. It involves a transformation to quasiparticle CAPs as explained in Section 5.3, and relates it to the properties of the ground state. The idea in short is to introduce an effective Hamiltonian for the superconducting state and bring it to a quasiparticle form by a transformation. The equation defining this transformation is the Bogoliubov–de Gennes (BdG) equation. It has solutions at eigenenergies, those can be seen as the energies of quasiparticles and simultaneously determine the energy of the ground state. Let us go step by step through this program.

   a. In Section 5.2 we have evaluated the total energy of a superconductor assuming $\langle \hat{\Delta}^\dagger \hat{\Delta} \rangle \approx \langle \hat{\Delta}^\dagger \rangle \langle \hat{\Delta} \rangle$. Show that under the same assumptions, this energy can be evaluated for an arbitrary state $|\psi\rangle$ as

   $$E = \min_\Delta \left\{ \langle \psi | \hat{H}_S | \psi \rangle + \frac{\mathcal{V}}{|U|} |\Delta|^2 \right\},$$

   where the effective superconducting Hamiltonian $\hat{H}_S$ is given by

   $$\hat{H}_S = \hat{H}^{(1)} + \sum_{\mathbf{k}} \left(\Delta \hat{c}_{\mathbf{k},\uparrow}^\dagger \hat{c}_{-\mathbf{k},\downarrow}^\dagger + \Delta^* \hat{c}_{-\mathbf{k},\downarrow} \hat{c}_{\mathbf{k},\uparrow} \right).$$

   Here, Δ is a complex *parameter*. Show that the optimal value of this parameter is consistent with the previous definition of Δ.

   b. Consider the generalization of the above Hamiltonian to the case of an inhomogeneous superconductor,

   $$\hat{H}_S = \sum_{n,m} \left(H_{nm} \hat{c}_n^\dagger \hat{c}_m + \frac{1}{2} \Delta_{nm} \hat{c}_n^\dagger \hat{c}_m^\dagger + \frac{1}{2} \Delta_{nm}^* \hat{c}_m \hat{c}_n \right).$$

   Explain why one can assume that the matrix Δ_{nm} is antisymmetric, $\Delta_{nm} = -\Delta_{mn}$. The Hamiltonian $\hat{H}_S$ contains products of two annihilation and creation operators. Let us bring it to a more conventional form, where annihilation and creation operators come in pairs. We can achieve this by *doubling* the set of operators. We introduce an index α which takes two values, e or h, and work with the set of operators $\hat{d}_{n,\alpha}$, defined by $\hat{d}_{n,e} = \hat{c}_n$, and $\hat{d}_{n,h} = \hat{c}_n^\dagger$. Show that the operator $\hat{H}_S$ can be written as

   $$\hat{H}_S = \frac{1}{2} \sum_{\alpha,\beta} \sum_{n,m} \hat{d}_{n,\alpha}^\dagger \mathcal{H}_{n,\alpha;m,\beta} \hat{d}_{m,\beta} + E_0,$$

   where the matrix $\mathcal{H}$ is best presented in the space $\{\alpha, \beta\}$,

   $$\mathcal{H} = \begin{pmatrix} \check{H} & \check{\Delta} \\ \check{\Delta}^\dagger & -\check{H}^T \end{pmatrix},$$

   the "checked" matrices still having the indices nm. Give an expression for E_0.

c. The matrix $\mathcal{H}$ is Hermitian and can therefore be presented in terms of its eigenfunctions $\psi_k(n, \alpha)$ and corresponding eigenvalues E_k, where the k index the eigenvalues. Demonstrate that this gives rise to a representation of $\hat{H}_S$ in the form

$$\hat{H} = \frac{1}{2} \sum_k \hat{\gamma}_k^\dagger E_k \hat{\gamma}_k + E_0,$$

where $\hat{\gamma}_k$ is obtained from $\hat{d}_{n,\alpha}$ with the help of ψ_k^*. Express $\hat{\gamma}_k$ in terms of the $\hat{d}$s.

d. Prove the following special property of $\mathcal{H}$: if $\psi_{n,\alpha} = (\psi_{n,e}, \psi_{n,h})$ is an eigenstate of $\mathcal{H}$ with eigenvalue E, the function $\bar{\psi}_{n,\alpha} \equiv (\psi_{n,h}^*, \psi_{n,e}^*)$ is an eigenstate of $\mathcal{H}$ with *opposite* energy $-E$. We label the eigenstate with opposite energy as $-k$, so that $E(k) = -E(-k)$, and $\bar{\psi}_k = \psi_{-k}$.

   Derive the relation between the operators $\hat{\gamma}_k$ at opposite energies. Making use of orthogonality relations for the eigenfunctions of $\mathcal{H}$, prove that the operators $\hat{\gamma}_k$ satisfy the canonical anticommutation relations

$$\{\hat{\gamma}_k, \hat{\gamma}_{k'}\} = 0 \quad \text{and} \quad \{\hat{\gamma}_k^\dagger, \hat{\gamma}_{k'}\} = \delta_{kk'},$$

provided $k \neq -k'$.

e. In the next step, we get rid of the $\hat{\gamma}_k$ corresponding to negative energy eigenvalues, thereby removing the doubling of the operator set introduced in (b). Prove that

$$\hat{H}_S = \sum_{k, E_k > 0} \hat{\gamma}_k^\dagger E_k \hat{\gamma}_k + E_g. \tag{5.51}$$

Give an explicit expression for E_g. Explain why (5.51) solves the problem of finding the ground state energy and excitation spectrum of the superconductor.

f. Write down and solve the BdG equations for a homogeneous superconductor, as described in the main text of this chapter, using $n = (\mathbf{k}, \sigma)$. Explain the relation between u and v and the eigenfunctions of $\mathcal{H}$. Reveal the spin structure of the eigenfunctions.

g. Write down the BdG equations in a coordinate representation for the effective Hamiltonian

$$\hat{H}_S = \int d\mathbf{r} \left(\sum_\sigma \hat{\psi}_\sigma^\dagger(\mathbf{r}) \left\{ -\frac{\hbar^2 \nabla^2}{2m} - \mu + V(\mathbf{r}) \right\} \hat{\psi}_\sigma(\mathbf{r}) \right.$$
$$\left. + \Delta(\mathbf{r}) \hat{\psi}_\uparrow^\dagger(\mathbf{r}) \hat{\psi}_\downarrow^\dagger(\mathbf{r}) + \Delta^*(\mathbf{r}) \hat{\psi}_\downarrow(\mathbf{r}) \hat{\psi}_\uparrow(\mathbf{r}) \right),$$

making use of the spin structure obtained in the previous step.

2. *Andreev bound states.* Let us consider a single electron level that is connected to two bulk superconductors, 1 and 2, where Δ is the same in modulus but differs in phase, $\Delta_1 = |\Delta| e^{i\phi_1}$, $\Delta_2 = |\Delta| e^{i\phi_2}$, and $\phi_1 \neq \phi_2$. The corresponding Hamiltonian resembles that of the Anderson model,

$$\hat{H} = \hat{H}_1 + \hat{H}_2 + \hat{H}_{\text{imp}} + \hat{H}_c^{(1)} + \hat{H}_c^{(2)},$$

$$\hat{H}_{\text{imp}} = \varepsilon_0 \left(2 \sum_\sigma \hat{a}_\sigma^\dagger \hat{a}_\sigma - 1 \right),$$

$$\hat{H}_c^{(i)} = \sum_{\mathbf{k},\sigma} t_{\mathbf{k}}^{(i)} \left(\hat{c}_{\mathbf{k},\sigma}^{\dagger(i)} \hat{a}_\sigma + \hat{a}_\sigma^\dagger \hat{c}_{\mathbf{k},\sigma}^{(i)} \right),$$

$$\hat{H}_i = \sum_{\mathbf{k},\sigma} \varepsilon_{\mathbf{k}} \hat{c}_{\mathbf{k},\sigma}^{\dagger(i)} \hat{c}_{\mathbf{k},\sigma}^{(i)} + \sum_{\mathbf{k}} \left(\Delta_i \hat{c}_{\mathbf{k},\uparrow}^{\dagger(i)} \hat{c}_{-\mathbf{k},\downarrow}^{\dagger(i)} + \Delta_i^* \hat{c}_{-\mathbf{k},\downarrow}^{(i)} \hat{c}_{\mathbf{k},\uparrow}^{(i)} \right).$$

The Hamiltonian $\hat{H}_{\text{imp}}$ describing the energy of the single electron level is written such that the unoccupied level has energy $-\varepsilon_0$ and the occupied level $+\varepsilon_0$. We further assume all coupling parameters $t_{\mathbf{k}}$ in $\hat{H}_c$ to be real. Note that the Hamiltonian does not contain interaction.

  a. Write down the corresponding BdG equation that encompasses the electron and hole amplitudes $\psi_{e,h}^{(i)}(\mathbf{k})$ of the extended states in both superconductors as well as the amplitudes $\Psi_{e,h}$ of the localized state.
  b. Assume that there is a bound state solution of the BdG equation with $|E| < |\Delta|$. Exclude the amplitudes of the extended states from the equation. Use the same notations $\Gamma_{1,2}$ as in Exercise 1 in Chapter 4.
  c. From the resulting equation, find the energy of the bound state assuming $\Gamma_{1,2} \gg \Delta$. What can you say about the dependence of this energy on the phase difference $\phi = \phi_1 - \phi_2$?

3. *Supercurrent.* Let us consider a coordinate-dependent superconducting order parameter $\Delta(\mathbf{r}) = \Delta e^{i\phi' x}$. The density of a supercurrent flowing in the x direction can be related to the ϕ'-dependence of the energy density $\mathcal{E}$ corresponding to this order parameter,

$$j = -\frac{2e}{\hbar} \frac{d\mathcal{E}}{d\phi'}.$$

  a. Write down and solve the BdG equations, find the eigenvalues.
  b. Compute the energy density $\mathcal{E}$ associated with the phase gradient ϕ' in the first non-vanishing order in ϕ' and express the supercurrent proportional to the phase gradient.
  c. Estimate the maximum current density possible in this superconductor.

4. *Electrons and phonons.* The electrons in a metal are coupled to the crystal lattice. The excitations of the lattice are phonons, that is, bosons. The frequency of such an excitation $\omega_{\mathbf{q}}$ depends on the wave vector of the phonon, and $\omega_{\mathbf{q}} = \omega_{-\mathbf{q}}$. The Hamiltonian reads

$$\hat{H} = \hat{H}_{\text{el}} + \hat{H}_{\text{ph}} + \hat{H}_{\text{el-ph}},$$

$$\hat{H}_{\text{el}} = \sum_{\mathbf{k},\sigma} \varepsilon_{\mathbf{k}} \hat{c}_{\mathbf{k},\uparrow}^\dagger \hat{c}_{\mathbf{k},\sigma}, \quad \hat{H}_{\text{ph}} = \sum_{\mathbf{q}} \hbar\omega_{\mathbf{q}} \hat{b}_{\mathbf{q}}^\dagger \hat{b}_{\mathbf{q}},$$

$$\hat{H}_{\text{el-ph}} = \lambda \sum_{\mathbf{k}_1,\mathbf{k}_2,\sigma} \sqrt{\frac{\hbar\omega_{\mathbf{k}_1-\mathbf{k}_2}}{\nu(0)\mathcal{V}}} \left(\hat{b}_{\mathbf{k}_2-\mathbf{k}_1}^\dagger + \hat{b}_{\mathbf{k}_1-\mathbf{k}_2} \right) \hat{c}_{\mathbf{k}_1,\sigma}^\dagger \hat{c}_{\mathbf{k}_2,\sigma}.$$

Here, $\hat{c}_{\mathbf{k},\sigma}$ and $\hat{b}_{\mathbf{q}}$ are the electron and phonon annihilation operators corresponding to the levels with quantized wave vectors $\mathbf{k}$ and $\mathbf{q}$, the index $\sigma = \uparrow, \downarrow$ labels the spin projection of the electron, and $\nu(0)$ denotes the electronic density of states at the Fermi energy. The last term, $\hat{H}_{\text{el-ph}}$, represents the electron–phonon interaction, $\lambda \ll 1$ being the dimensionless strength of the interaction. We show that a second-order perturbation theory in terms of λ can result in an effective attractive electron–electron interaction and can thus eventually lead to superconductivity. In the absence of electron–phonon interaction, the ground state of the metal $|g\rangle$ is such that all the electron states up to the energy level E_{F} are filled, all states with higher energy are empty, and there are no phonons. As usual, we count electron energies $\varepsilon_{\mathbf{k}}$ from E_{F}.

a. Let us consider the set of excited states $|p_{\mathbf{k}}\rangle = \hat{c}^{\dagger}_{\mathbf{k},\uparrow} \hat{c}^{\dagger}_{-\mathbf{k},\downarrow} |g\rangle$, with $|\mathbf{k}| > k_{\text{F}}$. They correspond to unbound electron pairs (Cooper pairs, in fact) with zero total momentum. Project the Schrödinger equation of the whole Hamiltonian for the eigenvalue E onto the states of the set and show that it can be reduced to a Schrödinger-type equation for the operator

$$\sum_{\mathbf{k}} 2\varepsilon_{\mathbf{k}} |p_{\mathbf{k}}\rangle \langle p_{\mathbf{k}}| + \sum_{\mathbf{k},\mathbf{k}'} H^{(\text{eff})}_{\mathbf{k}\mathbf{k}'}(E) |p_{\mathbf{k}}\rangle \langle p_{\mathbf{k}'}|,$$

where

$$H^{(\text{eff})}_{\mathbf{k}\mathbf{k}'}(E) = \langle p_{\mathbf{k}} | \hat{H}_{\text{el-ph}} \frac{1}{E - \hat{\tilde{H}}} \hat{H}_{\text{el-ph}} | p_{\mathbf{k}'} \rangle.$$

What is $\hat{\tilde{H}}$?

b. Compute the matrix elements of $\hat{H}^{(\text{eff})}$ in second order in $\hat{H}_{\text{el-ph}}$. Consider the limits $E \ll \hbar\omega_0$ and $E \gg \hbar\omega_0$, where ω_0 is the typical phonon frequency. Show that in the first limit, the matrix elements amount to an effective electron–electron interaction of the form (cf. (5.3))

$$\hat{H}^{(2)} = \frac{1}{\mathcal{V}} \sum_{\mathbf{k}_1, \mathbf{k}_2} \hat{Z}^{\dagger}_{\mathbf{k}_1} U(\mathbf{k}_1, \mathbf{k}_2) \hat{Z}_{\mathbf{k}_2},$$

where the sign of U corresponds to attraction. Give an expression for $U(\mathbf{k}_1, \mathbf{k}_2)$.

c. An essential feature of the attractive interaction derived is that it is strongly reduced if the electron energies are larger than the phonon energy involved. Since typical phonon energies are two orders of magnitude smaller than the typical E_{F}, there will only be attraction between electrons with very small energy, i.e. electrons close to the Fermi surface.

Let us use a simplified model, assuming

$$U(\mathbf{k}_1, \mathbf{k}_2) = -|U_0| \Theta(\hbar\omega_0 - |\varepsilon_{\mathbf{k}_1}|) \Theta(\hbar\omega_0 - |\varepsilon_{\mathbf{k}_2}|).$$

Solve the equations for the excitation energies and demonstrate that they permit a solution with negative energy. This contradicts the fact that they should be excitations and thus proves the transition to a qualitatively different ground state: that of a superconductor.

5. *Competition of repulsion and attraction.* In fact, there is also Coulomb interaction between the electrons in the metal, and this repulsive interaction can compete with the

attraction due to the phonons. Since the attraction is restricted to electron energies close to the Fermi surface, let us consider the following simple model for the total interaction:

$$U(\mathbf{k}_1, \mathbf{k}_2) = -U_a \Theta(\hbar\omega_0 - |\varepsilon_{\mathbf{k}_1}|)\Theta(\hbar\omega_0 - |\varepsilon_{\mathbf{k}_2}|) + U_r \Theta(E_c - |\varepsilon_{\mathbf{k}_1}|)\Theta(E_c - |\varepsilon_{\mathbf{k}_2}|).$$

Here $\hbar\omega_0 \ll E_c \ll E_F$, and $U_{a,r} > 0$.

   a. Write down the Schrödinger equation for Cooper pair excitations in this model.
   b. Analyze the possibility of negative energy solutions and derive the superconductivity criterion for the model given.

Solutions

1. *Bogoliubov–de Gennes equation.* While this is a demanding problem, there are enough hints given to make each step easy.

   a. From a mathematical point of view, this is a rather trivial exercise. We explicitly minimize the equation with respect to Δ^*. Differentiating the expression gives the optimal Δ,

   $$\Delta = -\frac{|U|}{\mathcal{V}} \sum_k \langle \psi | \hat{Z}_\mathbf{k} | \psi \rangle,$$

   which is indeed consistent with (5.5).
   Substituting this optimal Δ, we find

   $$E = \langle \psi | \hat{H}^{(1)} | \psi \rangle - \frac{\mathcal{V}}{|U|}|\Delta|^2 \approx \langle \psi | \left(\hat{H}^{(1)} + \hat{H}^{(2)} \right) | \psi \rangle,$$

   where we make use of the assumption $\langle \hat{\Delta}^\dagger \hat{\Delta} \rangle \approx \langle \hat{\Delta}^\dagger \rangle \langle \hat{\Delta} \rangle$ to arrive at the last equality.

   From a physical point of view, the transformation we use above is less trivial. Instead of evaluating an interaction term that is of fourth power in the CAPs, we evaluate a term quadratic in the CAPs and introduce the auxiliary parameter Δ. In a more general context, such a transformation is called a Hubbard–Stratonovich transformation and is instrumental for treating interactions of identical particles.

   b. Any symmetric part of Δ_{nm} would not enter the equation since the operators $\hat{c}_n$ and $\hat{c}_m$ anticommute.

   The equivalence of the two forms of $\hat{H}_S$ can be proven by explicitly substituting the definitions of the operators $\hat{d}$. The only term to take extra care of is the one with $\alpha = \beta = h$. The substitution in this case gives $-\frac{1}{2}\sum_{n,m} \hat{c}_n H_{mn} \hat{c}_m^\dagger$. We then use the anticommutation relations to change the order of the CAPs, so it becomes $\frac{1}{2}\sum_{n,m} H_{mn} \hat{c}_m^\dagger \hat{c}_n - \frac{1}{2}\sum_n H_{nn}$. Collecting all terms gives $E_0 = \frac{1}{2}\sum_n H_{nn}$.

   c. In terms of the eigenstates,

   $$\mathcal{H}_{n,\alpha;m,\beta} = \sum_k \psi_k(n,\alpha) E_k \psi_k^*(m,\beta),$$

which we substitute into the sum to obtain

$$\hat{H}_S - E_0 = \frac{1}{2} \sum_{\alpha,\beta} \sum_{n,m} \sum_k \hat{d}^\dagger_{n,\alpha} \psi_k(n,\alpha) E_k \psi_k^*(m,\beta) \hat{d}_{m,\beta}$$

$$= \frac{1}{2} \sum_k \hat{\gamma}_k^\dagger E_k \hat{\gamma}_k,$$

with $\hat{\gamma}_k = \sum_{n,\alpha} \psi_k^*(n,\alpha) \hat{d}_{n,\alpha}$.

d. To prove the property, we write down the eigenvalue equation for $\mathcal{H}$ and take its complex conjugate. Using that $H_{nm} = H_{mn}^*$ and $\Delta_{nm} = -\Delta_{mn}$, it becomes an eigenvalue equation for $-\mathcal{H}$. For the $\hat{\gamma}_k$ we find

$$\hat{\gamma}_k = \sum_n \psi_k^*(n,e)\hat{c}_n + \psi_k^*(n,h)\hat{c}_n^\dagger,$$

$$\hat{\gamma}_{-k} = \sum_n \psi_k(n,h)\hat{c}_n + \psi_k(n,e)\hat{c}_n^\dagger = \hat{\gamma}_k^\dagger.$$

To prove the anticommutation relations, let us consider the anticommutator of $\hat{\gamma}_k$ and $\hat{\gamma}_{k'}$. The only non-vanishing contributions come from products of $\hat{c}^\dagger$ and $\hat{c}$,

$$\{\hat{\gamma}_k, \hat{\gamma}_{k'}\} = \sum_n \left[\psi_k^*(n,e)\psi_{k'}^*(n,h) + \psi_{k'}^*(n,e)\psi_k^*(n,h) \right]$$

$$= \sum_n \left[\psi_k^*(n,e)\bar{\psi}_{k'}(n,e) + \psi_k^*(n,h)\bar{\psi}_{k'}(n,h) \right]$$

$$= \sum_n \left[\psi_k^*(n,e)\psi_{-k'}(n,e) + \psi_k^*(n,h)\psi_{-k'}(n,h) \right].$$

The latter expression is the inner product of the eigenfunctions with k and $-k'$, and is thus zero when $k \neq -k'$.

e. Using the anticommutation relations makes the task straightforward: $\hat{\gamma}_{-k}^\dagger \hat{\gamma}_{-k} = 1 - \hat{\gamma}_k \hat{\gamma}_k^\dagger$. Therefore,

$$E_g = E_0 - \frac{1}{2} \sum_k \Theta(E_k) E_k.$$

This can be explained by noticing that the $\hat{\gamma}_k$ at positive energies are common CAPs and describe quasiparticle excitations. Since all energies E_k are positive, the ground state is that of no excitations, its energy being E_g.

f. In this case, the only non-zero elements of the matrices are $H_{\mathbf{k},\sigma;\mathbf{k}',\sigma} = \varepsilon_{\mathbf{k}}$, and $\Delta_{\mathbf{k},\sigma;-\mathbf{k}',-\sigma} = \sigma\Delta$. This couples electrons and holes with *opposite spin*. For electrons with spin up and holes with spin down, the eigenvalue equation reads

$$\begin{pmatrix} \varepsilon_{\mathbf{k}} & \Delta \\ \Delta^* & -\varepsilon_{\mathbf{k}} \end{pmatrix} \begin{pmatrix} \psi_e \\ \psi_h \end{pmatrix} = E_k \begin{pmatrix} \psi_e \\ \psi_h \end{pmatrix}, \tag{5.52}$$

having the eigenenergies $E_k = \pm\sqrt{|\Delta|^2 + \varepsilon_{\mathbf{k}}^2}$.

Solving for the eigenvectors, and comparing this with the expressions for $u_\mathbf{k}$ and $v_\mathbf{k}$, we see that

$$\psi_e = u_\mathbf{k}, \ \psi_h = -v_\mathbf{k} \quad \text{for} \quad E = \sqrt{|\Delta|^2 + \varepsilon_\mathbf{k}^2},$$

$$\psi_e = v_\mathbf{k}, \ \psi_h = u_\mathbf{k} \quad \text{for} \quad E = -\sqrt{|\Delta|^2 + \varepsilon_\mathbf{k}^2}.$$

The other set of eigenfunctions corresponds to electrons with spin down and holes with spin up. The BdG Hamiltonian in this block differs by the sign of Δ. The eigenvectors then read

$$\psi_e = u_\mathbf{k}, \ \psi_h = v_\mathbf{k} \quad \text{for} \quad E = \sqrt{|\Delta|^2 + \varepsilon_\mathbf{k}^2},$$

$$\psi_e = -v_\mathbf{k}, \ \psi_h = u_\mathbf{k} \quad \text{for} \quad E = -\sqrt{|\Delta|^2 + \varepsilon_\mathbf{k}^2}.$$

This elucidates the spin structure.

g. We restrict ourselves to a single block, for instance, with electrons with spin up and holes with spin down. From the solution of the previous point, we see that the BdG equation can be written as

$$\left(-\frac{\hbar^2 \nabla^2}{2m} - \mu + V(\mathbf{r}) \right) \psi_e(\mathbf{r}) + \Delta(\mathbf{r}) \psi_h(\mathbf{r}) = E\psi_e(\mathbf{r}),$$

$$\left(\frac{\hbar^2 \nabla^2}{2m} + \mu - V(\mathbf{r}) \right) \psi_h(\mathbf{r}) + \Delta^*(\mathbf{r}) \psi_e(\mathbf{r}) = E\psi_h(\mathbf{r}).$$

6

Superfluidity

In the same series of experiments which led to the discovery of superconductivity, Kamerlingh Onnes was the first who succeeded in cooling helium below its boiling point of 4.2 K. In 1908, he was cooling this liquefied helium further in an attempt to produce solid helium. But since helium only solidifies at very low temperatures (below 1 K) and high pressures (above 25 bar), Kamerlingh Onnes did not succeed, and he turned his attention to other experiments.

It was, however, soon recognized that at 2.2 K (a temperature reached by Kamerlingh Onnes), the liquid helium underwent a transition into a new phase, which was called helium II. After this observation, it took almost 30 years to discover that this new phase is actually characterized by a complete absence of viscosity (a discovery made by Kapitza and Allen in 1937). The phase was therefore called *superfluid*. In a superfluid, there can be persistent frictionless flows – superflows – very much like supercurrents in superconductors. This helium II is, up to now, the only superfluid which is available for experiments.

The most abundant isotope of helium, ^{4}He, is a spinless boson. Given this fact, the origin of superfluidity was almost immediately attributed to Bose condensation, which was a known phenomenon by that time.[1] However, as we discuss in more detail below, a Bose condensate of non-interacting particles is *not* a superfluid; only the interaction makes it one. The theory that takes this interaction into account was developed in the 1940s by Landau (from a phenomenological approach) and Bogoliubov (from a microscopic consideration). Below, we outline the Bogoliubov theory.

More recent developments in the field concern Bose condensation of ultra-cold atoms in magnetic traps (first observed in 1995) and experiments that might indicate the existence of *supersolids*.

6.1 Non-interacting Bose gas

We start by considering a simple thing, a non-interacting Bose gas (NIB) of massive particles with energies $E(\mathbf{k}) = \hbar^2 k^2/2m$. The ground state of this gas is obvious. All particles

---

[1] The less abundant isotope ^{3}He is a fermion. Although Bose condensation cannot play a role here, ^{3}He also becomes a superfluid at very low temperatures (below 2.6 mK). In this case, the superfluidity is indeed of a different kind, more resembling the superconductivity discussed in the previous chapter: the fermionic ^{3}He atoms form pairs, much as the electrons in a superconductor form Cooper pairs, and these bosonic pairs then condense into a superfluid.

will tend to go to the lowest energy level available, that is, with $\mathbf{k} = 0$. Since the particles are bosons, they can all fit in, and the ground state of N particles is therefore

$$|N\rangle \equiv \frac{\left(\hat{b}_0^\dagger\right)^N}{\sqrt{N!}}|0\rangle = |N, 0, 0, 0, \ldots\rangle. \tag{6.1}$$

Let us now address a less obvious point. It is clear that states with different N have the same energy, namely zero. This implies that also any superposition of these states has zero energy and thus qualifies for the ground state as well. For instance, one can think of a superposition where all $|N\rangle$ are taken with the same weight,

$$|\phi\rangle = \sum_N |N\rangle. \tag{6.2}$$

This means that the model of the NIB, distinct from the model of the NIF, is too idealized to decide upon a true ground state. One has to take into account factors that make the Bose gas non-ideal, for instance interaction. This is what we do in the next sections.

We also note here that a state like $|\phi\rangle$ no longer has a fixed number of particles, which implies a broken gauge symmetry. We can indeed see this from the above construction of $|\phi\rangle$. Suppose we apply a gauge transform to the bosonic CAPs such that $\hat{b}_0^\dagger \to e^{i\chi}\hat{b}_0^\dagger$. For any gauge-symmetric state this would result in an unimportant overall phase factor. For the state $|\phi\rangle$, however, we see that the transform changes $|\phi\rangle$ in

$$|\phi\rangle \to |\phi'\rangle = \sum_N e^{iN\chi}|N\rangle, \tag{6.3}$$

which is *not* just an overall phase factor.

6.2 Field theory for interacting Bose gas

To remind the reader, in these chapters we are exemplifying the techniques of second quantization. In the previous two chapters, we considered magnetism and superconductivity, both involving fermions. Naturally, it is now time to treat an example for bosons. We could proceed with the same method that we used in the previous chapters and try a wave function for the ground state of the interacting Bose gas. However, it is both compelling and instructive to proceed in a different way.

We have already mentioned that it is possible to formulate a quantized field theory even without explicitly invoking a wave function. In principle, a closed and most concise formulation of a field theory is given by the set of evolution equations for the field operators. In the spirit of the Heisenberg picture, these equations are obtained from the Hamiltonian,

$$i\hbar\frac{\partial\hat{\psi}(\mathbf{r})}{\partial t} = [\hat{\psi}(\mathbf{r}), \hat{H}]. \tag{6.4}$$

We formulate our field theory for interacting bosons in this way. Although an exact solution cannot be obtained, we are able to do much with controllable approximate methods.

Lev Davidovich Landau (1908–1968)
Won the Nobel Prize in 1962 for "his pioneering theories for condensed matter, especially liquid helium."

Lev Landau was born to a Jewish family in Baku, now the capital of Azerbaijan. He entered Baku State University aged 14, and, after moving to Leningrad in 1924, he graduated from Leningrad University at the age of 19.

In 1929, Landau left the USSR to travel through Western Europe. After working for some time with Niels Bohr in Copenhagen, he considered himself a pupil of Bohr for the rest of his life. After 18 months, Landau returned to the USSR and moved to Kharkov Gorky State University (now Kharkiv, Ukraine), where he was made Head of the Physics School in 1935. In the following years he was extremely productive, producing on average one scientific paper every six weeks covering almost all disciplines of theoretical physics. Within a few years, Landau turned his School in Kharkov into the center of theoretical physics in the whole USSR.

Landau moved to Moscow in 1937 to become the Head of the Theoretical Physics Department of the Academy of Sciences of the USSR. In 1938 he was arrested together with two colleagues for allegedly writing an anti-Stalinistic leaflet, and Landau spent one year in prison. He was released on bail only after the famous experimental physicist Pyotr Kapitza wrote personally to Stalin and took responsibility for Landau.

Having intensively studied physics, mathematics, and chemistry as a student, Landau had an extremely broad basic knowledge. Besides, he also had an unusual deep intuition for physics. Almost any work in theoretical physics presented to Landau would be turned upside down on the spot and provided with a completely new, original, and intuitive understanding. During his years in Moscow, he made important contributions to fields ranging from fluid dynamics to quantum field theory, his theories for superconductivity and superfluidity being most well-known.

In January 1962, Landau was involved in a serious car accident and was severely injured. He spent several months in hospital during which he was declared clinically dead multiple times. Miraculously, however, he survived and even recovered to normal in most ways. Unfortunately, his scientific genius and creativity did not survive the accident, and in the last years of his life he did not produce any notable scientific output. Landau died six years later, in 1968, from complications of his injuries.

To find the right approximations to use, we take a corresponding classical field theory as a guideline. For the case of boson theories this works since bosonic field operators *do* have classical analogues, as opposed to fermionic ones. Indeed, let us consider a set of NIB ground states $|N\rangle$ as defined above, with $N \gg 1$. We know that the bosonic CAPs corresponding to the lowest energy level, $\hat{b}_0^\dagger$ and $\hat{b}_0$, obey

$$\langle N|\hat{b}_0^\dagger \hat{b}_0|N\rangle = N \quad \text{and} \quad \langle N|\hat{b}_0 \hat{b}_0^\dagger|N\rangle = N + 1. \tag{6.5}$$

But since $N \gg 1$,

$$\langle \hat{b}_0^\dagger \hat{b}_0 \rangle \approx \langle \hat{b}_0 \hat{b}_0^\dagger \rangle \gg \langle [\hat{b}_0, \hat{b}_0^\dagger] \rangle, \tag{6.6}$$

which implies that for this group of states we can neglect the commutator of the field operators in comparison with their typical values and thus treat the CAPs as *classical* commuting variables, i.e. as simple (complex) numbers. The bosonic field operator, as defined in (3.11), can thus be written as

$$\hat{\psi}(\mathbf{r}) = b_0 \psi_0(\mathbf{r}) + \sum_{\mathbf{k} \neq 0} \hat{b}_\mathbf{k} \psi_\mathbf{k}(\mathbf{r}), \tag{6.7}$$

where the CAPs for all states with $k \neq 0$ are still operators.

6.2.1 Hamiltonian and Heisenberg equation

We start with writing down the Hamiltonian in terms of field operators. From Chapter 3 we know that the non-interacting part of the Hamiltonian reads

$$\hat{H}^{(1)} = \int d\mathbf{r}\, \hat{\psi}^\dagger(\mathbf{r}) \left\{ -\frac{\hbar^2 \nabla^2}{2m} \right\} \hat{\psi}(\mathbf{r}), \tag{6.8}$$

while the term describing the interaction is

$$\hat{H}^{(2)} = \frac{1}{2} \int d\mathbf{r}_1 d\mathbf{r}_2 V^{(2)}(\mathbf{r}_1 - \mathbf{r}_2) \hat{\psi}^\dagger(\mathbf{r}_1) \hat{\psi}^\dagger(\mathbf{r}_2) \hat{\psi}(\mathbf{r}_2) \hat{\psi}(\mathbf{r}_1). \tag{6.9}$$

We again choose for a simple interaction by invoking the well-used contact potential. Then our model interaction simplifies to

$$\hat{H}^{(2)} = \frac{U}{2} \int d\mathbf{r}\, [\hat{\psi}^\dagger(\mathbf{r})]^2 [\hat{\psi}(\mathbf{r})]^2, \tag{6.10}$$

and is said to be *local* in space.

Control question. How do you explain this?

The next step is to use the relation (6.4) to produce the evolution equations for the field operators. We have already done this job for the non-interacting term. To deal with the interaction term $\hat{H}^{(2)}$, we have to compute $[\hat{\psi}(\mathbf{r}), \hat{\psi}^\dagger(\mathbf{r}')^2 \hat{\psi}(\mathbf{r}')^2]$. We find

$$[\hat{\psi}(\mathbf{r}), \hat{\psi}^\dagger(\mathbf{r}')^2 \hat{\psi}(\mathbf{r}')^2] = [\hat{\psi}(\mathbf{r}), \hat{\psi}^\dagger(\mathbf{r}')^2] \hat{\psi}(\mathbf{r}')^2$$

$$= \left([\hat{\psi}(\mathbf{r}), \hat{\psi}^\dagger(\mathbf{r}')] \hat{\psi}^\dagger(\mathbf{r}') - \hat{\psi}^\dagger(\mathbf{r}')[\hat{\psi}(\mathbf{r}), \hat{\psi}^\dagger(\mathbf{r}')] \right) \hat{\psi}(\mathbf{r}')^2 \tag{6.11}$$

$$= 2\delta(\mathbf{r} - \mathbf{r}')\hat{\psi}^\dagger(\mathbf{r}')\hat{\psi}(\mathbf{r}')^2.$$

So finally we obtain

$$i\hbar \frac{\partial \hat{\psi}(\mathbf{r})}{\partial t} = -\frac{\hbar^2 \nabla^2}{2m}\hat{\psi}(\mathbf{r}) + U\hat{\psi}^\dagger(\mathbf{r})\hat{\psi}(\mathbf{r})\hat{\psi}(\mathbf{r}). \tag{6.12}$$

This is a non-linear operator equation, physical solutions of which should satisfy the boson commutation relations. Still, the equation has many solutions. This is understandable since the full set of solutions describes all possible states of the boson field, including those very far from equilibrium. Such a situation is similar to classical non-linear dynamics where the general solutions can be complicated and unpredictable (a stormy ocean is a good example of a system described by classical non-linear dynamics). The way to analyze such a system is to find first a simplest solution (a still ocean) and then consider small deviations from this (small ripples). Let us try to implement this approach for our field theory.

6.3 The condensate

In our case, the still ocean is the condensate, i.e. the macroscopically occupied ground state. From (6.7) we use that for the condensate the field operator is mainly just a classical number,

$$\hat{\psi}(\mathbf{r}, t) \approx b_0 \psi_0(\mathbf{r}, t) \equiv \Psi(t). \tag{6.13}$$

This classical field corresponds to the state $\psi_0(\mathbf{r}, t)$ with momentum $\mathbf{k} = 0$ and therefore does not depend on $\mathbf{r}$. The Heisenberg equation (6.12) for $\Psi(t)$ then reduces to

$$i\hbar \frac{\partial \Psi(t)}{\partial t} = U|\Psi(t)|^2 \Psi(t). \tag{6.14}$$

This equation has a solution with a modulus not depending on time,

$$\Psi(t) = \Psi_0 e^{-\frac{i}{\hbar}\mu t} \quad \text{so that} \quad |\Psi(t)|^2 = |\Psi_0|^2. \tag{6.15}$$

Recalling the density operator in terms of the field operator, we see that the solution describes a stationary Bose condensate with a density $\rho(\mathbf{r}) = |\Psi_0|^2$. As to the parameter μ, by virtue of (6.14) it is directly related to the density, $\mu \equiv U|\Psi_0|^2 = U\rho$. The notation μ is no coincidence: the quantity coincides with the chemical potential of the condensate. Indeed, adding one more particle to the condensate absorbs the interaction energy U multiplied by the local density of bosons.

6.3.1 Broken symmetry

We have already hinted that in a superfluid the gauge symmetry is broken. Let us now consider this explicitly. Since a gauge transformation gives a phase factor to a single-particle wave function, $|\psi\rangle \to e^{i\chi}|\psi\rangle$, the elements in Fock space are transformed as $|\{n_k\}\rangle \to e^{iN\chi}|\{n_k\}\rangle$. Field operators are therefore transformed as $\hat{\psi}_k^\dagger \to e^{i\chi}\hat{\psi}_k^\dagger$. This of course also holds for the lowest momentum component of the field operator, meaning that the condensate wave function transforms as

$$\Psi \to \Psi e^{-i\chi}. \tag{6.16}$$

This implies that the macroscopic variable Ψ characterizing the condensate, changes upon a gauge transformation and thus manifests the broken symmetry. Therefore, a superfluid is in a macroscopic quantum-coherent state. The major difference from a superconductor is that the helium atoms are not charged. There are no electric manifestations of superfluidity.

6.4 Excitations as oscillations

So we have the still ocean, let us now describe small ripples – excitations of the stationary condensate. Again, we borrow a method from classical non-linear dynamics. We linearize the equations around the simplest solution to find the small-amplitude oscillations. We thus search for a solution in the form

$$\hat{\psi}(\mathbf{r}, t) \approx \Psi(\mathbf{r}, t) + \delta\hat{\psi}(\mathbf{r}, t)e^{-\frac{i}{\hbar}\mu t}. \tag{6.17}$$

We substitute this into the Heisenberg equation and keep only terms of first order in $\delta\hat{\psi}$. This immediately gives

$$i\hbar\frac{\partial(\delta\hat{\psi})}{\partial t} + \mu(\delta\hat{\psi}) = -\frac{\hbar^2\nabla^2}{2m}(\delta\hat{\psi}) + U\Psi_0^2(\delta\hat{\psi})^\dagger + 2U|\Psi_0|^2(\delta\hat{\psi}). \tag{6.18}$$

To simplify, let us first replace both $U\Psi_0^2$ and $U|\Psi_0|^2$ with μ, which is equivalent to assuming Ψ_0 real. The equation can then be rewritten as

$$i\hbar\frac{\partial(\delta\hat{\psi})}{\partial t} = -\frac{\hbar^2\nabla^2}{2m}(\delta\hat{\psi}) + \mu\left\{(\delta\hat{\psi}) + (\delta\hat{\psi})^\dagger\right\}. \tag{6.19}$$

We see that plane waves satisfy this equation. Expressing the corrections to the condensate field operator as in (6.7), this means that we can search for a solution of the form

$$\delta\hat{\psi}(\mathbf{r}, t) = \frac{1}{\sqrt{\mathcal{V}}}\sum_{\mathbf{k}}\hat{b}_{\mathbf{k}}e^{i\mathbf{k}\cdot\mathbf{r}}. \tag{6.20}$$

We recall here that the operators $\hat{\psi}(\mathbf{r}, t)$ should satisfy the boson commutation relations. This gives the familiar commutation relations for the $\hat{b}_{\mathbf{k}}$,

$$[\hat{b}_{\mathbf{k}}, \hat{b}_{\mathbf{k}'}^\dagger] = \delta_{\mathbf{k},\mathbf{k}'}. \tag{6.21}$$

This is a comforting idea. The operators $\hat{b}_{\mathbf{k}}^{(\dagger)}$ are, as expected, nothing but the familiar CAPs of non-interacting bosons. However, unfortunately reality is not that simple. If we substitute (6.20) into the Heisenberg equation (6.19) to obtain an equation of motion for $\hat{b}_{\mathbf{k}}$, we get

$$i\hbar\frac{\partial\hat{b}_{\mathbf{k}}}{\partial t} = \{E(\mathbf{k}) + \mu\}\hat{b}_{\mathbf{k}} + \mu\hat{b}_{-\mathbf{k}}^\dagger, \tag{6.22}$$

which contains a "weird" term with $\hat{b}^\dagger$. If this term were absent, we would recover the usual equation for non-interacting boson-like particles with energies $E(\mathbf{k}) + \mu$. The presence of this term makes it impossible to regard the operators $\hat{b}_{\mathbf{k}}^{(\dagger)}$ as describing actually existing bosons in the system. We thus need to find a workaround. What can we do?

6.4.1 Particles and quasiparticles

In 1946, Nikolay Bogoliubov confronted this problem. Before that, quantum mechanics was not his main field. He was an established mathematician working in what we call now "non-linear science" – the qualitative and quantitative analysis of classical non-linear dynamics. Perhaps this is why he was able to find a simple and elegant solution of this quantum problem. His idea was as follows. Equation (6.22) is weird since the $\hat{b}_{\mathbf{k}}$ are the wrong operators to work with. One should try to find the right operators $\hat{\gamma}_{\mathbf{k}}$ that satisfy the same boson commutation relations but also obey the "correct" equation of motion,

$$i\hbar \frac{\partial \hat{\gamma}_{\mathbf{k}}}{\partial t} = \epsilon_{\mathbf{k}} \hat{\gamma}_{\mathbf{k}}, \tag{6.23}$$

the same equation of motion as for non-interacting bosons. How to find these new operators? Since all equations are linear, it is natural to try to present the wrong operators in terms of the right ones. We thus write

$$\hat{b}_{\mathbf{k}} = \alpha_{\mathbf{k}} \hat{\gamma}_{\mathbf{k}} + \beta_{\mathbf{k}} \hat{\gamma}_{-\mathbf{k}}^{\dagger},$$
$$\hat{b}_{-\mathbf{k}}^{\dagger} = \alpha_{-\mathbf{k}}^{*} \hat{\gamma}_{-\mathbf{k}}^{\dagger} + \beta_{-\mathbf{k}}^{*} \hat{\gamma}_{\mathbf{k}}. \tag{6.24}$$

This is called a *Bogoliubov transform*. The requirement that both the wrong and right operators satisfy the same boson commutation relations gives as an extra constraint

$$|\alpha_{\mathbf{k}}|^2 - |\beta_{\mathbf{k}}|^2 = 1. \tag{6.25}$$

We have actually already encountered similar equations in Section 5.3 when we tried to find the operators describing the excitations in a superconductor. Then they also described the difference between the wrong CAPs (particles) and the right CAPs (quasiparticles). In the present case we get quasiparticles for bosons rather than for fermions. Besides, we did this by a different method, but this is of no fundamental importance. In fact, superconductivity can be solved with a Bogoliubov transform too.

Control question. Can you explain the difference between (6.25) and (5.39)?

To find the quasiparticle energies $\epsilon_{\mathbf{k}}$ along with the coefficients $\alpha_{\mathbf{k}}$ and $\beta_{\mathbf{k}}$, one substitutes the Bogoliubov transform into the evolution equation (6.22) and requires that it can be written as (6.23). One then obtains a linear algebra problem for an unknown vector $(\alpha_{\mathbf{k}}, \beta_{\mathbf{k}}^{*})$ with eigenvalues $\epsilon_{\mathbf{k}}$. The quasiparticle energies – the energies of the elementary excitations of the interacting Bose condensate – are given by

$$\epsilon_{\mathbf{k}} = \sqrt{E(\mathbf{k})\{E(\mathbf{k}) + 2\mu\}}. \tag{6.26}$$

We plot this quasiparticle energy spectrum in Figure 6.1 (solid line) versus $|\mathbf{k}|/k_B$, where $k_B \equiv \sqrt{2m\mu}/\hbar$. To compare, we also plot the (kinetic) energy of "usual" particles. The lower dotted line shows the free-particle spectrum $E(k) \propto k^2$. The upper dotted line shows the same spectrum but now taking into account interactions between the particles, this results in an offset of μ.

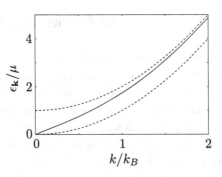

Fig. 6.1 The solid line gives the quasiparticle energy spectrum in a superfluid. The upper and lower dotted lines present the "usual" (kinetic) energy of particles with mass m, respectively with and without interaction with other particles, i.e. with and without taking into account the chemical potential μ.

The most astonishing fact about the quasiparticle spectrum is that the energy is *linear* in $|\mathbf{k}|$ at small $\mathbf{k}$. This indicates that the quasiparticles become quanta of *sound*! The sound velocity $v_p \equiv \frac{1}{\hbar} \lim_{\mathbf{k}\to 0} \partial\epsilon_{\mathbf{k}}/\partial\mathbf{k}$ is readily evaluated and reads

$$v_p = \sqrt{\frac{\mu}{m}}. \tag{6.27}$$

One can wonder if this sound, obtained by a quantum calculation, is the common sound observed in all gases and liquids or a sort of novel, quantum sound. Actually, it appears that this is a common sound. The same velocity v_p can be re-calculated with usual classical formulas involving the compressibility and the mass density of the liquid.[2] What cannot be obtained with classical formulas is that the sound excitations clearly become particles at higher energies. When $k \gg k_B$, the quasiparticle spectrum becomes quadratic in k, as is the case for free particles. More precisely, the spectrum approaches $E(\mathbf{k}) = \hbar^2 k^2/2m + \mu$, which is the energy of free particles with a chemical potential μ taking into account the repulsion a particle experiences from all other particles.

6.5 Topological excitations

In the previous section we described and discussed excitations of the superfluid that happened to be non-interacting quasiparticles. The procedure we used to deal with these resembles very much a complex version of a Taylor expansion, where we have restricted ourselves to the first few terms. However, we could go on considering higher order terms

---

[2] There was a long-lasting confusion about this. The reason for this confusion is the fact that at finite (non-zero) temperatures below the superfluid transitions, ^{4}He is a two-component liquid consisting of the superfluid condensate and a normal component that experiences some viscosity. The hydrodynamics of such a liquid are rich enough to support *four* different types of sound. However, both at zero temperature and above the transition, helium is a one-component liquid with a single common sound.

that should describe interactions of quasiparticles, and hope that we can account for all possible excitations in this way.

However, this hope is false. It turns out that many field theories possess excitations not captured by a Taylor expansion: *topological* excitations. They arise from the fact that the configurations of quantum fields in such theories can be (approximately) separated into classes such that no infinitesimally small change of the field configuration would cause a change from one class to another. Topology is a branch of mathematics dealing with the dichotomy between infinitesimally small and finite changes. This is why field configurations within the same class are called topologically connected, while those of distinct classes are topologically disconnected. Topological excitations are of interest because they can be regarded as emergent particles that are fundamentally distinct from the "original" elementary particles forming the ground state of the quantum system.

Superfluidity provides a good opportunity to illustrate these concepts. For the relevant configurations of the complex field $\langle \hat{\psi}(\mathbf{r}, t) \rangle$, the modulus (related to the particle density ρ) remains constant, while the phase can change rather freely. There exist topologically non-trivial configurations of the phase which are called *vortices*. They are emergent excitations, and depending on the dimensionality can be regarded as elementary excitations.

To simplify the discussion, we resort to a closed equation for $\langle \hat{\psi}(\mathbf{r}, t) \rangle \equiv \Psi(\mathbf{r}, t)$ that is a quasiclassical approximation for the full quantum field theory. To derive the equation, we take the Heisenberg equation (6.12), average it over the state and approximate $\langle \psi^{\dagger}(\mathbf{r}) \hat{\psi}(\mathbf{r}) \hat{\psi}(\mathbf{r}) \rangle = \langle \psi^{\dagger}(\mathbf{r}) \rangle \langle \hat{\psi}(\mathbf{r}) \rangle \langle \hat{\psi}(\mathbf{r}) \rangle = |\Psi(\mathbf{r})|^2 \Psi(\mathbf{r})$. It is also convenient to get rid of the time-dependent phase factor $e^{-\frac{i}{\hbar}\mu t}$, such that the field configuration of the ground state of the condensate is stationary. We thus include a factor $e^{\frac{i}{\hbar}\mu t}$ into $\Psi(\mathbf{r}, t)$ and write the Heisenberg equation

$$i\hbar \frac{\partial \Psi(\mathbf{r}, t)}{\partial t} = \left\{ -\frac{\hbar^2 \nabla^2}{2m} - \mu + U|\Psi(\mathbf{r}, t)|^2 \right\} \Psi(\mathbf{r}, t). \tag{6.28}$$

This equation is called the Gross–Pitaevskii equation, and describes the time evolution of the field configuration $\Psi(\mathbf{r}, t)$.

The quasiclassical approximation, as seen above, requires a large number of bosons. This number is determined by the density of the superfluid ρ and a space scale. We thus expect the Gross–Pitaevskii equation to be accurate at scales exceeding the interatomic distance $\simeq \rho^{-1/3}$, while not being able to resolve the details of $\Psi(\mathbf{r})$ at smaller distances. The characteristic length arising from (6.28) reads $\xi = \hbar/\sqrt{2mU\rho}$, also called the "healing length" of the superfluid. The Gross–Pitaevskii equation is thus consistent if $\xi \gg \rho^{-1/3}$, i.e. for weak interaction $mU \ll \hbar^2 \rho^{-1/3}$. For superfluid helium, $\xi \simeq 0.1$ nm is of the order of interatomic distances. The Gross-Pitaevskii equation is a classical Hamiltonian equation, it can be obtained by variation of the classical Hamiltonian

$$H_{\text{cl}} = \int d\mathbf{r} \left(\frac{\hbar^2}{2m} |\nabla \Psi(\mathbf{r})|^2 - \mu |\Psi(\mathbf{r})|^2 + \frac{U}{2} |\Psi(\mathbf{r})|^4 \right), \tag{6.29}$$

with respect to $\Psi^*(\mathbf{r})$ and $\Psi(\mathbf{r})$. Not surprisingly, $H_{\text{cl}} = \langle \hat{H} - \mu \hat{N} \rangle$. This implies conservation of total energy $E = H_{\text{cl}}$ in the process of time evolution.

Richard Feynman (1918–1988)
Shared the Nobel Prize in 1965 with Sin-Itiro Tomonaga and Julian Schwinger for "their fundamental work in quantum electrodynamics, with deep-ploughing consequences for the physics of elementary particles."

After Feynman received his Ph.D. degree in 1942 from Princeton University, he almost immediately left academia to join the Manhattan Project – the project for developing the atomic bomb. There at Los Alamos he met famous physicists such as Niels Bohr and Robert Oppenheimer, and the 24 year-old Feynman quickly attracted the attention of Bohr, who regularly invited him for one-to-one discussions.

In 1945, after the war had ended, Feynman accepted a professorship at Cornell University, where he stayed until 1950. The bombing of Hiroshima and Nagasaki had led him into a depression, and during his years at Cornell his scientific output was only meager. He nevertheless received many offers from other universities during these years, and in 1950 he decided to move to Caltech in California. As he claimed, this decision was made when he was trying to put the chains on his car during one of the many snowstorms in Ithaca: he then realized that "there must be a part of the world that doesn't have this problem."

Feynman stayed at Caltech until his death in 1988, producing numerous valuable contributions to theoretical physics. He also earned a reputation as an excellent teacher, claiming that any subject which could not be explained to first-year students was not fully understood yet.

Among his many discoveries, he was the first to come up with a full quantum mechanical theory of superfluidity, describing its ground state, but also its topological excitations. After its discovery in the 1930s, the first explanation for superfluidity was given by Landau's phenomenological theory, which provided an explanation for frictionless flow of a fluid, solely based on consideration of the energy and momentum of the fluid under Galilean transformations. In the 1940s, many qualitative theories and predictions were based on this work, but it was only in 1955 that Feynman came up with a microscopic explanation. Both the quantization of rotations in a superfluid, suggested by Onsager in the 1940s, and the existence of "rotons" as fundamental excitations, already conjectured by Landau, followed from Feynman's theory.

How can we relate the function $\Psi(\mathbf{r}, t)$ to more classical quantities characterizing a liquid? It is clear that that the modulus of Ψ is related to the local density of bosons,

$$\rho(\mathbf{r}, t) = \langle \hat{\psi}^{\dagger}(\mathbf{r}, t)\hat{\psi}(\mathbf{r}, t) \rangle \approx |\Psi(\mathbf{r}, t)|^2. \qquad (6.30)$$

To see if we can also attribute a classical "meaning" to the phase of Ψ, let us compute the time derivative of the density, $d\rho/dt$. From the Gross–Pitaevskii equation, we find that

$$\frac{d\rho(\mathbf{r},t)}{dt} = \frac{i\hbar}{2m}\nabla \cdot \{\Psi^*(\mathbf{r},t)\nabla\Psi(\mathbf{r},t) - \Psi(\mathbf{r},t)\nabla\Psi^*(\mathbf{r},t)\}, \tag{6.31}$$

where we have used the identity

$$\Psi\nabla^2\Psi^* - \Psi^*\nabla^2\Psi = \nabla \cdot (\Psi\nabla\Psi^* - \Psi^*\nabla\Psi). \tag{6.32}$$

Control question. Can you derive (6.32)?

We now see that if we define

$$\mathbf{j}(\mathbf{r},t) = -\frac{i\hbar}{2m}\{\Psi^*(\mathbf{r},t)\nabla\Psi(\mathbf{r},t) - \Psi(\mathbf{r},t)\nabla\Psi^*(\mathbf{r},t)\}, \tag{6.33}$$

we can write (6.31) as

$$\frac{d\rho(\mathbf{r},t)}{dt} + \nabla \cdot \mathbf{j}(\mathbf{r},t) = 0. \tag{6.34}$$

But this equation is exactly a continuity equation for the particle density! The change with time in the local particle density must equal the local divergence of the particle current density. The $\mathbf{j}(\mathbf{r},t)$ from (6.33) is thus nothing but the particle current density. It must equal the local particle density multiplied by the local velocity field, so we can extract the superfluid velocity as

$$\mathbf{v}_s(\mathbf{r},t) = \frac{\hbar}{m}\nabla\phi(\mathbf{r},t), \tag{6.35}$$

where we use the notation $\Psi(\mathbf{r},t) = \sqrt{\rho(\mathbf{r},t)}e^{i\phi(\mathbf{r},t)}$. The gradient of the phase ϕ of $\Psi(\mathbf{r},t)$ is thus proportional to the local velocity of the fluid.

Let us now illustrate the existence of topological excitations in a superfluid with the simplest setup. We consider a one-dimensional field theory, and confine the superfluid to a thin ring of cross-section s and radius R. We introduce the coordinate x describing position along the ring circumference. If the ring is thin enough, we can disregard the spatial dependence of field configurations in the perpendicular directions. We concentrate on solutions with a constant density, in which case the Gross–Pitaevskii equation reduces to the continuity equation and reads $v' = \phi'' = 0$. Except for a trivial solution (ϕ being a constant) the solutions have a constant (finite) gradient $\nabla\phi$, corresponding to the condensate moving with a constant velocity.

We now recall that the fluid is on a ring. This implies that the possible solutions for Ψ must be periodic in x with a period $2\pi R$. Since we assumed a constant density, we know that $\Psi = \sqrt{\rho}e^{i\phi(x)}$, and we see that periodicity is satisfied if the phase accumulated along the ring equals a multiple of 2π,

$$\oint dx\frac{\partial\phi}{\partial x} = 2\pi n. \tag{6.36}$$

The integer n is a *topological number*. Since it is a discrete number, it cannot be changed by a small variation of Ψ. Sometimes it is called a topological *charge*, by analogy with the discrete charge of elementary particles. Each field configuration with a non-zero constant

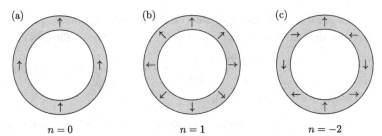

Fig. 6.2 Three topological states of a superfluid confined to quasi-one-dimensional ring. (a) The phase of the condensate wave function, indicated by the arrows, is constant along the ring, $n = 0$. (b) The phase winds once from zero to 2π when going around the ring, $n = 1$. (c) The phase winds from zero to 4π, and in the opposite direction, $n = -2$.

density in all points of the ring is characterized by a certain n, and those that differ in n belong to different topological sectors. The energy in a topological sector n is minimized by $\Psi = \sqrt{\rho}\exp(-inx/R)$ and equals $E_n = E_0 + n^2\pi\hbar\rho s/mR$. The true ground state corresponds to $n = 0$, while the states of the other sectors present topological excitations with respect to this ground state. In Fig. 6.2 we show three topological states: with $n = 0$, $n = 1$ and $n = -2$. The gray area represents the superfluid, which is confined to a quasi-one-dimensional ring. The superfluid density is assumed to be constant, but the phase of the condensate wave function, indicated by the arrows, can vary around the ring.

Control question. Can you sketch this drawing for $n = -1$? $n = 2$?

Relation (6.36) implies a quantization of the velocity of the condensate, $v_n = n\hbar/mR$. To understand this better, we compute the total angular momentum M of the superfluid with respect to the symmetry axis of the ring. Since the momentum of an infinitesimal element of the fluid reads $dp = s\rho mvdx$, the total angular momentum is given by

$$M_n = R\int dp = R\int_0^{2\pi R} dx\, \rho mv_n s = \hbar n s 2\pi R\rho = \hbar n N_p, \tag{6.37}$$

$N_p = s\rho 2\pi R$ being the total number of particles in the condensate. We see that in a topological sector n *each* particle of the superfluid acquires a quantized value of angular momentum $\hbar n$. This is different from the angular momentum quantization of non-interacting bosons and clearly manifests the collective nature of topological excitations.

6.5.1 Vortices

We found that in one dimension a topological excitation in a superfluid is anything but localized. Rather, it is global, being a property of the whole fluid confined to the ring. In a two-dimensional geometry, the same type of excitation looks more like a classical localized particle. Let us consider a thin film of a superfluid of thickness b, with (x, y) being the coordinates in the plane of the film. We imagine the film to be divided into concentric rings enclosing a single point $\mathbf{r}_0 \equiv (0, 0)$, which makes it convenient to describe coordinates in a polar system, $(x, y) \rightarrow (r \cos \theta, r \sin \theta)$. We assume the phase accumulation along the contour of a ring,

$$\Delta\phi = \oint d\mathbf{r} \cdot \nabla\phi = 2\pi, \qquad (6.38)$$

to be the same in all rings. We then see that a velocity distribution $\mathbf{v}_s(\mathbf{r}) = (\hbar n/mr)\hat{\theta}$, i.e. scaling with $1/r$ and always pointing in the tangential direction $\hat{\theta}$, solves (6.38). The corresponding phase is, apart from a constant, thus simply proportional to the angle θ. Eventually, if we deform the contour of integration in (6.38), we get the same result as long as the contour encloses the origin, which we illustrate in Fig. 6.3.

The resulting structure is called a *vortex*. Such structures readily occur in normal gases and liquids, like air and water. Examples of vortices which we all know are tornadoes, whirlwinds, and the small whirlpools which appear in your sink when you drain the water. All such structures have a characteristic velocity profile with the tangential velocity being proportional to r^{-1}. The distinction from normal liquids is that in a superfluid the velocity at each point in the fluid around a vortex can only take the quantized values $\hbar n/mr$. This results from the superfluid condensate being characterized by a phase, a quantity not present in conventional fluids. The integer n determines how often the phase of the wave function winds from zero to 2π when we go once around the vortex. Therefore, n is also called the *winding number* of the vortex.

The velocity of a vortex flow diverges for $r \rightarrow 0$, and the phase becomes undefined at this point. Something special must thus be going on close to the center of a vortex. For the "everyday" examples given above, we know that this is indeed the case: a tornado has in its center an "eye", a region where there is almost no airflow, and the whirlpool in the sink has an empty (waterless) column in its center. Most probably, also a vortex in a superfluid should have something similar in its center. Let us thus revert to the Gross–Pitaevskii equation, and see if we can obtain more information.

We thus seek for a cylindrically symmetric solution in the form $\Psi(\mathbf{r}) = \sqrt{\rho} f(r) \exp(in\theta)$, where ρ is the equilibrium density of the superfluid that is reached far away from the vortex center and $f(r)$ is a dimensionless function. The continuity of Ψ implies that $f(r) \rightarrow r^n$ at $r \rightarrow 0$, the density of the condensate must reach zero precisely in the center of the vortex.

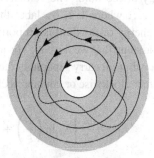

Fig. 6.3   Along all concentric circles around the origin the phase of the condensate wave function accumulates by the same amount of $2\pi n$ when we travel once around the origin. In fact, this phase accumulation is the same for any closed loop which encircles the origin, like the one indicated with the dashed line.

Control question. Do you see how?

The resulting equation for f can be made dimensionless by rescaling r to $u = r/\xi$, with $\xi = \hbar/\sqrt{2mU\rho_0}$ being the healing length, and reads

$$\frac{1}{u}\frac{d}{du}\left(u\frac{df}{du}\right) + \left(1 - \frac{n^2}{u^2}\right)f - f^3 = 0. \tag{6.39}$$

This equation indeed has a solution satisfying $f \to u^n$ at $u \to 0$ and $f \to 1$ at $u \to \infty$. This means that the depression of the density is restricted to the vortex *core*, being of the order of the healing length ξ. The velocities in the core are of the order of $\hbar/m\xi$, which is roughly equal to the sound velocity $v_p = \sqrt{\mu/m}$.

Let us also estimate the (kinetic) energy stored in a single vortex. Summing up the contributions of the individual rings, we obtain

$$E_n = \int d\mathbf{r} \, \frac{mv^2(\mathbf{r})}{2}\rho(\mathbf{r}) = \pi \frac{n^2\hbar^2}{m}\rho b \int \frac{dr}{r}. \tag{6.40}$$

The integral over r diverges both at the lower and upper limit and needs fixing. The lower limit is naturally set by the size of the vortex core where the density is suppressed. The upper limit is nothing but the typical size of the superfluid L. We thus obtain that the energy of the vortex weakly (logarithmically) depends on the system size,

$$E_n = n^2 b \frac{dE}{dl}\ln\left(\frac{L}{\xi}\right), \quad \text{with} \quad \frac{dE}{dl} = \pi\frac{\hbar^2\rho}{m}, \tag{6.41}$$

where we introduced dE/dl, being a measure of the total vortex energy density in the direction perpendicular to the plane of the thin film – the "energy per unit of thickness." An interesting observation is that the energy of the vortex scales with n^2. This means that it is energetically favorable to separate a single vortex with, for instance, $n = 2$ and thus energy $E_2 = 4b(dE/dl)\ln(L/\xi)$, into two vortices with winding number $n = 1$ and $E_1 = b(dE/dl)\ln(L/\xi)$. Generally, vortices with the lowest winding numbers $n = \pm1$ are thus preferred.

One can find the interaction energy of a *multi-vortex* configuration by simply calculating the kinetic energy of the fluid. The velocity profiles of different vortices add, $\mathbf{v}_{tot}(\mathbf{r}) = \mathbf{v}_1(\mathbf{r}) + \mathbf{v}_2(\mathbf{r}) + \ldots$, where $1, 2, \ldots$ label the different vortices (see Fig. 6.4(a)). Since the total kinetic energy of the fluid is proportional to v_{tot}^2, the vortex–vortex interaction is pairwise. Let us assume that we have N_v vortices, all with the same winding number $n = 1$ (or $n = -1$). They thus all carry separately the same energy E_1, and they all circulate in the same way, i.e. either clockwise or counterclockwise. The full energy including interactions then reads

$$E = b\frac{dE}{dl}\left\{N_v\ln\left(\frac{L}{\xi}\right) + \frac{1}{2}\sum_{i\neq j}\ln\left(\frac{L}{|\mathbf{r}_i - \mathbf{r}_j|}\right)\right\}, \tag{6.42}$$

$\mathbf{r}_i$ denoting the vortex coordinates. The interaction energy is apparently lowest when the vortices are far apart. This means that at a fixed concentration of vortices, they will try to be as far separated as possible from each other, thus forming a regular lattice.

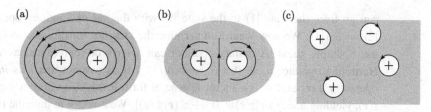

Fig. 6.4 Multi-vortex configurations. (a) The velocity profile and thus the angular momentum brought into the system by two vortices with equal winding numbers simply adds. Far away from the vortices the excitation looks like a single vortex with double winding number. (b) Far away from two vortices with opposite winding number, the condensate is in the ground state $n = 0$. (c) Of course, much more complicated combinations of vortices with different winding numbers are possible.

The vortices combine features of localized particles and of global excitations. On the one hand, they are characterized by coordinates like particles. On the other hand, the velocity distortion produced by each vortex falls off slowly with distance resulting in an energy that depends on the system size. An even more spectacular witness of their global character is the total angular momentum of the fluid. Since each vortex consists of concentric one-dimensional rings, (6.37) holds for the two-dimensional system as well. The total angular momentum thus equals $\hbar N_p N_v n$ ($n = \pm 1$ being the winding number of the vortices). That is, each particle of the fluid feels the presence of each vortex and acquires a quantum of angular momentum from it. From this it is also clear that the total topological charge of the fluid equals N_v (or $-N_v$, depending on sign of the winding number). All field configurations are thus separated into sectors with different N_v.

An interesting configuration is a pair of vortices with *opposite* winding numbers. The velocity profiles of the two vortices cancel each other at distances of the order of the distance d between the vortex centers, as illustrated in Fig. 6.4(b). The energy is $E \propto b(dE/dl)\ln(d/\xi)$, corresponding to *attraction* and not depending on the system size. This configuration corresponds to zero topological charge, and is topologically equivalent to the uniform ground state. Therefore, it can be seen as a superposition of elementary excitations – sound quanta – at the background of this state. The existence of this configuration implies that elementary excitations can actually also look like topological excitations. Of course, in the end they are not since they are topologically connected to the ground state.

6.5.2 Vortices as quantum states

Until now, we have worked with the quantum field $\hat{\psi}(\mathbf{r})$ as if it were a classical field $\Psi(\mathbf{r})$. It is important to understand that vortices are not just particles, but they are *quantum* particles. While we will not detail here the complex (quantum) dynamics of vortices, let us pinpoint a quantum state corresponding to a certain (topological) field configuration.

For an underlying bosonic field theory, such states are easy to produce. Let us consider two states, $|1\rangle$ and $|2\rangle$, corresponding to the field configurations $\langle\hat{\psi}(\mathbf{r})\rangle_{1,2} = \Psi_{1,2}(\mathbf{r})$. One can go from the state $|1\rangle$ to the state $|2\rangle$ with the aid of a unitary operator $\hat{S}_{21}$, so that $|2\rangle = \hat{S}_{21}|1\rangle$. We now want to determine the structure of this operator, and for this we use a simple guess. A unitary operator can always be represented as an exponent of a Hermitian operator $\hat{h}$, so that $\hat{S} = \exp(i\hat{h})$. Let us assume that $\hat{h}$ is *linear* in the field operators $\hat{\psi}(\mathbf{r})$ and $\hat{\psi}^\dagger(\mathbf{r})$. Its most general form involves an arbitrary complex function $z(\mathbf{r})$, yielding $\hat{h} = \int d\mathbf{r}\{z(\mathbf{r})\hat{\psi}^\dagger(\mathbf{r}) + z^*(\mathbf{r})\hat{\psi}(\mathbf{r})\}$. We now try to tune the function $z(\mathbf{r})$ such that we produce the desired operator $\hat{S}_{21}$.

Using the commutation relation of the field operators we find

$$[\hat{\psi}(\mathbf{r}), \hat{S}] = iz(\mathbf{r})\hat{S}. \tag{6.43}$$

Control question. Can you prove the above relation?

With this, we compute

$$\Psi_2(\mathbf{r}) = \langle 2|\hat{\psi}(\mathbf{r})|2\rangle = \langle 1|\hat{S}_{21}^{-1}\hat{\psi}(\mathbf{r})\hat{S}_{21}|1\rangle = \langle 1|\hat{S}_{21}^{-1}\hat{S}_{21}\hat{\psi}(\mathbf{r})|1\rangle + \langle 1|\hat{S}_{21}^{-1}[\hat{\psi}(\mathbf{r}),\hat{S}]|1\rangle$$
$$= \Psi_1(\mathbf{r}) + iz(\mathbf{r}), \tag{6.44}$$

so we have to set $iz(\mathbf{r}) = \Psi_2(\mathbf{r}) - \Psi_1(\mathbf{r})$. The unitary operator we need reads therefore

$$\hat{S}_{21} = \exp\left\{ \int d\mathbf{r}\big[\{\Psi_2(\mathbf{r}) - \Psi_1(\mathbf{r})\}\hat{\psi}^\dagger(\mathbf{r}) - \{\Psi_2^*(\mathbf{r}) - \Psi_1^*(\mathbf{r})\}\hat{\psi}(\mathbf{r})\big]\right\}. \tag{6.45}$$

Here we cannot replace operators with numbers, as we frequently have in this chapter. If we do this, the operator $\hat{S}_{12}$ would become just a multiplication with an unimportant phase factor and would not give rise to a new state. We indeed really need the non-commuting parts of the bosonic CAPs in this case.

Let us apply this result to a single-vortex configuration with a vortex centered at the point $\mathbf{r}_0$. The state $|1\rangle$ is the uniform ground state of the condensate $|g\rangle$ with $\Psi_1 = \sqrt{\rho}$, while the state $|2\rangle$ is the quantum state with a single vortex $|\mathbf{r}_0\rangle$. We seek for the operator connecting the two states, $|\mathbf{r}_0\rangle = \hat{S}_v(\mathbf{r}_0)|g\rangle$, and using the above result we see that it reads

$$\hat{S}_v(\mathbf{r}_0) = \exp\left\{ \sqrt{\rho} \int d\mathbf{r} \left(\frac{x+iy}{r} - 1 \right) \hat{\psi}^\dagger(\mathbf{r} + \mathbf{r}_0) - \text{H.c.}\right\}, \tag{6.46}$$

which can be interpreted as a *vortex creation operator*.

Control question. Can you trace the origin of the above expression?

With this, one can construct a quantum field theory for vortices. A word of caution is required: the operators $\hat{S}_v(\mathbf{r}_0)$ and $\hat{S}_v^\dagger(\mathbf{r}_0)$ do not obey the standard commutation relations at different $\mathbf{r}_0$ since the states at different $\mathbf{r}_0$ do not automatically form an orthogonal basis, $\langle \mathbf{r}_0|\mathbf{r}_0'\rangle \neq 0$ and have to be orthogonalized before use. Similarly, one can find the state corresponding to the vortex–antivortex configuration. As explained in the previous section, this state is topologically trivial and one finds that it can be presented as a superposition of sound quanta (see Exercise 4).

6.5.3 Vortex lines

Let us now finally turn to the more realistic three-dimensional liquid. There, vortices are not point-like excitations as in the thin superfluid film, but the vortex cores are long lines that must start and end at the boundaries of the superfluid and penetrate the whole volume of the fluid, as illustrated in Fig. 6.5. This statement sounds a bit dogmatic. Why could a vortex line not just end in the bulk of the fluid? The answer is the topological condition (6.38). If we find a certain circulation along a contour in the superfluid, we must find it along any other contour that is obtained by continuous deformation of the original contour, provided no vortex lines are intersected in course of the deformation. Suppose now that a vortex line ends somewhere in the fluid. We could then start with a contour that encloses the line far from the end, but then pull it off from the line as grandma pulls a loop off a knitting needle. The accumulated phases along the contour before and after deforming should differ, but topology forbids this and therefore the vortex lines can never end (in the superfluid). The energy of a line is proportional to its length and therefore to the system size, $E_1 = L(dE/dl)\ln(L/\xi)$. The line can therefore not be regarded as a particle, but it is rather a *macroscopic* excitation.

In principle, producing these vortex lines in a real superfluid should be relatively easy: one only needs to set the superfluid into rotation. Before turning to superfluids, let us first consider some basics about "ordinary" rotation. Suppose we have a normal rigid body (let's say an apple) rotating with an angular frequency ω about a fixed axis. Everyday experience tells us that under these conditions the apple in principle does not tend to fall apart, so every small part of it moves with the same angular frequency around the axis. In terms of the velocity field describing the movement of the apple, we can say that $\mathbf{v} = \omega \times \mathbf{r}$, where we have set the origin of the (cylindrical) coordinate system onto the rotation axis.

We can also do the reverse. Given a (complicated) velocity field $\mathbf{v}(x, y, z)$, we can decide if this field, or parts of it, have a rotating component. For this, we need to calculate the curl of the velocity field, $\nabla \times \mathbf{v}$, which yields its local *vorticity* – the "density of rotation"

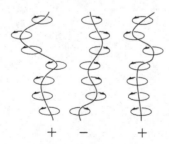

+   −   +

Fig. 6.5   In three dimensions the vortex core is a line inside the superfluid.

in the field. This can be appreciated by considering the integral of this curl over a surface Σ in the field. According to Stokes' theorem, this surface integral $\int_{\Sigma} (\nabla \times \mathbf{v}) \cdot d\Sigma$ equals the closed path integral $\oint_{\partial \Sigma} \mathbf{v} \cdot d\mathbf{r}$ along the boundary of the surface. Clearly this contour integral is a measure for the net amount of circulation of the field $\mathbf{v}$ along this contour. For the above example of a rotating rigid body, we find that $\nabla \times \mathbf{v} = 2\omega \hat{z}$, where $\hat{z}$ is the unit vector along the axis of rotation. The vorticity of a rotating rigid object is thus simply the angular frequency of the rotation.

How about the vorticity of superflows? We have shown above that in a superfluid $\mathbf{v}_s = (\hbar/m)\nabla \phi$. By virtue of the topological condition (6.38), the vorticity through a surface is set by the number of vortices N_v penetrating the surface, and reads $(2\pi\hbar/m)N_v$. The vorticity of a superflow without vortices is thus always zero! Also the total angular momentum M of the fluid is only due to vortices, $M = \hbar N_p N_v$. A superfluid without vortices is thus *irrotational*. If we were to try to rotate a vessel containing a superfluid, it would just slip along the vessel walls.

However, there are ways to enforce rotation. For instance, one can rotate a vessel filled with liquid helium in the normal state, which of course acquires a finite angular momentum. When this liquid is then gradually cooled further into the superfluid state, the angular momentum cannot disappear. After the transition to a superfluid has taken place, there thus must be a lattice of vortex lines formed in the liquid, the number of lines corresponding to the initial angular momentum.

Table 6.1 Summary: Superfluidity

Non-interacting Bose gas

Hamiltonian: $\hat{H}^{(1)} = \int d\mathbf{r}\, \hat{\psi}^\dagger(\mathbf{r}) \left\{ -\frac{\hbar^2}{2m} \nabla^2 \right\} \hat{\psi}(\mathbf{r})$

lowest energy: all N particles in the ground state, $|N\rangle \equiv \dfrac{(\hat{b}_0^\dagger)^N}{\sqrt{N!}} |0\rangle$

all $|N\rangle$ are degenerate $\rightarrow$ need interactions to decide on true ground state

Add interactions

model interaction: $U(r) = U\delta(r)$, contact interaction

Hamiltonian: $\hat{H}^{(2)} = \frac{1}{2} U \int d\mathbf{r}\, [\hat{\psi}^\dagger(\mathbf{r})]^2 [\hat{\psi}(\mathbf{r})]^2$

Field theory for bosons

equation of motion: $i\hbar \dfrac{\partial \hat{\psi}(\mathbf{r})}{\partial t} = [\hat{\psi}(\mathbf{r}), \hat{H}] = -\frac{\hbar^2}{2m} \nabla^2 \hat{\psi}(\mathbf{r}) + U \hat{\psi}^\dagger(\mathbf{r}) \hat{\psi}(\mathbf{r}) \hat{\psi}(\mathbf{r})$

use: macroscopic occupation of the ground state, $\langle \hat{b}_0^\dagger \hat{b}_0 \rangle \approx \langle \hat{b}_0 \hat{b}_0^\dagger \rangle$

so the field operator is $\hat{\psi}(\mathbf{r}) \approx b_0 \psi_0(\mathbf{r}) + \sum_{\mathbf{k} \neq 0} \hat{b}_\mathbf{k} \psi_\mathbf{k}(\mathbf{r})$

condensate: $\hat{\psi}(\mathbf{r}, t) \approx b_0 \psi_0(\mathbf{r}, t) \equiv \Psi(t)$, with equation of motion $i\hbar \dfrac{\partial \Psi(t)}{\partial t} = U |\Psi(t)|^2 \Psi(t)$

$\rightarrow \Psi(t) = \Psi_0 e^{-\frac{i}{\hbar}\mu t}$, where $\mu = U\rho$ (chemical potential)

interpretation of Ψ_0: $|\Psi_0|^2 = \rho$ (density) and $\frac{\hbar}{m} \nabla \arg(\Psi_0) = \mathbf{v}_s(\mathbf{r}, t)$ (superfluid velocity)

Excitations

ripples $\delta\hat{\psi}(\mathbf{r}, t) \propto \sum_\mathbf{k} \hat{b}_\mathbf{k} e^{i\mathbf{k}\cdot\mathbf{r}}$, but as $|\Psi(t)\rangle$ is not a number state, $\hat{b}^{(\dagger)}$ are the *wrong* operators

instead quasiparticle: $|q_\mathbf{k}\rangle = \left(\lambda_\mathbf{k} \hat{b}_\mathbf{k}^\dagger + \mu_\mathbf{k} \hat{b}_{-\mathbf{k}} \right) |\Psi(t)\rangle$, with energy $\epsilon_\mathbf{k} = \sqrt{E(k)\{E(k) + 2\mu\}}$

for $k \rightarrow 0$ linear spectrum with sound velocity $v_p = \sqrt{\mu/m}$

Topological excitations

vortex: Ψ_0 must be single-valued everywhere, so $\oint d\mathbf{r} \cdot \nabla \arg(\Psi_0) = 2\pi n$ along closed contour

if $n \neq 0$, a vortex is enclosed and ρ must vanish at the vortex core

configurations with different n are disconnected $\rightarrow$ *topological* excitations

energy of a vortex $E_n = \pi n^2 b\rho \frac{\hbar^2}{m} \ln\left(\frac{L}{\xi}\right)$, with $L \times b$ system size

Broken symmetry

the superfluid ground state has no definite particle number: gauge symmetry is broken

Exercises

1. *Bose–Einstein condensate in a harmonic trap* (solution included). We consider a system of N identical bosons, confined in a 3-dimensional potential well. The Hamiltonian for this system reads

$$\hat{H} = \sum_{i=1}^{N} \left\{ \frac{\hat{p}_i^2}{2m} + \frac{m\omega^2}{2}(\hat{x}_i^2 + \hat{y}_i^2 + \hat{z}_i^2) \right\},$$

which has a discrete spectrum with the energies

$$E_{n_x n_y n_z} = \hbar\omega(n_x + n_y + n_z + \tfrac{3}{2}),$$

the numbers n_x, n_y, and n_z being non-negative integers.

   a. We assume zero temperature and no interactions. In that case, all bosons simply occupy the ground level with energy E_{000}. Give the wave function $\psi_{000}(\mathbf{r})$ corresponding to the ground level.

   Suddenly, the potential trap is removed and the particles can move freely in space. After some time t a "photograph" is made of the expanding cloud of particles, thereby imaging the spatial distribution (density profile) of the cloud. Let us assume that the size of the original trap is negligible compared to the size of the expanded cloud. The measured density profile is then actually proportional to the original distribution of momenta in the trap: if a particle is detected at position $\mathbf{r}$, its momentum was $m\mathbf{r}/t$.

   b. Calculate the single-boson wave function in momentum space $\psi_{000}(\mathbf{p})$. How does the density profile look after a waiting time t?

2. *Bose–Einstein condensation in a harmonic trap at finite temperature.* In reality temperature is of course never exactly zero, and other states besides the ground state are populated. Let us investigate the role of finite temperature, but still neglect the interaction between the bosons.

   a. We start with a system consisting of a single bosonic level $|k\rangle$ with energy ε_k. The system is in contact with a reservoir with which it can exchange particles, so the allowed configurations of the system are $\propto (\hat{b}_k^\dagger)^N |\text{vac}\rangle$ with energies $N\varepsilon_k$, where N is an integer determining the number of bosons occupying the level. The system is in equilibrium at chemical potential μ.

   Use the Boltzmann factors mentioned in Section 1.10 to show that the average thermal occupation $\langle N(\varepsilon_k, \mu) \rangle$ of this level is given by

$$\langle N(\varepsilon_k, \mu) \rangle \equiv n_B(\varepsilon_k, \mu) = \frac{1}{e^{(\varepsilon_k - \mu)/k_B T} - 1},$$

   the well-known Bose–Einstein distribution function.

   b. The energy spectrum of the trap is $E_{n_x n_y n_z} = \hbar\omega(n_x + n_y + n_z + \tfrac{3}{2})$, the numbers n_x, n_y, and n_z being non-negative integers. Give the relation between the total number of particles in the trap N and the chemical potential μ.

   Usually, to convert the sum found in (b) into an explicit expression for $N(\mu)$, one assumes that all levels are so finely spaced that one can approximate the sum over the levels by an integral over energy, $\sum_k \rightarrow \int d\varepsilon\, D(\varepsilon)$, where $D(\varepsilon)$ is the density of states as a function of energy.

c. Calculate the density of states $D(\varepsilon)$ in a parabolic trap defined by the potential

$$V_{\text{trap}}(\mathbf{r}) = \frac{m\omega^2}{2}(x^2 + y^2 + z^2).$$

From here on we assume for convenience that $k_B T \gg \hbar\omega$, so that for most relevant levels also $\varepsilon \gg \hbar\omega$. Practically this means that you can forget about the ground state energy $\frac{3}{2}\hbar\omega$ in your integrals.

Since we allow for a macroscopic occupation of the ground state, we cannot, as suggested above, convert the sum we have found in (b) into an integral. We have to take special care of the ground state.

d. What is $N_0(\mu)$, the number of particles in the ground state as a function of the chemical potential?

Since we assume that $k_B T \gg \hbar\omega$, we see that a macroscopic occupation of the ground state, $N_0 \gg 1$ corresponds to a small chemical potential, $\mu/k_B T \ll 1$. In fact, we see that $N_0 \approx k_B T/(\frac{3}{2}\hbar\omega - \mu)$.

e. We have taken care of the ground state separately and found that it is occupied by $N_0 \approx k_B T/(\frac{3}{2}\hbar\omega - \mu) \gg 1$ particles. Now calculate the expected total number of bosons in the well, forgetting about the special status of the ground state. Show that this number is

$$N_{\text{th}} \approx \zeta(3)\left(\frac{k_B T}{\hbar\omega}\right)^3. \tag{6.47}$$

For a trap in which a significant fraction of the number of particles has condensed into the ground state, we have thus found that the relation between the total number of particles and the chemical potential is roughly

$$N = N_{\text{th}} + N_0 \approx \zeta(3)\left(\frac{k_B T}{\hbar\omega}\right)^3 + \frac{k_B T}{\frac{3}{2}\hbar\omega - \mu}.$$

Another way of looking at this relation is saying that for a given temperature, a number N_{th} of particles "fits" in the well without having a significant occupation of the ground state. All "extra" particles added occupy the ground state and ultimately raise its occupation number to a macroscopic level. The relation (6.47) thus defines for a given total number of particles a critical temperature below which the occupation of the ground state will start rising to a macroscopic level.

f. Suppose we do again the experiment of Exercise 2. We trap a certain number of bosons in a parabolic well. Then we suddenly remove the well, and after some time t we make an image of the particle density profile. However, this time we take the finite temperature of the system into account. We do not cool the system to absolute zero, but to $\frac{1}{2}T_c$, where T_c is the critical temperature as defined above, $k_B T_c = \hbar\omega\sqrt[3]{N/\zeta(3)}$. Calculate what fraction N_0/N of the particles was condensed in the ground state. How does the density profile look in this case?

3. *Bose–Einstein condensate with interactions.* While understanding the phenomenon of superfluidity, we noticed that the interaction between the particles plays a crucial role.

Let us try to include the effect of weak interactions into our simple description of the trapped Bose–Einstein condensate, and see what changes.

If we include interactions, the wave function of the condensed particles is no longer simply $\psi_{000}(\mathbf{r})$ – the solution of the Schrödinger equation for a particle in a harmonic potential well. Instead, the condensate wave function is governed by the Gross–Pitaevskii equation (6.28). Let us write a stationary version of it,

$$\left\{ -\frac{\hbar^2 \nabla^2}{2m} - \mu + V_{\text{trap}}(\mathbf{r}) + U|\Psi(\mathbf{r})|^2 \right\} \Psi(\mathbf{r}) = 0,$$

where, as usual, $V_{\text{trap}}(\mathbf{r}) = \frac{1}{2}m\omega^2 r^2$. We include a chemical potential μ since we are dealing with interacting particles.

This non-linear equation is hard to solve. Let us therefore consider one specific limit, the so-called Thomas–Fermi limit, in which the number of particles is so large that the mean field energy $U|\Psi(\mathbf{r})|^2$ is much larger than the kinetic energy.

a. Calculate the density profile $|\Psi(\mathbf{r})|^2$ in the Thomas–Fermi limit.

b. Calculate the chemical potential μ as a function of the total number of particles N.

c. Find the momentum density in the condensate, $|\Psi(\mathbf{k})|^2$.

  Hint. You might need the integral

$$\int_0^a \sin(kr)\sqrt{r^2 - \frac{r^4}{a^2}} = \frac{\pi a}{2k} J_2(ka),$$

  where $J_2(z)$ is the second Bessel function of the first kind.

d. What is roughly the width of this momentum density profile? Estimate the number of particles needed for the Thomas–Fermi approximation to hold.

4. *Vortex–antivortex pair.* We consider an infinite two-dimensional superfluid in which there are two vortices: one with winding number $n = 1$ centered around position $\mathbf{r}_0$, and one with winding number $n = -1$ centered around $\mathbf{r}_0'$. As discussed in the text, this combination of a vortex and antivortex is topologically trivial, and we can write it as a superposition of bosonic plane wave states. Let us derive this explicitly.

a. We define $\mathbf{r}_d = \mathbf{r}_0' - \mathbf{r}_0$ to be the distance between the vortex and antivortex. Give an explicit expression for the vortex–antivortex pair creation operator $\hat{S}_{av}(\mathbf{r}_0, \mathbf{r}_0')$ in terms of $\mathbf{r}_d$.

b. We assume the superfluid to have a low density ρ, which suggests that we expand the vortex–antivortex creation operator found at (a) in powers of $\sqrt{\rho}$. Perform this expansion to first order and write the result in terms of bosonic CAPs in momentum space $\hat{b}_{\mathbf{k}}^{(\dagger)}$.

c. Let this operator act on the ground state and give the resulting vortex–antivortex state $|av\rangle$.

d. The result found at (c) can be written

$$|av\rangle = |g\rangle + \sum_{\mathbf{k} \neq 0} c_{\mathbf{k}} \hat{b}_{\mathbf{k}}^\dagger |g\rangle,$$

where the squared moduli $|c_{\mathbf{k}}|^2$ give the probabilities for finding a plane wave bosonic mode with wave vector $\mathbf{k}$.

Let us concentrate on the behavior of the superfluid on length scales greatly exceeding r_d, or in other words, we investigate the occurrence of plane waves with small wave vectors $k \ll r_d^{-1}$. Find the distribution of plane waves with small k and discuss their directions.

Solutions

1. *Bose–Einstein condensation in a harmonic trap.*
    a. This is a generalization of the solution we know for the ground state of the one-dimensional harmonic potential well. Since the Hamiltonian can be split in three parts only depending on a single coordinate, $\hat{H} = \hat{H}_x(\hat{p}_x, \hat{x}) + \hat{H}_y(\hat{p}_y, \hat{y}) + \hat{H}_z(\hat{p}_z, \hat{z})$, the eigenstates of $\hat{H}$ are product states of one-dimensional solutions. For the ground level we can thus write

    $$\psi_{000}(\mathbf{r}) = \left(\frac{m\omega}{\pi\hbar}\right)^{\frac{3}{4}} \exp\left\{-\frac{m\omega}{2\hbar}r^2\right\},$$

    where $r^2 = x^2 + y^2 + z^2$.
    b. We rewrite the differential equation defining $\psi_{000}(\mathbf{r})$ (cf. (1.92)) in the momentum representation, using that in this representation $\hat{\mathbf{r}} = i\hbar\nabla_{\mathbf{p}}$,

    $$\left(\mathbf{p} + \hbar\omega m \frac{d}{d\mathbf{p}}\right)\psi_{000}(\mathbf{p}) = 0,$$

    and we can immediately write down the solution

    $$\psi_{000}(\mathbf{p}) = \left(\frac{1}{\pi\hbar\omega m}\right)^{\frac{3}{4}} \exp\left\{-\frac{p^2}{2\hbar\omega m}\right\}.$$

    So, the measured density profile is Gaussian and the cloud has a typical size of $\sim t\sqrt{\hbar\omega/m}$.

PART III

FIELDS AND RADIATION

In classical physics one can make a clear distinction between particles and fields. Particles are point-like objects which are at any time fully characterized by their position and momentum. The state of a particle in three-dimensional space is thus described by six independent variables, or, in other words, the particle lives in a six-dimensional phase space. Fields, on the other hand, have in general infinitely many degrees of freedom, and are described by (multi-dimensional) continuous functions. Examples of classical fields are the pressure field $p(\mathbf{r}, t)$ inside a container filled with gas (a scalar field), or the velocity field $\mathbf{v}(\mathbf{r}, t)$ of a flowing liquid (a vector field).

It is the goal of the coming chapters to develop methods to include fields and the interaction between particles and fields into our quantum mechanical description of the world. Although fields are continuous quantities, we show that it is possible to quantize a field, and to treat it in the end in a very particle-like way. In the quantum world, particles and fields are not that different.

The first step along this route is to realize that all classical fields are very similar: their dynamics are governed by linear equations, which can usually be reduced to simple wave equations describing the field configurations. We show that, as a consequence, most fields can be equivalently described as large sets of oscillator modes. In Chapter 8 we then demonstrate how an oscillator mode can be quantized, and then we are done: with the same successive steps we can quantize any classical field. With this knowledge at hand, we then quantize the electromagnetic field. In Chapter 9 we show how to describe the interaction between matter and radiation. This chapter contains several worked-out examples of processes involving emission and absorption of radiation: we discuss the radiative decay of excited atomic states, as well as Cherenkov radiation and Bremsstrahlung, and we give a simplified picture of how a laser works. This third part is concluded with a short introduction to *coherent states*: a very general concept, but one in particular that is very important in the field of quantum optics.

7 Classical fields

All classical fields actually look alike, and can be represented as a set of oscillators. The most interesting field is the electromagnetic field in vacuum and many examples of it are discussed in later chapters. Since this field and its classical description are rather involved, we start the chapter with a simpler example of a classical field theory, that of the deformation field of an elastic string. We derive the equation of motion for the field, and show that it can indeed be reduced to a set of oscillator modes. Only after this do we refresh the classical theory of the electromagnetic field and show that it can be treated along the exact same lines, allowing us to present an arbitrary solution of the Maxwell equations in vacuum in terms of oscillator modes. At the very end of the chapter, we briefly discuss a very constructive finite-element approach to the electromagnetic field: electric circuit theory. See Table 7.1 for a summary of the chapter's content.

7.1 Chain of coupled oscillators

The simplest example of a classical field we give in this chapter is the deformation field of an elastic string. However, before we turn to the continuous string, as a sort of warm-up exercise, we first consider a linear chain of coupled oscillators, as depicted in Fig. 7.1. A number N of particles with mass m sits on a straight line, and all neighboring particles are coupled to each other by massless springs with spring constant K. We keep the problem fully one-dimensional, so we only allow movement of the particles along the direction of the chain, as illustrated in the figure. As we recognize in the next section, the way to describe and solve this discrete problem is basically identical to our approach to the continuous problem, the elastic string.

Since we are facing a problem of discrete particles, we know how to solve it. The Hamiltonian of the full chain is simply the sum of the kinetic energy of all particles and the potential energy stored in the springs. We can thus write

$$H = \sum_n \left\{ \frac{p_n^2}{2m} + \frac{1}{2}K(x_{n+1} - x_n)^2 \right\}, \tag{7.1}$$

where p_n is the momentum of particle n and x_n its position. The equations of motion for this system follow from Hamilton's equations,

$$\frac{dx_n}{dt} = \frac{\partial H}{\partial p_n} \quad \text{and} \quad \frac{dp_n}{dt} = -\frac{\partial H}{\partial x_n}, \tag{7.2}$$

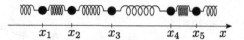

Fig. 7.1 Part of a linear chain of coupled oscillators. All coupled particles have equal mass m, and are coupled to their nearest neighbors by two massless springs with spring constant K. The state of the chain can at any moment be fully characterized by all coordinates x_n and momenta p_n of the particles.

which yield for our chain

$$\frac{d^2x_n}{dt^2} = \frac{K}{m}(x_{n+1} - 2x_n + x_{n-1}). \tag{7.3}$$

The form of this equation suggests that we try a plane wave solution

$$x_n = u_k \exp\{i(kn - \omega_k t)\} + u_k^* \exp\{-i(kn - \omega_k t)\}. \tag{7.4}$$

If we insert this ansatz into (7.3), we find the dispersion relation

$$-\omega_k^2 = 2\frac{K}{m}(\cos k - 1) \quad \Rightarrow \quad \omega_k = 2\sqrt{\frac{K}{m}}|\sin(k/2)|, \tag{7.5}$$

which for small k is approximately linear, $\omega_k \approx k\sqrt{K/m}$. Note that both the plane waves and the dispersion relation are periodic in k with period 2π. Solutions with k and $k + 2\pi$ are identical, and their parameterization by k spanning from $-\infty$ to ∞ is ambiguous. To remove this ambiguity, we must restrict the values of k to an interval of length 2π, for instance to $(-\pi, \pi)$.

The motion of a chain of coupled oscillators can thus indeed be described in terms of a superposition of propagating plane waves,

$$x_n = \sum_k \left\{ u_k e^{i(kn - \omega_k t)} + u_k^* e^{-i(kn - \omega_k t)} \right\}, \tag{7.6}$$

the frequency of each component being given by (7.5).

Let us finally address the question what are the allowed values for k. We assume that the chain is long enough that the particular choice of boundary conditions (i.e. are the two outer particles fixed, or maybe free?) does not make a significant difference. The most convenient choice is then to assume periodic boundary conditions: $x_{n+N} = x_n$. We see that this together with (7.4) implies that $e^{ikN} = 1$, leading to a discrete set of allowed wave numbers $k = 2\pi l/N$, where l is an integer. Since k is restricted to $(-\pi, \pi)$, l is restricted to the interval $(-N/2, N/2)$. The number of discrete values of k thus equals the number of degrees of freedom in the chain.

7.2 Continuous elastic string

Let us now turn to a deformed continuous elastic string, as illustrated in Fig. 7.2. To keep things simple, we allow the string to deform only in one transverse direction, i.e. in the plane of the drawing. The thick line shows a possible deformation of (a part of) the string at

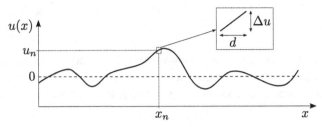

Fig. 7.2 A deformed elastic string at a certain moment of time. The equilibrium position of the string is indicated with the dashed line. The scalar field $u(x, t)$ describes the deviations from this equilibrium.

time t, away from its equilibrium position indicated with the dashed line. The deformation can thus be characterized by the scalar field $u(x, t)$, giving the local displacement from equilibrium as a function of position and time.

7.2.1 Hamiltonian and equation of motion

It is the goal of this section to derive an equation of motion for the field $u(x, t)$, so let us thus try to find the Hamiltonian describing the elastic string. We are now dealing with a continuous system, which means that in principle we cannot talk about a collection of separate particles and/or springs and then just sum up their contributions to the total Hamiltonian. We could proceed by guessing a reasonable expression for something like a Hamiltonian *density* and then integrate that over the length of the string. If we have made a good guess, we can probably end up with the right Hamiltonian and equation of motion, but let us see if we can do better than guessing.

Let us discretize the system first, since we know how to deal with discrete objects, and then see if the results can unambiguously be extended to the continuous limit. We thus cut the string into a large number of discrete pieces of length d, one of them is shown in the zoom-in in Fig. 7.2. The position of this small separated piece of string is given by $u_n(t)$, and since it has a mass ρd, where ρ is the mass density of the string, that is, the mass per unit length, its momentum simply reads $p_n = \rho d \partial_t u_n$. The potential energy in the Hamiltonian is due to the deformation of the string, and we treat the small isolated piece as a small spring with spring constant κ. Its length is given by $l = \sqrt{d^2 + (\Delta u)^2}$, and the associated potential energy thus reads $\frac{1}{2}\kappa l^2 = \frac{1}{2}\kappa(\Delta u)^2$, where we drop the part not depending on the deformation. We add the two contributions, and construct the Hamiltonian of the whole string as a sum of the separate pieces,

$$H = \sum_n d \left[\frac{1}{2} \frac{p_n^2}{\rho d^2} + \frac{1}{2}\kappa d \left(\frac{u_{n+1} - u_n}{d} \right)^2 \right], \tag{7.7}$$

where we approximate $\Delta u \approx u_{n+1} - u_n$. This is a good approximation when the deformation of the string is smooth on the scale of d, which is satisfied when $\Delta u \ll d$ for all pieces. The Hamiltonian is basically identical to the Hamiltonian (7.1) derived in the previous section for the chain of coupled oscillators, which was to be expected since the discretized string *is* a chain of coupled oscillators. For the discretized string we thus find the same set of equations of motion

$$\frac{d^2 u_n}{dt^2} = \frac{\kappa d}{\rho} \frac{u_{n+1} - 2u_n + u_{n-1}}{d^2}, \tag{7.8}$$

as we had for the the chain of oscillators before.

We are now ready to go to the continuous limit, letting $d \to 0$. Since $\kappa \propto 1/d$, the product κd goes to a constant K which represents the tension in the string. We introduce the momentum density $p(x)$ being the limit of p_n/d at $d \to 0$, and we regard $p(x)$ as a continuous function of x. This allows us to write the Hamiltonian for the continuous elastic string,

$$H = \int dx \left[\frac{1}{2\rho} p^2(x) + \frac{K}{2} \left(\frac{\partial u}{\partial x} \right)^2 \right]. \tag{7.9}$$

We see that the description of the continuous string indeed makes use of a Hamiltonian density, as we guessed before. This density integrated over the whole length of the string yields the full Hamiltonian. To get the equations of motion, we compute the variations of the Hamiltonian with respect to $u(x)$ and $p(x)$,

$$\frac{\partial u}{\partial t} = \frac{\delta H}{\delta p(x)} = \frac{p(x)}{\rho}, \tag{7.10}$$

$$\frac{\partial p}{\partial t} = -\frac{\delta H}{\delta u(x)} = K \frac{d^2 u(x,t)}{dx^2}. \tag{7.11}$$

Combining these two equations into a single equation of motion, like (7.8) but now for $u(x)$, we obtain in the continuous limit

$$\frac{d^2 u(x,t)}{dt^2} = \frac{K}{\rho} \frac{d^2 u(x,t)}{dx^2}. \tag{7.12}$$

This is a linear (wave) equation and describes oscillations in a specific medium, in this case a one-dimensional elastic string. In other media, natural or artificial, we can find numerous examples of fields and dynamical equations that are either intrinsically linear or approximately linear in the limit of small deviations from equilibrium. We have mentioned surface ocean waves already, but one could also think of sound waves in solids, spin waves in ferromagnets, etc. Due to this fundamental similarity, everything we learn in the next few chapters about description and quantization of classical fields can (with minor adjustments) be extended to all these systems.

Control question. Give three other examples of classical oscillating fields.

7.2.2 Solution of the equation of motion

We have thus found – not too surprisingly – that the dynamics of the deformation profile of the continuous elastic string are governed by a wave equation. As we found in the previous section for its discrete counterpart, this equation is satisfied by plane waves $u_k(x,t)$,

$$u_k(x,t) = u_k e^{i(kx - \omega_k t)} + u_k^* e^{-i(kx - \omega_k t)}, \tag{7.13}$$

where k is the wave vector, and the corresponding frequency is $\omega_k = k\sqrt{K/\rho}$. We have again combined the two complex conjugated terms to make sure that the displacement is real. Since the string is continuous, so to say, each segment of it consists of an infinite number of atoms, k is not restricted to any interval and spans from $-\infty$ to ∞.

To obtain the most general solution of the wave equation (7.12) we construct a superposition of all allowed particular solutions $u_k(x, t)$. In order to make sure we do not miss anything, we work with a discrete set of wave vectors as we did for particles. To do so, we consider the field configurations in a long but finite string of length L, and we impose periodic boundary conditions on the opposite ends of the string, $u(x, t) = u(x + L, t)$.

Control question. What are the allowed k in this case?

A general solution of (7.12) can then be written as

$$u(x, t) = \sum_k \left\{ u_k e^{i(kx - \omega_k t)} + u_k^* e^{-i(kx - \omega_k t)} \right\}, \tag{7.14}$$

where the sum is over all k allowed by the periodic boundary conditions. We chose our coordinate system such that the average displacement of the string is zero, i.e. $u_0 = 0$.

Another way of arriving at (7.14) is to recognize that the most general solution of (7.12) is $u(x, t) = u(x - vt)$ with $v = \sqrt{K/\rho}$. Equation (7.14) just expresses this solution in terms of its Fourier components. This is a discrete set since we assumed periodic boundary conditions, so the function $u(x, t)$ is periodic with period L.

7.2.3 The elastic string as a set of oscillators

Let us now have a closer look at the Hamiltonian (7.9) we derived above. We know that the function $u(x, t)$ can be written as a set of plane waves, so let us express the Hamiltonian in terms of the Fourier components u_k of the solution. We express momentum density using (7.14), substitute everything into the Hamiltonian and introduce the renormalized Fourier components $d_k \equiv u_k\sqrt{2\rho L \omega_k}$, where L is the length of the string. We then arrive at the remarkably elegant expression

$$H = \sum_k \omega_k d_k^* d_k. \tag{7.15}$$

Control question. The Hamiltonian contains second powers of the field. Generally, one thus expects double sums of the kind $\sum_{\mathbf{k}, \mathbf{k}'}$. Why is there only a single sum in the above equation? Why does it not depend on time?

One who recalls the single-particle Hamiltonian in second quantization understands that by writing the field Hamiltonian in this particular way, the quantization of the field is already half-done. In the next chapter we show how to do the next steps.

One can view the coefficients d_k as *normal coordinates* presenting arbitrary configurations of the field $u(x, t)$. The time-dependence of the field can be assigned to the d_k,

$$d_k(t) \equiv d_k e^{-i\omega_k t} \quad \text{and} \quad \ddot{d}_k(t) \equiv -\omega_k^2 d_k(t). \tag{7.16}$$

A minor disadvantage of these new coordinates is that they are complex. We can introduce real variables in the following way,

$$Q_k = \frac{1}{\sqrt{2\omega_k}}(d_k + d_k^*) \quad \text{and} \quad P_k = -i\sqrt{\frac{\omega_k}{2}}(d_k - d_k^*), \qquad (7.17)$$

and rewrite the Hamiltonian (7.15) in terms of these variables as

$$H = \sum_k \frac{1}{2}(P_k^2 + \omega_k^2 Q_k^2). \qquad (7.18)$$

But this is exactly the Hamiltonian for particles with mass $m = 1$ in a parabolic potential! The dynamics of the elastic string are thus fully equivalent to the dynamics of a set of harmonic oscillators, labeled by their wave vector k.

Based on this similarity to a mechanical oscillator, we say that Q_k and P_k are the generalized coordinate and momentum of the displacement field $u(x, t)$. Generalized is crucial here: the quantities do not even have dimensions of coordinate and momentum. The meaning and use of the notation is to stress the following fact: all linear oscillators in Nature – electromagnetic waves, sound, pendula, a steel ball hanging on a spring – are very similar to each other, and can be regarded in a unified way.

In terms of the mechanical analogy, the first term in (7.18) pictures the kinetic energy of the particle while the second gives the potential energy. This Hamiltonian is a classical Hamiltonian function, so the equations of motion read

$$\dot{Q}_k = \frac{\partial H}{\partial P_k} \quad \text{and} \quad \dot{P}_k = -\frac{\partial H}{\partial Q_k}. \qquad (7.19)$$

These equations describe fully the dynamics of the elastic string, and can be equivalently used instead of (7.12).

7.3 Classical electromagnetic field

Let us now consider the classical electromagnetic field, of which the description is based on the *Maxwell equations*. We focus on the solutions of the Maxwell equations in vacuum, and show in that case all possible field configurations can again be interpreted as an infinite set of oscillators.

We start by refreshing the Maxwell equations, and supplement them by a series of useful relations for forces arising from the fields, and energy and energy flow of the fields. We then review the concept of scalar and vector potentials for the fields and understand their use better. We come again to gauge transformations at this point, and consider two convenient gauge choices. Finally, we then find all solutions of the Maxwell equations in vacuum, and present the field as a set of oscillators.

7.3.1 Maxwell equations

As explained before, a classical field is an object defined in each point $(\mathbf{r}, t)$ of space and time. Electromagnetism deals with two three-component fields – an electric field $\mathbf{E}$ and a magnetic field $\mathbf{B}$ – which are a vector and a pseudovector respectively. The fields obey the following Maxwell equations,

$$\nabla \times \mathbf{E} = -\frac{\partial \mathbf{B}}{\partial t}, \tag{7.20}$$

$$\nabla \times \mathbf{B} = \mu_0\left(\mathbf{j} + \varepsilon_0 \frac{\partial \mathbf{E}}{\partial t}\right), \tag{7.21}$$

$$\nabla \cdot \mathbf{E} = \frac{\rho}{\varepsilon_0}, \tag{7.22}$$

$$\nabla \cdot \mathbf{B} = 0. \tag{7.23}$$

These are four equations. Actually, this number is a tribute to tradition depending on the choice of presentation of the electric and magnetic fields. In his original book, Maxwell gave 20 equations for 20 variables. We actually soon end up with two equations, and in relativistic theory there is just a single Maxwell equation. Equations (7.20) and (7.21) express the time derivatives of the fields in terms of their current configurations, and are therefore evolution equations. Equations (7.22) and (7.23) are in fact just constraints imposed on the fields.

As one can see from the equations, the sources of the fields are electric charges (ρ being the charge density) and currents ($\mathbf{j}$ the current density). In principle, this makes the Maxwell equations incomplete, since the dynamics of charged particles are again affected by the fields. A complete description of a physical system including charges and currents would include (a model for) the feedback of the fields on the charged particles. However, the Maxwell equations as presented above are complete for a system without any charges and currents, and thus are able to describe electromagnetic waves in vacuum.

Note that, as in the case of the elastic string, we have a set of linear equations describing the dynamics of the fields. Indeed, the Maxwell equations form another example of linear equations describing oscillations in a specific medium, in this case oscillations of $\mathbf{E}$ and $\mathbf{B}$ in the physical vacuum.

As a side-remark, we mention here that we have written the equations in a certain system of units called Système International (SI), which is the basis of modern metrology, widespread and even made obligatory by law in many countries. This does not mean that this system is logical and/or convenient. The problem is that the equations contain ε_0 and μ_0, the permittivity and the permeability of vacuum, which does not make much sense nowadays. By definition, $\mu_0 = 4\pi \times 10^7$ H/m, and the permittivity follows from the permeability via a true physical constant, c, the speed of light in vacuum,

$$\mu_0 \varepsilon_0 c^2 = 1. \tag{7.24}$$

In an alternative system of units, the so-called CGS, the Maxwell equations take a more logical form and contain the speed of light only,

James Clerk Maxwell (1831–1879)

By the end of the first half of the 19th century, many different phenomena involving electricity, magnetism, and light were known. The existence of a relation between electric and magnetic fields was already discovered, and was expressed through phenomenological laws such as Ampère's circuital law and Faraday's law of induction. Light was still regarded as a separate phenomenon, the fundamental research on it mainly being focused on the question whether it consisted of waves or particles. By 1850, the evidence for light being a wave-like excitation of some medium (dubbed the *ether*) was overwhelming, and it finally seemed that the wave theory of light was correct.

In 1861 and 1862, James Clerk Maxwell published a series of papers in which he cast the known equations of electrodynamics in a new form, introducing the displacement current as a correction to Ampère's law. While contemplating on the set of equations he had arrived at, he realized that they allowed for a wave-like solution in vacuum with a propagation speed which was remarkably close to the speed of light. He wrote: "we can scarcely avoid the inference that light consists in the transverse undulations of the same medium which is the cause of electric and magnetic phenomena." The insight that light is an electromagnetic wave is maybe the most important discovery of all 19th century physics, and for this many regard Maxwell, next to Newton and Einstein, as one of the greatest physicists of all time.

Maxwell was born to a rich Scottish family, and already as a very young boy he displayed a great curiosity for the world around him. He was educated first at Edinburgh University and later at Trinity College, Cambridge. At age 25, he returned to Scotland to take up a professorial position at the Marischal College in Aberdeen. He stayed in Aberdeen for four years, and in 1860 he moved to King's College, London. In London, he had the most productive years of his career; there he even produced the world's first color photograph. In 1871, he became the first Cavendish Professor of Physics at the University of Cambridge, where he stayed until his death in 1879 at the young age of 48.

$$\nabla \times \mathbf{E} = -\frac{1}{c}\frac{\partial \mathbf{B}}{\partial t}, \tag{7.25}$$

$$\nabla \times \mathbf{B} = \frac{1}{c}\left(4\pi \mathbf{j} + \frac{\partial \mathbf{E}}{\partial t}\right), \tag{7.26}$$

$$\nabla \cdot \mathbf{E} = 4\pi\rho, \tag{7.27}$$

$$\nabla \cdot \mathbf{B} = 0. \tag{7.28}$$

However, we stick to the SI in this book.

7.3.2 Useful relations

The Maxwell equations are difficult to apply without several useful relations which we need here and in later chapters. The feedback of the electromagnetic fields on a particle with a charge q is given by the corresponding force,

$$\mathbf{F} = q[\mathbf{E} + (\mathbf{v} \times \mathbf{B})]. \tag{7.29}$$

Considering these forces, one finds the energy *density* U associated with the electric and magnetic fields,

$$U = \frac{1}{2}\varepsilon_0(\mathbf{E}^2 + c^2\mathbf{B}^2). \tag{7.30}$$

This presents the energy accumulated between the plates of an electric capacitor or inside a magnetic coil. The density of flux of this energy, or *Poynting vector*, is given by

$$\mathbf{S} = \frac{1}{\mu_0}\mathbf{E} \times \mathbf{B}. \tag{7.31}$$

In the absence of any charges, the energy of the electromagnetic field is conserved. This is expressed by

$$\nabla \cdot \mathbf{S} + \frac{\partial U}{\partial t} = 0, \tag{7.32}$$

in the differential form.

Control question. Can you prove this from the Maxwell equations?

If one takes the divergence of (7.21) and uses (7.22) to exclude the divergence of the electric field, one obtains

$$\nabla \cdot \mathbf{j} + \frac{\partial \rho}{\partial t} = 0, \tag{7.33}$$

which *expresses* the conservation of electric charge. Some say it *proves* the conservation of electric charge. However, strictly speaking the charge and current configurations are not the business of the Maxwell equations: they just give the configuration of the fields at any given input of charges and currents. The relation (7.33) is a message from the equations: "We won't work if you give us a non-conserving charge, you'd better not!".

7.3.3 Vector and scalar potentials

Instead of working with electric and magnetic fields, it is often convenient to work with their respective *vector* and *scalar potentials*, $\mathbf{A}$ and φ, sometimes called the magnetic and electric potential. A mathematical observation is that the substitution

$$\mathbf{B} = \nabla \times \mathbf{A} \quad \text{and} \quad \mathbf{E} = -\nabla\varphi - \frac{\partial \mathbf{A}}{\partial t}, \tag{7.34}$$

automatically satisfies (7.20) and (7.23). The remaining two become

$$\nabla^2 \varphi + \frac{\partial}{\partial t}(\nabla \cdot \mathbf{A}) = -\frac{\rho}{\varepsilon_0}, \tag{7.35}$$

$$\nabla^2 \mathbf{A} - \frac{1}{c^2}\frac{\partial^2 \mathbf{A}}{\partial t^2} - \nabla\left(\nabla \cdot \mathbf{A} + \frac{1}{c^2}\frac{\partial \varphi}{\partial t}\right) = -\mu_0 \mathbf{j}, \tag{7.36}$$

which together form a complete expression of the Maxwell equations. As promised earlier, we have thus reduced the number of equations to two. The price paid is that the equations now contain derivatives of higher orders.

The vector and scalar potentials as defined above *cannot* be measured in principle, their values are ambiguous. One can only measure their derivatives, the magnetic and electric fields, by their effect on charged particles. A reasonable question is: why would we use these "unphysical" variables? Well, there are several reasons for this. First of all, the potentials can be convenient for simplifying the math of the Maxwell equations. We indeed have managed to reduce the number of equations. Secondly, the potentials are also required to build up a Hamiltonian for charged particles in the presence of an electromagnetic field. The true reason, however, is that the ambiguity itself is not a "bad" property. It is actually a "good" property that allows us to understand deep features of the theory which are not obvious at the level of the fields.

Let us concentrate on this ambiguity. The observables, the electric and magnetic fields, are invariant with respect to *gauge transformations* of the potentials. A gauge transformation is defined by a scalar function $\chi(\mathbf{r}, t)$ as follows

$$\mathbf{A} \rightarrow \mathbf{A}' = \mathbf{A} + \nabla\chi \quad \text{and} \quad \varphi \rightarrow \varphi' = \varphi - \frac{\partial \chi}{\partial t}. \tag{7.37}$$

Control question. Do you see why the physical fields $\mathbf{E}$ and $\mathbf{B}$ do not change upon such a transformation?

This gives rise to many interesting results, but first of all it gives us some personal freedom. We can choose the χ in the most convenient way, one that best suits the problem at hand. The physical results do not depend on our subjective choice by virtue of the gauge invariance.

7.3.4 Gauges

A particular choice of $\chi(\mathbf{r}, t)$ or, equivalently, a choice of a constraint that implicitly defines χ is called a *gauge*. Two gauges are particularly popular for practical applications: the Lorenz[1] gauge and the Coulomb gauge.

The Lorenz gauge is given by the constraint

$$\nabla \cdot \mathbf{A} + \frac{1}{c^2}\frac{\partial \varphi}{\partial t} = 0. \tag{7.38}$$

---

[1] Ludvig Lorenz (1829–1891) was a Dane.

One can start with an arbitrary configuration $\mathbf{A}'$ and φ' of the potentials and find a gauge transform that brings the potentials to the Lorenz gauge. To this end, one takes the relation (7.37) and substitutes it into (7.38), which then becomes an equation for χ to solve.

The advantage of the Lorenz gauge is that it is Lorentz covariant,[2] that is, it satisfies the symmetry of a relativistic theory. If one rewrites the Maxwell equations (7.35) and (7.36) using the Lorenz constraint, one observes a remarkable symmetry between the scalar and vector potentials,

$$\Box \varphi = \frac{\rho}{\varepsilon_0},\tag{7.39}$$

$$\Box \mathbf{A} = \mu_0 \mathbf{j}.\tag{7.40}$$

The symbol $\Box$ here is the D'Alembert (or wave) operator,

$$\Box \equiv \frac{1}{c^2}\frac{\partial^2}{\partial t^2} - \nabla^2.\tag{7.41}$$

Actually, it was a reflection upon this symmetry which brought about the theory of relativity. Both potentials in this formulation are *retarded*: their values in a given point do not react instantly to the motion of charges in another point since the effect of such motion cannot propagate faster than light. There is more about relativity in Chapter 13.

Another choice, that works best for non-relativistic particles, is the Coulomb gauge where the vector potential satisfies the constraint

$$\nabla \cdot \mathbf{A} = 0.\tag{7.42}$$

In this gauge, the Maxwell equations read

$$\nabla^2 \varphi = -\frac{\rho}{\varepsilon_0},\tag{7.43}$$

$$\nabla^2 \mathbf{A} - \frac{1}{c^2}\frac{\partial^2 \mathbf{A}}{\partial t^2} = -\mu_0 \mathbf{j} + \frac{1}{c^2}\frac{\partial \nabla \varphi}{\partial t}.\tag{7.44}$$

Conveniently, the electrostatic potential is not retarded anymore.

Control question. Does this imply that the electric field is not retarded?

7.3.5 Electromagnetic field as a set of oscillators

Let us now finally perform a similar procedure as in Section 7.2 for the elastic string, and show how one can equivalently describe the electromagnetic field in vacuum in terms of a set of independent oscillators. We work in the Coulomb gauge, and since there are no charges, the scalar potential is absent. The Maxwell equations reduce to one single equation for the vector potential,

$$\nabla^2 \mathbf{A} - \frac{1}{c^2}\frac{\partial^2 \mathbf{A}}{\partial t^2} = 0,\tag{7.45}$$

---

[2] Hendrik A. Lorentz was a very famous Dutch physicist. Most textbooks eventually ascribe the gauge to him.

and the electric and magnetic fields are expressed in terms of $\mathbf{A}$,

$$\mathbf{B} = \nabla \times \mathbf{A} \quad \text{and} \quad \mathbf{E} = -\frac{\partial \mathbf{A}}{\partial t}.$$

But the remaining Maxwell equation (7.45) is a simple wave equation: the very same wave equation as (7.12), which describes the displacement field of the elastic string! We thus know how to deal with it. The equation is satisfied by plane electromagnetic waves with wave vector $\mathbf{k}$ and frequency $\omega_{\mathbf{k}} = ck$,

$$\mathbf{A}_{\mathbf{k},\alpha}(\mathbf{r}, t) = \mathbf{e}_\alpha a_{\mathbf{k}} \exp\{i(\mathbf{k} \cdot \mathbf{r} - \omega_{\mathbf{k}}t)\} + \mathbf{e}_\alpha a_{\mathbf{k}}^* \exp\{-i(\mathbf{k} \cdot \mathbf{r} - \omega_{\mathbf{k}}t)\}, \tag{7.46}$$

where $\mathbf{e}_\alpha$ is a (generally complex) vector of *polarization*. Since the Coulomb gauge implies $\nabla \cdot \mathbf{A} = 0$, the polarization is always perpendicular to the wave vector, $\mathbf{e} \cdot \mathbf{k} = 0$. This leaves for each wave vector two independent polarizations indexed by $\alpha = 1, 2$. It is convenient to make them real and orthonormal,

$$\mathbf{e}_\alpha \cdot \mathbf{e}_\beta = \delta_{\alpha\beta}. \tag{7.47}$$

We again work with a discrete set of wave vectors, as we did for the elastic string. We thus consider field configurations in a large but finite cube of length L and thus volume $\mathcal{V} = L^3$, and we impose periodic boundary conditions on the opposite faces of the cube.

Control question. What is the spacing between the discrete values of $\mathbf{k}$?

Now we are ready to write the most general solution of the Maxwell equations in vacuum. It is a linear combination of all possible partial solutions,

$$\mathbf{A}(\mathbf{r}, t) = \sum_{\mathbf{k}} \sum_{\alpha=1,2} \frac{1}{\sqrt{2\varepsilon_0 \omega_{\mathbf{k}} \mathcal{V}}} \left[d_{\mathbf{k}\alpha} \mathbf{e}_\alpha \exp\{i(\mathbf{k} \cdot \mathbf{r} - \omega_{\mathbf{k}}t)\} + \text{c.c.} \right], \tag{7.48}$$

where we have introduced the new coefficients $d_{\mathbf{k}\alpha} = a_{\mathbf{k}\alpha} \sqrt{2\varepsilon_0 \omega_{\mathbf{k}} \mathcal{V}}$. The use of this renormalization with a strange square root factor becomes transparent when we calculate the full energy and full flux corresponding to the solution using relations (7.30) and (7.31),

$$H_{\text{e.m.}} = \int d\mathbf{r} \frac{\varepsilon_0}{2}(\mathbf{E}^2 + c^2 \mathbf{B}^2) = \sum_{\mathbf{k},\alpha} \omega_{\mathbf{k}} d_{\mathbf{k}\alpha}^* d_{\mathbf{k}\alpha}, \tag{7.49}$$

$$\mathbf{S}_{\text{tot}} = \int d\mathbf{r} \frac{1}{\mu_0}(\mathbf{E} \times \mathbf{B}) = \sum_{\mathbf{k},\alpha} c^2 \mathbf{k} \, d_{\mathbf{k}\alpha}^* d_{\mathbf{k}\alpha}. \tag{7.50}$$

It provides simple expressions for energy and energy flux. We also see that the Hamiltonian of the field (7.49) now looks identical to the Hamiltonian (7.15) describing the displacement field of the elastic string, so we can treat it in the exact same way to arrive at a description in terms of oscillator modes.

From the normal coordinates $d_{\mathbf{k}\alpha}$ we again construct the generalized coordinate and momentum,

$$Q_{\mathbf{k}\alpha} = \frac{1}{\sqrt{2\omega_{\mathbf{k}}}}(d_{\mathbf{k}\alpha} + d_{\mathbf{k}\alpha}^*) \quad \text{and} \quad P_{\mathbf{k}\alpha} = -i\sqrt{\frac{\omega_{\mathbf{k}}}{2}}(d_{\mathbf{k}\alpha} - d_{\mathbf{k}\alpha}^*), \tag{7.51}$$

and rewrite the Hamiltonian as

$$H_{\text{e.m.}} = \sum_{\mathbf{k},\alpha} \frac{1}{2}(P_{\mathbf{k}\alpha}^2 + \omega_k^2 Q_{\mathbf{k}\alpha}^2), \tag{7.52}$$

again a perfect copy of the Hamiltonian for a set of mechanical oscillators with mass $m = 1$. If we were to apply Hamilton's equations and derive the equations of motion from this Hamiltonian, we would obtain the Maxwell equations as well as the oscillations of the particle – a feature of the unified description. Analogously to the elastic string, we have thus shown how to describe the electromagnetic field as a set of oscillators, in this case labeled by their wave vector $\mathbf{k}$ and their polarization α.

To get acquainted better with this set, let us do some counting of oscillator modes. We would like to learn how many modes there are in a small volume of wave vector space around a point $\mathbf{k}$ and how many allowed frequencies there are in a given interval $d\omega$. The first question is answered just by counting the discrete values of $\mathbf{k}$ in each direction. The number of oscillators is thus given by

$$2_p \frac{dk_x L}{2\pi} \frac{dk_y L}{2\pi} \frac{dk_z L}{2\pi} = V\frac{d\mathbf{k}}{4\pi^3}, \tag{7.53}$$

where the factor 2_p takes into account that there are two allowed polarizations per $\mathbf{k}$. The second question is answered by calculating the volume in $\mathbf{k}$-space corresponding to the frequency interval $d\omega$. We use the fact that $d\mathbf{k} = 4\pi k^2 dk$ and then find the number of frequencies in $d\omega$ to be

$$\frac{\omega^2 d\omega}{\pi^2 c^3}. \tag{7.54}$$

7.3.6 The LC-oscillator

The above discussion concerned the Maxwell equations in vacuum. Similar equations describe electromagnetism in (artificial) media, homogeneous as well as inhomogeneous. The equations are in principle for coordinate-dependent $\mathbf{E}(\mathbf{r}, t)$ and $\mathbf{B}(\mathbf{r}, t)$. It is known that many practical problems of electromagnetism are solved with electric circuit theory: a finite-element approach to the Maxwell equations. Within the approach, the coordinate dependence of fields is abstracted out. Instead, the theory is formulated in terms of voltage drops and currents ascribed to each element. While it is not a goal of this book to review circuit theory, we give one example of a simple electromagnetic oscillator, and show how it fits in our picture of the Maxwell equations.

The oscillator we consider consists of a capacitor with capacitance C and an inductor with inductance L, as illustrated in Fig. 7.3. This is a circuit without dissipation. The energy,

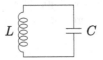

A simple electromagnetic oscillator: a capacitor with capacitance C connected to an inductor with inductance L.

once stored in either capacitor or inductor, stays in the circuit although it oscillates between the elements. There are two variables characterizing the circuit: the voltage difference V and the current I. They are connected in such a way that both current and voltage are the same for both elements. The energy reads

$$H = \frac{1}{2}CV^2 + \frac{1}{2}LI^2. \tag{7.55}$$

The dynamics of the circuit are given by

$$\dot{q} = I = C\dot{V} \quad \text{and} \quad \dot{I} = -V/L. \tag{7.56}$$

The first equation describes the current through the capacitor, the second equation is the inductive voltage produced by the current change in the inductor. The solution of these equations is an oscillation with frequency

$$\omega_0 = \frac{1}{\sqrt{LC}}. \tag{7.57}$$

Control question. Do electrons actually traverse the capacitor? What type of current is there in the capacitor and which Maxwell equation is the most relevant to describe this situation?

Let us try to derive the generalized momentum and coordinate for this oscillator, similarly as for the electromagnetic field in vacuum. We do it as follows. In terms of the generalized momentum and coordinate, the energy of the oscillator should read

$$H = \frac{1}{2}(P^2 + \omega_0^2 Q^2). \tag{7.58}$$

This suggests that the momentum is proportional to the current and the coordinate to the voltage. If we compare (7.58) and (7.55), we see that the right substitution is

$$P = \sqrt{L}I \quad \text{and} \quad Q = C\sqrt{L}V. \tag{7.59}$$

Reverting to (7.51), we then obtain the normal coordinate

$$d = \frac{1}{\sqrt{2\omega_0}}\left(i\sqrt{L}I + \sqrt{C}V\right). \tag{7.60}$$

We could go even further with this circuit theory and represent each oscillator mode of the electromagnetic field in vacuum with an LC-oscillator. In that case, the effective inductance of these oscillators depends on $\mathbf{k}$,

$$C_k = \varepsilon_0 \sqrt[3]{\mathcal{V}} \quad \text{and} \quad L_k = \frac{\mu_0}{k^2 \sqrt[3]{\mathcal{V}}}. \tag{7.61}$$

This simple example shows that oscillators in artificial electric circuits are quite similar to the oscillators representing the electromagnetic waves in vacuum.

Table 7.1 Summary: Classical fields

Generalized coordinate and momentum

dynamics of fields obey usual linear wave equations, e.g.

discrete 1D chain: $\qquad \dfrac{d^2 u_n(t)}{dt^2} = k\{u_{n+1}(t) - 2u_n(t) + u_{n-1}(t)\}$

continuous 1D chain: $\qquad \dfrac{d^2 u(x,t)}{dt^2} = \dfrac{K}{\rho}\dfrac{d^2 u(x,t)}{dx^2}$

general solution: $\qquad u(x,t) = \sum_k \left\{ u_k e^{i(kx-\omega_k t)} + u_k^* e^{-i(kx-\omega_k t)} \right\}$

Hamiltonian: $\qquad$ can be written $H = \sum_k \frac{1}{2}(P_k^2 + \omega_k^2 Q_k^2) = \sum_k \omega_k d_k^* d_k$

$\qquad\qquad$ with $Q_k = \dfrac{1}{\sqrt{2\omega_k}}(d_k + d_k^*)$ and $P_k = -i\sqrt{\dfrac{\omega_k}{2}}(d_k - d_k^*)$, where $d_k \propto u_k$

Q_k and P_k: $\qquad$ *generalized* coordinate and momentum

Classical electromagnetic field: electric and magnetic field

Maxwell equations: $\quad \nabla \times \mathbf{E} = -\dfrac{\partial \mathbf{B}}{\partial t}, \nabla \times \mathbf{B} = \mu_0\left(\mathbf{j} + \varepsilon_0 \dfrac{\partial \mathbf{E}}{\partial t}\right), \nabla \cdot \mathbf{E} = \dfrac{\rho}{\varepsilon_0}$, and $\nabla \cdot \mathbf{B} = 0$

e.m. force: $\qquad \mathbf{F} = q[\mathbf{E} + (\mathbf{v} \times \mathbf{B})]$

energy flux: $\qquad \mathbf{S} = \frac{1}{\mu_0}\mathbf{E} \times \mathbf{B}$

energy density: $\qquad U = \frac{1}{2}\varepsilon_0(\mathbf{E}^2 + c^2\mathbf{B}^2)$

Classical electromagnetic field: vector and scalar potential

introduce: $\qquad \mathbf{B} = \nabla \times \mathbf{A}$ and $\mathbf{E} = -\nabla\varphi - \dfrac{\partial \mathbf{A}}{\partial t}$

Maxwell equations: $\quad \nabla^2\varphi + \dfrac{\partial}{\partial t}(\nabla \cdot \mathbf{A}) = -\dfrac{\rho}{\varepsilon_0}, \nabla^2\mathbf{A} - \dfrac{1}{c^2}\dfrac{\partial^2 \mathbf{A}}{\partial t^2} - \nabla\left(\nabla \cdot \mathbf{A} + \dfrac{1}{c^2}\dfrac{\partial\varphi}{\partial t}\right) = -\mu_0\mathbf{j}$

gauge transformation: $\mathbf{A} \to \mathbf{A} + \nabla\chi$ and $\varphi \to \varphi - \dfrac{\partial\chi}{\partial t}$: same e.m. field $\to \chi$ free to choose

examples of gauges: $\quad \nabla \cdot \mathbf{A} + \dfrac{1}{c^2}\dfrac{\partial\varphi}{\partial t} = 0$: Lorenz gauge, Lorentz covariant

$\qquad\qquad\qquad \nabla \cdot \mathbf{A} = 0$: Coulomb gauge, scalar potential not retarded

Classical electromagnetic field in vacuum

Maxwell equations: in Coulomb gauge: $\nabla^2\mathbf{A} - \dfrac{1}{c^2}\dfrac{\partial^2 \mathbf{A}}{\partial t^2} = 0 \to$ wave equation

general solution: $\quad \mathbf{A}(\mathbf{r},t) = \sum_{\mathbf{k},\alpha} \dfrac{1}{\sqrt{2\varepsilon_0\omega_\mathbf{k}\mathcal{V}}}\left[d_{\mathbf{k}\alpha}\mathbf{e}_\alpha e^{i(\mathbf{k}\cdot\mathbf{r}-\omega_\mathbf{k} t)} + \text{c.c.}\right]$

$\qquad\qquad\qquad$ with $\mathbf{e}_\alpha$ polarization vector (two directions) and $\omega_\mathbf{k} = ck$

introduce: $\qquad Q_{\mathbf{k}\alpha} = \dfrac{1}{\sqrt{2\omega_\mathbf{k}}}(d_{\mathbf{k}\alpha} + d_{\mathbf{k}\alpha}^*)$ and $P_{\mathbf{k}\alpha} = -i\sqrt{\dfrac{\omega_\mathbf{k}}{2}}(d_{\mathbf{k}\alpha} - d_{\mathbf{k}\alpha}^*)$

energy flux: $\qquad \mathbf{S}_{\text{tot}} = \sum_{\mathbf{k},\alpha} c^2\mathbf{k}\, d_{\mathbf{k}\alpha}^* d_{\mathbf{k}\alpha}$

Hamiltonian: $\qquad H_{\text{e.m.}} = \int d\mathbf{r}\,\frac{1}{2}\varepsilon_0(\mathbf{E}^2 + c^2\mathbf{B}^2) = \sum_{\mathbf{k},\alpha} \omega_\mathbf{k} d_{\mathbf{k}\alpha}^* d_{\mathbf{k}\alpha} = \sum_{\mathbf{k},\alpha} \frac{1}{2}(P_{\mathbf{k}\alpha}^2 + \omega_k^2 Q_{\mathbf{k}\alpha}^2)$

Exercises

1. *Bending oscillations of a thin rod* (solution included). Let us consider a thin rod of length L that cannot be compressed but can bend. Its shape can be parameterized by a unit vector $\mathbf{n}(l)$ that gives the direction of the rod axis at the point situated at the distance l from the left end of the rod (see Fig. 7.4). The bending energy reads

$$E = \frac{C}{2} \int dl \left(\frac{d\mathbf{n}}{dl}\right)^2,$$

where C is a constant with dimension J·m. The mass per unit length is ρ. The left end is clamped and the right end is free. In its equilibrium state, the rod is straight, and lies along the x-axis. We consider small deflections of the rod in the y-direction.

a. Write down the Hamiltonian in terms of $y(x)$ and the momentum $p(x)$ assuming small deflections.

b. Derive the equation of motion for $y(x)$. Give the typical scale Ω of the oscillation frequency.

c. Solve the equation of motion taking the boundary conditions into account.
   Hint. Those read $y(0) = 0$, $y'(0) = 0$ at the clamped end and $y''(L) = 0$, $y'''(L) = 0$ at the free end.
   Show that the frequencies of the oscillations can be expressed in terms of Ω and the roots κ_n of the equation $\cosh \kappa \cos \kappa = -1$.

d. The general solution of the equation can be given in the following form:

$$y(x, t) = \sqrt{L} \sum_n \phi_n(x) \left\{ y_n \exp(-i\omega_n t) + \text{H.c.} \right\},$$

where $\phi_n(x)$ is an orthonormal set of solutions corresponding to the frequencies ω_n. (The $\phi_n(x)$ can be straightforwardly found, but their concrete expressions are rather involved, so we don't use them here.) Rewrite the Hamiltonian in the standard form

$$H = \sum_n \omega_n d_n^* d_n,$$

and find the correspondence between d_n and y_n and express $y(x, t)$ in terms of d_n.

2. *Normal modes of a monatomic three-dimensional crystal.* Let us first consider a very relevant extension of the linear chain of oscillators described in Section 7.1: traveling waves in a three-dimensional crystal. Of course, this is not a field theory in the strict

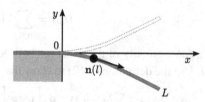

Fig. 7.4   Oscillations of a thin rod clamped at the left end.

sense, since it deals with discrete particles (lattice atoms) all having a definite position and momentum. It is however an instructive exercise, and it can be extended to the continuous case thereby connecting to elasticity theory for crystals.

When the crystal is in equilibrium, the atoms occupy equilibrium positions denoted by the lattice site vectors $\mathbf{R}$. Deviations from this equilibrium are caused by atoms being displaced from their lattice site, having positions $\mathbf{r} \equiv \mathbf{R} + \mathbf{u}(\mathbf{R})$. The vector $\mathbf{u}(\mathbf{R})$ thus gives the displacement away from its equilibrium position of the atom at lattice site $\mathbf{R}$.

a. Since the internal potential energy of the crystal must be minimal in equilibrium, we can most generally write the deformation energy in the crystal to leading order using the matrix $\Phi(\mathbf{R})$ defined as

$$\Phi_{\alpha\beta}(\mathbf{R} - \mathbf{R}') = \frac{\partial^2 U}{\partial u_\alpha(\mathbf{R}) \partial u_\beta(\mathbf{R}')}, \tag{7.62}$$

where the derivative is taken in the point $\mathbf{u}(\mathbf{R}) = \mathbf{u}(\mathbf{R}') = 0$ and $\alpha, \beta = x, y, z$. In this expression, U gives the potential energy of the crystal exactly as a function of all $\mathbf{u}(\mathbf{R})$. To this (harmonic) order, the deformation energy then reads

$$U_{\text{def}} = \frac{1}{2} \sum_{\mathbf{R}\mathbf{R}'} \mathbf{u}(\mathbf{R})\Phi(\mathbf{R} - \mathbf{R}')\mathbf{u}(\mathbf{R}').$$

Show that (i) $\Phi(\mathbf{R}) = \Phi^T(-\mathbf{R})$, (ii) $\sum_{\mathbf{R}} \Phi(\mathbf{R}) = 0$, and (iii) if the crystal has inversion symmetry $\Phi(\mathbf{R}) = \Phi(-\mathbf{R})$.

b. Write down the equation of motion for $\mathbf{u}(\mathbf{R})$ and show that the plane wave ansatz $\mathbf{u}(\mathbf{R}) = \mathbf{e} \exp\{i(\mathbf{k} \cdot \mathbf{R} - \omega t)\}$, with $\mathbf{e}$ again a vector of polarization, leads to the eigenvalue equation

$$m\omega^2 \mathbf{e} = \Phi(\mathbf{k})\mathbf{e},$$

where m is the mass of the atoms and $\Phi(\mathbf{k}) = \sum_{\mathbf{R}} \Phi(\mathbf{R})e^{-i\mathbf{k}\cdot\mathbf{R}}$ is the Fourier transform of Φ.

From properties (i) and (ii) derived in (a), we see that $\Phi(\mathbf{k})$ is symmetric and real, so that for each $\mathbf{k}$ it has three orthonormal eigenvectors $\mathbf{e}_{1,2,3}(\mathbf{k})$ with corresponding eigenvalues $\lambda_{1,2,3}(\mathbf{k})$. We thus have for each $\mathbf{k}$ three normal modes with

$$\omega_n(\mathbf{k}) = \sqrt{\frac{\lambda_n(\mathbf{k})}{m}},$$

with $n = 1, 2, 3$.

c. Show that $\Phi(\mathbf{k}) = -2 \sum_{\mathbf{R}} \Phi(\mathbf{R}) \sin^2(\frac{1}{2}\mathbf{k} \cdot \mathbf{R})$.

d. The main contribution to U_{def} comes from the interaction between neighboring atoms. We assume that the potential can be written as

$$U_{\text{def}} = \sum_{\text{n.n.}} V(|\mathbf{r}_i - \mathbf{r}_j|),$$

it is a function only of the distances between all pairs of nearest-neighbor atoms i and j. Use the definition (7.62) to show that

$$\Phi_{\alpha\beta}(\mathbf{k}) = 2 \sum_{\mathbf{R}} \sin^2(\tfrac{1}{2}\mathbf{k} \cdot \mathbf{R}) \left\{ V'(R)\left[\frac{\delta_{\alpha\beta}}{R} - \frac{R_\alpha R_\beta}{R^3}\right] + V''(R)\frac{R_\alpha R_\beta}{R^2} \right\},$$

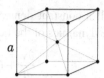

a

Fig. 7.5 The unit cell of a body-centered cubic lattice.

where the sum is over all nearest neighbors of the atom located at $\mathbf{R} = 0$. We defined $R \equiv |\mathbf{R}|$ and the derivatives of the interaction energy as $V'(R) \equiv [\partial V(r)/\partial r]_{r=R}$, etc.

e. We now consider a so-called body-centered cubic crystal with unit cell size a. In such a crystal the unit cell has a basic cubic shape, but there is an extra atom located in the center of the cube (see Fig. 7.5). Show that for waves traveling in the diagonal direction $\mathbf{k} = k\{\frac{1}{2}\sqrt{2}, \frac{1}{2}\sqrt{2}, 0\}$ there is one longitudinal mode with frequency

$$\omega_l = \sqrt{\frac{8}{3}\frac{3C_1 + 2C_2}{m}} \sin\left(\frac{ak}{2\sqrt{2}}\right),$$

and two transverse modes with frequencies

$$\omega_{t1} = \sqrt{\frac{8}{3}\frac{3C_1 + C_2}{m}} \sin\left(\frac{ak}{2\sqrt{2}}\right) \quad \text{and} \quad \omega_{t2} = 2\sqrt{\frac{2C_1}{m}} \sin\left(\frac{ak}{2\sqrt{2}}\right),$$

polarized respectively in the directions $\{0, 0, 1\}$ and $\{-1, 1, 0\}$. The constants C_1 and C_2 are defined as

$$C_1 = \frac{V'(R)}{R} \quad \text{and} \quad C_2 = V''(R) - \frac{V'(R)}{R}.$$

3. *Gravity waves in water.* We now consider the "most common" waves we can imagine – the small ripples on the surface of a liquid – and see if we can derive a wave equation describing their dynamics. We characterize the liquid by the following fields: a mass density $\rho(\mathbf{r}, t)$, a pressure field $p(\mathbf{r}, t)$, and a velocity field $\mathbf{v}(\mathbf{r}, t)$ describing the flow of the liquid. In the framework of this description, we assume the liquid to be *continuous* on all relevant length scales. Microscopically, this means that any infinitesimally small volume dV we consider still contains a large (macroscopic) number of particles, so that it makes sense to speak about the mass density or pressure inside this volume element.

a. Derive a continuity equation expressing the conservation of mass. Show that for an incompressible liquid, assuming that the mass density $\rho(\mathbf{r}, t) = \rho$ is constant, this equation reads $\nabla \cdot \mathbf{v} = 0$.

Hint. You will need Gauss' theorem relating the divergence of a vector field inside a volume to its flux through the surface bounding this volume,

$$\int_V (\nabla \cdot \mathbf{v}) dV = \oint_S (\mathbf{v} \cdot \mathbf{n}) dS,$$

where $\mathbf{n}$ is the unit vector in each point normal to the surface S.

b. Assuming still a constant density, use Newton's laws to derive Euler's equation describing the dynamics of the liquid,

$$\frac{\partial \mathbf{v}}{\partial t} + (\mathbf{v} \cdot \nabla)\mathbf{v} = -\frac{1}{\rho}\nabla p + \mathbf{g}, \tag{7.63}$$

where $\mathbf{g}$ is the gravitational acceleration.

c. Give the solution for the pressure field p when the liquid is at rest.

d. We assume that the fluid is rotation-free so that $\nabla \times \mathbf{v} = 0$ everywhere (see Section 6.5.3). This implies that the velocity field can be written as the derivative of some scalar potential, $\mathbf{v} = \nabla\varphi$.

When focusing on small ripples, i.e. small deviations from the equilibrium state $\mathbf{v} = 0$, we can neglect the term in (7.63) which is quadratic in $\mathbf{v}$. Consider the dynamics *at* the surface of the liquid, and rewrite (7.63) as a wave equation for φ at the surface.

e. Let us forget about boundary conditions: we assume the liquid to be infinitely deep and its surface to be infinitely large, and we try to find a plane wave solution propagating in the x-direction. A good guess is a solution of the form

$$\varphi(\mathbf{r}, t) = f(z)\cos(kx - \omega t).$$

Determine $f(z)$, give the velocity profile $\mathbf{v}(\mathbf{r}, t)$, derive the dispersion relation of the waves $\omega(k)$, and give the velocity of the propagation of the waves $v_k = \partial\omega/\partial k$.

4. *Oscillations of magnetization.* In Chapter 4 we have shown how the interactions between electrons can cause a metal to have a ferromagnetic ground state. We then investigated the elementary excitations in a ferromagnet, and found low-energy excitations corresponding to slow spatial variations of the magnetization. In this exercise, we address the spatial oscillations of the magnetization in a more general way, and derive a classical Hamiltonian describing these oscillations.

a. We characterize the magnetization of a ferromagnet with the field $\mathbf{m}(\mathbf{r})$. Let us first focus on a small region in the magnet with a uniform magnetization $\mathbf{m}$, which we assume to be of unit magnitude $|\mathbf{m}| = 1$. In the presence of a magnetic field $B_{\text{ext}} = B\hat{z}$ the energy of this magnetization is $E = -Bm_z$. Express this energy for small deviations from $\mathbf{m} = \hat{z}$ in terms of m_x and m_y.

b. Since the magnetization of the magnet can be associated with the local orientation of the electron spins, the equations of motion for $\mathbf{m}$ in B_{ext} read

$$\dot{m}_x = Bm_y \quad \text{and} \quad \dot{m}_y = -Bm_x.$$

Identify the above components of the magnetization with a generalized coordinate Q and momentum P, and derive a Hamiltonian from the energy found at (a).

Also, in the absence of a magnetic field, the energy of the ferromagnet depends on $\mathbf{m}(\mathbf{r})$. The lowest energy corresponds to a uniform magnetization. If the magnetization is not perfectly uniform but varies slightly as a function of coordinate, there must be a small increase in energy in the magnet. This implies that the local energy density due to the non-uniformity of the magnetization must be some function involving the local spatial derivatives of $\mathbf{m}$. A constraint on this function is that it must be invariant under rotations of **m** with respect to the coordinate system: rotating the magnetization field as a whole should not make a difference to the energy stored in it due to local oscillations of this field. With this constraint, we find that the simplest expression (involving the smallest number of derivatives) one could write for the energy, is given by

$$E = C \sum_{\alpha,\beta} \int d\mathbf{r} \, \frac{\partial m_\alpha}{\partial r_\beta} \frac{\partial m_\alpha}{\partial r_\beta},$$

where C is some constant and $\alpha, \beta \in \{x, y, z\}$ label the coordinates.

c. Write the corresponding Hamiltonian in terms of the generalized coordinate and momentum found at (b), still assuming that $|\mathbf{m}(\mathbf{r})| = 1$. Simplify this Hamiltonian assuming that the deviations from the uniform magnetization $\mathbf{m}(\mathbf{r}) = \hat{z}$ are small.

d. Let us introduce $m^\pm \equiv m_x \pm im_y$ and their Fourier components,

$$m^+(\mathbf{r}) = \frac{1}{\sqrt{\mathcal{V}}} \sum_{\mathbf{k}} e^{-i\mathbf{k}\cdot\mathbf{r}} m_{\mathbf{k}} \quad \text{and} \quad m^-(\mathbf{r}) = \frac{1}{\sqrt{\mathcal{V}}} \sum_{\mathbf{k}} e^{i\mathbf{k}\cdot\mathbf{r}} m_{\mathbf{k}}^*.$$

Show that this allows the Hamiltonian to be written in terms of the modes **k** as

$$H = \frac{C}{\mathcal{V}} \sum_{\mathbf{k}} k^2 m_{\mathbf{k}}^* m_{\mathbf{k}}.$$

Solutions

1. *Bending oscillations of a thin rod.*

a. For small deflections we have $l \approx x$, $n_x \approx 1$, and $n_y \approx dy/dx$. Therefore, the momentum is in the y-direction, $p_y \equiv p(x)$ being the relevant momentum density. The Hamiltonian thus reads

$$H = \frac{1}{2} \int dx \left(\frac{p^2}{\rho} + C(y'')^2 \right).$$

b. Varying the Hamiltonian and excluding $p(x)$, we obtain

$$\ddot{y} + \frac{C}{\rho} y^{IV} = 0,$$

the superscript Roman IV standing for the fourth derivative. For an estimation of the lowest frequency, we assume that $y(x)$ changes at a typical scale L. From this, $y^{IV} \simeq L^{-4} y$ and $\Omega = (1/L^2)\sqrt{C/\rho}$.

c. If we substitute $y(x) = \exp\{i(kx - \omega t)\}$, we find the dispersion relation $\omega^2 = (C/\rho)k^4$. For given ω, there are four values of k satisfying this equation, $k = \pm\sqrt{\omega}(C/\rho)^{-1/4}$ or $k = \pm i\sqrt{\omega}(C/\rho)^{-1/4}$. In other terms, if we define k as a positive real number, then there are four independent solutions of the equation: $\sin(kx)$,

$\cos(kx)$, $\sinh(kx)$, and $\cosh(kx)$. Let us define a set of four equivalent convenient functions,

$$F_1(x) = \sin(kx) - \sinh(kx),$$
$$F_2(x) = F_1'/k = \cos(kx) - \cosh(kx),$$
$$F_3(x) = F_2'/k = -\sin(kx) - \sinh(kx),$$
$$F_4(x) = F_3'/k = -\cos(kx) - \cosh(kx),$$

with $F_1 = F_4'/k$. The functions $F_{1,2}(x)$ satisfy the boundary conditions at the left end, $y(0) = 0$, and $y'(0) = 0$. We thus seek for a solution in the form $y(x) = AF_1(x) + BF_2(x)$. The boundary conditions at the free end take the form

$$0 = \begin{pmatrix} y''(L)/k^2 \\ y'''(L)/k^3 \end{pmatrix} = \begin{pmatrix} F_3(L) & F_4(L) \\ F_4(L) & F_1(L) \end{pmatrix} \begin{pmatrix} A \\ B \end{pmatrix}.$$

There is a solution for A and B provided the determinant of the 2×2 matrix is zero, that is, $F_1(L)F_3(L) = F_4^2(L)$. The latter equation reduces to $\cos(kL)\cosh(kL) = -1$. The roots of this equation give an infinite series of discrete values of k. In distinction from the elementary examples considered in the chapter, these values are not equidistant. They are approximately equidistant at $kL \gg 1$ when the equation is reduced to $\cos(kL) = 0$. Introducing a dimensionless $\kappa = kL$, we find that the frequencies of the oscillations are given by

$$\frac{\omega_n}{\Omega} = \kappa_n^2 \quad \text{where} \quad \cos(\kappa_n)\cosh(\kappa_n) = -1.$$

d. For this task, we need to substitute the expression for $y(x, t)$ and the momentum density $p(x, t) = \rho \dot{y}(x, t)$ into the Hamiltonian. When doing so, it is instructive to note that

$$\int dx (y'')^2 = \int dx \, y \, y^{IV} = (\rho/C) \int dx \, y \, \ddot{y}.$$

Using the orthonormality of $\phi(x)$, we arrive at

$$H = 2\rho L \sum \omega_n^2 y_n^* y_n.$$

From this, $d_n = \sqrt{2\rho L \omega_n} y_n$. With this, we can re-express $y(t, x)$ in terms of d_n.

In this chapter we explain the procedure for quantizing classical fields, based on the preparatory work we did in the previous chapter. We have seen that classical fields can generally be treated as an (infinite) set of oscillator modes. In fact, for all fields we considered in the previous chapter, we were able to rewrite the Hamiltonian in a way such that it looked identical to the Hamiltonian for a generic (mechanical) oscillator, or a set of such oscillators. We therefore understand that, if we can find an unambiguous way to quantize the mechanical oscillator, we get the quantization of all classical fields for free!

We thus start considering the mechanical harmonic oscillator and refresh the quantum mechanics of this system. The new idea is to associate the oscillator with a level for bosonic particles representing the quantized excitations of the oscillator. We then turn to a concrete example and show how the idea can be used to quantize the deformation field of the elastic string discussed in Section 7.2. In this case, the bosonic particles describing the excitations of the field are called phonons: propagating wave-like deformations of the string. As a next example, we revisit the low-lying spinfull excitations in a ferromagnet, already described in Section 4.5.3. Here, we start from a classical field theory for the magnetization of the magnet, and show how the quantization procedure leads to the same bosonic magnons we found in Chapter 4. Finally, we implement the quantization technique for the electromagnetic field, leading us to the concept of photons – the quanta of the electromagnetic field.

In the last part of the chapter, we treat the quantized electromagnetic field in more detail. We write down field operators for the electric and magnetic fields in terms of the photonic creation and annihilation operators. This then allows us to investigate several interesting quantum mechanical properties of the fields: the uncertainty relations between the electric and magnetic fields, the zero-point energy associated with the vacuum, and fluctuations of the field in vacuum. Finally, to complete the connection with the previous chapter, we quantize the simplest element of a circuit-theoretical description of the electromagnetic field: the LC-circuit discussed in Section 7.3.6. The content of the chapter is summarized in Table 8.1.

8.1 Quantization of the mechanical oscillator

Let us review and refresh the properties of a system that plays a central role in both classical and elementary quantum mechanics: the mechanical harmonic oscillator, already discussed in Section 1.9. To focus on the essentials, we consider a particle of mass $m = 1$ moving in a parabolic potential $\frac{1}{2}\omega^2 Q^2$, where Q is the coordinate of the particle.

On a classical level, the harmonic oscillator can be described in a variety of ways. We choose a way which has the best correspondence with the quantum description, so we write its Hamiltonian,

$$H = \frac{1}{2}(P^2 + \omega^2 Q^2), \tag{8.1}$$

where P and Q are numbers representing the momentum and the coordinate of the particle. By definition of the Hamiltonian function, the equations of motion for these variables read

$$\dot{Q} = \frac{\partial H}{\partial P} \quad \text{and} \quad \dot{P} = -\frac{\partial H}{\partial Q}, \tag{8.2}$$

which for our concrete case reduce to

$$\dot{Q} = P \quad \text{and} \quad \dot{P} = -\omega^2 Q. \tag{8.3}$$

We easily verify that these two coupled differential equations have a straightforward solution: they describe harmonic oscillations with frequency ω of P and Q around their equilibrium values $P = Q = 0$.

Suppose for now that we did not know how to write the correct quantum mechanical Hamiltonian for the oscillator. Naively, the fastest way to get to a quantum description of the oscillator would be to add "hats" to the variables making them operators,

$$\hat{H} = \frac{1}{2}(\hat{P}^2 + \omega^2 \hat{Q}^2). \tag{8.4}$$

This is, however, not enough, and we need something else. To understand what we need, let us turn to the Heisenberg picture. In this picture, the evolution equations for operators are formed by commuting the operators with the Hamiltonian operator,

$$i\hbar\dot{\hat{Q}} = [\hat{Q}, \hat{H}] \quad \text{and} \quad i\hbar\dot{\hat{P}} = [\hat{P}, \hat{H}]. \tag{8.5}$$

Since the Hamiltonian contains the operators $\hat{P}$ and $\hat{Q}$, we need to specify a rule for the commutation of the two. Let us assume that they obey the *canonical commutation relation*

$$[\hat{Q}, \hat{P}] = i\hbar. \tag{8.6}$$

This assumption has the status of a hypothesis at the moment. To check our hypothesis, we evaluate (8.5) using the above commutation rule. We obtain

$$\dot{\hat{Q}} = \hat{P} \quad \text{and} \quad \dot{\hat{P}} = -\omega^2 \hat{Q}. \tag{8.7}$$

These quantum equations look precisely like the classical equations (8.3). We thus achieve correspondence with classical mechanics and are sure that the hypothesis – the quantization postulate – makes sense.

The Heisenberg picture is sometimes not very convenient for practical calculations since it does not specify a wave function. Let us thus turn to the Schrödinger picture and write the

Schrödinger equation in the coordinate representation. The operators in this representation become

$$\hat{Q} \to q \quad \text{and} \quad \hat{P} \to -i\hbar\frac{\partial}{\partial q}, \tag{8.8}$$

obviously satisfying the quantization postulate. The Schrödinger equation for a stationary wave function $\psi(q)$ thus reads

$$E\psi = \hat{H}\psi = \left(-\frac{\hbar^2}{2}\frac{\partial^2}{\partial q^2} + \frac{\omega^2 q^2}{2}\right)\psi(q). \tag{8.9}$$

In Section 1.9 we have shown that the eigenenergies of this equation form a discrete equidistant spectrum $E_n = \hbar\omega(n + \frac{1}{2})$ while the eigenfunctions can be expressed in terms of Hermite polynomials.

For the purposes of this chapter it is important that we consider the equation in a more abstract but more comprehensive way. Let us again express the Hamiltonian in terms of the auxiliary operator $\hat{a}$ defined as

$$\hat{a} = \frac{\omega\hat{Q} + i\hat{P}}{\sqrt{2\hbar\omega}} \quad \text{and} \quad \hat{a}^\dagger = \frac{\omega\hat{Q} - i\hat{P}}{\sqrt{2\hbar\omega}}. \tag{8.10}$$

As already mentioned in Section 1.9 the operators $\hat{a}$ and $\hat{a}^\dagger$ obey the commutation relations

$$[\hat{a}, \hat{a}^\dagger] = 1 \quad \text{and} \quad [\hat{a}, \hat{a}] = [\hat{a}^\dagger, \hat{a}^\dagger] = 0. \tag{8.11}$$

In the framework of the elementary quantum mechanics of Chapter 1 this implied nothing: writing the Hamiltonian in terms of the operators $\hat{a}^{(\dagger)}$ merely served as a convenient mathematical tool of simplification of the problem. But now, being experienced in second quantization, we get a jolt: we recognize *bosons*! Indeed, we have already found that we can write the Hamiltonian as

$$\hat{H} = \hbar\omega(\hat{a}^\dagger\hat{a} + \frac{1}{2}) = \hbar\omega(\hat{n} + \frac{1}{2}), \tag{8.12}$$

having the eigenenergies $E_n = \hbar\omega(n + \frac{1}{2})$. The operator $\hat{n}$ is thus nothing but the number operator for the bosons described by the operators $\hat{a}^{(\dagger)}$. The harmonic oscillator is in all respects identical to a single level with energy $\hbar\omega$ that can host any number of boson particles. All the states of the oscillator are states with different integer number of particles, and all possible energies of the oscillator are sums of single-particle energies.

8.1.1 Oscillator and oscillators

What is valid for a single oscillator, must be valid for an (infinite) set of oscillators as well. Let us just repeat the above reasoning for an oscillator set labeled by k. A mechanical analogy consists of particles indexed by k with mass $m = 1$ and coordinates Q_k and momenta P_k moving in corresponding parabolic potentials $\frac{1}{2}\omega_k^2 Q_k^2$ (Fig. 8.1). The Hamiltonian function for the set reads

$$H = \frac{1}{2}\sum_k (P_k^2 + \omega_k^2 Q_k^2), \tag{8.13}$$

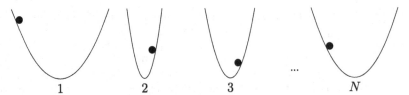

1   2   3   ...   N

Fig. 8.1   A set of N uncoupled oscillators, each with a distinct eigenfrequency ω_k.

and the evolution equations are independent for each oscillator of the set,

$$\dot{Q}_k = \frac{\partial H}{\partial P_k} = P_k \quad \text{and} \quad \dot{P}_k = -\frac{\partial H}{\partial Q_k} = -\omega_k^2 Q_k. \tag{8.14}$$

Let us now quantize the set. We start again by naively assigning hats to the variables in the Hamiltonian,

$$\hat{H} = \frac{1}{2} \sum_k (\hat{P}_k^2 + \omega_k^2 \hat{Q}_k^2). \tag{8.15}$$

We then need to specify the commutation rules. It is natural to assume that operators corresponding to independent oscillators do commute, and we thus guess the canonical commutation relations to read

$$[\hat{Q}_k, \hat{P}_{k'}] = i\hbar \delta_{kk'} \quad \text{and} \quad [\hat{Q}_k, \hat{Q}_{k'}] = [\hat{P}_k, \hat{P}_{k'}] = 0. \tag{8.16}$$

Our guess seems to be right again: the quantum evolution equations obtained by commuting the operators with the Hamiltonian are precisely like the classical ones. The next step is to introduce the operators

$$\hat{a}_k = \frac{\omega_k \hat{Q}_k + i\hat{P}_k}{\sqrt{2\hbar\omega_k}} \quad \text{and} \quad \hat{a}_k^\dagger = \frac{\omega_k \hat{Q}_k - i\hat{P}_k}{\sqrt{2\hbar\omega_k}}, \tag{8.17}$$

that again satisfy the commutation relations for bosonic CAPs,

$$[\hat{a}_k, \hat{a}_{k'}^\dagger] = \delta_{kk'} \quad \text{and} \quad [\hat{a}_k, \hat{a}_{k'}] = [\hat{a}_k^\dagger, \hat{a}_{k'}^\dagger] = 0. \tag{8.18}$$

This means that the set of oscillators can be regarded as a set of levels for non-interacting bosons.

The Hamiltonian in terms of the CAPs reads

$$\hat{H} = \sum_k \hbar\omega_k(\hat{a}_k^\dagger \hat{a}_k + \tfrac{1}{2}) = \sum_k \hbar\omega_k(\hat{n}_k + \tfrac{1}{2}). \tag{8.19}$$

The eigenstates of the full Hamiltonian can thus be represented by a set of numbers $\{n_k\}$, giving the occupation numbers of all levels indexed with k. There is one special state, the vacuum, which is defined as

$$\hat{a}_k|0\rangle = 0, \quad \text{for any } k. \tag{8.20}$$

All other states can be obtained from the vacuum by applying the proper sequence of creation operators,

$$|\{n_k\}\rangle = \prod_k \frac{(\hat{a}^\dagger)^{n_k}}{\sqrt{n_k!}}|0\rangle, \tag{8.21}$$

and the energy corresponding to a certain eigenstate $\{n_k\}$ of the Hamiltonian is the sum of the single-particle energies $\hbar\omega_k$,

$$E_{\{n_k\}} = \sum_k \hbar\omega_k(n_k + \tfrac{1}{2}). \tag{8.22}$$

With this, our task is completed. We already knew how to represent a classical field by a set of oscillators, and now we have learned how to quantize this set of oscillators. We thus have all tools at hand to quantize any field and describe its excitations in terms of bosonic particles.

This also complements our activities in the first chapters. There, we started from the concept of particles, we used the symmetry postulate to figure out the Fock space of all possible particle configurations, we introduced CAPs in this space, and we finally managed to relate these operators to *fields* describing the particles. We actually found that in the case of bosons these fields also have classical analogues. The quest of the previous and the present chapters was exactly the other way around. We started with classical fields, presented them as a set of oscillators, we introduced bosonic CAPs describing the excitations of the fields, and then we are there: we ended up with particles. So now we have mastered a unified description of particles and fields from *both sides*.

8.2 The elastic string: phonons

Let us now illustrate the procedure outlined above, and give several examples of the quantization of a classical field theory. We start implementing these last steps of quantization for the small oscillations in the elastic string. We did all the preparatory work in Section 7.2, and in fact we are almost done. As we see from (7.18), we have already brought the Hamiltonian into the form of (8.13), and we thus know how to define bosonic CAPs, which excite and de-excite the oscillator modes in the system. Using the definition (7.17) of the coordinates Q_k and momenta P_k for this case, we straightforwardly derive the Hamiltonian describing the excitations

$$\hat{H} = \sum_k \hbar\omega_k(\hat{a}_k^\dagger \hat{a}_k + \tfrac{1}{2}), \tag{8.23}$$

where we use $\hat{a}_k = \hat{d}_k/\sqrt{\hbar}$ and $\hat{a}_k^\dagger = \hat{d}_k^\dagger/\sqrt{\hbar}$, which we can express in terms of the Fourier components of the displacement field as

$$\hat{a}_k = \sqrt{\frac{2\rho L\omega_k}{\hbar}}\hat{u}_k \quad \text{and} \quad \hat{a}_k^\dagger = \sqrt{\frac{2\rho L\omega_k}{\hbar}}\hat{u}_k^\dagger. \tag{8.24}$$

The energies of the bosonic levels k are given by the phonon dispersion relation $\hbar\omega_k = \hbar k v$, where $v = \sqrt{K/\rho}$ is the group velocity of the propagating waves (see Section 7.2.2).

The creation operator $\hat{a}_k^\dagger$ thus excites plane wave modes in the string with wave number k, and the annihilation operator $\hat{a}_k$ de-excites the same modes. To see this even more explicitly, let us write down the resulting expression for the position operator $\hat{u}(x,t)$,

$$\hat{u}(x,t) = \sum_k \sqrt{\frac{\hbar}{2\rho L \omega_k}} \left\{ \hat{a}_k(t) e^{ikx} + \hat{a}_k^\dagger(t) e^{-ikx} \right\}, \tag{8.25}$$

where the time-dependence $e^{\pm i\omega_k t}$ has been absorbed into the CAPs. We see that the CAPs indeed can be directly identified with plane waves traveling through the wire. These quantized bosonic excitations created and annihilated by $\hat{a}^\dagger$ and $\hat{a}$ are called *phonons*.

We can investigate some properties of these phonons in more detail. Let us, as a simple exercise, create a state with one phonon present, $|1_k\rangle = \hat{a}_k^\dagger|0\rangle$, and see if we can find the expected displacement field of the line in this state. We thus calculate

$$\langle 1_k | \hat{u}(x,t) | 1_k \rangle = \langle 0 | \hat{a}_k \hat{u}(x,t) \hat{a}_k^\dagger | 0 \rangle = \langle 0 | \hat{a}_k \sum_{k'} \sqrt{\frac{\hbar}{2\rho L \omega_{k'}}} \left\{ \hat{a}_{k'} e^{ikx} + \hat{a}_{k'}^\dagger e^{-ikx} \right\} \hat{a}_k^\dagger | 0 \rangle.$$
$$\tag{8.26}$$

But since $\langle 0 | \hat{a}_k \hat{a}_{k'} \hat{a}_k^\dagger | 0 \rangle = \langle 0 | \hat{a}_k \hat{a}_{k'}^\dagger \hat{a}_k^\dagger | 0 \rangle = 0$ the expectation value of $\hat{u}(x,t)$ is zero everywhere! Oops. That does not really look like the traveling wave we imagined to be created by $\hat{a}_k^\dagger$. What did we do wrong?

In fact we did not do anything wrong: the above expectation value is indeed zero. The reason is that states created by the bosonic CAPs, having well defined numbers of phonons in all modes, are eigenstates of the Hamiltonian, with accordingly a well defined energy. In quantum mechanics, energy eigenstates often bear not very much resemblance to classical states. Usually one needs a significant uncertainty in the energy of a state before it starts behaving like a classical state. In fact, there are special quantum states which optimally resemble their classical analogues. They are called *coherent states* and are discussed in detail in Chapter 10.

We can illustrate this for the case of phonons in the elastic string (Fig. 8.2). Let us naively introduce uncertainty in the energy by creating the state $|s\rangle = \frac{1}{\sqrt{2}}\{|1_k\rangle + |2_k\rangle\}$, i.e. an equal superposition of the states with one and two phonons present in the mode k. We again calculate the expectation value of $\hat{u}(x,t)$,

$$\langle s | \hat{u}(x,t) | s \rangle = \frac{1}{2} \left\{ \langle 1_k | + \langle 2_k | \right\} \hat{u}(x,t) \left\{ |1_k\rangle + |2_k\rangle \right\} = 2\sqrt{\frac{\hbar}{\rho L \omega_k}} \cos(kx - \omega_k t), \tag{8.27}$$

which indeed looks much more like a classical traveling wave.

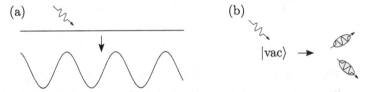

Fig. 8.2   Phonons in an elastic string. (a) After the string has been excited from outside (we for instance hit it with our hands), the atoms in the string perform wave-like oscillations. (b) Another way of viewing it is saying that in the initial vacuum (the string in equilibrium) particles, *phonons*, have been created.

A remaining question is what this tells us about the state $|1_k\rangle$. Does it mean that a string with exactly one phonon in it does not move at all? Obviously, the answer is no. The explanation is that the creation operator $\hat{a}_k^\dagger$ creates a traveling wave with a well-defined wave number k and energy ω_k, but with an undefined phase. In classical terms, displacement fields proportional to $\cos(kx - \omega_k t + \alpha)$ are allowed for any α. When calculating the expectation value of $\hat{u}(x, t)$ all possible phases are averaged and the result is of course zero. To show that the state $|1_k\rangle$ indeed does not correspond to no displacement at all, we can calculate the expectation value of $[\hat{u}(x, t)]^2$, where the exact phase α should play no role. Indeed, a straightforward evaluation yields

$$\langle 1_k |[\hat{u}(x, t)]^2|1_k\rangle = \frac{\hbar}{2\rho L \omega_k}, \tag{8.28}$$

and not zero as we feared.

8.3 Fluctuations of magnetization: magnons

In Section 4.5 we discussed several possibilities to create excitations in ferromagnets. Apart from the obvious possibility to simply inject extra electrons or holes into the magnet, we also investigated excitations built out of electron–hole pairs. We found that one can create separate electron–hole excitations carrying arbitrarily small energy and momentum as long as the excitations are *spinless*. Indeed, when both the electron and hole exist in the same band (the ↑-band or ↓-band), then the energy splitting between the two bands does not play a role and one can simply promote an electron from just below the Fermi level to just above it. However, as soon as we tried to create *spin-carrying* electron–hole excitations (by adding a hole to one band and an electron to the other), we found that due to the exchange splitting it was impossible to have electron–hole excitations with both small energy and momentum (recall Fig. 4.5(c) and 4.5(d)).

The simplest spinfull low-lying excitations turned out to be the Goldstone modes which are present due to breaking of the rotational symmetry by **P**, the polarization. The excitations consist of superpositions of many spin-carrying electron–hole excitations, the resulting state having a polarization which varies slowly in space. We called those excitations *magnons*, and in section 4.5.3 we derived the corresponding magnon creation and annihilation operators and the energy spectrum of the magnons. In this section we show how we can arrive at the same magnons, now starting from a quasi-classical description of the magnetization of the magnet.

Let us consider a ferromagnet which has a finite but non-uniform polarization. In classical language, we would say that the magnet has a coordinate dependent magnetization $\mathbf{m}(\mathbf{r})$. The magnitude of this vector $|\mathbf{m}(\mathbf{r})|$ is a measure of the local degree of polarization and its direction gives the direction along which the polarization is built up. For simplicity we focus on the case of a uniform degree of polarization $|\mathbf{m}(\mathbf{r})| = 1$, but allow for small fluctuations in the *direction* of $\mathbf{m}(\mathbf{r})$.

In Exercise 4 of the previous chapter, we derived that in this case the (classical) Hamiltonian describing small deviations from the equilibrium magnetization has the form

$$H = C \sum_{\mathbf{k}} k^2 m_{\mathbf{k}}^* m_{\mathbf{k}}, \tag{8.29}$$

where $m_{\mathbf{k}}$ denotes the Fourier transform of $m^+(\mathbf{r}) = m_x(\mathbf{r}) + i m_y(\mathbf{r})$. With an appropriate definition of generalized coordinates and momenta $Q_{\mathbf{k}}$ and $P_{\mathbf{k}}$ we can rewrite this Hamiltonian again in the form (8.13).

Control question. Can you express $Q_{\mathbf{k}}$ and $P_{\mathbf{k}}$ in terms of $m_{\mathbf{k}}$?

As above for the case of phonons, we can thus write

$$\hat{H} = \sum_{\mathbf{k}} \hbar \omega_{\mathbf{k}} (\hat{a}_{\mathbf{k}}^\dagger \hat{a}_{\mathbf{k}} + \tfrac{1}{2}), \tag{8.30}$$

if we define

$$\hat{a}_{\mathbf{k}} = \frac{1}{\sqrt{\hbar}} \hat{m}_{\mathbf{k}}, \quad \hat{a}_{\mathbf{k}}^\dagger = \frac{1}{\sqrt{\hbar}} \hat{m}_{\mathbf{k}}^\dagger, \quad \text{and} \quad \omega_{\mathbf{k}} = C k^2. \tag{8.31}$$

We see that $\hat{a}_{\mathbf{k}}^\dagger$ can be written in terms of a "magnetization operator" as

$$\hat{a}_{\mathbf{k}}^\dagger = \frac{1}{\sqrt{\hbar}} \int d\mathbf{r}\, \hat{m}^-(\mathbf{r}) e^{-i\mathbf{k}\cdot\mathbf{r}}, \tag{8.32}$$

for unit volume. Let us now compare the expression found here with the one we derived in Section 4.5.3 for the magnon creation operator $\hat{m}_{\mathbf{k}}^\dagger$. This operator was shown to read in terms of real-space electronic field operators

$$\hat{m}_{\mathbf{k}}^\dagger \propto \int d\mathbf{r}\, e^{-i\mathbf{k}\cdot\mathbf{r}} \hat{c}_\downarrow^\dagger(\mathbf{r}) \hat{c}_\uparrow(\mathbf{r}), \tag{8.33}$$

see Eq. (4.36). If we now compare the two expressions, we see that they are the same if we identify $\hat{m}^-(\mathbf{r}) \leftrightarrow \hat{c}_\downarrow^\dagger(\mathbf{r})\hat{c}_\uparrow(\mathbf{r})$. In terms of electronic spin density operators, we find

$$\hat{m}^-(\mathbf{r}) = \hat{m}_x(\mathbf{r}) - i\hat{m}_y(\mathbf{r}) \quad \leftrightarrow \quad \hat{c}_\downarrow^\dagger(\mathbf{r})\hat{c}_\uparrow(\mathbf{r}) = \hat{s}_x(\mathbf{r}) - i\hat{s}_y(\mathbf{r}). \tag{8.34}$$

We see that the two creation operators create the same excitation if we associate the magnetization operator with the local spin density operator, which is consistent with the idea that the manifestation of magnetism is in fact due to the alignment of the spins of the electrons in a material. This means that the bosons, which can be associated with the CAPs defined in (8.31), are indeed the same magnons we have already found in Chapter 4.

Also, the spectrum of the excitations we found here is the same as the one for the magnons: it is quadratic in the wave vector, $\omega_{\mathbf{k}} \propto k^2$. The prefactor in the spectrum can of course be different. In Chapter 4 we focused on the (freely moving) conduction electrons in a metal, the arguments in the present section are more general and also hold, e.g., for ferromagnetic insulators.

8.4 Quantization of the electromagnetic field

Now that we have practiced quantizing the deformation field in an elastic string and the magnetization fluctuations in a ferromagnet, it is time to turn our attention to the electromagnetic field. Again, we recognize that actually all the work has been done already. In Section 7.3.5 we showed how one can represent the electromagnetic field in vacuum as a large set of oscillators. Combining this with the results of Section 8.1, we know how to write down the correct bosonic CAPs describing the elementary excitations of the field. With this we thus finalize the quantization of the classical electromagnetic field.

8.4.1 Photons

We recall that the oscillators representing the electromagnetic field are labeled by their wave vector and polarization index, $k = (\mathbf{k}, \alpha)$. With our knowledge of the quantization of oscillators, we can now establish what is a *quantum* state of the electromagnetic field. In general, it is a state in a bosonic Fock space with basis vectors corresponding to certain occupation numbers of *particles*,

$$|\{n_k\}\rangle = \prod_k \frac{(\hat{a}^\dagger)^{n_k}}{\sqrt{(n_k)!}} |0\rangle. \tag{8.35}$$

The operators $\hat{a}_k^\dagger$ that generate these states from the vacuum are again readily obtained from the generalized momenta P_k and coordinates Q_k of the corresponding oscillator. The particles we have just obtained are called *photons*, the quanta of light. The elementary properties of photons are probably already known to you, if not from high school, then from an introductory physics course of the first year of university. Let us refresh this knowledge here with a more solid background.

Photons thus "live" in levels labeled by wave vector and polarization $(\mathbf{k}, \alpha)$. To figure out the energy and the momentum of the photons, let us recall the presentation of the classical energy and integrated Poynting vector in terms of the normal coordinates $d_{\mathbf{k}\alpha}$,

$$H_{\text{e.m.}} = \int d\mathbf{r} \frac{\varepsilon_0}{2} (\mathbf{E}^2 + c^2 \mathbf{B}^2) = \sum_{\mathbf{k},\alpha} \omega_{\mathbf{k}} d_{\mathbf{k}\alpha}^* d_{\mathbf{k}\alpha}, \tag{8.36}$$

$$\mathbf{S}_{\text{tot}} = \int d\mathbf{r} \frac{1}{\mu_0} (\mathbf{E} \times \mathbf{B}) = \sum_{\mathbf{k},\alpha} c^2 \mathbf{k} \, d_{\mathbf{k}\alpha}^* d_{\mathbf{k}\alpha}. \tag{8.37}$$

These classical expressions are already highly suggestive: using our experience from the previous sections, we immediately recognize that their quantization is achieved by replacing $d \to \sqrt{\hbar}\hat{a}$. Therefore

$$\hat{H}_{\text{e.m.}} = \sum_{\mathbf{k},\alpha} \hbar\omega_{\mathbf{k}} (\hat{n}_{\mathbf{k}\alpha} + \tfrac{1}{2}), \tag{8.38}$$

$$\hat{\mathbf{S}}_{\text{tot}} = \sum_{\mathbf{k},\alpha} c^2 \hbar \hat{n}_{\mathbf{k}\alpha} \mathbf{k}. \tag{8.39}$$

We know from the previous chapter that $\omega_{\mathbf{k}} = ck$, so the photon *energy* is $\hbar ck$. To get the momentum from the last equation, we note that the photons propagate with velocity c. From the energy flux we can derive the momentum density by dividing it by the velocity squared, c^2. From this it follows that the *momentum* of an individual photon is $\mathbf{P}_{\mathbf{k}\alpha} = \hbar\mathbf{k}$.

The *statistics* of photons are bosonic by construction: indeed, we cannot make anything fermionic from a simple oscillator. Further, since photons are elementary particles, they should have *spin*, and this spin has to be integer to comply with their bosonic statistics. However, there seems to be a problem here. Photons with a given wave vector $\mathbf{k}$ come with two polarizations, which suggests a correspondence to spin $s = 1/2$, which is not integer. In fact, it turns out that the spin of a photon is $s = 1$. The two polarizations correspond to the two projections of the spin on the wave vector with spin ± 1. If we define the z-axis along the $\mathbf{k}$ we thus have $S_z = \pm 1$. The third possible projection, $S_z = 0$, is absent since photons move with the speed of light. This is a relativistic effect that guarantees that the electric field in electromagnetic waves is always transverse.

The same relativistic effects dictate that the *coordinate* of a moving photon is not quite a well-defined object. In fact, there is hardly a way to measure the coordinate of a photon without absorbing it. In this case, the coordinate can be determined, yet the photon is gone. In the relativistic framework, a photon is a *massless* particle and can never be at rest. Relativistic theory suggests the following line of separation between matter and irradiation: the classical limit of massless particles corresponds to fields, waves and oscillations, while the classical limit of massive particles corresponds to matter where particles move slowly and can have both coordinate and momentum.

8.4.2 Field operators

To investigate the quantum properties of the electromagnetic field, as a next step we express the formerly classical field variable of the vector potential $\mathbf{A}$ and the fields $\mathbf{E}$ and $\mathbf{B}$ in terms of photon CAPs. In fact, this procedure is a reverse of what we did to quantize the electromagnetic field. We have started from the classical vector potential and ended with the CAPs,

$$\mathbf{A}(\mathbf{r}, t) \Rightarrow d_{\mathbf{k}\alpha} \Rightarrow P_{\mathbf{k}\alpha}, Q_{\mathbf{k}\alpha} \Rightarrow \hat{a}^\dagger_{\mathbf{k}\alpha}, \hat{a}_{\mathbf{k}\alpha}, \tag{8.40}$$

but now we do the reverse to get operator expressions for the field variables,

$$\hat{a}^\dagger_{\mathbf{k}\alpha}(t), \hat{a}_{\mathbf{k}\alpha}(t) \Rightarrow \hat{P}_{\mathbf{k}\alpha}, \hat{Q}_{\mathbf{k}\alpha} \Rightarrow \hat{d}_{\mathbf{k}\alpha} \Rightarrow \hat{\mathbf{A}}(\mathbf{r}, t). \tag{8.41}$$

The fastest way to do this is to use the expression for $\mathbf{A}$ in terms of normal coordinates $d_{\mathbf{k}\alpha}$ and replace $d_{\mathbf{k}\alpha} \to \sqrt{\hbar}\hat{a}_{\mathbf{k}\alpha}$. We find

$$\hat{\mathbf{A}}(\mathbf{r}, t) = \sum_{\mathbf{k}} \sum_{\alpha=1,2} \sqrt{\frac{\hbar}{2\varepsilon_0\omega_{\mathbf{k}}\mathcal{V}}} \left[\hat{a}_{\mathbf{k}\alpha}(t)\mathbf{e}_\alpha \exp\{i\mathbf{k}\cdot\mathbf{r}\} + \text{H.c.}\right], \tag{8.42}$$

Paul Dirac (1902–1984)

Shared the Nobel Prize in 1933 with Erwin Schrödinger for "the discovery of new productive forms of atomic theory."

Paul Dirac was born in Bristol as the second child of Charles Dirac, a Swiss immigrant and French teacher in Bristol, and Florence Dirac, a librarian. In Bristol, Dirac suffered a difficult childhood. His father bullied his wife and forced his three children to address him in French, a language that Dirac never really mastered. Although his father supposedly never physically abused anyone, Dirac was severely humiliated on a daily basis. This, together with a mild form of autism, made Dirac a very introvert and silent person. His approach to conversation was mostly monosyllabic: his colleagues once jokingly introduced the unit of a Dirac, one word per hour.

After studying engineering at the University of Bristol, Dirac continued for a Bachelor's degree in mathematics. With his graduation he won a scholarship which enabled him to go to Cambridge to obtain a Ph.D. in mathematics. In these years, Schrödinger and Heisenberg put forward their theories of wave mechanics and matrix mechanics. Dirac, by then a skilled mathematician, recognized the similarity between Heisenberg's quantization rules and the Poisson brackets in classical Hamiltonian mechanics. He showed that quantum mechanics could be fully formulated in terms of canonically conjugated variables obeying specific commutation relations, a scheme in which both Schrödinger's and Heisenberg's theory fitted.

In the course of this work, Dirac was also the first to present a quantized theory for the electromagnetic field. His most famous work however, was his derivation of the Dirac equation: a relativistic (Lorentz covariant) version of the Schrödinger equation, see Chapter 13. Based on this equation, he predicted in 1928 the existence of antimatter, which was experimentally observed only four years later.

During the war Dirac was invited to come to the US to work on the Manhattan Project. Dirac, however, refused because he was afraid to leave his daily routine in Cambridge. He stayed in Cambridge until 1968, when he moved to Florida to be close to his daughter Mary. For the last 14 years of his life, Dirac was a professor at Florida State University.

and from this we construct operators for **E** and **B** by taking derivatives,

$$\hat{\mathbf{E}}(\mathbf{r}, t) = i \sum_{\mathbf{k}} \sum_{\alpha=1,2} \mathbf{e}_\alpha \sqrt{\frac{\hbar \omega_{\mathbf{k}}}{2\varepsilon_0 \mathcal{V}}} \left[\hat{a}_{\mathbf{k}\alpha}(t) \exp\{i\mathbf{k} \cdot \mathbf{r}\} - \text{H.c.} \right], \tag{8.43}$$

$$\hat{\mathbf{B}}(\mathbf{r}, t) = -i \sum_{\mathbf{k}} \sum_{\alpha=1,2} (\mathbf{e}_\alpha \times \mathbf{k}) \sqrt{\frac{\hbar}{2\varepsilon_0 \omega_{\mathbf{k}} \mathcal{V}}} \left[\hat{a}_{\mathbf{k}\alpha}(t) \exp\{i\mathbf{k} \cdot \mathbf{r}\} - \text{H.c.} \right]. \tag{8.44}$$

This, in a way, completes our quantization of the electromagnetic field: we have found out how to describe the field in terms of bosonic particles – photons – and we have derived operator expressions for the field observables $\mathbf{E}$ and $\mathbf{B}$ in terms of photon creation and annihilation operators.

8.4.3 Zero-point energy, uncertainty relations, and vacuum fluctuations

Let us now study some immediate consequences of quantization of the electromagnetic field we have just performed. We see that, since the energy of the set of oscillators reads

$$E = \sum_{\mathbf{k},\alpha} \hbar\omega_{\mathbf{k}}(n_{\mathbf{k}\alpha} + \tfrac{1}{2}), \tag{8.45}$$

there is a finite energy associated with the field configuration containing no photons at all. This energy is called the zero-point energy of the oscillator, and it exists because the oscillator is in fact not at rest in its ground state: it exerts zero-point motion owing to the fact that its momentum and coordinate cannot be simultaneously set to the energetically favorable values $P_{\mathbf{k}} = Q_{\mathbf{k}} = 0$.

The first question coming to mind is of course whether we can use this energy and/or detect the associated zero-point motion. One cannot do this immediately, although many pseudo-scientists would gladly do so. Indeed, the zero-point energy is actually the ground state energy of the vacuum. To release this energy, one would have to be able to break the vacuum, a rich idea that promises infinite outcomes. Indeed, let us calculate the total zero-point energy stored in a volume $\mathcal{V}$,

$$E = \sum_{\mathbf{k},\alpha} \tfrac{1}{2}\hbar\omega_{\mathbf{k}} = \frac{\mathcal{V}}{(2\pi)^3} \int d\mathbf{k}\, \hbar\omega_{\mathbf{k}} = \frac{\hbar\mathcal{V}}{2\pi^2 c^3} \int d\omega\, \omega^3 = \mathcal{V} \cdot \infty! \tag{8.46}$$

The energy per unit volume is infinite! But since we cannot break the vacuum, we can probably forget about extracting this energy from the vacuum. However, there is still a problem combining zero-point energy and gravitation. According to Einstein, any energy is a mass and any mass does gravitate. So even if we could not profit from the zero-point energy, the gravitational field produced by it would smash us. How do we survive? To our knowledge, this problem has not been solved yet. Zumino noted in 1975 that the total energy density of vacuum is zero if there is a symmetry between fermions and bosons, called *supersymmetry*. However, this supersymmetry is not exact below $\omega \simeq 10^{26}$ Hz. If one takes this as an upper cut-off in the above integral, one still has to cope with an unreasonably large energy density.

Let us turn, however, to earthly tasks and evaluate the zero-point energy at scales where we can modify the properties of the vacuum for electromagnetic waves. How can we modify its properties? We could take, for instance, a transparent isolator with dielectric constant $\varepsilon = \varepsilon_r \varepsilon_0$ instead of the vacuum. This change reduces the energies of the oscillators from $\hbar\omega_k$ to $\hbar\omega_k/\sqrt{\varepsilon_r}$. An important point, however, is that in fact not all oscillators change their energy. The isolator consists of atoms, and their reaction time is finite. The atoms cannot react to the fastest oscillations, and these oscillator modes thus do not "feel" the presence of the isolator, and experience no change in dielectric constant. Let us make some estimations here. The effective speed of light in a dielectric medium is $c^* = c/\sqrt{\varepsilon_r}$. We substitute into (8.46) this speed for $\omega < \omega_c$, where ω_c is the upper frequency at which the atoms can still react to the electromagnetic field. With this, the change in zero-point energy density can be estimated as

$$\frac{\Delta E_0}{\mathcal{V}} = \frac{\hbar}{8\pi^2 c^3} \left(\varepsilon_r^{3/2} - 1 \right) \omega_c^4. \tag{8.47}$$

The cut-off frequency thus must be related to the atomic energy scale: the typical distance between atomic levels, $\hbar\omega_c \simeq E_{\text{at}}$. How can we estimate E_{at}? An estimate is obtained by equating the potential and kinetic energy of a hydrogen atom,

$$E_{\text{at}} \simeq \text{Ry} \simeq \frac{e^2}{8\pi\varepsilon_0 a_B}, \quad \text{where} \quad a_B \simeq \frac{4\pi\varepsilon_0\hbar^2}{e^2 m_e}, \tag{8.48}$$

a_B, the Bohr radius, being the typical size of an atom. To get a clear picture, let us estimate ΔE_0 for the volume of an atom,

$$\Delta E_0 \simeq \frac{a_B^3 E_{\text{at}}^4}{\hbar^3 c^3} \simeq E_{\text{at}} \left(\frac{e^2}{4\pi\varepsilon_0\hbar c} \right)^3. \tag{8.49}$$

This is the first time in this book that we encounter the celebrated fine structure constant α,

$$\alpha \equiv \frac{e^2}{4\pi\varepsilon_0\hbar c} \approx \frac{1}{137.03}, \tag{8.50}$$

which is a small dimensionless number related to the strength of quantum electromagnetic effects. We see that the eventual zero-point energy at the atomic scale is small, at least six orders of magnitude smaller than the atomic energy. This energy change is responsible for Casimir forces between two metallic bodies. With some reservations, the observation of Casimir forces is in fact an observation of zero-point energy, see Exercise 1.

Let us now turn to the *uncertainty relations* and *vacuum fluctuations*, since they are related. The most common uncertainty relation is a consequence of the canonical commutation relation between the operators of coordinate and momentum,

$$[\hat{p}, \hat{x}] = -i\hbar, \quad \text{so} \quad (\Delta p)(\Delta x) \simeq \hbar. \tag{8.51}$$

In a harmonic oscillator, as already mentioned above, this gives rise to fluctuations of both coordinate and momentum in the ground state of the oscillator – vacuum fluctuations. Since we have presented the electromagnetic fields as a set of oscillators, we can write similar uncertainty relations and establish the vacuum fluctuations of the field.

Now it is time for a little algebra. To simplify the derivation, we do it first for rather abstract Hermitian operators that are linear in bosonic CAPs. We take

$$\hat{X} = \sum_k x_k \hat{a}_k + x_k^* \hat{a}_k^\dagger \quad \text{and} \quad \hat{Y} = \sum_k y_k \hat{a}_k + y_k^* \hat{a}_k^\dagger. \tag{8.52}$$

Later, we use the expressions for the field operators (8.43) and (8.44) instead.

To find the uncertainty relations, we calculate $[\hat{X}, \hat{Y}]$, which involves in principle a double sum. Since we know that pairs of creation operators and pairs of annihilation operators always commute, we keep only terms with $\hat{a}$ and $\hat{a}^\dagger$ to obtain

$$[\hat{X}, \hat{Y}] = \sum_{k,l} \left(x_k y_l^* [\hat{a}_k, \hat{a}_l^\dagger] + x_k^* y_l [\hat{a}_k^\dagger, \hat{a}_l] \right). \tag{8.53}$$

Now we implement the commutation relations $[\hat{a}_k, \hat{a}_l^\dagger] = \delta_{kl}$ and get

$$[\hat{X}, \hat{Y}] = \sum_{k,l} \left(x_k y_k^* - x_k^* y_k \right), \tag{8.54}$$

a purely imaginary expression, by the way.

The vacuum fluctuations can then be presented as the average of the symmetrized product of the two, $\frac{1}{2}\langle \hat{X}\hat{Y} + \hat{Y}\hat{X} \rangle$, which in fact can be written as $\frac{1}{2}\langle 0|\{\hat{X}, \hat{Y}\}|0\rangle$. Again, terms involving pairs of creation operators as well as pairs of annihilation operators do not contribute to the sum since the average of such a pair is certainly zero. We thus keep only the terms with $\hat{a}$ and $\hat{a}^\dagger$,

$$\frac{1}{2}\langle 0|\{\hat{X}, \hat{Y}\}|0\rangle = \frac{1}{2}\sum_{k,l} \left(x_k y_l^* \langle 0|[\hat{a}_k, \hat{a}_l^\dagger]|0\rangle + x_k^* y_l \langle 0|[\hat{a}_k^\dagger, \hat{a}_l]|0\rangle \right). \tag{8.55}$$

Now we use the fact that $\langle 0|[\hat{a}_k, \hat{a}_l^\dagger]|0\rangle = \delta_{kl}$.

Control question. Can you prove the above short formula?

So then we are left with a single sum,

$$\frac{1}{2}\langle 0|\{\hat{X}, \hat{Y}\}|0\rangle = \frac{1}{2}\sum_{k,l} \left(x_k y_k^* + x_k^* y_k \right), \tag{8.56}$$

and if $\hat{X} = \hat{Y}$, this reduces to

$$\langle 0|\hat{X}^2|0\rangle = \sum_k |x_k|^2. \tag{8.57}$$

Now we can turn our attention to the electromagnetic field. If we look again at the operator expressions for the electric and magnetic fields (8.43) and (8.44), we see that they have the same structure as the abstract operators $\hat{X}$ and $\hat{Y}$ we just considered. The uncertainty relation between the different components of electric and magnetic field, for instance $[\hat{B}_y(\mathbf{r}, t), \hat{E}_x(\mathbf{r}', t)]$, can thus readily be obtained. Combining (8.54), (8.43), and (8.44), we compute

$$\begin{aligned}
[\hat{B}_y(\mathbf{r}, t), \hat{E}_x(\mathbf{r}', t)] &= \frac{i\hbar}{\varepsilon_0 \mathcal{V}}\mathrm{Im}\sum_{\mathbf{k},\alpha} e^{i\mathbf{k}\cdot(\mathbf{r}-\mathbf{r}')}(\mathbf{k}\times \mathbf{e}_\alpha)_y e_\alpha^x \\
&= \frac{i\hbar}{\varepsilon_0 \mathcal{V}}\mathrm{Im}\sum_{\mathbf{k},\alpha} e^{i\mathbf{k}\cdot(\mathbf{r}-\mathbf{r}')}(k_z e_\alpha^x - k_x e_\alpha^z) e_\alpha^x.
\end{aligned} \tag{8.58}$$

A tricky point is to evaluate the polarization vectors since their components do depend on the direction of $\mathbf{k}$. We use the relation

$$\sum_\alpha e_\alpha^a e_\alpha^b = \delta_{ab} - \frac{k_a k_b}{k^2}, \tag{8.59}$$

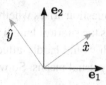

Fig. 8.3 This equality can readily be seen if one expresses the orthonormal vectors $\hat{x}, \hat{y}$ in a new coordinate system, spanned by $\mathbf{e}_1, \mathbf{e}_2$, and $\mathbf{k}/k$ (here illustrated for two dimensions). Orthogonality between $\hat{x}, \hat{y}$, and $\hat{z}$ in the new basis yields the relation.

where $a, b \in \{x, y, z\}$. This relation directly follows from the orthonormality of the set of two polarizations and their orthogonality to $\mathbf{k}$, see Fig. 8.3. Then we obtain

$$\operatorname{Im} \sum_{\mathbf{k},\alpha} e^{i\mathbf{k}\cdot(\mathbf{r}-\mathbf{r}')} k_z = \frac{\mathcal{V}}{(2\pi)^3} \operatorname{Im} \int d\mathbf{k}\, k_z e^{i\mathbf{k}\cdot(\mathbf{r}-\mathbf{r}')} = -\frac{\mathcal{V}}{(2\pi)^3} \operatorname{Re}\frac{\partial}{\partial z} \int d\mathbf{k}\, e^{i\mathbf{k}\cdot(\mathbf{r}-\mathbf{r}')}, \quad (8.60)$$

so that

$$[\hat{B}_y(\mathbf{r}, t), \hat{E}_x(\mathbf{r}', t)] = -\frac{i\hbar}{\varepsilon_0} \frac{\partial}{\partial z} \delta(\mathbf{r} - \mathbf{r}'). \quad (8.61)$$

In words this means that $\hat{B}_y$ and $\hat{E}_x$ commute, except in the same point, where the uncertainty is eventually infinite. To make sense out of this, let us average the fields over a cubic volume of size Δl. The relation becomes

$$\Delta B \cdot \Delta E \simeq \frac{\hbar}{\varepsilon_0 (\Delta l)^4}. \quad (8.62)$$

That is, the uncertainty grows with decreasing volume: quantum effects on the fields become more pronounced at smaller space scale.

Let us now turn to the vacuum fluctuations. Combining (8.43), (8.44), and (8.57), we find for the square of the electric field

$$\langle 0 | \hat{\mathbf{E}} \cdot \hat{\mathbf{E}} | 0 \rangle = \frac{\hbar}{2\varepsilon_0 \mathcal{V}} \sum_{\mathbf{k},\alpha} \mathbf{e}_\alpha \cdot \mathbf{e}_\alpha = \frac{\hbar c}{2\pi^2 \varepsilon_0} \int k^3 dk = \infty, \quad (8.63)$$

and a similar relation holds for the magnetic field,

$$\langle 0 | \hat{\mathbf{B}} \cdot \hat{\mathbf{B}} | 0 \rangle = \frac{\hbar}{2\pi^2 \varepsilon_0 c^2} \int k^3 dk = \infty = \frac{1}{c^2} \langle 0 | \hat{\mathbf{E}} \cdot \hat{\mathbf{E}} | 0 \rangle. \quad (8.64)$$

The vacuum fluctuations thus diverge at short length scales (large frequencies), as the zero-point energy does. In fact, the fluctuations and the zero-point energy are related since the vacuum energy density can be presented as

$$U_{\text{vac}} = \frac{\langle 0 | \hat{H}_{\text{e.m.}} | 0 \rangle}{\mathcal{V}} = \frac{\varepsilon_0}{2} \left(\langle 0 | \hat{\mathbf{E}}^2 | 0 \rangle + c^2 \langle 0 | \hat{\mathbf{B}}^2 | 0 \rangle \right) = \frac{\hbar c}{2\pi^2} \int k^3 dk = \infty.$$

Let us make sense of the fluctuations by skipping the short length scales. If we cut the integral over k at $k \simeq 1/(\Delta l)$, the fluctuations of the electric field become

$$\langle 0 | \hat{\mathbf{E}}^2 | 0 \rangle \simeq \frac{\hbar c}{\varepsilon_0 (\Delta l)^4}, \quad (8.65)$$

which is finite. The question arises whether these fluctuations are large or small. Let us use the atomic scale a_B as the shortest length scale. The idea is that electromagnetic waves with a shorter wave length would hardly affect the motion of electrons in atoms and therefore cannot be detected by usual means. So we estimate the fluctuations as

$$\langle 0|\hat{\mathbf{E}}^2|0\rangle \simeq \frac{\hbar c}{\varepsilon_0 a_B^4}. \tag{8.66}$$

It would be interesting to compare the typical magnitude of these vacuum fluctuations of **E** with the typical electric field created by the charges of the electrons and nucleus inside the atom. The latter can be estimated as

$$\mathbf{E}_{\text{at}}^2 \simeq \left(\frac{e}{\varepsilon_0 a_B^2}\right)^2, \tag{8.67}$$

roughly the field of an elementary charge at a distance a_B. Comparing the two, we obtain

$$\frac{\mathbf{E}_{\text{fluc}}^2}{\mathbf{E}_{\text{at}}^2} \simeq \frac{\varepsilon_0 \hbar c}{e^2} \simeq \frac{1}{\alpha} \simeq 137. \tag{8.68}$$

We thus (finally) come to a truly astonishing result: the electrons in atoms feel a field of vacuum fluctuations that exceeds by at least one order of magnitude the field of the nuclei and other electrons.

We have apparently done something wrong. The fluctuations we have taken into account are in fact instant and mostly contributed by very fast oscillations with frequencies c/a_B. These oscillations are too fast to affect the electrons. As mentioned, efficient fluctuations have frequencies not exceeding the atomic scale $E_{\text{at}}/\hbar$ and live at a much longer space scale $\zeta \simeq c\hbar/\text{Ry} \gg \Delta l$. The fluctuations of their field are small,

$$\langle 0|\hat{\mathbf{E}}^2|0\rangle_{\text{eff}} \simeq \frac{\hbar c}{\varepsilon_0 \zeta^4} \quad \Rightarrow \quad \frac{\mathbf{E}_{\text{fluc}}^2}{\mathbf{E}_{\text{at}}^2} \simeq \left(\frac{e^2}{4\pi\varepsilon_0 \hbar c}\right)^3 = \alpha^3, \tag{8.69}$$

in concurrence with our estimations of the zero-point energy at this scale.

We still have not answered the question whether there is a way to observe these vacuum fluctuations. In fact, there is no direct way, the fluctuations are a property of the ground state and one can extract no energy from them. There are, however, several indirect ways. For instance, the rate of spontaneous emission from excited atomic states is related to vacuum fluctuations of electric field at the frequency corresponding to the energy difference between the excited and ground state. The fluctuations also cause small but measurable shifts of the positions of the energy levels in atoms. We discuss these topics in more detail in Chapter 9.

8.4.4 The simple oscillator

So we have managed to do an important job. We have quantized the electromagnetic field in vacuum, and found several quantum effects for the field (the uncertainty relations and vacuum fluctuations). Let us now see how we can do the same job for the simple LC-oscillator

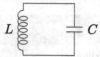

Fig. 8.4   A simple electric circuit, consisting of a capacitor C and an inductance L.

(Fig. 8.4) with which we also concluded the previous chapter. In Section 7.3.6 we reduced its Hamiltonian to the standard form

$$H = \frac{1}{2}(P^2 + \omega_0^2 Q^2), \tag{8.70}$$

where

$$\omega_0 = \frac{1}{\sqrt{LC}}, \quad P = \sqrt{L}I, \quad \text{and} \quad Q = C\sqrt{L}V. \tag{8.71}$$

We are by now so well trained in quantizing oscillators that we can skip all intermediate steps and immediately construct the "field operators" for the electric variables V and I in terms of the bosonic CAPs,

$$\hat{V} = (\hat{a}^\dagger + \hat{a})\sqrt{\frac{\hbar}{2C^{3/2}L^{1/2}}} \quad \text{and} \quad \hat{I} = i(\hat{a}^\dagger - \hat{a})\sqrt{\frac{\hbar}{2C^{1/2}L^{3/2}}}, \tag{8.72}$$

where $\hat{a}^\dagger$ and $\hat{a}$ excite and de-excite the oscillator.

With this in hand, we can consider similar quantum effects as we did above for the electromagnetic field. The zero-point energy is simply that of a single oscillator,

$$E_0 = \frac{\hbar\omega_0}{2}, \tag{8.73}$$

and the vacuum fluctuations of the current and voltage are found to be

$$\langle 0|\hat{V}^2|0\rangle = \frac{\hbar\omega_0}{2C} \quad \text{and} \quad \langle 0|\hat{I}^2|0\rangle = \frac{\hbar\omega_0}{2L}. \tag{8.74}$$

Again, we can ask ourselves whether these fluctuations are large or small. This turns out to depend on the circumstances. Let us try to solve the problem in a "fundamental" way. We have the fundamental unit of charge, e, and compute the fluctuations of the capacitor charge $q = CV$ in these units,

$$\frac{\langle 0|\hat{q}^2|0\rangle}{e^2} = \frac{\hbar\omega_0 C}{2e^2} = \frac{1}{\pi Z G_Q}, \tag{8.75}$$

where we have defined the effective impedance $Z \equiv \sqrt{L/C}$ of the oscillator and the so-called conductance quantum $G_Q = e^2/\pi\hbar \approx (12909\ \Omega)^{-1}$. Depending on the actual impedance of the oscillator, the vacuum fluctuations of the charge can thus be both much larger or much smaller than e.

Control question. Your mobile phone should contain (a complicated form of) an LC-oscillator to receive and transmit signals. Can you estimate the vacuum fluctuations of the charge in this oscillator?

Table 8.1 Summary: Quantization of fields

Quantization of one oscillator

classical: $H = \frac{1}{2}(P^2 + \omega^2 Q^2)$, $\dot{Q} = \frac{\partial H}{\partial P} = P$, $\dot{P} = -\frac{\partial H}{\partial Q} = -\omega^2 Q$

quantize: $\hat{H} = \frac{1}{2}(\hat{P}^2 + \omega^2 \hat{Q}^2)$, with $[\hat{Q}, \hat{P}] = i\hbar$ reproduces correct equations of motion

introduce: $\hat{a} = \dfrac{\omega\hat{Q} + i\hat{P}}{\sqrt{2\hbar\omega}}$ and $\hat{a}^\dagger = \dfrac{\omega\hat{Q} - i\hat{P}}{\sqrt{2\hbar\omega}}$ $\rightarrow$ $[\hat{a}, \hat{a}^\dagger] = 1$, $[\hat{a}, \hat{a}] = [\hat{a}^\dagger, \hat{a}^\dagger] = 0$

$\hat{a}$ and $\hat{a}^\dagger$ describe bosons, rewrite $\hat{H} = \hbar\omega(\hat{a}^\dagger\hat{a} + \frac{1}{2}) = \hbar\omega(\hat{n} + \frac{1}{2})$

Quantization of many oscillators

if oscillators are uncoupled: same procedure, same results

classical: $H = \frac{1}{2}\sum_k (P_k^2 + \omega_k^2 Q_k^2)$, $\dot{Q}_k = \frac{\partial H}{\partial P_k} = P_k$, $\dot{P}_k = -\frac{\partial H}{\partial Q_k} = -\omega_k^2 Q_k$

quantize: $\hat{H} = \frac{1}{2}\sum_k (\hat{P}_k^2 + \omega_k^2 \hat{Q}_k^2)$, with $[\hat{Q}_k, \hat{P}_{k'}] = i\hbar\delta_{kk'}$ reproduces correct equations of motion

introduce: $\hat{a}_k = \dfrac{\omega_k\hat{Q}_k + i\hat{P}_k}{\sqrt{2\hbar\omega_k}}$ and $\hat{a}_k^\dagger = \dfrac{\omega_k\hat{Q}_k - i\hat{P}_k}{\sqrt{2\hbar\omega_k}}$ $\rightarrow$ $[\hat{a}_k, \hat{a}_{k'}^\dagger] = \delta_{kk'}$, $[\hat{a}_k, \hat{a}_{k'}] = [\hat{a}_k^\dagger, \hat{a}_{k'}^\dagger] = 0$

$\hat{a}_k$ and $\hat{a}_k^\dagger$ describe bosons in level k, rewrite $\hat{H} = \sum_k \hbar\omega_k(\hat{a}_k^\dagger\hat{a}_k + \frac{1}{2}) = \sum_k \hbar\omega_k(\hat{n}_k + \frac{1}{2})$

Examples

phonons: $\hat{a}_k = \frac{1}{\sqrt{\hbar}}\sqrt{2\rho\omega_k L}\,\hat{u}_k$ and $\hat{a}_k^\dagger = \frac{1}{\sqrt{\hbar}}\sqrt{2\rho\omega_k L}\,\hat{u}_k^\dagger$

where $\hat{u}_k$ is Fourier transform of displacement operator $\hat{u}(x, t)$

magnons: $\hat{a}_\mathbf{k} = \frac{1}{\sqrt{\hbar}}\int d\mathbf{r}\,\hat{m}^+(\mathbf{r})e^{i\mathbf{k}\cdot\mathbf{r}}$ and $\hat{a}_\mathbf{k}^\dagger = \frac{1}{\sqrt{\hbar}}\int d\mathbf{r}\,\hat{m}^-(\mathbf{r})e^{-i\mathbf{k}\cdot\mathbf{r}}$

where $\hat{m}^\pm(\mathbf{r}) = \hat{m}_x(\mathbf{r}) \pm i\hat{m}_y(\mathbf{r})$ are local magnetization operators

---

Quantization of the electromagnetic field

photons: $\hat{a}_{\mathbf{k}\alpha} = \frac{1}{\sqrt{\hbar}}\sqrt{2\varepsilon_0\omega_\mathbf{k}\mathcal{V}}\,\hat{\mathbf{A}}_\mathbf{k}\cdot\mathbf{e}_\alpha$ and $\hat{a}_{\mathbf{k}\alpha}^\dagger = \frac{1}{\sqrt{\hbar}}\sqrt{2\varepsilon_0\omega_\mathbf{k}\mathcal{V}}\,\hat{\mathbf{A}}_\mathbf{k}^\dagger\cdot\mathbf{e}_\alpha$

where $\hat{\mathbf{A}}_\mathbf{k}$ is Fourier transform of vector potential operator

and $\mathbf{e}_{1,2}$ are allowed polarizations, $\mathbf{e}_\alpha\cdot\mathbf{k} = 0$

with $\mathbf{e}_\alpha$ polarization vector (two directions) and $\omega_\mathbf{k} = ck$

Hamiltonian: $\hat{H}_{\text{e.m.}} = \sum_{\mathbf{k},\alpha} \hbar\omega_\mathbf{k}(\hat{a}_{\mathbf{k}\alpha}^\dagger\hat{a}_{\mathbf{k}\alpha} + \frac{1}{2}) = \sum_{\mathbf{k},\alpha} \hbar\omega_\mathbf{k}(\hat{n}_{\mathbf{k}\alpha} + \frac{1}{2})$ $\rightarrow$ photon energy is $\hbar ck$

energy flux: $\hat{\mathbf{S}}_{\text{tot}} = \sum_{\mathbf{k},\alpha} c^2\hbar\hat{n}_{\mathbf{k}\alpha}\mathbf{k}$ $\rightarrow$ photon momentum is $\hbar\mathbf{k}$

field operators: $\hat{\mathbf{A}}(\mathbf{r}, t) = \sum_{\mathbf{k},\alpha} \sqrt{\hbar/2\varepsilon_0\omega_\mathbf{k}\mathcal{V}}\left[\hat{a}_{\mathbf{k}\alpha}(t)\mathbf{e}_\alpha\exp\{i\mathbf{k}\cdot\mathbf{r}\} + \text{H.c.}\right]$

$\hat{\mathbf{E}}(\mathbf{r}, t) = i\sum_{\mathbf{k},\alpha} \mathbf{e}_\alpha\sqrt{\hbar\omega_\mathbf{k}/2\varepsilon_0\mathcal{V}}\left[\hat{a}_{\mathbf{k}\alpha}(t)\exp\{i\mathbf{k}\cdot\mathbf{r}\} - \text{H.c.}\right]$

$\hat{\mathbf{B}}(\mathbf{r}, t) = -i\sum_{\mathbf{k},\alpha} (\mathbf{e}_\alpha\times\mathbf{k})\sqrt{\hbar/2\varepsilon_0\omega_\mathbf{k}\mathcal{V}}\left[\hat{a}_{\mathbf{k}\alpha}(t)\exp\{i\mathbf{k}\cdot\mathbf{r}\} - \text{H.c.}\right]$

Zero-point energy, uncertainty relations, and vacuum fluctuations

zero-point energy: in the vacuum $E_0 = \langle 0|\hat{H}_{\text{e.m.}}|0\rangle = \mathcal{V}\cdot\infty$ is infinite, but inaccessible

one *can* detect changes $\Delta E_0/\mathcal{V}$ in this energy: Casimir effect, Van der Waals force, ...

estimate of ΔE_0 due to presence of an atom: cut-off energy $\hbar\omega_c = \text{Ry}$, then $\Delta E_0 \sim \alpha^3\,\text{Ry}$

fine structure constant: $\alpha \equiv \dfrac{e^2}{4\pi\varepsilon_0\hbar c} \approx \dfrac{1}{137}$, relative strength of quantum electromagnetic effects

uncertainty relations: generally $(\Delta A)(\Delta B) \gtrsim |\langle[\hat{A}, \hat{B}]\rangle|$

two components of e.m. field $[\hat{B}_y(\mathbf{r}, t), \hat{E}_x(\mathbf{r}', t)] = -\frac{i\hbar}{\varepsilon_0}\partial_z\delta(\mathbf{r} - \mathbf{r}')$

average over volume $(\Delta l)^3$ $\rightarrow$ $(\Delta B_y)(\Delta E_x) \gtrsim \hbar/\varepsilon_0(\Delta l)^4$

vacuum fluctuations: $\mathbf{E}_{\text{fluc}}^2 = \langle 0|\hat{\mathbf{E}}\cdot\hat{\mathbf{E}}|0\rangle = c^2\langle 0|\hat{\mathbf{B}}\cdot\hat{\mathbf{B}}|0\rangle = E_0/\varepsilon_0\mathcal{V}$, related to E_0

estimate of $\mathbf{E}_{\text{fluc}}^2$ "felt" by an atom: cut-off energy $\hbar\omega_c = \text{Ry}$, then $\mathbf{E}_{\text{fluc}}^2 \sim \alpha^3\mathbf{E}_{\text{at}}^2$

where $\mathbf{E}_{\text{at}}$ is the typical electric field of an elementary charge at the atomic distance $a_B \equiv \hbar/\alpha mc$

Exercises

1. *The Casimir effect* (solution included). When two uncharged perfectly conducting parallel plates are very close to each other, the zero-point energy of the quantized electromagnetic field between the two plates becomes a sensitive function of the distance between the plates. The dependence of the energy on the distance can manifest itself as an attractive force between the plates. This effect was predicted in 1948 by the Dutch physicists Hendrik Casimir and Dirk Polder, and experimentally confirmed in 1958 by Marcus Sparnaay.

   We consider an uncharged conducting plate with dimensions $W \times W$ which divides a box of size $L \times W \times W$ into two, as depicted in Fig. 8.5. The walls of the box are also assumed to be uncharged and perfectly conducting. The task of this exercise is to calculate the dependence of the zero-point energy of the electromagnetic field on the separation a.

   a. The allowed frequencies for the electromagnetic field inside a box of size $a \times W \times W$ with perfectly conducting walls are

   $$\omega_{nlm} = c|\mathbf{k}| = c\sqrt{\frac{n^2\pi^2}{a^2} + \frac{l^2\pi^2}{W^2} + \frac{m^2\pi^2}{W^2}},$$

   where $\mathbf{k}$ denotes the wave vector of the mode. The numbers n, l, and m are non-negative integers numbering the allowed modes of the field. Write down an expression for the zero-point energy of the field in the box. You can assume W to be large, so that the sums over l and m can be written as integrals.

   b. The expression found at (a) diverges, as we know. However, modes with very high energies will not provide an important contribution, since they correspond to frequencies too high to notice any of the conducting walls at all. To introduce a cut-off, we can include a factor $e^{-\eta k/\pi}$ where k is the length of the wave vector and η a small factor which we hopefully can set to zero at the end of the calculation.

   Write the integral in terms of the coordinates $(x, y) = \frac{a}{W}(l, m)$. Then introduce polar coordinates (r, ϕ) with which to represent the coordinates (x, y) and show that the zero-point energy found at (a) can be written as

   $$E_0(a, \eta) = \frac{\hbar c\pi^2 L^2}{2a^3} \sum_{n=1}^{\infty} \int_0^{\infty} dr\, r\sqrt{n^2 + r^2}\, e^{-\frac{\eta}{a}\sqrt{n^2+r^2}}.$$

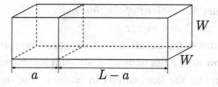

Fig. 8.5   A box of size $L \times W \times W$, divided into two by a plate at distance a from the left wall.

Note that we left the term with $n = 0$ out of the sum: it is not important for our calculation since it yields in the end a contribution not depending on a.

c. We define $z = \sqrt{n^2 + r^2}$, so that $r\,dr = z\,dz$. Write the integral over r as an integral over z and show that the integrand can be written as a double derivative with respect to η. Writing the integral like this, one can immediately see how to perform it. Do first the integral and then the sum, and show that the result reads

$$E_0(a, \eta) = \frac{\hbar c \pi^2 W^2}{2a} \frac{\partial^2}{\partial \eta^2} \frac{a}{\eta} \frac{1}{e^{\eta/a} - 1}.$$

d. Expand the η-dependent part of the above expression as a series in η, keeping only the terms which will survive when we take the limit $\eta \to 0$. Show that the result can be written as

$$E_0(a, \eta) = \hbar c \pi^2 W^2 \left(\frac{3a}{\eta^4} - \frac{1}{2\eta^2} - \frac{1}{720a^3} + \mathcal{O}[\eta] \right).$$

e. We can write the net force on the middle plate due to the dependence of the zero-point energy in the two parts of the box on the position a of the plate as

$$F = F_\mathrm{L} + F_\mathrm{R} = \lim_{\eta \to 0} \left\{ -\frac{\partial}{\partial a} E_0(a, \eta) - \frac{\partial}{\partial a} E_0(L - a, \eta) \right\}. \qquad (8.76)$$

Show that for $L \gg a$ this yields

$$F = -\frac{\hbar c \pi^2 W^2}{240 a^4},$$

the famous Casimir force. For what distance a does the Casimir force become equivalent to a pressure of 1 bar?

2. *Force in an LC-oscillator.* The capacitance of an LC-oscillator is proportional to the distance between the plates. This leads to a Casimir-like force between the two plates. Calculate this force as a function of the distance a between the plates.

3. *Casimir forces in an elastic string.* Let us revisit the one-dimensional elastic string considered in this chapter. We attach two small extra masses m_1 and m_2 to the string. The masses are separated by the distance b. The goal is to compute the interaction force between the masses. We do this by computing the correction to the energy of the ground state.
   a. What is the perturbation term in the Hamiltonian?
   b. How can it be expressed in terms of phonon CAPs?
   c. What are the virtual states involved in the correction? What does the b-dependent part of the correction look like?
   d. Give the expression for the force.

4. *Quantum fluctuations of the thin rod's end.* Let us revisit the thin rod setup considered in Exercise 1 of Chapter 7.
   a. Estimate the quantum fluctuations of the vertical position of the free end of the rod. How do the fluctuations scale with increasing the rod's length?
   b. Express the fluctuations in terms of the normalized solutions $\phi_n(x)$ discussed in Exercise 1 of Chapter 7.

Solutions

1. The Casimir effect.

a. The zero-point energy in a box with dimensions $a \times W \times W$ is

$$E_0 = 2\frac{1}{2}\hbar c \sum_{n=0}^{\infty}{}' \int_0^{\infty} dl\, dm \sqrt{\frac{n^2\pi^2}{a^2} + \frac{l^2\pi^2}{W^2} + \frac{m^2\pi^2}{W^2}},$$

where the factor 2 takes into account the two possible polarizations. In fact, the mode with $n = 0$ has only one allowed polarization, so for this term there should be included a factor $\frac{1}{2}$ which is indicated by the prime at the sum.

b. We first write

$$E_0 = \hbar c \sum_{n=1}^{\infty} \int_0^{\infty} dl\, dm \frac{\pi}{a} \sqrt{n^2 + \frac{l^2 a^2}{W^2} + \frac{m^2 a^2}{W^2}}\, e^{-\frac{\eta}{a}\sqrt{n^2 + \frac{l^2 a^2}{W^2} + \frac{m^2 a^2}{W^2}}}.$$

Then we introduce new coordinates $\frac{la}{L} \to x$ and $\frac{ma}{L} \to y$, yielding

$$E_0 = \frac{\hbar c \pi W^2}{a^3} \sum_{n=1}^{\infty} \int_0^{\infty} dx\, dy \sqrt{n^2 + x^2 + y^2}\, e^{-\frac{\eta}{a}\sqrt{n^2 + x^2 + y^2}}.$$

Then we write the integral in the xy-plane in polar coordinates (r, ϕ),

$$E_0 = \frac{\hbar c \pi W^2}{a^3} \sum_{n=1}^{\infty} \int_0^{\infty} r\, dr \int_0^{\pi/2} d\phi \sqrt{n^2 + r^2}\, e^{-\frac{\eta}{a}\sqrt{n^2 + r^2}}.$$

Performing the integral over ϕ yields the result desired.

c. We find

$$E_0 = \frac{\hbar c \pi^2 W^2}{2a^3} \sum_{n=1}^{\infty} \int_n^{\infty} dz\, z^2 e^{-\frac{\eta}{a} z},$$

which indeed can be written as

$$E_0 = \frac{\hbar c \pi^2 W^2}{2a} \sum_{n=1}^{\infty} \int_n^{\infty} dz\, \frac{\partial^2}{\partial \eta^2} e^{-\frac{\eta}{a} z}.$$

We take the differentiation operator to the left and find

$$E_0 = \frac{\hbar c \pi^2 W^2}{2a} \frac{\partial^2}{\partial \eta^2} \sum_{n=1}^{\infty} \int_n^{\infty} dz\, e^{-\frac{\eta}{a} z} = \frac{\hbar c \pi^2 W^2}{2a} \frac{\partial^2}{\partial \eta^2} \sum_{n=1}^{\infty} \frac{a}{\eta} e^{-\frac{\eta}{a} n}.$$

The sum over n can now be performed, yielding the expression given in the exercise. We now also see that the term with $n = 0$ is not important: differentiating the summand with respect to η makes all terms proportional to n^2.

d. Expand

$$\frac{a}{\eta} \frac{1}{e^{\eta/a} - 1},$$

in terms of η up to power η^2. Higher powers vanish after differentiating twice with respect to η and then letting $\eta \to 0$.

e. Evaluating (8.76) yields

$$F = -\hbar c \pi^2 W^2 \left\{ \frac{1}{240a^4} - \frac{1}{240(L-a)^4} \right\},$$

which reduces for $L \gg a$ to the expression given in the exercise. The pressure, or force, on the plate per unit of area, reads

$$\frac{F}{W^2} = -\frac{\hbar c \pi^2}{240a^4},$$

which becomes equal to 10^5 N/m for $a \approx 11$ nm.

9

Radiation and matter

Most introductory courses on physics mention the basics of atomic physics: we learn about the ground state and excited states, and that the excited states have a finite life-time decaying eventually to the ground state while emitting radiation. You might have hoped for a better understanding of these decay processes when you started studying quantum mechanics. If this was the case, you must have been disappointed. Courses on basic quantum mechanics in fact assume that all quantum states have an *infinitely long* life-time. You learn how to employ the Schrödinger equation to determine the wave function of a particle in various setups, for instance in an infinitely deep well, in a parabolic potential, or in the spherically symmetric potential set up by the nucleus of a hydrogen atom. These solutions to the Schrödinger equation are explained as being stationary, the only time-dependence of the wave function being a periodic phase factor due to the energy of the state. Transitions between different quantum states are traditionally not discussed and left for courses on advanced quantum mechanics. If you have been disappointed by this, then this chapter will hopefully provide some new insight: we learn how matter and radiation interact, and, in particular, how excited atomic states decay to the ground state.

Our main tool to calculate emission and absorption rates for radiative processes is Fermi's golden rule (see Section 1.6.1). We thus start by refreshing the golden rule and specifying its proper use. Before we turn to explicit calculations of transition rates, we use the first part of the chapter for some general considerations concerning emission and absorption, those being independent of the details of the interaction between matter and electromagnetic radiation. In this context we briefly look at master equations that govern the probabilities to be in certain states and we show how to use those to understand properties of the equilibrium state of interacting matter and radiation. We show that it is even possible to understand so-called black-body radiation without specifying the exact interaction Hamiltonian of matter and radiation.

Later in the chapter, we turn to concrete examples. First, we derive a Hamiltonian describing the interaction between matter and radiation. This finally allows us to evaluate all matrix elements needed to calculate explicit transition rates. We then apply these results to several examples. First, we evaluate the rate of spontaneous emission of photons by an excited atom. Using the dipole approximation we can arrive at a rough estimate for this rate. Then we go into more detail, discuss selection rules, and finally apply everything we learned to the simplest example of an excited hydrogen atom. The second example is the emission of radiation by free electrons. Such emission is impossible in vacuum but is allowed for electrons moving faster than the speed of light in a dielectric medium. This so-called Cherenkov radiation is responsible for the blue glow one can observe when high-energy electrons coming from nuclear reactors travel through water. A third example concerns *Bremsstrahlung*: electrons can scatter off an external potential while emitting a photon. We end the chapter with a simple discussion of the laser. We briefly outline the processes in gaseous lasers and derive master equations that describe the lasing process. Table 9.1 summarizes the content of the chapter.

9.1 Transition rates

Fermi's golden rule is a nice, simple, and powerful tool to evaluate transition rates in quantum systems, and this is why we included it in the introductory chapter on elementary quantum mechanics. Suppose that we have a system which is at some given time in an initial state $|i\rangle$, and there is a small perturbation $\hat{H}_{\text{int}}$ present which couples $|i\rangle$ to other states $|f\rangle$. We have shown in Chapter 1, using first-order time-dependent perturbation theory, that for short times the probability of finding the system in one of the other states $|f\rangle$ grows with a rate

$$\Gamma = \frac{2\pi}{\hbar} \sum_f |\langle f|\hat{H}_{\text{int}}|i\rangle|^2 \delta(E_i - E_f). \tag{9.1}$$

Therefore Γ can be interpreted as the decay rate of the state $|i\rangle$.

We already hinted in Chapter 1 at some problems concerning the golden rule. We see that for a discrete spectrum the delta function (which takes care of the conservation of energy) makes this decay rate always zero, except for the infinitesimally unlikely case that one level f is perfectly aligned with the level i, and then the rate becomes infinite. This is indeed rather unphysical, and in Chapter 1 we have already argued that the golden rule only makes sense when applied to a continuum of states. Indeed, if there are really only two states, we can solve the Schrödinger equation exactly, and we find that the system will never transit into $|f\rangle$. Rather, the wave function oscillates between $|i\rangle$ and $|f\rangle$. The same happens if we have any finite number of final states: the wave function oscillates, and sooner or later the system can be found back in its initial state. The situation, however, changes qualitatively if the number of final states is *infinite* and they form a continuous energy spectrum. In this case, the system never returns to the same initial state, and then we can indeed speak about a real *transition*.

The problem can also be formulated on a more fundamental level, reasoning from a puristic quantum mechanical point of view, which deals with the wave function only. Like the equations of classical mechanics, the time-dependent Schrödinger equation is time-reversible. Given a wave function at some moment of time, we can evaluate the wave function not only for future times, but we can also figure out what it was in the past. Already this is incompatible with the mere concept of a transition between two states of a system, either classical or quantum. Our intuitive notion of a transition implies that once a transition has occurred, the system does not remember anymore where it has been in the past (before the transition). This is in clear contradiction of our assumption that the Schrödinger equation gives a complete description of the time-dependent wave function for both directions of time. We discuss this issue in more detail in Chapters 11 and 12, the "dissipative" part of the book.

In this chapter, we do not worry about this problem, and exclusively calculate transition rates involving continuous spectra. The problem we extensively work on is that of radiative decay of excited atoms. The initial state of the system is thus an excited state of an atom together with the vacuum of the electromagnetic field. The final states we consider are the ground state of the atom and one of the states of the field in which one oscillator mode is excited – in the language of particles, one single photon has been emitted by the atom. These possible final states are labeled by the wave vector $\mathbf{k}$ of the emitted photon. Further, as argued above, a meaningful answer for the decay rate can only be obtained upon replacing the sum over $\mathbf{k}$ by an integration,

$$\sum_{\mathbf{k}} \quad \rightarrow \quad \mathcal{V} \int \frac{d\mathbf{k}}{(2\pi)^3}. \tag{9.2}$$

The main ingredients still missing before we can simply apply Fermi's golden rule and calculate explicit photon emission and absorption rates for atoms, are the matrix elements $\langle f|\hat{H}_{\text{int}}|i\rangle$ coupling initial and final states. The first thing we therefore have to figure out is $\hat{H}_{\text{int}}$, the Hamiltonian describing the coupling between the atomic states and the radiative field. Having an expression for this Hamiltonian, we can evaluate the matrix elements, and thus identify which transitions between atomic states are allowed and evaluate the corresponding rates. Before we do this, however, we first make a small detour, and give some more general considerations about emission and absorption.

9.2 Emission and absorption: General considerations

There is something to learn about emission and absorption that is general for all systems radiating, irrespective of their nature. We should learn this before turning to concrete systems. To do so, let us calculate transition rates at a rather abstract level, similar to what we did for the vacuum fluctuations in the previous chapter. We represent the interaction Hamiltonian as a sum over operator products $\hat{M}^{(a)}\hat{X}^{(a)}$, where the operators $\hat{M}$ deal with matter only and the operators $\hat{X}$ act on the bosonic field of radiation,

$$\hat{H}_{\text{int}} = \sum_{a} \hat{M}^{(a)}\hat{X}^{(a)}. \tag{9.3}$$

The field operators $\hat{X}^{(a)}$ are assumed to be linear in the CAPs,

$$\hat{X}^{(a)} = \sum_{q} (x_q^{(a)}\hat{a}_q + x_q^{*(a)}\hat{a}_q^{\dagger}). \tag{9.4}$$

We now consider two states of matter, $|A\rangle$ and $|B\rangle$, and study radiative transitions between these states. For definiteness, let us take $E_A > E_B$. The quantum states of the whole system are in principle direct products of matter states and states of radiation,

$$|A\rangle \otimes |\{n_q\}\rangle \quad \text{and} \quad |B\rangle \otimes |\{n_q'\}\rangle. \tag{9.5}$$

We concentrate solely on processes involving a certain mode k. In this case, we can disregard the occupation numbers of all modes except k, since all other occupations are not changing in the course of the transition. We then see that there are two processes which we can investigate: (i) absorption of radiation (a transition from B to A, with increasing energy), and (ii) emission of radiation (from A to B, with decreasing energy). For the case of absorption, the initial and final states are

$$|i\rangle = |B\rangle \otimes |n_k\rangle, \tag{9.6}$$

$$|f\rangle = |A\rangle \otimes |n_k - 1\rangle, \tag{9.7}$$

while for emission

$$|i\rangle = |A\rangle \otimes |n_k\rangle, \tag{9.8}$$

$$|f\rangle = |B\rangle \otimes |n_k + 1\rangle, \tag{9.9}$$

where we use shorthand notation for a state of the field, writing n_k only.

To use Fermi's golden rule, we need the transition matrix elements. For absorption,

$$
\begin{aligned}
H_{\text{abs}} = \langle f|\hat{H}_{\text{int}}|i\rangle &= \sum_a \langle n_k - 1|\hat{X}^{(a)}|n_k\rangle \langle A|\hat{M}^{(a)}|B\rangle \\
&= \sum_a M_{AB}^{(a)} x_k^{(a)} \langle n_k - 1|\hat{a}|n_k\rangle \\
&= \sum_a M_{AB}^{(a)} x_k^{(a)} \sqrt{n_k},
\end{aligned}
\tag{9.10}
$$

and the square of the matrix element that enters the transition rate is

$$|H_{\text{abs}}|^2 = n_k \left| \sum_a M_{AB}^{(a)} x_k^{(a)} \right|^2. \tag{9.11}$$

For emission we find similarly,

$$
\begin{aligned}
H_{\text{em}} = \langle f|\hat{H}_{\text{int}}|i\rangle &= \sum_a \langle n_k + 1|\hat{X}^{(a)}|n_k\rangle \langle B|\hat{M}^{(a)}|A\rangle \\
&= \sum_a M_{AB}^{*(a)} x_k^{*(a)} \langle n_k + 1|\hat{a}^\dagger|n_k\rangle \\
&= \sum_a M_{AB}^{*(a)} x_k^{*(a)} \sqrt{n_k + 1},
\end{aligned}
\tag{9.12}
$$

and the square is

$$|H_{\text{em}}|^2 = (n_k + 1) \left| \sum_a M_{AB}^{(a)} x_k^{(a)} \right|^2. \tag{9.13}$$

Using Fermi's golden rule, we can thus write for the absorption and emission rates involving the mode k only

$$\Gamma_{\text{abs}}(k) = \Gamma_k n_k, \tag{9.14}$$

$$\Gamma_{\text{em}}(k) = \Gamma_k(n_k + 1), \tag{9.15}$$

where

$$\Gamma_k = \frac{2\pi}{\hbar}\left|\sum_a M_{AB}^{(a)} x_k^{(a)}\right|^2 \delta(\hbar\omega_k - E_A + E_B). \tag{9.16}$$

We immediately notice a remarkable property of these two rates: their ratio depends only on the occupation number of the corresponding mode,

$$\frac{\Gamma_{\text{abs}}(k)}{\Gamma_{\text{em}}(k)} = \frac{n_k}{n_k + 1}, \tag{9.17}$$

and does not depend on the concrete parameters of the interaction.

To compute the total rates (not the partial rates involving only one specific mode k), one simply sums over all modes satisfying $\hbar\omega_k = E_A - E_B$. We note that there are usually many discrete modes satisfying this, each having its own occupation number n_k. We thus replace the actual occupation numbers by the average occupation number $\bar{n}$ of all modes satisfying $\hbar\omega_k = E_A - E_B$, and we assume that $\bar{n}$ depends on frequency only. Then for the total rates we obtain

$$\Gamma_{\text{abs}} = \Gamma\bar{n}, \tag{9.18}$$

$$\Gamma_{\text{em}} = \Gamma(\bar{n} + 1), \tag{9.19}$$

where

$$\Gamma = \sum_k \Gamma_k = \sum_k \frac{2\pi}{\hbar}\left|\sum_a M_{AB}^{(a)} x_k^{(a)}\right|^2 \delta(\hbar\omega_k - E_A + E_B). \tag{9.20}$$

We see that also the ratio of the total rates does not depend on the details of the interaction and is expressed in terms of $\bar{n}$,

$$\frac{\Gamma_{\text{abs}}}{\Gamma_{\text{em}}} = \frac{\bar{n}}{\bar{n} + 1}. \tag{9.21}$$

Let us investigate the *emission* rate (9.19) in more detail. The first term, $\Gamma\bar{n}$, is proportional to the occupation number of the relevant modes, and is called the *stimulated* emission rate. The second term, simply Γ, is called the *spontaneous* emission rate. The idea is that in the absence of any radiation, when $\bar{n} = 0$, there is still some emission from an excited state of matter just "by itself." The presence of other photons then further increases the rate, in other words, "stimulates" it. This picture, although widely used, does not immediately explain why the *absorption* rate and stimulated emission rate are actually the same.

Another way of looking at these emission and absorption rates is to regard them as a result of time-dependent fluctuations, to be more specific, of the noise of the variables $\hat{X}^{(a)}$. Transitions between $|A\rangle$ and $|B\rangle$ are in fact proportional to the spectral intensity of this noise at the "right" frequency $(E_A - E_B)/\hbar$, very much like a sound coming from a radio-set tuned to a certain frequency. This noise can be separated into a "classical" and a "quantum" part. We have already seen that in the limit of large occupation numbers bosons can be treated as classical fields. Applying this to the rates, we recognize that for large occupation numbers, both absorption and stimulated emission come from a single source: the classical noise. Both rates are the same since the states $|A\rangle$ and $|B\rangle$ are brought into resonance and there is an equal probability to loose or to receive energy from a classical field. This is in contrast with the "quantum" noise arising from the vacuum fluctuations of $\hat{X}^{(a)}$. Since the vacuum fluctuations are a property of the ground state, they cannot supply any energy but can only consume it. This gives rise to spontaneous emission, to decay of the excited state $|A\rangle$. So, as mentioned, radiative decay to some extent probes the vacuum fluctuations of the electromagnetic field. We review this picture in more detail in Section 11.3 and in Chapter 12.

9.2.1 Master equations

What are these transition rates actually good for? They allow a bridge between a quantum and a (classical) statistical description of many-body systems. To see the difference between the descriptions in general, let us talk about quantum mechanics in a basis of wave functions labeled by $|i\rangle$. The pure quantum description operates with the exact wave function in this basis, $|\psi\rangle = \sum_i \psi_i |i\rangle$. The classical statistical description, however, deals only with probabilities to be in a certain state of this basis, p_i. Suppose we figured out (using quantum mechanics) the transition rates from a state $|i\rangle$ to other states $|j\rangle$ – let us call these rates $\Gamma_{i \to j}$. We can then use these rates to create a set of equations governing the dynamics of the (classical) occupation probabilities. Such equations are called *master equations* and in general look like

$$\frac{dp_i}{dt} = -p_i \sum_j \Gamma_{i \to j} + \sum_j p_j \Gamma_{j \to i}. \tag{9.22}$$

This equation is nothing but a probability balance. The change per unit time of the probability of being in the state $|i\rangle$ is contributed by two groups of terms. The terms of the first group encompass the transitions *from* $|i\rangle$ to all other states. This reduces the probability p_i, so these terms come in (9.22) with a minus sign. The factor p_i accounts for the fact that the system must be in the state $|i\rangle$ before such a transition can take place. The second group encompasses transitions *to* $|i\rangle$ from all other states. They increase p_i, so they enter (9.22) with a plus sign. In this group, each individual rate comes with a factor p_j that accounts for the fact that the system must be in the state $|j\rangle$ for such a transition to happen.

Let us now turn again to the system we discussed before: only two states of matter and many boson modes. The general probabilities to find one of the states look terribly complicated since they in principle depend on the occupation numbers of all the modes, $P(A; \{n_k\})$ and $P(B; \{n_k\})$. Fortunately, the field modes and states of matter can be regarded as independent, so the probabilities factorize,

$$P(A; \{n_k\}) = P_A \prod_k p_k(n_k) \quad \text{and} \quad P(B; \{n_k\}) = P_B \prod_k p_k(n_k). \tag{9.23}$$

Here, $P_{A(B)}$ is the probability of being in the state $A(B)$, and $p_k(n)$ is the probability of having n photons in the mode k.

For the matter part of the system, the master equation is defined by the total rates Γ_{abs} and Γ_{em},

$$\frac{dP_A}{dt} = -\Gamma_{\text{em}}P_A + \Gamma_{\text{abs}}P_B, \qquad (9.24)$$

$$\frac{dP_B}{dt} = -\Gamma_{\text{abs}}P_B + \Gamma_{\text{em}}P_A. \qquad (9.25)$$

Substituting the rates we derived above, we find

$$\frac{dP_A}{dt} = -\bar{n}\Gamma P_A + (\bar{n} + 1)\Gamma P_B, \qquad (9.26)$$

$$\frac{dP_B}{dt} = -(\bar{n} + 1)\Gamma P_B + \bar{n}\Gamma P_A. \qquad (9.27)$$

(To remind you, $\bar{n}$ is the average occupation number of the relevant modes.)

Let us now turn to the master equation for the field mode k. Each process of emission increases the number of photons by 1, and each absorption process decreases it by 1. Besides, the rates Γ_{em} and Γ_{abs} depend on the number of photons in the mode. So we write for the mode k

$$\frac{dp_k(n)}{dt} = \Gamma_{\text{em}}(k, n-1)P_A p_k(n-1) + \Gamma_{\text{abs}}(k, n+1)P_B p_k(n+1)$$
$$- [\Gamma_{\text{em}}(k, n)P_A + \Gamma_{\text{abs}}(k, n)P_B]p_k(n). \qquad (9.28)$$

Substituting the rates (9.14) and (9.15), we find

$$\frac{dp_k(n)}{dt} = \Gamma_k n P_A p_k(n-1) + \Gamma_k(n+1)P_B p_k(n+1) - \Gamma_k[(n+1)P_A + nP_B]p_k(n). \quad (9.29)$$

9.2.2 Equilibrium and black-body radiation

The simplest thing one can do with master equations is to find their stationary solutions, where all probabilities are in *equilibrium*. We do this simply by setting $dP/dt = 0$ and solving the resulting equations. To start with, we assume that the matter is brought to a thermal equilibrium state by some other source, not by the radiation we are considering. This implies that the probabilities $P_{A,B}$ conform to the Boltzmann distribution with temperature T,

$$P_{A,B} \propto \exp\{-E_{A,B}/k_B T\}, \quad \text{so that} \quad \frac{P_A}{P_B} = \exp\{-(E_A - E_B)/k_B T\}. \qquad (9.30)$$

If we now use the master equations (9.26) and (9.27), we obtain

$$\frac{P_B}{P_A} = \frac{\Gamma_{\text{em}}}{\Gamma_{\text{abs}}} = \exp\left\{\frac{E_A - E_B}{k_B T}\right\} = \exp\left\{\frac{\hbar\omega}{k_B T}\right\} = \frac{\bar{n} + 1}{\bar{n}}. \qquad (9.31)$$

From this we can solve for $\bar{n}$ at a certain energy $\hbar\omega = E_A - E_B$. We find

$$\bar{n} = \frac{1}{\exp\{\hbar\omega/k_BT\} - 1} = n_B(\omega), \tag{9.32}$$

which is the well-known Bose distribution corresponding to the temperature T!

Let us also work out the equilibrium condition for the master equation (9.29). Here we can use a detailed balance condition. In equilibrium all probabilities $p_k(n)$ are stationary, which implies that the *net* rate for transitions from a state with n photons to one with $n + 1$ photons must vanish. For the two underlying processes – going from $n + 1$ to n by absorption of a photon by the matter, and going from n to $n + 1$ by emission – this means that they must exactly cancel each other. If we set these two rates equal, we find

$$\Gamma_k(n+1)P_B p_k(n+1) = \Gamma_k(n+1)P_A p_k(n), \quad \text{so that} \quad \frac{p_k(n+1)}{p_k(n)} = \frac{P_A}{P_B}.$$

The ratio of two probabilities for neighboring n is constant and determined by the temperature of the matter. From this we infer that

$$p_k(n) = \frac{e^{-\hbar\omega_k n/k_BT}}{1 - e^{-\hbar\omega_k/k_BT}}, \tag{9.33}$$

where the denominator comes from the normalization condition $\sum_n p_k(n) = 1$. We see that this is nothing but the Boltzmann distribution corresponding to the energy $E(n) = \hbar\omega_k n$.

Radiation which is in equilibrium with emitting and absorbing matter is called *black-body radiation*. Sunlight is an example of (almost) black-body radiation with a temperature corresponding to the temperature of the solar photosphere (5780 K). A typical application of the above theory is to evaluate the energy flux of black-body radiation out of the body. To calculate it, let us consider the region in space between two infinite black planes. The left plane is kept at a temperature T and the right one is kept cold, having a temperature $T_1 \ll T$. All photons that travel to the right ($k_z > 0$) have been emitted from the left plane[1] and thus bear occupation numbers $n_B(\omega_k)$ corresponding to the temperature T.

Recalling the definition of energy flux – the Poynting vector of the electromagnetic field – we find for the flux per unit area

$$\frac{S_z}{A} = 2_p \int \frac{d\mathbf{k}}{(2\pi)^3} n_B(\omega_k)\Theta(k_z)c^2\hbar k_z. \tag{9.34}$$

The factor $\Theta(k_z)$ singles out photons traveling to the right. We integrate this expression over all directions of $\mathbf{k}$ and replace $k \to \omega/c$ to find the flux per frequency interval,

$$\frac{S_z}{A} = \int d\omega\, n_B(\omega)\frac{\hbar\omega^3}{4\pi^2c^2}, \tag{9.35}$$

and we arrived at the famous Planck law (solid line in Fig. 9.1). It was first guessed by Planck to fit experimental data accumulated by the end of the 19th century. The theory

---

[1] At this point "black" is important. The planes should be black, that is, ideally absorbing, since otherwise photons could reflect from the planes and we would no longer know for sure that photons traveling to the right originated from the left plane.

Max Planck (1858–1947)
Won the Nobel Prize in 1918 "in recognition of the services he rendered to the advancement of Physics by his discovery of energy quanta."

Max Planck was born into an intellectual family, his father being Professor of Law at the University of Kiel. Planck was educated at the Universities of Munich and Berlin, and graduated in 1879 on a thesis about the second law of thermodynamics. After his graduation, he spent several years as an unpaid lecturer in Munich, until he was appointed Professor at the University of Kiel in 1885. He stayed there for only four years, after which he succeeded Kirchhoff at Berlin University. He stayed at this university until his retirement in 1926.

In Berlin, Planck started working on the problem of black-body radiation. The theories existing at the time were unable to explain the measured spectra of radiation and predicted divergences for low or high frequencies. Planck's new law, presented in 1900, was based on the hypothesis that radiation could only be emitted in fixed quanta of energy hf, where f is the frequency and h Planck's constant. Planck regarded this quantization merely as a formal assumption, and did not realize its profound physical implications. Although thus being the father of quantum theory, Planck remained skeptical toward the standard interpretation of quantum mechanics put forward by Bohr and Heisenberg in the 1920s.

After the age of 50, Planck was struck by many personal tragedies. His first wife died suddenly of tuberculosis in 1909, his oldest son died in the first world war at the front in Verdun, and his two twin daughters both died giving birth in 1917 and 1919. He seems however to have endured all these losses relatively stoically. After the Nazis came to power in 1933, Planck saw many of his Jewish friends and colleagues flee the country. He himself, however, felt it his duty to stay in Nazi Germany, although he strongly opposed the politics, especially the persecution of the Jews. He did not hide his views, which at some point even led to a serious investigation of his ancestry, resulting in an official quantification of Planck as being a one-sixteenth Jew. After his house in Berlin had been destroyed by allied bombings in 1944 and his second son Erwin was executed in 1945 by the Nazis after an attempt to assassinate Hitler, all the will to live had left him. He died in in Göttingen in 1947, aged 89.

of that time gave, as we would call it now, a classical result, which can be obtained from Planck's law by taking the limit $\hbar \to 0$ (dotted line in the figure),

$$\frac{S_z}{A} = \int d\omega \, k_B T \frac{\omega^2}{4\pi^2 c^2}. \tag{9.36}$$

The problem with this expression is that the integral diverges at large frequencies, yielding an infinite light intensity: an *ultraviolet catastrophe*. Needless to say, this contradicted the

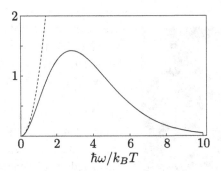

Fig. 9.1 Planck's law for the energy flux in black-body radiation (solid line). In the limit $\hbar \to 0$ one recovers the result from classical theory (dotted line), resulting in an ultraviolet catastrophe.

experiments. Planck's law effectively cuts the integral at $\omega \simeq k_B T / \hbar$, and the maximum flux corresponds to $\hbar\omega \approx 2.8 \, k_B T$. The integral over frequencies in (9.35) is finite, and yields

$$\frac{S_z}{A} = \frac{\pi^2 \hbar}{60 c^2} \left(\frac{k_B T}{\hbar} \right)^4,$$ (9.37)

which is called Stefan's law and was first discovered experimentally.

9.3 Interaction of matter and radiation

Let us now turn to the more concrete problem of an actual atom interacting with the radiative field. As we already noted at the end of Section 9.1, the main ingredient still missing to evaluate explicit transition rates is the Hamiltonian describing the *interaction* between matter and radiation.

As a first step in figuring out the nature of this interaction, let us try to derive the Schrödinger equation for a charged particle in the presence of a (classical) electromagnetic field. Here we could proceed as usual: look at the classical equations describing the particle in the field, get inspired by these equations and make an educated guess for the corresponding quantum mechanical Hamiltonian, and then hopefully show that our guess turns out to make sense. Let us, however, proceed in a different way.

In Section 7.3.3, we showed that the description of the electromagnetic field in terms of a vector and scalar potential still has one degree of freedom too many. Indeed, the gauge transformation $\mathbf{A} \to \mathbf{A} + \nabla \chi$ and $\varphi \to \varphi - \partial_t \chi$ turned out not to change any physical property of the "real" electric and magnetic fields. In Chapter 7 we then merely used this gauge freedom to manipulate the Maxwell equations to a form most convenient for describing the fields in vacuum – we just used it as a tool to simplify the maths. But in fact this gauge freedom has a deeper meaning, some aspects of which become clear below.

If we think again about particles, we realize that we have a similar freedom of gauge. Multiplying the wave function of a particle with an arbitrary phase factor $e^{i\alpha}$ does not change any physical property of the particle. Inspired by this similarity, we make here an important claim, at this point a mere hypothesis: the Hamiltonian of a single particle in the presence of an electromagnetic field can be obtained if one *identifies* the gauge transformation of the wave function with the gauge transformation of the potentials.

Let us see what this statement can bring us. The single-particle Schrödinger equation (assuming zero vector and scalar potentials),

$$i\hbar \frac{\partial \psi}{\partial t} = -\frac{\hbar^2}{2m} \frac{\partial^2}{\partial \mathbf{r}^2} \psi, \tag{9.38}$$

remains the same if we multiply ψ with a phase factor, for instance $\psi \to e^{\frac{i}{\hbar}e\chi}\psi$ where χ does not depend on time and coordinates. Let us now instead multiply it with a phase factor which *does* depend on time and coordinates,

$$\psi(\mathbf{r}, t) \to e^{\frac{i}{\hbar}e\chi(\mathbf{r},t)}\psi(\mathbf{r}, t). \tag{9.39}$$

The Schrödinger equation for the new wave function becomes

$$i\hbar \frac{\partial \psi}{\partial t} + e \frac{\partial \chi}{\partial t}\psi = -\frac{\hbar^2}{2m} \left(\frac{\partial}{\partial \mathbf{r}} - i\frac{e}{\hbar}\nabla\chi \right)^2 \psi. \tag{9.40}$$

We now *identify* the χ in the phase factor with the χ in the gauge transformation of the electromagnetic field. This means that the multiplication of a wave function by the phase factor $e^{\frac{i}{\hbar}e\chi(\mathbf{r},t)}$ should have the same effect as performing the gauge transformation $\mathbf{A} \to \mathbf{A} + \nabla\chi$ and $\varphi \to \varphi - \partial_t\chi$. The original vector and scalar potentials were zero, and after this gauge transformation they thus read $\mathbf{A} = \nabla\chi$ and $\varphi = -\partial_t\chi$. We can thus substitute in (9.40)

$$\nabla\chi \to \mathbf{A} \quad \text{and} \quad \frac{\partial \chi}{\partial t} \to -\varphi, \tag{9.41}$$

to obtain the Schrödinger equation in the presence of the potentials,

$$i\hbar \frac{\partial \psi}{\partial t} = \left\{ -\frac{\hbar^2}{2m} \left(\frac{\partial}{\partial \mathbf{r}} - i\frac{e}{\hbar}\mathbf{A} \right)^2 + e\varphi \right\} \psi. \tag{9.42}$$

We recognize this effective Hamiltonian in the presence of electric and magnetic fields from classical mechanics, and it can be justified by a classical argument. The gradient of the scalar potential reproduces the electric force acting on the charged particle, while the Lorentz force due to the magnetic field is reproduced by making the momentum "long" with the aid of the vector potential,

$$\mathbf{p} \to \mathbf{p} - e\mathbf{A}, \tag{9.43}$$

which in quantum mechanics should be changed to

$$-i\hbar\nabla \to -i\hbar\nabla - e\mathbf{A}(\mathbf{r}, t). \tag{9.44}$$

With (9.42) we have thus arrived at the same expression we probably would have guessed starting from classical theory. The new element here is the way – the gauge way – we have derived it. By the way, this gauge way does not seem flawless. Formally, the derivation is done for potentials that do not give rise to any electric and magnetic field. Logically, the equation does not have to hold for potentials that do produce fields. However, it *does*.

This is why the electromagnetic field is said to be a *gauge field*. A gauge field is associated with a certain continuous group of transformations of the particle field operators. If the parameters characterizing a particular transformation do not depend on time and coordinates, then the field equations do not change. In this case, the transformation is an element of the symmetry group. However, if the parameters do depend on coordinates and time, the transformation produces extra terms in the field equations. The (time-)gradients of the parameters of the transformation are associated with the gauge field, and these gradients represent the interaction mediated by the gauge field. We have just run this procedure for the electromagnetic field. In this case, the transformation was a multiplication with a phase factor (the so-called $U(1)$ group), the function $\chi(\mathbf{r}, t)$ was the parameter of the transformation, and its (time-)gradient introduced the vector and scalar potentials into the Schrödinger equation, thus describing the interaction between the particle and the field.

Gauge fields were invented by discontented individuals who could not just accept that particles and fields interact, but wondered *why* they do so. The best they could come up with is the "principle of local gauge invariance" just described. It is now believed that at least three of the four fundamental interactions – the electromagnetic, weak and strong interaction – are gauge fields. They all can be described together assuming the gauge group $SU(3) \times SU(2) \times U(1)$ in the framework of the Standard Model of particle physics.

Now that we have seen how the electromagnetic field enters the Hamiltonian of charged particles, we can specify the interaction between matter and irradiation. Writing the full Hamiltonian for charged particles in a field, we combine the above results with those of the previous chapter,

$$\hat{H} = \int d\mathbf{r}\, \hat{\psi}^\dagger \left\{ \frac{(-i\hbar\nabla - e\hat{\mathbf{A}})^2}{2m} + e\varphi \right\} \hat{\psi} + \hat{H}_{\text{e.m.}}, \qquad (9.45)$$

with $\hat{H}_{\text{e.m.}} = \sum_{\mathbf{k}\alpha} \hbar\omega_{\mathbf{k}}(\hat{n}_{\mathbf{k}\alpha} + \frac{1}{2})$ describing the electromagnetic field, and the first term describing the particles in the presence of this field.

Since the scalar potential φ is the result of a finite charge density ρ, it is only present in (9.45) due to the particles themselves. We can thus regard this term as describing interaction *between* the particles. Instead of keeping φ in the above expression, we can add an interaction term to the Hamiltonian,

$$\hat{H}^{(2)} = \frac{1}{2} \int d\mathbf{r}_1 d\mathbf{r}_2\, U(\mathbf{r}_1 - \mathbf{r}_2)\hat{\psi}^\dagger(\mathbf{r}_1)\hat{\psi}^\dagger(\mathbf{r}_2)\hat{\psi}(\mathbf{r}_2)\hat{\psi}(\mathbf{r}_1), \qquad (9.46)$$

where the pairwise potential is given by the Coulomb law

$$U(\mathbf{r}) = \frac{e^2}{4\pi\varepsilon_0\, |\mathbf{r}|}. \qquad (9.47)$$

With this substitution, the Hamiltonian (9.45) is rewritten as

$$\hat{H} = \int d\mathbf{r}\, \hat{\psi}^\dagger \frac{(-i\hbar\nabla - e\hat{\mathbf{A}})^2}{2m} \hat{\psi} + \hat{H}_{\text{e.m.}} + \hat{H}^{(2)}. \qquad (9.48)$$

We can understand this Hamiltonian better by separating it into three blocks: one for matter, one for radiation, and a third block for the interaction,

$$\hat{H} = \underbrace{\hat{H}_0 + \hat{H}^{(2)}}_{\text{matter}} + \underbrace{\hat{H}_{\text{e.m.}}}_{\text{radiation}} + \underbrace{\int d\mathbf{r}\, \hat{\psi}^\dagger \left(\frac{e}{2m} i\hbar\{\nabla, \hat{\mathbf{A}}\} + \frac{(e\hat{\mathbf{A}})^2}{2m_i} \right) \hat{\psi}}_{\text{interaction}}, \qquad (9.49)$$

where $\hat{H}_0 = -\int d\mathbf{r}\, \hat{\psi}^\dagger (\hbar^2\nabla^2/2m)\hat{\psi}$ is the Hamiltonian for free particles. This separation looks a bit artificial, and indeed it is. Nature itself does not separate matter and radiation in the way we humans do, quanta of radiation are actually particles and therefore matter. So this separation has been performed just for our convenience.

It also looks as if we have missed something. As a matter of fact, we know that there is a magnetic moment associated with the spin of a charged particle, and an energy associated with the orientation of this moment in a magnetic field. The Hamiltonian (9.49) apparently does not account for this effect. The reason is that this interaction is in fact of relativistic origin. We come back to this in Chapter 13.

We now concentrate on the interaction term. Its second-quantized form for a single particle is given above, and for many identical particles we can write

$$\hat{H}_{\text{int}} = \sum_n \left\{ -\frac{e}{m}\hat{\mathbf{A}}(\mathbf{r}_n) \cdot \hat{\mathbf{p}}_n + \frac{e^2}{2m}\hat{\mathbf{A}}^2(\mathbf{r}_n) \right\}, \qquad (9.50)$$

where n labels the particles.

Control question. Do you understand why we can simplify $i\hbar\{\nabla, \hat{\mathbf{A}}\} \to -2\hat{\mathbf{A}} \cdot \hat{\mathbf{p}}$?

We always treat the interaction term as a small perturbation and restrict ourselves to first-order corrections. This means that we should keep only terms of first power in $\hat{\mathbf{A}}$,

$$\hat{H}_{\text{int}} = -\int d\mathbf{r}\, \hat{\mathbf{A}}(\mathbf{r}) \cdot \hat{\mathbf{j}}(\mathbf{r}), \qquad (9.51)$$

where $\hat{\mathbf{j}}(\mathbf{r})$ is the operator of electric current density. It reads

$$\hat{\mathbf{j}}(\mathbf{r}) = -\frac{i\hbar e}{2m} \left\{ \hat{\psi}^\dagger(\mathbf{r})[\nabla\hat{\psi}(\mathbf{r})] - [\nabla\hat{\psi}^\dagger(\mathbf{r})]\hat{\psi}(\mathbf{r}) \right\}, \qquad (9.52)$$

in terms of field operators, or

$$\hat{\mathbf{j}}(\mathbf{r}) = \sum_n e\frac{\hat{\mathbf{p}}_n}{m} \delta(\mathbf{r} - \mathbf{r}_n), \qquad (9.53)$$

for the N-particle Hamiltonian.

Now that we have derived a Hamiltonian describing the interaction between matter and radiation, we can finally calculate explicit transition rates for systems emitting or absorbing electromagnetic radiation. We start with the simplest setup, that of a single atom interacting with the electromagnetic field.

9.4 Spontaneous emission by atoms

The dense hot gas of the solar photosphere emits light in a continuous spectrum almost identical to the black-body spectrum we derived in Section 9.2.2. In 1814, Joseph von Fraunhofer discovered that within this continuous spectrum, there are in fact many tiny dark lines. Later, these lines were attributed to the absorption of solar radiation by sparse colder gases far away from the sun. When the atoms of such a gas are sufficiently far away from each other, they absorb and emit radiation only at a set of discrete frequencies given by the differences of energies of their discrete states. In 1854, the emission spectrum of such a sparse gas was measured by David Alter and it was shown to match exactly the dark absorption lines observed in the spectrum of sunlight. This then paved the way for the development of the theory of atoms and ultimately for the birth of quantum mechanics.

Let us concentrate here on spontaneous *emission* and try to calculate the spontaneous decay rate of an atom in an excited state. Here we first briefly remind you of the method with which we evaluate this rate. As we know, the word "spontaneous" in the context of emission means that there are no photons in the initial state of the system. The initial state is therefore a direct product of an excited atomic state $|A\rangle$, and the photon vacuum, $|i\rangle = |A\rangle \otimes |0\rangle$. The atom can emit a photon and decay to state $|B\rangle$ which has a lower energy. For such a process, the possible final photon states are labeled by the index of the mode where the photon has been created, $|f_k\rangle = |B\rangle \otimes |1_k\rangle$. The total rate of such radiative decay can be calculated with Fermi's golden rule and is contributed to by all possible final states allowed by energy conservation. The modes are indexed by the photon wave vector $\mathbf{k}$ and two possible polarizations α, the label k thus actually is $(\mathbf{k}, \alpha)$. The decay rate reads

$$\Gamma = \frac{2\pi}{\hbar} \sum_{\mathbf{k},\alpha} |H_{em}(k)|^2 \delta(E_A - E_B - \hbar\omega_{\mathbf{k}}), \tag{9.54}$$

where H_{em} is the matrix element of the interaction Hamiltonian between the initial and final state, $H_{em}(k) = \langle f_k | \hat{H}_{int} | i \rangle$. As we know, in order to get a sensible result, the summation over the discrete $\mathbf{k}$ has to be replaced by an integration,

$$\Gamma = \frac{2\pi}{\hbar} \mathcal{V} \sum_\alpha \int \frac{d\mathbf{k}}{(2\pi)^3} |H_{em}(k)|^2 \delta(E_A - E_B - \hbar\omega_{\mathbf{k}}). \tag{9.55}$$

9.4.1 Dipole approximation

The transition matrix element $H_{em}(k)$ that appears in the above expressions depends on both $\mathbf{k}$ and the polarization vector $\mathbf{e}_\alpha$. For atoms, one may evaluate it in a rather simplifying approximation, the *dipole* approximation. To explain this approximation, we assume that the atom is at rest and we put the origin of the coordinate system at the position of its nucleus. As we have shown in Section 9.3, the interaction between matter and radiation may be presented as

$$\hat{H}_{int} = -\int d\mathbf{r}\, \hat{\mathbf{A}}(\mathbf{r}) \cdot \hat{\mathbf{j}}(\mathbf{r}), \tag{9.56}$$

where $\hat{\mathbf{j}}$ consists of matter (electron) variables and $\hat{\mathbf{A}}$ represents the radiation. Since the electrons in the atom are confined to distances of $a_B \approx 0.1$ nm from the nucleus, the operator $\hat{\mathbf{j}}(\mathbf{r})$ is non-zero only at $r \simeq a_B$. As to the radiation operator, its typical space scale is set by the typical photon wave vectors, k, involved. As mentioned in Section 8.4.3 when we discussed vacuum fluctuations, this typical wave vector is $k \simeq E_{\mathrm{at}}/\hbar c$, and the corresponding space scale (the wave length of the emitted light) is $\lambda = 2\pi/k \simeq \hbar c/E_{\mathrm{at}} \gg a_B$. The conclusion is that the vector potential can be regarded as a constant operator at the atomic scale, and the interaction operator reads

$$\hat{H}_{\mathrm{int}} = -\int d\mathbf{r}\, \hat{\mathbf{A}}(0) \cdot \hat{\mathbf{j}}(\mathbf{r}). \qquad (9.57)$$

Using the expressions derived in Sections 9.3 and 8.4.2, we replace

$$\int d\mathbf{r}\, \hat{\mathbf{j}}(\mathbf{r}) \rightarrow \sum_n e\frac{\hat{\mathbf{p}}_n}{m}, \qquad (9.58)$$

$$\hat{\mathbf{A}}(0) \rightarrow \sum_{\mathbf{k},\alpha} \sqrt{\frac{\hbar}{2\varepsilon_0\omega_{\mathbf{k}}\mathcal{V}}}\, \mathbf{e}_\alpha\big[\hat{a}_{\mathbf{k}\alpha} + \hat{a}^\dagger_{\mathbf{k}\alpha}\big]. \qquad (9.59)$$

The operator of current is now the sum over the velocities of all electrons in the atom multiplied by the electronic charge.

To evaluate the spontaneous emission rate, we first of all need the matrix elements of this operator between the atomic states $|A\rangle$ and $|B\rangle$. There is a general relation between matrix elements of velocity and coordinate that comes from the fact that the velocity is just the time derivative of the coordinate. For one electron, we derive

$$\langle A|\frac{\hat{\mathbf{p}}}{m}|B\rangle = \langle A|\dot{\hat{\mathbf{r}}}|B\rangle = \frac{i}{\hbar}\langle A|[\hat{H},\hat{\mathbf{r}}]|B\rangle = \frac{i}{\hbar}(E_A - E_B)\langle A|\hat{\mathbf{r}}|B\rangle \equiv i\omega_{AB}\mathbf{r}_{AB}, \qquad (9.60)$$

where $\omega_{AB} \equiv (E_A - E_B)/\hbar$. For many electrons, one simply sums over all the electrons. Therefore, the matrix element needed is that of the operator of the atomic *dipole moment* $\sum_n e\,\mathbf{r}_n$, and this is why it is called the dipole approximation. It is based on a small parameter $a_B/\lambda \simeq \alpha$, with α being the fine structure constant.

Summarizing all transformations we made, we find an expression for the square of the transition matrix element in the dipole approximation,

$$\mathcal{V}|H_{\mathrm{em}}|^2 = |\mathbf{r}_{AB} \cdot \mathbf{e}_\alpha|^2 \frac{e^2\hbar\omega_{AB}}{2\varepsilon_0}. \qquad (9.61)$$

To obtain an order-of-magnitude estimate for the matrix element, we note that both $\hbar\omega_{AB}$ and $e^2/\varepsilon_0 a_B$ are of the atomic energy scale and $\mathbf{r}_{AB} \simeq a_B$. This gives

$$\mathcal{V}|H_{\mathrm{em}}|^2 \simeq E_{\mathrm{at}}^2 a_B^3. \qquad (9.62)$$

9.4.2 Transition rates

Now we have everything we need to evaluate the transition rates characterizing the radiative decay. To begin with, let us write Fermi's golden rule in a differential form. We look at a partial transition rate $d\Gamma$ that arises from the emission of photons with polarization α and wave vectors in an infinitesimally small volume element $d\mathbf{k}$ of wave vector space. Using the expression for $\mathcal{V}|H_{em}|^2$ we found above, we have

$$d\Gamma = \frac{2\pi}{\hbar}|\mathbf{r}_{AB} \cdot \mathbf{e}_\alpha|^2 \frac{e^2\hbar\omega_{AB}}{2\varepsilon_0}\delta(\hbar\omega_{AB} - \hbar ck)\frac{d\mathbf{k}}{(2\pi)^3}. \tag{9.63}$$

As usual, we separate the integration over $\mathbf{k}$ into an integration over its modulus and over all directions, $d\mathbf{k} \to k^2\, dk\, d\Omega$, where $d\Omega$ is an element of solid angle. We then integrate over the modulus of $\mathbf{k}$ to obtain the partial rate of emission of photons with polarization α into the element of solid angle $d\Omega$. This integration absorbs the δ-function and gives

$$d\Gamma = \frac{2\pi}{\hbar}|\mathbf{r}_{AB} \cdot \mathbf{e}_\alpha|^2 \frac{e^2\hbar\omega_{AB}}{2\varepsilon_0}\frac{\omega_{AB}^2}{8\pi^3\hbar c^3}d\Omega. \tag{9.64}$$

To see the physical significance of this, let us place some excited atoms at the origin of the coordinate system and detect their radiation with a detector which has an area A and is located at the point $\mathbf{R}$. Provided the size of the detector is much smaller than the distance R between the detector and the atoms, it collects their radiation emitted in the small solid angle $d\Omega = A/R^2$ in the direction $\mathbf{R}/R$, thereby measuring $d\Gamma$.

Control question. Given the number of atoms in the excited state, N_A, what is the energy flux collected by the detector?

At first sight, this rate does not depend on direction but rather on the polarization of the light emitted, $\mathbf{e}_\alpha$. However, as we know, not all directions of polarization are allowed. Given a wave vector $\mathbf{k}$, only polarization vectors orthogonal to $\mathbf{k}$ are allowed. We can thus choose one of the vectors, for instance $\mathbf{e}_2$, in such a way that it is orthogonal to both $\mathbf{r}_{AB}$ and $\mathbf{k}$ (see Fig. 9.2). Since the transition matrix element for emission of a photon with polarization α is proportional to $\mathbf{r}_{AB} \cdot \mathbf{e}_\alpha$, emission with polarization along $\mathbf{e}_2$ does not take place. The remaining polarization direction $\mathbf{e}_1$ is then orthogonal to both $\mathbf{k}$ and $\mathbf{e}_2$. If θ is the angle between the directions of $\mathbf{r}_{AB}$ and $\mathbf{k}$, then we have $\mathbf{r}_{AB} \cdot \mathbf{e}_1 \propto \sin\theta$. The intensity of the light emitted with no regard for polarization is therefore given by

$$d\Gamma = \frac{e^2\omega_{AB}^3}{8\pi^2\varepsilon_0\hbar c^3}|\mathbf{r}_{BA}|^2\sin^2\theta\, d\Omega. \tag{9.65}$$

There is no emission in the direction of $\mathbf{r}_{AB}$, while the emission reaches a maximum for $\mathbf{k} \perp \mathbf{r}_{AB}$. The angular dependence of the intensity of emission is schematically sketched in Fig. 9.3.

Fig. 9.2   Directions of the polarizations of the light emitted.

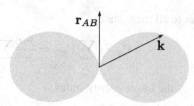

Fig. 9.3   The emission intensity in a certain direction is proportional to the length of the gray interval drawn in this direction.

So now we know how light is emitted if the dipole moment $\mathbf{r}_{AB}$ is given. We discuss what determines the actual direction of this vector in the next subsection. Let us first evaluate the rate of spontaneous emission in all possible directions: the total rate. This rate determines the life-time of the excited state. To find this rate, we integrate the partial rate (9.65) over the solid angle. We see that

$$\int d\Omega \, \sin^2\theta = \int_0^{2\pi} d\phi \int_0^{\pi} d\theta \, \sin^3\theta = 2\pi \frac{4}{3} = \frac{8\pi}{3}. \tag{9.66}$$

Therefore, the total rate of spontaneous emission from state A to state B reads

$$\Gamma_{\mathrm{sp}} = \frac{4\alpha\omega_{AB}^3}{3c^2}|\mathbf{r}_{BA}|^2. \tag{9.67}$$

Here we again make use of the fine structure constant $\alpha \equiv e^2/4\pi\varepsilon_0\hbar c$.

Let us also provide an order-of-magnitude estimate of Γ_{sp}. The ratio of Γ_{sp} and the emitted frequency characterizes an inverse quality factor of the excited state. Good quality corresponds to a long life-time of the state, while poor quality states do not even give rise to a well-defined energy. By virtue of the Heisenberg uncertainty relation, the energy uncertainty of the excited state is of the order of $\hbar\Gamma_{\mathrm{sp}}$. The dipole matrix element and the emitted energy are both estimated by their atomic scale values a_B and E_{at}. Therefore,

$$\frac{\Gamma_{\mathrm{sp}}}{\omega_{AB}} \simeq \alpha \left(\frac{\omega_{AB}}{c}|\mathbf{r}_{BA}|\right)^2 \simeq \alpha^3. \tag{9.68}$$

The quality factor of an excited atomic state is therefore roughly α^{-3}, or about a million. The typical values of spontaneous emission rates are $10^8 - 10^9$ Hz.

We note that this 10^{-6} comes from two small factors in the above relation, the fine structure constant α and $(\omega_{AB}r_{AB}/c)^2 \simeq (a_B/\lambda)^2$. The second factor shows again that the atoms are small in comparison with the wavelength and justifies the use of the dipole approximation. What would happen if the atoms were larger? It seems that an increase of the matrix element would lead to larger and larger rates. However, if r_{AB} were to become of the order of λ the dipole approximation fails, and a further increase of r_{AB} would not increase the rate. Indeed, classical electromagnetism states that an optimal size for an emitter meant to emit waves with wave length λ is of the order of λ. For an optimal dipole (that is, if $r_{AB} \simeq \lambda$), the decay rate is still small in comparison with the frequency of the emitted radiation, $\Gamma_{\mathrm{sp}}/\omega_{AB} \simeq \alpha$.

For a sufficiently highly excited state, the lower energy state B is not unique. It can be the ground state or yet another excited state. In this case, the total rate is the sum of the emission rates to all these states,

$$\Gamma_{sp} = \sum_j \Gamma_{sp,j} = \frac{4\alpha}{3c^2} \sum_j (\omega_{AB_j})^3 |\mathbf{r}_{AB_j}|^2, \tag{9.69}$$

where j indexes the lower energy states.

9.4.3 Selection rules

Atoms are known to be round, or in scientific terms, the Hamiltonian of an atom has a spherical symmetry. The transition matrix element is a vector, so it has a certain direction, thereby breaking the spherical symmetry of the atomic potential. The eigenstates of a symmetric Hamiltonian do not necessarily possess the same symmetry as the Hamiltonian. However, they can be classified by their symmetry types. These symmetry types are called irreducible representations of the symmetry group. Group theory can tell us if a matrix element $\langle A|\hat{\mathbf{r}}|B\rangle$ is zero, provided the symmetry types of $|A\rangle$, $|B\rangle$, and $\mathbf{r}$ are given. It thus can forbid transitions between certain states and selects pairs of states $|A\rangle$ and $|B\rangle$ that can be involved in a transition. These rules are called *selection rules*. For a spherical symmetry, the irreducible representations correspond to different integer values of total angular momentum l. Each value of the total momentum l gives rise to $2l + 1$ degenerate states. These states are indexed by the projection m of the angular momentum on a certain axis (usually the z-axis), $-l \leq m \leq l$, called the azimuthal quantum number. There are in principle many states with the same l and m but with different energies. An arbitrary atomic state can therefore be indexed by three quantum numbers: $|nlm\rangle$.

The selection rules for atomic states are such that the transition matrix element $\langle l'm'|\hat{\mathbf{r}}|lm\rangle$ is non-zero only if l' and l differ by one, $|l - l'| = 1$. In addition to this, there is a selection rule with respect to the azimuthal quantum number,

$$\begin{aligned}
\langle l'm'|z|lm\rangle &\neq 0 &&\text{only if } m' = m, \\
\langle l'm'|x + iy|lm\rangle &\neq 0 &&\text{only if } m' = m + 1, \\
\langle l'm'|x - iy|lm\rangle &\neq 0 &&\text{only if } m' = m - 1.
\end{aligned} \tag{9.70}$$

Let us exemplify these rules for the atomic s ($l = 0$) and p ($l = 1$) states. The s states are isotropic, so no vector matrix elements are possible between these states. There are three degenerate p states that differ in azimuthal quantum number, $|m\rangle$ with $m \in \{-1, 0, 1\}$. These three independent states can be chosen in a more convenient way as

$$|z\rangle = |0\rangle, \ |x\rangle = \frac{1}{\sqrt{2}}\{|1\rangle - i|-1\rangle\}, \text{ and } |y\rangle = \frac{1}{\sqrt{2}}\{|1\rangle + i|-1\rangle\}. \tag{9.71}$$

These three states transform under rotation in precisely the same way as a three-component vector $\mathbf{r}$. By symmetry, the only non-zero matrix elements are

$$\langle s|\hat{x}|x\rangle = \langle s|\hat{y}|y\rangle = \langle s|\hat{z}|z\rangle. \tag{9.72}$$

Now we are ready to answer the postponed question: what determines the direction of the atomic dipole $\mathbf{r}_{AB}$? Since the p states are degenerate, the excited atom with energy E_A can be in any superposition of the three states, most generally $\alpha_x|x\rangle + \alpha_y|y\rangle + \alpha_z|z\rangle$.

Recalling (9.72), we see that $\mathbf{r}_{AB}$ is parallel to $\boldsymbol{\alpha}$. If we manage to prepare the excited atom in a certain superposition,[2] the emission is anisotropic in agreement with (9.65). However, commonly the excited atoms are found with equal probability in all three states, which corresponds to a randomly distributed $\boldsymbol{\alpha}$. In this case, the emitted radiation is isotropic.

The simplest example to illustrate the selection rules with, would be a hydrogen atom containing only one single electron. The energy levels of the atom are known to depend only on the main quantum number n,

$$E_n = -\frac{\mathrm{Ry}}{n^2} = \frac{m_e c^2 \alpha^2}{2n^2}, \tag{9.73}$$

where the m_e in this equation denotes the mass of the electron. This implies that, for instance, the $2s$ and $2p$ states have the same energy, and transitions between these states do not involve electromagnetic radiation. A transition between the lowermost p and s states, the $2p \to 1s$ transition, has an energy difference of $\hbar\omega = \frac{3}{4}\mathrm{Ry}$, and can occur spontaneously, accompanied by the emission of a photon with the same energy. The transition rate can be calculated relatively straightforwardly (see Exercise 1), and is

$$\Gamma = \left(\frac{2}{3}\right)^8 \alpha^5 \frac{m_e c^2}{\hbar} \simeq 6.27 \times 10^8 \ \mathrm{s}^{-1}. \tag{9.74}$$

One might wonder what would happen if a hydrogen atom were to be brought to an isotropic excited state, for instance $2s$. From this initial state, the selection rules seem to forbid a radiative transition to the only lower lying state, the ground state. However, this prohibition applies only to a simplest decay process involving only the emission of a single photon. It has been shown that a $2s \to 1s$ transition in fact occurs with emission of two photons with total energy $E_2 - E_1$. This process was identified as the main source of the continuous spectrum emission from nebulae. It is clear that the rate of two-photon processes should be much smaller than the corresponding single-photon processes. A rough estimate for this rate reads $\Gamma_{2\text{-ph}}/\omega \simeq (\Gamma_{1\text{-ph}}/\omega)^2$, which yields $\Gamma_{2\text{-ph}} \simeq \alpha^6 \omega$. An exact calculation gives

$$\Gamma_{2s \to 1s} \simeq 8.2 \ \mathrm{s}^{-1}. \tag{9.75}$$

The quality factor of the $2s$ state is therefore $\simeq 10^{14}$, which makes it the best natural resonator known.

9.5 Blue glow: Cherenkov radiation

As a next example of the interaction between matter and radiation, we consider the "blue glow" discovered by Pavel Cherenkov in 1934. A freely moving electron could in principle reduce its kinetic energy and momentum by emitting photons. It does not do so in vacuum, since there emission is forbidden by the principle of relativity. Indeed, relativity tells us

---

[2] For instance, we could excite it with polarized light with polarization $\mathbf{e} \parallel \boldsymbol{\alpha}$.

Pavel Alekseyevich Cherenkov (1904–1990)
Shared the Nobel Prize in 1958 with Ilya Frank and Igor Tamm for "the discovery and the interpretation of the Cherenkov effect."

Pavel Cherenkov was born in the rural district of Voronezh to a peasant family. He did his undergraduate studies at Voronezh University, where, in 1928, he earned a degree in mathematics and physics. After teaching at a high school for a short time, he got his Ph.D. in 1935 in Leningrad and moved to Moscow for a research position at the P. N. Lebedev Physical Institute, where he stayed until his death in 1990.

From the early 1930s, Cherenkov studied the absorption of radioactive radiation by different fluids. The absorption was almost always accompanied by weak luminescence of the fluid, which in most cases could be explained as being simple fluorescence: the radiation excites the molecules in the fluid and then they relax back to their ground state while emitting photons. Cherenkov noticed however that beams of high energy electrons passing through a fluid always produced a faint blue glow, which was impossible to interpret as fluorescence. Lacking highly sensitive photon detectors, Cherenkov made most observations with the naked eye. To prepare for his experiments, Cherenkov usually started his working day by spending an hour in a totally dark room. Then, with extreme patience and persistence, he would try for hours to see and quantify the blue glow.

Many colleagues were skeptical about Cherenkov's results, being concerned about the simplicity of the peasant son's methods. Despite this, he thoroughly investigated the phenomenon, which was theoretically explained in 1937 by his colleagues Ilya Frank and Igor Tamm (with whom Cherenkov shared the Nobel Prize).

that the physics must be the same in all inertial reference frames. If we go to the reference frame where the electron is at rest, it cannot emit anything. Therefore, it also does not emit photons when moving with a constant velocity.

This is different if the electron moves in a dielectric medium with an electrical permittivity $\varepsilon = \varepsilon_0 \varepsilon_r$. The speed of light light c' in the medium is always smaller than that in vacuum, $c' = c/\sqrt{\varepsilon_r}$. High-energy electrons may thus move with a velocity v exceeding this light barrier, and such "faster-than-light" electrons do emit radiation. This phenomenon was discovered by Pavel Cherenkov in 1934 in the course of tedious experimenting, but natural radiation sources hardly produced any detectable light. The situation improved in the mid-1940s when the first pool-type nuclear reactors came into operation. The reactor core emits β-particles – high-energy electrons – that propagate faster than light in water. They produce an ominous glow that seems bluish to the human eye.

Let us begin with the conservation laws for photon emission. We denote the wave vector of the electron before and after the emission as $\mathbf{k}$ and $\mathbf{k}'$, and the wave vector of the photon

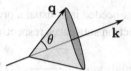

Fig. 9.4 The cone of Cherenkov radiation. An electron with wave vector k emits a photon with wave vector q, the angle θ between two vectors being fixed by $\cos\theta = c'/v$.

as $\mathbf{q}$. Conservation of momentum gives

$$\hbar\mathbf{k} = \hbar\mathbf{k}' + \hbar\mathbf{q}, \tag{9.76}$$

and conservation of energy

$$E(k) = E(k') + \hbar c' q, \tag{9.77}$$

where $E(k) = \sqrt{m^2c^4 + (\hbar k)^2 c^2}$ is the energy of the relativistic electron. The wave vector of a relativistic electron exceeds by far the wave vector of a visible photon, so we can safely assume $q \ll k, k'$. We substitute $\mathbf{k}' = \mathbf{k} - \mathbf{q}$ in the energy conservation relation and expand $E(k')$ to first order in q to obtain

$$\frac{\partial E}{\partial \mathbf{k}} \cdot \mathbf{q} = \hbar c' q. \tag{9.78}$$

Since the electron velocity $\mathbf{v} = \frac{1}{\hbar}\partial E/\partial \mathbf{k}$, we have

$$\mathbf{v} \cdot \mathbf{q} = c' q, \quad \text{and thus} \quad \cos\theta = \frac{c'}{v}, \tag{9.79}$$

where θ is the angle between the direction of electron propagation and the direction of emission (see Fig. 9.4). We thus have proven that Cherenkov radiation propagates at the surface of a cone centered at the moving electron with cone angle $\arccos(c'/v)$. Since $\cos\theta < 1$, we see that radiation is indeed only possible for electrons moving faster than light. The propagation of Cherenkov radiation closely resembles the propagation of a sonic boom – the acoustic shock wave coming from a supersonic aircraft.

9.5.1 Emission rate and spectrum of Cherenkov radiation

Let us evaluate the transition rate that determines the intensity of the radiation. The initial and final states for the photons are as usual $|0\rangle$ and $|1_{q\alpha}\rangle$, α labeling the polarization of the photon. The initial and final states of the electron are plane waves with corresponding wave vectors, $|\mathbf{k}\rangle$ and $|\mathbf{k}'\rangle = |\mathbf{k} - \mathbf{q}\rangle$. This means that the electron is delocalized and we cannot implement the dipole approximation we used before. So we proceed directly with the general interaction term

$$\hat{H}_{\text{int}} = -\int d\mathbf{r}\, \hat{\mathbf{A}}(\mathbf{r}) \cdot \hat{\mathbf{j}}(\mathbf{r}), \tag{9.80}$$

where for non-relativistic electrons

$$\hat{\mathbf{j}}(\mathbf{r}) = -\frac{i\hbar e}{2m}\left\{ \hat{\psi}^\dagger(\mathbf{r})[\nabla\hat{\psi}(\mathbf{r})] - [\nabla\hat{\psi}^\dagger(\mathbf{r})]\hat{\psi}(\mathbf{r}) \right\}. \tag{9.81}$$

The matrix element needed is as usual a product of the matrix elements of $\hat{\mathbf{A}}(\mathbf{r})$ and $\hat{\mathbf{j}}(\mathbf{r})$ in the photon and electron subspaces respectively,

$$\langle 1_{\mathbf{q}\alpha}|\hat{\mathbf{A}}(\mathbf{r})|0\rangle = \mathbf{e}_\alpha \sqrt{\frac{\hbar}{2\varepsilon\omega_{\mathbf{q}}\mathcal{V}}} e^{-i\mathbf{q}\cdot\mathbf{r}}, \tag{9.82}$$

$$\langle 0|\hat{c}_{\mathbf{k}-\mathbf{q}}\hat{\mathbf{j}}(\mathbf{r})\hat{c}_{\mathbf{k}}^\dagger|0\rangle = e^{i\mathbf{q}\cdot\mathbf{r}}\frac{e\hbar}{2m\mathcal{V}}(\mathbf{k}-\mathbf{q}+\mathbf{k}) = e^{i\mathbf{q}\cdot\mathbf{r}}\frac{e}{2\mathcal{V}}\big[\mathbf{v}(\mathbf{k}-\mathbf{q})+\mathbf{v}(\mathbf{k})\big]. \tag{9.83}$$

Since $q \ll k$, we find that $\mathbf{v}(\mathbf{k}-\mathbf{q}) \approx \mathbf{v}(\mathbf{k})$ in the last expression. Under these assumptions, the expression is valid for a relativistic electron, even if we cannot use the non-relativistic expression $\hat{\mathbf{p}}/m$ for the velocity. This is because the current operator is related to the electron velocity, and the plane waves are eigenfunctions of the velocity operator, $\hat{\mathbf{v}}|\mathbf{k}\rangle = \mathbf{v}(\mathbf{k})|\mathbf{k}\rangle$. With this, the matrix element between initial and final states reads

$$\langle f|\hat{H}_{\text{int}}|i\rangle = -e\sqrt{\frac{\hbar}{2\varepsilon\omega_{\mathbf{q}}\mathcal{V}}} \, \mathbf{v}(\mathbf{k})\cdot\mathbf{e}_\alpha. \tag{9.84}$$

As we did for atoms, we look at the partial rate of emission to the small element of photon wave vector space $d\mathbf{q}$. We use the above expression for the transition matrix element to arrive at

$$d\Gamma_{\mathbf{k},\mathbf{q}} = \frac{2\pi}{\hbar}\frac{e^2\hbar}{2\varepsilon\omega_{\mathbf{q}}}(\mathbf{v}\cdot\mathbf{e}_\alpha)^2\delta(E_i-E_f)\frac{d\mathbf{q}}{(2\pi)^3}. \tag{9.85}$$

The analysis of the contribution of the two different polarizations is similar to that for atomic transitions, the vector $\mathbf{v}$ now plays the role of the dipole moment $\mathbf{r}_{AB}$. One of the two polarization directions, let us call it $\mathbf{e}_2$, is perpendicular to both $\mathbf{v}$ and $\mathbf{q}$. Since this implies that $\mathbf{v}\cdot\mathbf{e}_2 = 0$, photons with this direction of polarization are not emitted. For the other polarization direction we have $\mathbf{v}\cdot\mathbf{e}_1 = v\sin\theta$, where θ is (as before) the angle between the direction of propagation of the electron $\mathbf{k}$ and the direction of photon emission $\mathbf{q}$. Making use of this, replacing $d\mathbf{q} \to q^2 dq\, d\varphi\, d(\cos\theta)$, and taking the energy conservation law in the form (9.79), we obtain

$$d\Gamma_{\mathbf{k},\mathbf{q}} = \frac{e^2\pi}{\varepsilon\omega_{\mathbf{q}}}v^2\sin^2\theta\,\delta(\hbar vq\cos\theta - \hbar c'q)\frac{q^2 dq\, d\varphi\, d(\cos\theta)}{(2\pi)^3}. \tag{9.86}$$

From here on, we cannot proceed in the same way as we did for atoms. In the case of atoms, emission occurred at a certain frequency fixed by energy conservation, and then we integrated over $q = \omega/c$ to remove the δ-function. Now the emission spectrum is continuous and energy conservation does not fix the modulus of the wave vector, but rather the angle θ. So in this case, the next step is to integrate over angles rather than over the modulus of $\mathbf{q}$. What we then get is the rate per frequency interval,

$$d\Gamma = \alpha\frac{v}{c'}\left(1 - \frac{c'^2}{v^2}\right)d\omega, \tag{9.87}$$

giving the number of photons per second and per energy interval emitted by a faster-than-light electron. Interestingly, the rate does not depend on frequency.

This, of all things, tells us something about human vision. A spectrum which does not depend on frequency is usually associated with white: a noise spectrum with a constant intensity for a large range of frequencies is called white noise. So why is Cherenkov radiation when observed not white but rather blue? The answer is that the amount of energy emitted to a small frequency interval is proportional to the frequency, $dE \simeq \alpha\hbar\omega\, d\omega$, meaning that there is more energy emitted at higher frequencies. Our eyes are sensitive to energy rather than to the number of photons absorbed so the Cherenkov light seems blue to us.

This also implies that the integral over frequencies diverges. To see how to deal with this divergence, let us recall our treatment of vacuum fluctuations and understand that the deviation of ε from ε_0 is restricted to frequencies smaller than $E_{at}/\hbar$. This gives an upper cut-off in the integral over frequencies, and a total rate $\Gamma \simeq \alpha E_{at}/\hbar$.

Control question. Since the electron emits energy, it decelerates. Can you deduce from the above estimate along what kind of path the electron travels before it decelerates below c' and the emission ceases?

9.6 Bremsstrahlung

So, free electrons cannot emit photons owing to conservation of momentum and energy, unless they move faster than light. Another possibility for an electron to emit a photon is to break momentum conservation by combining the emission of a photon with scattering at another charged particle while transferring some momentum to it. The electron is typically decelerated in the course of such an event, and the corresponding emission process is called *Bremsstrahlung* (German for "braking radiation").

Let us compute the rate of Bremsstrahlung assuming an electron moving in a stationary potential field $U(\mathbf{r})$. This allows us to disregard the dynamics of the charged particle and is an excellent approximation for scattering at nuclei. We take into account both $U(\mathbf{r})$ and $\hat{H}_{int}$ in the lowest order of perturbation theory. With this, the Bremsstrahlung is due to a second-order process with a rate given by the general formula (1.63). Let us assume that the electron has momentum $\hbar\mathbf{k}$ in the initial state. In the final state the situation is as follows: (i) there is a photon with momentum $\hbar\mathbf{q}$, (ii) the electron momentum is $\hbar(\mathbf{k} - \mathbf{q} - \mathbf{k}')$, and (iii) the momentum $\hbar\mathbf{k}'$ has been transferred to the potential field. Energy conservation during this process demands that

$$E_f - E_i = E(\mathbf{k} - \mathbf{q} - \mathbf{k}') + \hbar c q - E(\mathbf{k}) = 0. \tag{9.88}$$

There are two possible virtual states for this second order process, differing by the order of potential scattering and photon emission as can be seen in Fig. 9.5. In case (a) the virtual state $|A\rangle$ is obtained from the initial state by potential scattering, so that the virtual electron momentum is $\mathbf{k} - \mathbf{k}'$. The energy difference with the initial (or final) state is therefore $E_A - E_i = E(\mathbf{k} - \mathbf{k}') - E(\mathbf{k})$. In case (b) the virtual state is obtained by photon emission, so that the virtual electron momentum is $\mathbf{k} - \mathbf{q}$, and the energy difference is $E_B - E_i = \hbar\omega_\mathbf{q} + E(\mathbf{k} - \mathbf{q}) - E(\mathbf{k}) = E(\mathbf{k} - \mathbf{q}) - E(\mathbf{k} - \mathbf{q} - \mathbf{k}')$. The latter equality follows from energy conservation, see (9.88).

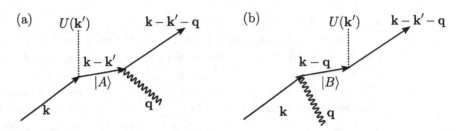

Fig. 9.5 The two possible virtual states of the Bremsstrahlung process. (a) The state $|A\rangle$ describes the electron with wave vector $\boldsymbol{k} - \boldsymbol{k'}$, part of its initial momentum having been absorbed by the potential field. (b) The state $|B\rangle$ describes the electron with wave vector $\boldsymbol{k} - \boldsymbol{q}$ and a single photon with wave vector $\boldsymbol{q}$.

We assume the electron to be non-relativistic, $v \ll c$. Energy conservation then implies that $q \simeq (v/c)(k - k') \ll k, k'$ and we can disregard the photon wave vector in comparison with the electron wave vectors. Under this assumption, the two energy denominators of the virtual states are opposite, $E_A - E_i = -\hbar c q = -(E_B - E_i)$. Since the matrix element of potential scattering, $U(\mathbf{k'})$, is the same for the contributions of both virtual states, the contributions tend to cancel each other. With this given, one should attentively analyze the matrix elements of $\hat{H}_{\text{int}}$ entering both contributions. Adjusting (9.83) to the current situation and using again that $q \ll k$, we obtain

$$\langle i|\hat{H}_{\text{int}}|B\rangle = -e\sqrt{\frac{\hbar}{2\varepsilon_0\omega_\mathbf{q}\mathcal{V}}}\ \mathbf{v}(\mathbf{k}) \cdot \mathbf{e}_\alpha,$$

$$\langle A|\hat{H}_{\text{int}}|f\rangle = -e\sqrt{\frac{\hbar}{2\varepsilon_0\omega_\mathbf{q}\mathcal{V}}}\ \mathbf{v}(\mathbf{k} - \mathbf{k'}) \cdot \mathbf{e}_\alpha. \tag{9.89}$$

The effective matrix element is proportional to the difference of the two, and therefore to the difference of the electron velocities in the initial and final states. Its square reads

$$|M_{if}|^2 = 2\pi\alpha\frac{|U(\mathbf{k'})|^2}{\mathcal{V}}\frac{1}{q^3}\left(\frac{\mathbf{v}_i - \mathbf{v}_f}{c} \cdot \mathbf{e}_\alpha\right)^2, \tag{9.90}$$

where $\mathbf{v}_i$ and $\mathbf{v}_f$ denote the initial and final electron velocity respectively.

To estimate the magnitude of the Bremsstrahlung rate, we compare the matrix element with the square of the potential scattering matrix element, $|U(\mathbf{k'})|^2/\mathcal{V}^2$, which gives the rate of electron collisions. The ratio of the two corresponds to the chance to emit a photon with momentum $\hbar\mathbf{q}$ in the course of a collision. The total number of photon states available for emission is estimated as $q^3\mathcal{V}$. With this, the total chance to emit a photon can be estimated as $\alpha(|\mathbf{v}_i - \mathbf{v}_f|/c)^2$. The estimation is composed of two small factors: $(v/c)^2$, related to low electron velocities, and α the fine structure constant. The latter guarantees that the Bremsstrahlung rate is still slow even for relativistic electrons with $v \simeq c$.

The polarization and angular distribution of the Bremsstrahlung is governed by the factor $[(\mathbf{v}_i - \mathbf{v}_f) \cdot \mathbf{e}_\alpha]^2$ and is therefore similar to that of Cherenkov radiation. The difference is that the polarization vector now is the difference of the velocities before and after the collision, rather than the velocity itself. The spectral distribution of the emitted radiation is, naturally enough, bound by the kinetic energy of the incoming electron. The concrete form of the distribution depends on the shape of the potential field. For scattering from nuclei, $U(\mathbf{k}') \simeq (k')^{-2}$. In this case, the spectral intensity does not depend on the photon energy at $\hbar\omega_\mathbf{q} \ll E(\mathbf{k})$.

An interesting interpretation of the Bremsstrahlung process is obtained if we consider electron motion in a potential that varies slowly in comparison with the electron wave length, implying $k' \ll k$. In the limit $\mathbf{k}' \to 0$, the potential scattering amplitude in the matrix element can be replaced by the acceleration $\mathbf{a}$ of the electron,

$$(\mathbf{v}_f - \mathbf{v}_i)U(\mathbf{k}') \approx -\frac{\hbar\mathbf{k}'}{m}U(\mathbf{k}') \quad \to \quad -i\frac{\hbar}{m}\nabla U = i\hbar\mathbf{a}. \tag{9.91}$$

The energy of the emitted photon is fixed by energy conservation to $\hbar\omega_\mathbf{q} = \hbar(\mathbf{v} \cdot \mathbf{k}')$. We integrate over the wave vectors of the photon to obtain the emission rate. The rate Γ_B in fact diverges in the limit of small $\mathbf{k}'$. A sensible quantity to evaluate is the power emitted by the electron per unit time, $P_B \equiv \hbar\omega_\mathbf{q}\Gamma_B$. It is given by

$$P_B = \frac{2}{3}\alpha\hbar\frac{\mathbf{a}^2}{c^2} = \frac{e^2}{6\pi\varepsilon_0 c}\frac{\mathbf{a}^2}{c^2}. \tag{9.92}$$

The latter equality proves that the emitted power in fact does not depend on $\hbar$ and should correspond to the result of a classical calculation. Indeed, one can solve the Maxwell equations for the field produced by an accelerating charge and demonstrate that the emitted power is given by (9.92).

9.7 Processes in lasers

Another example for which we can use what we have learned about emission and absorption of photons concerns lasers. Lasers were invented about 50 years ago and are nowadays more than widely available in everyday life. It is, however, not evident that they should appear in a book about advanced *quantum* mechanics. In the very beginning, the research field dealing with lasers was called "quantum electronics." Today this is completely out of fashion, and a modern laser textbook calls this a misunderstanding and stresses that "no preliminary knowledge of quantum mechanics is required." Indeed, lasers are mostly about human ingenuity rather than about the fundamentals of the Universe, and many things about lasers can be equally well understood from both classical and quantum points of view. For the present book, the quantum aspect is of interest and extreme simplification of the laser physics is necessary.

Officially, the word laser stands for "light amplification by stimulated emission of radiation". The widely used simple explanation of its operation is as follows. Suppose we have some atoms in an excited state. As we know, for each atom the emission rate to a given mode can be written as $\gamma(n + 1)$, where n is the number of photons in the mode. In the beginning, there are no photons at all in the laser. Then, at some instance, one atom emits a photon spontaneously. The photon emitted then increases the rate of emission for all the

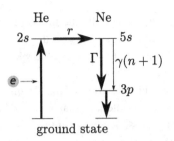

Fig. 9.6 Processes in a He-Ne laser. Free electrons excite He atoms to the 2s state. This state is resonant with the 5s state of Ne, so the 5s state gets populated with a rate r. The excited Ne atoms can decay via a p state to the ground state emitting photons in the visible range.

atoms by a factor of 2 so it is likely to stimulate another photon to go. These two photons then increase the rate even more, resulting in an avalanche. We can say that the first spontaneous emission has been amplified by the laser.

This explanation gives the requirements that have to be fulfilled to make a laser. The first requirement is to constantly have enough excited atoms inside the laser. To achieve a constant laser output, one should thus continuously provide excited atoms by pumping energy into the system. The second requirement is to keep the photons confined in a single mode. To stimulate emission, they should stay in the laser for a while and should not leave it too soon. Usually the confinement is very selective: only the photons in a few specific modes are strongly confined, whereas all photons emitted in other modes leave the laser.

To see how this works, let us give a more detailed description, pictured in Fig. 9.6. We thus turn to a concrete example of a laser, the He-Ne gaseous laser, which descended from the common neon signs. A high voltage is applied to a mixture of the two gases confined in a glass tube. The voltage is high enough to ignite discharges in the gas so that free electrons are running from the anode to the cathode. These electrons collide with He atoms, exciting some of them into the isotropic excited state 2s. As we have seen, this state cannot decay with the emission of a single photon and the He atoms are trapped in this state for a relatively long time. The "engineering" trick implemented is that the energy of the 2s state of a He atom is very close to that of the excited 5s state of a Ne atom. Collisions between excited He atoms and ground state Ne atoms are thus resonant and excite some Ne atoms to the 5s state. This is how the voltage source supplies excited Ne atoms. The efficiency of this process can be characterized by a rate r, the number of Ne atoms excited per unit time. The Ne atoms in the 5s state mostly lose energy by emitting radiation and going to p states of lower energy, and this is the process creating the visible light. Subsequent (faster) transitions bring the atoms from these p states back to the ground state.

In fact, the light radiated during the $5s \to p$ transition of the Ne atoms is emitted in all directions, and most light leaves the tube very soon. Exceptions are formed by a few modes which have a wave vector strictly parallel to the tube axis. These modes are confined with the help of two mirrors at the ends of the tube. The back mirror reflects all light while the front mirror transmits 1% of the light. Therefore, a photon in a confined mode makes it on average 100 times through the cavity before it leaves the laser through the front mirror. It is customary to characterize the photon life-time in the mode by the quality factor of the mode, $\tau = Q/\omega$, where ω is the frequency of the laser light. For a tube of length L, the lifetime is thus $\tau \simeq 100 \cdot (2L/c)$. For typical parameters $L = 0.5$ m and $\omega = 10^{15}$ Hz we have $Q \simeq 0.3 \times 10^9$. We assume that there is only one single confined mode and that the photons are accumulated in this mode only. The output power of the laser is then roughly $P_{out} = n\hbar\omega/\tau = n\hbar\omega^2/Q$, where n is the number of photons in the confined mode.

Control question. Do you see where the above relation comes from?

The rate of radiative decay to all non-confined modes, Γ, is of the order of our estimate for the decay rate of atoms to *all* modes, say $\Gamma \simeq 10^9$ Hz. Since photons are not accumulated in these modes, this rate is different from the emission rate to the *confined* mode, $\gamma(n+1)$. The ratio of the prefactors, that is, the ratio of the spontaneous rate to all modes to the spontaneous rate to a single mode, Γ/γ, is roughly equal to the number of modes available for emission. This number is proportional to the volume of the laser cavity and is astronomically large, say 10^{10}. This means that the spontaneous rate to the lasing mode is quite small, $\gamma \simeq 10^{-1}$ Hz, or once every 10 seconds.

9.7.1 Master equation for lasers

The above description defines a simple model for the laser which can be formulated in terms of master equations. The first master equation we can write is for the number of atoms excited,

$$\frac{dN_{ex}}{dt} = r - N_{ex}\Gamma - N_{ex}\gamma(\bar{n}+1). \tag{9.93}$$

The actual number N_{ex} is thus determined from the competition between the pumping rate r and the total decay rate $\Gamma + \gamma(\bar{n}+1)$, where $\bar{n}$ is the average number of photons in the *lasing* mode. This number is determined from the second master equation,

$$\frac{d\bar{n}}{dt} = N_{ex}\gamma(\bar{n}+1) - \frac{\omega}{Q}\bar{n}. \tag{9.94}$$

Here, the competition takes place between emission to the lasing mode, at a rate $N_{ex}\gamma(\bar{n}+1)$, and the emission of the laser light through the front mirror.

A more detailed master equation holds for the probability p_n to have n photons in the lasing mode,

$$\frac{dp_n}{dt} = -p_n\left[N_{ex}\gamma(n+1) + \frac{\omega}{Q}n\right] + p_{n-1}N_{ex}\gamma n + \frac{\omega}{Q}(n+1)p_{n+1}, \tag{9.95}$$

where we take into account that each emission process from the excited atoms adds one photon and each leak through the front mirror subtracts one photon from the mode.

Let us find the stationary solution of (9.93) and (9.94). The stationary number of excited atoms is given by

$$N_{ex} = \frac{r}{\Gamma + \gamma(\bar{n}+1)}. \tag{9.96}$$

We substitute this into (9.94) and assume that $\bar{n} \gg 1$, yielding

$$\frac{d\bar{n}}{dt} = \bar{n}\left(\frac{r\gamma}{\Gamma + \gamma\bar{n}} - \frac{\omega}{Q}\right). \tag{9.97}$$

At sufficiently low pumping rates, this time derivative is always negative. The linear gain $r\gamma/\Gamma$ from the stimulated emission cannot compensate for the photon loss through the mirror ω/Q. This means that at low r, the average photon number in the lasing mode is very low, and the laser will not work. The "laser threshold" is achieved at a pumping rate $r_{\text{th}} = \omega\Gamma/\gamma Q$, which is $r_{\text{th}} \approx 10^{16}$ s^{-1} for the example numbers we chose above. Above this threshold, the derivative $d\bar{n}/dt$ is positive if one starts with small $\bar{n}$. The number of photons grows, and the light is amplified. The number of photons is stabilized at

$$\bar{n} = \frac{rQ}{\omega} - \frac{\Gamma}{\gamma}. \tag{9.98}$$

The reason that it will stabilize and not grow on forever is that the increased number of photons also increases the decay rate and therefore reduces the number of excited atoms. The total emission of photons into the lasing mode,

$$\Gamma_{\text{em}} = \frac{r\gamma(\bar{n}+1)}{\Gamma + \gamma(\bar{n}+1)} < r, \tag{9.99}$$

never exceeds the pumping rate r. It saturates at r in the limit of a large number of photons $\bar{n} \gg \Gamma/\gamma$. In a well-saturated laser, almost all pumped power goes into the output laser light, and we have

$$P_{\text{out}} = \hbar\omega r. \tag{9.100}$$

For our example parameters, $P_{\text{out}} \approx 1$ mW for $r \simeq 10^{16}$ s^{-1}.

9.7.2 Photon number distribution

Let us now look at the distribution of the number of photons in a laser. We restrict ourselves to the case of a well-saturated laser where the emission rate to the lasing mode $N_{\text{ex}}\gamma n$ equals the pumping rate r. Then (9.95) simplifies to

$$\frac{dp_n}{dt} = -p_n\left[r + \frac{\omega}{Q}n\right] + p_{n-1}r + \frac{\omega}{Q}(n+1)p_{n+1}. \tag{9.101}$$

To solve the above equation, we implement the detailed balance reasoning which we also used in the discussion of the black-body irradiation,

$$p_{n+1}\frac{\omega}{Q}(n+1) = p_n r, \quad \text{so that} \quad \frac{p_{n+1}}{p_n} = \frac{\bar{n}}{n+1}. \tag{9.102}$$

In order to derive the last equation, we used the fact that $\bar{n} = rQ/\omega$ for a well-saturated laser. The properly normalized distribution following from this detailed balance equation reads

$$p_n = e^{-\bar{n}}\frac{\bar{n}^n}{n!}, \tag{9.103}$$

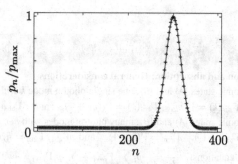

Fig. 9.7 Poisson distribution of the number of photons for $\bar{n} = 300$. The solid line presents the Gaussian approximation.

which is the famous Poisson distribution widely used in probability theory! Generally, it expresses the probability of the occurrence of a specified number of events in a fixed period of time, using that one knows the average rate for the occurrence of the events, and assuming that occurrences are independent of the time elapsed since the last event. For large $\bar{n}$, the Poisson distribution is concentrated around $\bar{n}$ (see Fig. 9.7). The variance of the distribution is $\langle (n - \bar{n})^2 \rangle = \bar{n}$, that is, the relative deviation of number of photons from the average value is $\simeq 1/\sqrt{\bar{n}}$, and the distribution can be approximated by a Gaussian distribution with the same average and variance. This is in contrast to the Boltzmann distribution we found for the black-body radiation.

Seemingly, our derivation implies that at a given moment the lasing mode is in a state with a certain number of photons. From time to time, it randomly jumps between states with slightly more or slightly fewer photons in it (photons are emitted to and from the mode at random) and this number thus fluctuates. Since the variance of the photon number is $\bar{n}$, the number of quantum states involved is of the order $\sqrt{\bar{n}} \gg 1$. In fact, everything in this section is correct except the last implication. In the next chapter, we learn that the laser, under some conditions, can be approximated by a *single* quantum state and emissions to and from the mode are in fact not random.

Table 9.1 Summary: Radiation and matter

Emission and absorption: General considerations

two atomic states $|A\rangle$ and $|B\rangle$, one single photon mode k with $\hbar\omega_k = E_A - E_B$

rates: $\Gamma_{\text{abs}}(k) = \Gamma_k n_k$, $\Gamma_{\text{em}}(k) = \Gamma_k(n_k + 1)$,   and $\Gamma_{\text{abs}}(k)/\Gamma_{\text{em}}(k) = n_k/(n_k + 1)$

two atomic states $|A\rangle$ and $|B\rangle$, many photon modes with $\hbar\omega_k = E_A - E_B$

rates: $\Gamma_{\text{abs}} = \Gamma\bar{n}$, $\Gamma_{\text{em}} = \Gamma(\bar{n} + 1)$,   and $\Gamma_{\text{abs}}/\Gamma_{\text{em}} = \bar{n}/(\bar{n} + 1)$,   with $\bar{n}$ average occupation

master equations: $\dfrac{dp_i}{dt} = -p_i \sum_j \Gamma_{i\to j} + \sum_j p_j \Gamma_{j\to i}$,   with probabilities p and rates Γ

equilibrium: solve $dp_i/dt = 0$ for all probabilities

example: radiating atoms in thermal equilibrium $\to \bar{n} = n_B(E_A - E_B)$, Bose distribution

Planck's law: radiating black body (many different atoms) $\to \dfrac{S_z}{A} = \int d\omega\, n_B(\omega) \dfrac{\hbar\omega^3}{4\pi^2 c^2}$

Interaction of matter and radiation

identify gauge transformations $\psi(\mathbf{r}, t) \to e^{\frac{i}{\hbar}e\chi(\mathbf{r},t)}\psi(\mathbf{r}, t)$ and $\mathbf{A} \to \mathbf{A} + \nabla\chi$ and $\varphi \to \varphi - \partial_t\chi$

then: Schrödinger equation *with* e.m. field: $i\hbar\dfrac{\partial\psi}{\partial t} = \left\{ -\dfrac{\hbar^2}{2m}\left(\nabla - i\dfrac{e}{\hbar}\mathbf{A}\right)^2 + e\varphi \right\}\psi$

where φ can be absorbed into electron–electron interaction term $\hat{H}^{(2)}$

interaction: $\hat{H}_{\text{int}} = -\int d\mathbf{r}\, \hat{\mathbf{A}}(\mathbf{r}) \cdot \hat{\mathbf{j}}(\mathbf{r})$,   to leading order in $\mathbf{A}$, and with $\hat{\mathbf{j}}(\mathbf{r}) = \sum_n \dfrac{e}{m}\hat{\mathbf{p}}_n\, \delta(\mathbf{r} - \mathbf{r}_n)$

Example: Spontaneous emission by atoms, transitions from $|A\rangle \otimes |0\rangle$ to $|B\rangle \otimes |1_k\rangle$

dipole approximation: $\hat{\mathbf{A}}(\mathbf{r}) \approx \hat{\mathbf{A}}(0) \;\to\; \hat{H}_{\text{int}} = -ie\omega_{AB}\langle A|\hat{\mathbf{r}}|B\rangle \cdot \hat{\mathbf{A}}(0) \equiv -ie\omega_{AB}\mathbf{r}_{AB} \cdot \hat{\mathbf{A}}(0)$

angular dependent rate: $d\Gamma_{\text{sp}} = \dfrac{\alpha\omega_{AB}^3}{2\pi c^2}|\mathbf{r}_{BA}|^2 \sin^2\theta\, d\Omega$, where θ is angle between $\mathbf{r}_{AB}$ and $\mathbf{k}$

averaged over angles: $\Gamma_{\text{sp}} = \dfrac{4\alpha\omega_{AB}^3}{3c^2}|\mathbf{r}_{BA}|^2 \simeq \dfrac{4\alpha\omega_{AB}^3}{3c^2}a_B^2$

selection rules: select allowed $|A\rangle$ and $|B\rangle$, for spherical potential only $|l_A - l_B| = 1$

Example: Cherenkov radiation, transitions from $|\mathbf{k}\rangle \otimes |0\rangle$ to $|\mathbf{k}'\rangle \otimes |1_q\rangle$

electrons moving "faster than light" in a dielectric medium can emit photons and decelerate

conservation of energy and momentum $\to \cos\theta = c'/v$, with θ angle of emission and $c' = c/\sqrt{\varepsilon_r}$

emission rate: using $q \ll k$, one finds $d\Gamma = \alpha\dfrac{v}{c'}\left(1 - \dfrac{c'^2}{v^2}\right)d\omega$, with a fixed angle θ

Example: Bremsstrahlung, transitions from $|\mathbf{k}\rangle \otimes |0\rangle$ to $|\mathbf{k}'\rangle \otimes |1_q\rangle$

electrons moving in an external potential can emit photons while scattering and decelerate

$\to$ second order process in which momentum is transferred to the potential

estimate: chance to emit photon while scattering $\simeq \alpha(\Delta v/c)^2$

Example: Laser, stimulated emission in a single mode k

atoms get excited with rate r, decay to many non-confined modes with rate Γ, and decay to a single confined mode k with rate $\gamma(\bar{n}_k + 1)$ (stimulated emission)

equilibrium: $N_{\text{ex}} = \dfrac{r}{\Gamma + \gamma(\bar{n}_k + 1)}$, $\bar{n} = r\tau - \dfrac{\Gamma}{\gamma}$,   with τ life time in mode k

output power: $P_{\text{out}} = \hbar\omega r$,   for saturated laser

distribution for n_k: $p_n = e^{-\bar{n}}\dfrac{\bar{n}^n}{n!}$,   Poisson distribution   (from detailed balance condition)

Exercises

1. *Spontaneous emission from the hydrogen atom* (solution included). Let us illustrate both the selection rules and the calculation of transition matrix elements with the simple example of the hydrogen atom. For a single electron in a hydrogen atom, we can write the wave functions $|nlm\rangle$ explicitly. In spherical coordinates (r, θ, ϕ) they read

$$\varphi_{nlm}(\mathbf{r}) = R_{nl}(r)Y_l^m(\theta, \phi).$$

The angular dependence of these functions is given by the spherical harmonics

$$Y_l^m(\theta, \phi) = \sqrt{\frac{(2l+1)}{4\pi}\frac{(l-|m|)!}{(l+|m|)!}} e^{im\phi} P_l^{|m|}(\cos\theta),$$

where $P_l^m(x)$ are the associated Legendre polynomials,

$$P_l^m(x) = \frac{(-1)^m}{2^l l!}(1-x^2)^{m/2}\frac{d^{l+m}}{dx^{l+m}}(x^2-1)^l.$$

a. Use the recurrence relations for the associated Legendre polynomials,

$$(l-m+1)P_{l+1}^m(x) = (2l+1)xP_l^m(x) - (l+m)P_{l-1}^m(x),$$

$$\sqrt{1-x^2}P_l^m(x) = \frac{1}{2l+1}\left[P_{l-1}^{m+1}(x) - P_{l+1}^{m+1}(x)\right],$$

$$\sqrt{1-x^2}P_l^m(x) = \frac{1}{2l+1}\left[(l-m+1)(l-m+2)P_{l+1}^{m-1}(x) - (l+m-1)(l+m)P_{l-1}^{m-1}(x)\right],$$

to derive recurrence relations expressing $Y_l^m(\theta, \phi)$ respectively in terms of $Y_{l\pm1}^m(\theta, \phi)$, $Y_{l\pm1}^{m+1}(\theta, \phi)$, and $Y_{l\pm1}^{m-1}(\theta, \phi)$.

b. Use the recurrence relations found at (a) to calculate the spontaneous transition rate from the state $|nlm\rangle$ to the state $|n'l'm'\rangle$. We have not specified yet the radial functions $R_{nl}(r)$, so express your answer in terms of the matrix elements

$$R_{nl}^{n'l'} \equiv \int dr\, r^3 R_{nl}(r)R_{n'l'}(r).$$

You must use the fact that the spherical harmonics are orthonormal, i.e.

$$\int d\Omega \left[Y_{l'}^{m'}(\theta, \phi)\right]^* Y_l^m(\theta, \phi) = \delta_{l',l}\delta_{m',m}.$$

c. To calculate explicit emission rates, we also need the radial functions $R_{nl}(r)$. They can be defined in terms of associated Laguerre polynomials, but for the purpose of this exercise the exact definition is not important. Let us focus on one particular transition: that between the lowermost p and s states, $2p \to 1s$. We assume the atoms in the $2p$ state to be randomly polarized, i.e. all three quantum numbers

$m = 0, \pm 1$ can be found equally likely. Calculate the average emission rate, using the radial functions for the hydrogen atom

$$R_{10}(r) = 2a_B^{-3/2} e^{-r/a_B},$$

$$R_{21}(r) = \frac{1}{\sqrt{24}} a_B^{-3/2} \frac{r}{a_B} e^{-r/2a_B},$$

where $a_B = \hbar/(m_e c \alpha)$ is the Bohr radius, m_e being the electron mass and α the fine structure constant. Whereas the angular functions $Y_l^m(\theta, \phi)$ are the same for any spherically symmetric potential, the radial functions $R_{nl}(r)$ are specific for the $1/r$ potential set up by the proton in the hydrogen atom.

2. *Stark effect* (solution included). In an electric field, a neutral object like a hydrogen atom can become polarized and obtain a finite dipole moment. This effect yields small corrections to the spectrum of the atomic states.

   a. We apply an electric field of magnitude E in the z-direction, which we describe by the scalar potential $\varphi = -Ez$. Fix the origin of the coordinate system to the position of the proton. What is the Hamiltonian $\hat{H}_E$ describing the effect of this field on the atom?

   b. We treat the field Hamiltonian as a small perturbation. Calculate the first order correction to the energy of the ground state due to the presence of the field.

   c. The four states with $n = 2$, that is three $2p$ states and one $2s$ state, are degenerate and all have an energy $\frac{3}{4}$Ry above the ground state. In this subspace, $\hat{H}_E$ cannot be treated as a perturbation. Diagonalize $\hat{H}_E$ in this subspace.
   Hint. $R_{21}^{20} = -3\sqrt{3}a_B$.

   d. Due to the electric field, the $n = 2$ states thus become mixed, and all four states acquire a finite dipole coupling to the $1s$ state. An observable effect is that the spectral line corresponding to the $1s$–$2p$ radiative transition splits. What are the possible frequencies a photon can have when it is emitted during $1s$–$2p$ radiative decay? How does the emission intensity for each frequency depend on the direction of emission? If one of the frequencies corresponds to multiple $n = 2$ states, assume that the observed radiation with this frequency comes from a random mixture of these states.

3. *Lamb shift.* In Exercise 2 we calculated corrections to the $1s$–$2p$ spectral lines linear in the applied electric field E. In the absence of a field, in the electromagnetic vacuum, the expectation value $\langle \hat{\mathbf{E}} \rangle$ vanishes, and these corrections are absent. We know, however, that in the vacuum the expectation value $\langle \hat{\mathbf{E}}^2 \rangle$ does *not* vanish, it is finite due to the vacuum fluctuations of the field. There are thus second order corrections to the spectrum of hydrogen, proportional to $\mathbf{E}^2$, and due to the interaction of the atoms with the vacuum fluctuations of the electromagnetic field. The most spectacular manifestation of these corrections is the energy splitting of $2s$ and $2p$ levels that is called the Lamb shift. The theoretical calculation and experimental verification of this shift has played an important role in the development of quantum electrodynamics.

   In principle, the calculation of the Lamb shift is a lengthy exercise in second-order perturbation theory. The big number of various virtual states involved and the necessity to go beyond the dipole approximation when computing the matrix elements make this exercise rather tedious. Here, we restrict ourselves to the correction to the energy difference $E_{1s} - E_{2p}$ between the ground s state and one of p-states, say, p_z, and take into account only transitions between these two states. This gives an order-of-magnitude estimation of the shift.

a. Give the second order corrections to the energies E_{1s} and E_{2p} in terms of sums over the wave vector $\mathbf{q}$ of the virtual photon. Assume the validity of the dipole approximation when evaluating the matrix elements. Demonstrate that these two contributions to $E_{1s} - E_{2p}$ cancel each other at large q.

b. Express the correction in terms of an integral over the photon energies E and the spontaneous emission rate Γ from p to s.

c. Note the logarithmic divergence of this integral at large energies. Consider the validity of the dipole approximation at large energies and evaluate the correction with logarithmic accuracy.

4. *Emission from a one-dimensional quantum dot.* An electron is confined in a quasi-one-dimensional potential well defined by the potential (assuming $a \ll L$)

$$V(\mathbf{r}) = \begin{cases} 0 & \text{when } 0 < x, y < a, \text{ and } 0 < z < L, \\ \infty & \text{otherwise.} \end{cases}$$

a. Give a general expression for the wave functions and energies of the electronic levels in the well.

b. We focus on the lowest three electronic levels, which we denote by $|111\rangle$, $|112\rangle$, and $|113\rangle$. Suppose the electron is initially in the highest state of the three, $|113\rangle$, and we are interested in the radiative decay rate from this state to the ground state. For what L can we use the dipole approximation?

c. Write down the matrix element M_{if} coupling the initial state, the electron in $|113\rangle$ and the photon vacuum, to the final state, the electron in $|111\rangle$ and one photon with wave vector $\mathbf{q}$ and polarization α. Use the dipole approximation.

d. Calculate from this matrix element the radiative decay rate from $|113\rangle$ to the ground state.

e. Direct decay $|113\rangle \rightarrow |111\rangle$ is not possible, as proven at (d). Decay thus takes place in two steps: radiative decay from $|113\rangle$ to $|112\rangle$ and successive radiative decay from $|112\rangle$ to $|111\rangle$. We first focus on the first transition, from $|113\rangle$ to $|112\rangle$. Give the matrix elements connecting the initial and possible final states.

f. Calculate the decay rate from $|113\rangle$ to $|112\rangle$.

g. Give the typical time it takes to decay from $|113\rangle$ to the ground state.

5. *Classical emission of an electron trapped by an ion.* Consider an electron trapped by an ion in a highly excited bound state, such that its (negative) energy E is much smaller than the atomic energy scale. In this case, the motion of the electron can be regarded as classical, and we assume that the electron moves along a circular orbit.

a. Implement (9.92) for the power of Bremsstrahlung to describe the time-dependence of the electron energy.

b. Describe qualitatively the crossover to the quantum regime.

Solutions

1. *Spontaneous emission from the hydrogen atom.*

   a. After some basic manipulations, you find

   $$Y_l^m \cos\theta = Y_{l+1}^m \sqrt{\frac{(l+1)^2 - m^2}{4(l+1)^2 - 1}} + Y_{l-1}^m \sqrt{\frac{l^2 - m^2}{4l^2 - 1}},$$

   $$Y_l^m \sin\theta e^{i\phi} = -Y_{l+1}^{m+1} \sqrt{\frac{(l+m+1)(l+m+2)}{4(l+1)^2 - 1}} + Y_{l-1}^{m+1} \sqrt{\frac{(l-m)(l-m-1)}{4l^2 - 1}},$$

   $$Y_l^m \sin\theta e^{-i\phi} = Y_{l+1}^{m-1} \sqrt{\frac{(l-m+1)(l-m+2)}{4(l+1)^2 - 1}} - Y_{l-1}^{m-1} \sqrt{\frac{(l+m)(l+m-1)}{4l^2 - 1}}.$$

   b. The decay rate for the case $l' = l + 1$ reads

   $$\frac{4}{3}\frac{\alpha\omega^3}{c^2} |R_{n,l}^{n',l+1}|^2 \left\{ \frac{(l+1)^2 - m^2}{4(l+1)^2 - 1}\delta_{m',m} + \frac{(l \pm m + 1)(l \pm m + 2)}{8(l+1)^2 - 1}\delta_{m',\,m\pm1} \right\},$$

   and for the case $l' = l - 1$

   $$\frac{4}{3}\frac{\alpha\omega^3}{c^2} |R_{n,l}^{n',l-1}|^2 \left\{ \frac{l^2 - m^2}{4l^2 - 1}\delta_{m',m} + \frac{(l \mp m)(l \mp m - 1)}{8l^2 - 1}\delta_{m',\,m\pm1} \right\}.$$

   For all other combinations of l and l', the rate is zero. Note that the expressions found here indeed comply with the selection rules explained in Section 9.4.3

   c. First we calculate

   $$R_{21}^{10} = \frac{2}{\sqrt{24}}\frac{1}{a_B^4} \int_0^\infty dr\, r^4 e^{-3r/2a_B}$$

   $$= \lim_{\eta \to 1} \frac{2}{\sqrt{24}}\frac{1}{a_B^4} \int_0^\infty dr \frac{\partial^4}{\partial\eta^4}\left(\frac{2a_B}{3}\right)^4 e^{-3r\eta/2a_B} = \frac{2^7}{3^4}\sqrt{\frac{2}{3}}a_B.$$

   Then we simply use the decay rates found at (b) to calculate the transition rates for $|2,1,-1\rangle \to |1,0,0\rangle$, for $|2,1,0\rangle \to |1,0,0\rangle$, and for $|2,1,1\rangle \to |1,0,0\rangle$. To account for the random polarization of the atoms, we take the average of the three rates, yielding finally

   $$\Gamma_{2p \to 1s} = \left(\frac{2}{3}\right)^8 \alpha^5 \frac{m_e c^2}{\hbar}.$$

2. *Stark effect.*

   a. The Hamiltonian reads

   $$\hat{H}_E = -eE\hat{z},$$

   where $\hat{z}$ is the z-component of the electronic position operator.

   b. This amounts to calculating $-eE\langle 100|\hat{z}|100\rangle$, but since the $1s$ ground state is spherically symmetric, the expectation value of $\hat{z}$ vanishes. The answer is zero.

c. From the selection rules we know that the only non-vanishing matrix element of the Hamiltonian is

$$-eE \langle 2p_z|\hat{z}|2s\rangle = -eER_{21}^{20} \int d\Omega \, \cos\theta \left[Y_1^0(\theta,\phi)\right]^* Y_0^0(\theta,\phi)$$

$$= -eER_{21}^{20}\sqrt{\frac{1}{3}} = -3eEa_B.$$

All other elements of the Hamiltonian are zero, so we can diagonalize

$$\hat{H}_E = -3eEa_B\{|+\rangle\langle+| - |-\rangle\langle-|\},$$

with $|\pm\rangle = \frac{1}{\sqrt{2}}\{|2p_z\rangle \pm |2s\rangle\}$.

d. We found at (c) that the $2p_z$ state and the $2s$ mix, and split by an amount of $6eEa_B$. The frequencies observed are thus (leaving out the factor of $\hbar$)

$$\tfrac{3}{4}\mathrm{Ry} \quad \text{and} \quad \tfrac{3}{4}\mathrm{Ry} \pm 3eEa_B.$$

The lowest and highest spectral lines are due to a decay process with the dipole moment $\mathbf{r}_{AB} = \langle 1s|\hat{\mathbf{r}}|2p_z\rangle$. This vector only has a z-component, the intensity of the radiation is thus proportional to $\sin^2\theta$, where θ is the angle between the direction of emission and the z-axis (see Section 9.4.2). The central line represents a mix of transitions from $|2p_x\rangle$ and $|2p_y\rangle$ to the ground state. The intensity of the transitions from $|2p_x\rangle$ is proportional to $\sin^2\alpha_x$, where α_x is the angle between the direction of emission and the x-axis. In spherical coordinates, we find $\mathbf{n} \cdot \mathbf{x} = \sin\theta\cos\phi = \cos\alpha_x$, where $\mathbf{n}$ and $\mathbf{x}$ are unit vectors in the direction of emission and along the x-axis respectively. The intensity is thus proportional to $1 - \sin^2\theta\cos^2\phi$. Similarly, the intensity from the other transition (from $|2p_y\rangle$) is proportional to $1 - \sin^2\theta\sin^2\phi$. An equal mixture of the two is thus proportional to $2 - \sin^2\theta$.

10

Coherent states

This chapter is devoted to coherent states. Generally speaking, a wave function of a system of identical particles does not have to have a certain number of particles. It can be a superposition of states with different numbers of particles. We have already encountered this in Chapters 5 and 6 when we studied superconductivity and superfluidity. Another important example of states with no well-defined number of particles is given by *coherent states*. In a way, these states are those which most resemble classical ones: the uncertainty in conjugated variables (such as position and momentum) is minimal for both variables, and their time-evolution is as close to classical trajectories as one can get.

Coherent states of radiation are relatively easy to achieve. They arise if we excite the electromagnetic field with classical currents. We look in detail at the coherent state of a single-mode oscillator and all its properties. The coherent state turns out to be an eigenfunction of the annihilation operator, the distribution of photon numbers is Poissonian, and it provides an optimal wave function describing a classical electromagnetic wave. We come back to our simple model of the laser and critically revise it with our knowledge of coherent states, and estimate the time at which the laser retains its optical coherence. We then derive Maxwell–Bloch equations that combine the master equation approach to the lasing (as outlined in Chapter 9) with the concept of coherence.

Next, we turn to coherent states of matter. They can be realized with the super-phenomena we have studied before – superconductivity and superfluidity. We learn about phase-number uncertainty and the difference in wave functions of a bulk and isolated condensates. We thoroughly illustrate the uncertainties with an example of a macroscopic quantum device: a Cooper pair box. At the end of the chapter, we then achieve an unexpected synthesis: the Cooper pair box may exhibit both coherent states of particles and coherent states of radiation. Table 10.1 summarizes the content of this chapter.

10.1 Superpositions

The notion that radiation is emitted and absorbed in quanta of $\hbar\omega$ is mostly due to Albert Einstein. In 1905, he very successfully described the photoelectric effect with this concept and was later awarded a Nobel prize for this discovery. Yet he doubted quantum mechanics and remained critical of its postulates until his death in 1955. One of the phrases he was famous for was: "The sale of beer in pint bottles does not imply that beer exists only in indivisible pint portions." That is, from the fact that radiation is emitted and absorbed in quanta one cannot deduce that radiation obeys the laws of quantum mechanics and is composed of photons.

Quantum mechanics appears to be sufficiently flexible to adequately reflect this beer paradox. As we have seen, the basis states of the electromagnetic field are those with a fixed number of photons $n_{\mathbf{k}\alpha}$ in each mode, $|\{n_{\mathbf{k}\alpha}\}\rangle$. That is, each state is characterized by a set of numbers telling you how many pints there are in each mode. However, the basic principle of quantum mechanics – even more basic than the quantization itself – is the superposition principle. It states that any linear superposition of basis states is also a legitimate quantum state. An arbitrary quantum state of the field therefore reads

$$|\psi(t)\rangle = \sum_{\{n_{\mathbf{k}\alpha}\}} c(\{n_{\mathbf{k}\alpha}\}, t)|\{n_{\mathbf{k}\alpha}\}\rangle, \qquad (10.1)$$

where the summation goes over all possible sets of occupation number and c is the coefficient of the superposition. If the photons do not interact, the basis states are stationary states of definite energy, while a superposition is not. The coefficients of the superposition therefore do depend on time,

$$c(\{n_{\mathbf{k}\alpha}\}, t) = c(\{n_{\mathbf{k}\alpha}\}) \exp\left\{ i \sum_{\{n_{\mathbf{k}\alpha}\}} n_{\mathbf{k}\alpha} \omega_{\mathbf{k}\alpha} t \right\}. \qquad (10.2)$$

There is a class of superpositions that are especially important in the quantum theory of radiation and actually also just in quantum *theory*. These states were extensively investigated and popularized by Roy Glauber about 50 years ago, the laser being the aim of his research. He wrote: "Such states must have indefinite numbers of quanta present. Only in that way can they remain unchanged when one quantum is removed." His work brought him the Nobel prize in 2005, which symbolically closed a century cycle of research on (non-)quantized radiation.

10.2 Excitation of an oscillator

Since all modes of the electromagnetic field are independent, it suffices to concentrate on a single mode only. So we take the simplest electrical oscillator for our study of coherent states: the LC-circuit already described in Chapters 7 and 8. An important new element we need is a voltage source included in the circuit (see Fig. 10.1). With the source, we apply a time-dependent voltage to *excite* the oscillator.

Let us first sort out the excitation process at a classical level. The equations describing the oscillator with the source read

$$I = \dot{q} = C\dot{V} \quad \text{and} \quad -L\dot{I} = V - V_{\text{ex}}(t), \qquad (10.3)$$

the voltage $V_{\text{ex}}(t)$ of the source contributes to the sum of the voltage drops along the circuit. We assume that initially the oscillator is at rest, $I = V = V_{\text{ex}} = 0$. We then excite the oscillator with a pulse of external voltage, $V_{\text{ex}}(t) = V$ for $0 < t < T$ and zero otherwise.

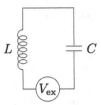

Fig. 10.1   The simplest electrical oscillator, an LC-circuit, with a voltage source.

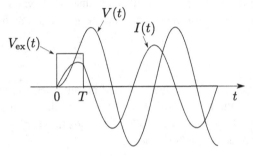

Fig. 10.2   Excitation of the oscillator with a voltage pulse of $V_{ex}(t) = V$ for $0 < t < T$ and zero otherwise. The resulting current $I(t)$ and voltage $V(t)$ remain oscillating after the pulse.

The resulting current and voltage are plotted versus time in Fig. 10.2. After the pulse has ended, the oscillator remains in an excited state characterized by an amplitude and phase,

$$V(t) = V_0 \cos(\omega_0 t + \phi) \quad \text{and} \quad I = C\dot{V}. \tag{10.4}$$

For our idealized oscillator, the excitation persists forever. A more realistic oscillator, i.e. including dissipation, is considered in the next chapter.

Instead of two variables, V and I, it is advantageous to use a single complex coordinate, as we did in Chapter 7. We thus use $d(t)$, the normal coordinate of the mode, defined as

$$d = \frac{1}{\sqrt{2\omega_0}} \left(i\sqrt{L}I + \sqrt{C}V \right). \tag{10.5}$$

The equation of motion for d is simply

$$\dot{d} = -i\omega_0 d + iv(t), \quad \text{where} \quad v(t) \equiv V_{ex}(t)\sqrt{\frac{\omega_0 C}{2}}. \tag{10.6}$$

Control question. Can you find a general solution of the above equation?

The solution for such a pulse excitation is depicted in Fig. 10.3 as a trajectory of d in the complex plane. The circle corresponds to the persistent excitation after the pulse, $d(t) = |d| \exp\{-i\omega_0 t\}$. In fact, it would be even better to plot $d(t) \exp\{i\omega_0 t\}$ instead of $d(t)$: the persistent excitation would then correspond to a single point in the plot.

Let us now turn to the quantum description of this problem. First of all, we need a Hamiltonian. It is obtained from the original Hamiltonian with a voltage shift,

$$\hat{H} = \frac{C\hat{V}^2}{2} + \frac{L\hat{I}^2}{2} - C\hat{V}V_{ex}(t). \tag{10.7}$$

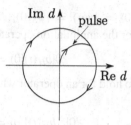

Fig. 10.3   Response of the oscillator to a voltage pulse, plotted as a trajectory of the normal coordinate $d(t)$.

We can convince ourselves that this Hamiltonian is correct by deriving the equations of motion from it and comparing those with (10.3). Next, we rewrite the Hamiltonian in terms of the bosonic CAPs $\hat{b}^\dagger$ and $\hat{b}$ of the mode. We obtain

$$\hat{H} = \hbar\omega_0 \hat{b}^\dagger \hat{b} - \sqrt{\hbar}v(t)\hat{b}^\dagger - \sqrt{\hbar}v^*(t)\hat{b}. \tag{10.8}$$

The role of the voltage source is thus restricted to the last two terms, which are linear in the CAPs.

A problem is that the Hamiltonian explicitly depends on time. It makes no sense to search for its stationary eigenfunctions, they do not exist. Although we might be able to find the time-dependent solutions we need, it is simpler to turn to the Heisenberg picture first. We try to solve for the time-dependent Heisenberg annihilation operator $\hat{b}(t)$ that satisfies $i\hbar\dot{\hat{b}} = [\hat{b}, \hat{H}(t)]$. Carrying out all commutations, we arrive at the equation

$$\dot{\hat{b}} = -i\omega_0 \hat{b} + i\frac{v(t)}{\sqrt{\hbar}}. \tag{10.9}$$

This operator equation is easy to solve, it is enough to recognize its similarity with (10.6). The solution reads

$$\hat{b}(t) = \hat{b}_S e^{-i\omega_0 t} + \frac{d(t)}{\sqrt{\hbar}}, \tag{10.10}$$

where $d(t)$ is the solution of the classical equation (10.6) and $\hat{b}_S$ the common annihilation operator in the Schrödinger picture. We find that the Heisenberg annihilation operator in the presence of an external source gets a "classical addition" – a constant term that corresponds to the solution of the classical equation. We use this result extensively in the part about dissipative quantum mechanics.

Still we want to find a solution for the state of the system, that is, we should turn back to the Schrödinger picture. To find the corresponding state, we implement the following trick that nicely illustrates the relation between the two pictures. We know that for any operator $\hat{A}$ and wave function $|\psi\rangle$ the expectation value does not depend on the picture,

$$\langle\psi_H|\hat{A}(t)_H|\psi_H\rangle = \langle\psi_S(t)|\hat{A}_S|\psi_S(t)\rangle, \tag{10.11}$$

where we use the indices H and S to distinguish between operators and states in the Heisenberg and Schrödinger picture respectively. There is a moment in time where the operators and wave functions are identical. For us it is convenient to choose this moment right before the pulse, at $t = 0$. At this moment, the oscillator is not excited so it is in its ground state, $|0\rangle = |\psi_H\rangle = |\psi_S(0)\rangle$. The relation (10.11) holds at any time for any operator, so also for the annihilation operator,

$$\langle 0|\hat{b}_H(t)|0\rangle = \langle \psi(t)|\hat{b}_S|\psi(t)\rangle. \qquad (10.12)$$

In fact, it must also hold for an operator which is an arbitrary function f of the annihilation operator,

$$\langle 0|f[\hat{b}_H(t)]|0\rangle = \langle \psi(t)|f[\hat{b}_S]|\psi(t)\rangle. \qquad (10.13)$$

For the annihilation operator given by (10.10), the left-hand side of the above relation can be evaluated as

$$\langle 0|f[\hat{b}_H(t)]|0\rangle = \langle 0|f[\hat{b}_S e^{-i\omega_0 t} + \frac{d(t)}{\sqrt{\hbar}}]|0\rangle = f[\frac{d(t)}{\sqrt{\hbar}}]. \qquad (10.14)$$

The last equality follows from the definition of the vacuum, $\hat{b}_S|0\rangle = 0$, that is, one can safely replace $\hat{b}_S$ by 0 in all operator expressions acting on the vacuum. So the excited state sought satisfies

$$f[\frac{d(t)}{\sqrt{\hbar}}] = \langle \psi(t)|f[\hat{b}_S]|\psi(t)\rangle, \qquad (10.15)$$

for an arbitrary function f. Since this function can be arbitrary (it can also contain higher powers of $\hat{b}_S$), the state $|\psi(t)\rangle$ must be an *eigenstate* of the annihilation operator,

$$\hat{b}|\psi(t)\rangle = \lambda|\psi(t)\rangle, \qquad (10.16)$$

where obviously we must have the eigenvalue $\lambda = d(t)/\sqrt{\hbar}$.

We have thus found that upon an external excitation the oscillator goes from the vacuum state to an eigenstate of the annihilation operator. The complex eigenvalue of the eigenstate is given by the solution of the classical equations for the excitation of the oscillator. Such a state is called a *coherent* state.

10.3 Properties of the coherent state

The coherent state has many interesting properties that can be derived directly from its definition,

$$\hat{b}|\psi\rangle = \lambda|\psi\rangle. \qquad (10.17)$$

First of all, since the annihilation operator does not commute with the operator of particle number, $\hat{n} \equiv \hat{b}^\dagger \hat{b}$, the coherent state is *not* a state with a fixed number of photons. It is not difficult to find an expression for the coherent state in the basis of states with fixed particle numbers, $|n\rangle$. We search for it in the form

$$|\psi\rangle = \sum_n \psi_n|n\rangle. \qquad (10.18)$$

By definition of the annihilation operator,

$$\hat{b}|n\rangle = \sqrt{n}|n - 1\rangle, \qquad (10.19)$$

and therefore

$$\sum_n \sqrt{n}\psi_n|n-1\rangle = \lambda \sum_n \psi_n|n\rangle, \tag{10.20}$$

and

$$\sqrt{n+1}\psi_{n+1} = \lambda\psi_n, \quad \text{so that} \quad \psi_{n+1} = \frac{\lambda}{\sqrt{n+1}}\psi_n. \tag{10.21}$$

So we can express all ψ_n by induction, starting from the coefficient ψ_0 for the vacuum state,

$$\psi_n = \frac{\lambda^n}{\sqrt{n!}}\psi_0. \tag{10.22}$$

This coefficient is fixed by the normalization condition,

$$1 = \sum_n |\psi_n|^2 = |\psi_0|^2 \sum_n \frac{|\lambda|^{2n}}{n!} = |\psi_0|^2 e^{|\lambda|^2}, \tag{10.23}$$

and we therefore obtain an explicit form of the coherent state

$$|\psi\rangle = e^{-\frac{1}{2}|\lambda|^2} \sum_n \frac{\lambda^n}{\sqrt{n!}}|n\rangle. \tag{10.24}$$

Let us evaluate the average number of photons in the coherent state. By direct calculation, we obtain

$$\bar{n} = \sum_n n|\psi_n|^2 = e^{-|\lambda|^2} \sum_n n\frac{|\lambda|^{2n}}{n!} = |\lambda|^2. \tag{10.25}$$

The average number of photons thus fixes the modulus of the coherent state parameter λ, but does not fix its phase. In general, we thus have $\lambda = \sqrt{\bar{n}}e^{i\varphi - i\omega_0 t}$, where we have explicitly written the trivial time-dependent phase $\omega_0 t$.

To understand the coherent state in more intuitive terms, it is constructive to make plots in the plane of complex λ (such as in Fig. 10.4). Since $\lambda(t) = d(t)/\sqrt{\hbar}$, the classical plot in Fig. 10.3 gives the time evolution of the coherent state upon excitation. Initially the oscillator was in the vacuum state (coherent state with $\lambda = 0$). During the pulse, the state evolves with $|\lambda|$ increasing. This corresponds to an increase of the average photon number

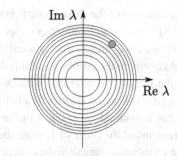

Fig. 10.4 Representation of coherent states in the plane of complex λ. The concentric circles show states with a fixed number of photons but indefinite phase. The small circle with diameter $\frac{1}{2}$ shows the coherent state.

(and associated average energy $\bar{E} = \hbar\omega\bar{n}$) in the process of excitation. After the pulse, the coherent state parameter orbits the plane of complex λ along a circle. The average number of photons and energy remain constant, as they should since the excitation process has ended, and the phase of λ evolves periodically with $e^{-i\omega_0 t}$. In the same plane of complex λ, states with a fixed number of photons n do not have any parameter that would account for a definite phase. Those states can be pictured as thin concentric circles with radius $\sqrt{n}$.

To put the coherent state in this picture, we get rid of the trivial time-dependent phase $\omega_0 t$. In the resulting coordinates, the classical solution $d(t)$ (after excitation) is represented by a point rather than an orbit. The corresponding coherent state is thus also centered around a stationary point. In the plot, we show it as a circle with a diameter $\frac{1}{2}$. With this finite diameter, we give a pictorial presentation of the fact that coherent states do not form an orthogonal set: they are not distinct from each other if the difference between their parameters λ is sufficiently small. To quantify this, we can look at the overlap between two states with λ and λ',

$$\langle \psi_{\lambda'} | \psi_\lambda \rangle = e^{-\frac{1}{2}|\lambda'|^2} \sum_n \frac{(\lambda'^*)^n (\lambda)^n}{n!} e^{-\frac{1}{2}|\lambda|^2} = \exp\{-\tfrac{1}{2}|\lambda'|^2 + \lambda'^* \lambda - \tfrac{1}{2}|\lambda|^2\}, \quad (10.26)$$

$$|\langle \psi_{\lambda'} | \psi_\lambda \rangle|^2 = \exp\{-|\lambda - \lambda'|^2\}. \quad (10.27)$$

The states are thus distinct if the distance between λ' and λ becomes of the order of 1, which shows that our choice of $\frac{1}{2}$ for the diameter of the circle representing the coherent state roughly makes sense.

This pictorial presentation of the uncertainty in λ also gives us an idea of the uncertainty in the number of photons in the coherent state. An uncertainty of $\Delta|\lambda| = \frac{1}{2}$ (which is relatively small as long as $|\lambda| \gg 1$) would translate to

$$\Delta|\lambda| = \Delta\sqrt{\bar{n}} \approx \frac{\Delta n}{2\sqrt{\bar{n}}} = \frac{1}{2}, \quad \text{so} \quad \Delta n \approx \sqrt{\bar{n}}. \quad (10.28)$$

We can actually arrive at the same result by investigating the probability distribution for the number of photons in the coherent state. The probability $p(n)$ to have n photons is given by the coefficients

$$p(n) = |\psi_n|^2 = \frac{\bar{n}^n}{n!} e^{-\bar{n}}, \quad (10.29)$$

where we again recognize the Poisson distribution. We know that this distribution has a variance $\langle (n - \bar{n})^2 \rangle = \bar{n}$, which indeed corresponds again to $\Delta n = \sqrt{\bar{n}}$.

We recall that we have already encountered the Poisson distribution while investigating lasers. We could compare the properties of this distribution with the Boltzmann distribution we found for the number of photons in black-body radiation. If the average number of photons $\bar{n}$ is large, the Poisson distribution is highly concentrated around this value since relative error $\Delta n/\bar{n}$ is small $\simeq 1/\sqrt{\bar{n}}$. Contrastingly, for the Boltzmann distribution the relative error remains of the order of 1, since there $\Delta n/\bar{n} \approx 1/\sqrt{2}$.

The Poisson distribution signals independent processes. In the context of the excitation of an oscillator it demonstrates that all photons are excited independently – no wonder, since they are non-interacting particles.

Another important property of the coherent state is that it is in some way an optimal state. This property was addressed by Schrödinger in the early days of quantum mechanics. At that time he was searching for a single-particle wave function that approximates a classical state in a most accurate way. To quantify this search, he demanded that this optimal function correspond to the minimum of the uncertainty relation

$$\Delta p \, \Delta x \geq \frac{\hbar}{2}. \tag{10.30}$$

This then sets a variational equation for the wave function $\psi(x)$. The solution of this variational equation with $\langle p \rangle = \langle x \rangle = 0$ is a Gaussian function,

$$\psi(x) = \frac{1}{\sqrt{\sigma \sqrt{2\pi}}} \exp \left\{ -\frac{x^2}{4\sigma^2} \right\}. \tag{10.31}$$

We in fact know that this function describes the vacuum state of an oscillator (see Section 1.9). A solution with non-zero expectation values $\langle p \rangle = p_0$ and $\langle x \rangle = x_0$ is obtained with shifts in x and p, yielding

$$\psi(x) = \frac{1}{\sqrt{\sigma \sqrt{2\pi}}} \exp \left\{ -\frac{(x - x_0)^2}{4\sigma^2} + i\frac{p_0 x}{\hbar} \right\}. \tag{10.32}$$

If we expand this function in the basis of eigenstates of the oscillator, we find that it indeed is a coherent state. The parameter of the state is found from $\lambda = \langle \hat{b} \rangle$ and the expression of the annihilation operator in terms of $\hat{p}$ and $\hat{x}$ is as we extensively discussed in Chapters 7 and 8. To conclude, we have identified the coherent state to be the one which optimizes the uncertainty relation and provides thereby the best approximation to a classical state having well-defined position *and* momentum.

We have already shown how the presentation of the coherent state in Fig. 10.4 allows us to read off the uncertainty in the number of photons. The same thing can of course easily be done for the uncertainty in the phase, the other coordinate in our description of the oscillator. Using again that $|\lambda| = \sqrt{\bar{n}} \gg 1$, and that the coherent state is pictured as a circle with diameter $\frac{1}{2}$, we find for the uncertainty in the phase

$$\Delta\phi \approx \frac{\Delta|\lambda|}{|\lambda|} \approx \frac{1}{2\sqrt{\bar{n}}}, \tag{10.33}$$

see Fig. 10.5. Combining this with the result that $\Delta n \approx \sqrt{\bar{n}}$, we see that for our coherent state we have $\Delta n \, \Delta \phi \approx \frac{1}{2}$. Since this coherent state optimizes the uncertainty relation, we know that generally we must have

$$\Delta n \, \Delta \phi \geq \frac{1}{2}. \tag{10.34}$$

In the spirit of Chapter 8, we now extend our conclusions for a single mode to the case of (infinitely) many modes, for instance, to all photon modes available in the physical vacuum. The general coherent state is then a direct product of the coherent states in each mode, and each state is characterized by its own parameter λ_k,

$$|\psi\rangle = \prod_k |\psi_{\lambda_k}\rangle, \quad \text{with} \quad \hat{b}_k |\psi_{\lambda_k}\rangle = \lambda_k |\psi\rangle. \tag{10.35}$$

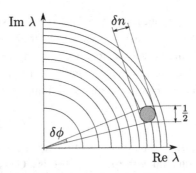

The uncertainty relation for phase and particle number can be deduced from this simple plot. The optimal (coherent) state is a circle with radius $\frac{1}{2}$. The spread of this state in the radial coordinate shows the uncertainty in n, and the azimuthal spread the uncertainty in ϕ.

There is in fact a direct and simple correspondence between any classical field configuration and the parameters of the coherent states that approximate this classical state best. This can easily be understood if we look at a simple example. In Section 8.2 we quantized the deformation field of an elastic string, and we were able to describe the system in terms of phonon creation and annihilation operators, let us for consistency call them here $\hat{b}^{\dagger}$ and $\hat{b}$. We then found for the position operator describing the local displacement of the string

$$\hat{u}(x,t) = \sum_k \sqrt{\frac{\hbar}{2\rho L \omega_k}} \left\{ \hat{b}_k e^{i(kx-\omega t)} + \hat{b}_k^{\dagger} e^{-i(kx-\omega t)} \right\}. \tag{10.36}$$

We immediately noted that this operator behaved somewhat counterintuitively at first sight: the expectation value of $\hat{u}(x,t)$ for the state with one singly excited mode $|1_k\rangle$ turned out to be zero. We then showed that introducing uncertainty in the energy, that is, creating a superposition of states with different numbers of phonons in the mode, helped. Already the state $\frac{1}{\sqrt{2}}\{|1_k\rangle + |2_k\rangle\}$ yielded a traveling-wave-like periodic amplitude $\langle \hat{u}(x,t)\rangle$, as we would expect from a classical point of view. We also hinted that the superpositions which resemble the classical physics best are the coherent states. Indeed, if we use what we learned in this chapter, and we calculate the expected displacement in a coherent state $|\psi_{\lambda_k}\rangle$, we find

$$\begin{aligned} \langle \psi_{\lambda_k} | \hat{u}(x,t) | \psi_{\lambda_k} \rangle &= \sqrt{\frac{\hbar}{2\rho L \omega_k}} \left\{ \lambda_k e^{i(kx-\omega t)} + \lambda_k^* e^{-i(kx-\omega t)} \right\} \\ &= \sqrt{\frac{2\hbar \bar{n}_k}{\rho L \omega_k}} \cos(kx - \omega t + \phi), \end{aligned} \tag{10.37}$$

where we use that we can write $\lambda_k = \sqrt{\bar{n}_k} e^{i\phi}$. We thus find a traveling wave with a well-defined number of phonons $\bar{n}_k$ and well-defined phase ϕ, which is the best correspondence to a classical traveling wave that one can achieve.

The same can also be done for the electromagnetic field. Given a classical field configuration $\mathbf{A}(\mathbf{r},t)$, we can deduce the coherent state with the best correspondence from the expression for field operators, using again that $\langle \hat{b}_k \rangle = \lambda_k$,

$$A(\mathbf{r},t) \quad\Leftrightarrow\quad \sum_{\mathbf{k},\alpha}\sqrt{\frac{\hbar}{2\varepsilon_0\omega_{\mathbf{k}}\mathcal{V}}}\left[\lambda_{\mathbf{k}\alpha}(t)\mathbf{e}_\alpha\exp\{i\mathbf{k}\cdot\mathbf{r}\}+\text{c.c.}\right]. \tag{10.38}$$

In analogy to the single oscillator described above, such coherent states result from excitation of the vacuum by time-dependent currents $\mathbf{j}(\mathbf{r},t)$. To find the relation between the currents and the λ_k one solves the classical Maxwell equations with current sources $\mathbf{j}(\mathbf{r},t)$ and then makes use of the correspondence (10.38). The Hamiltonian that describes the interaction between the currents and radiation is of familiar form,

$$\hat{H}_{\text{int}} = -\int d\mathbf{r}\,\hat{A}(\mathbf{r})\cdot\hat{\mathbf{j}}(\mathbf{r}).$$

10.4 Back to the laser

Let us now revise the simple model of the laser, introduced in the previous chapter, using our insight into coherent states. There is a spectacular feature of lasers we intentionally failed to mention in the previous chapter. If a single-mode laser is properly operated, its emitted light is optically coherent in space as well as in time with astonishing precision. In the time domain, this means that the electric field of the laser light in a given point of space oscillates at a given frequency with a very stable phase ϕ, that is, $E(t)=E\cos(\omega t+\phi)$. The phase drifts at the scale of the optical coherence time τ_{oc} which, under favorable circumstances, can be of the order of hours. In the frequency domain, this determines the line-width of the laser $\Delta\omega \simeq 1/\tau_{\text{oc}} \simeq 10^{-4}$ Hz. The laser is unbelievably monochromatic.

The approach of the previous chapter implied that at a given moment the lasing mode is in a state with a fixed number of photons, and this fixed number fluctuates in time. The phase-number uncertainty relation however (10.34) tells us that this cannot be the case for a perfectly monochromatic laser. Such a state would have a big phase uncertainty, not compatible with the optical coherence observed. So, the question is in what kind of state an operating laser actually is. We have just learned that a coherent state with a large average number of photons has a well-defined phase, as the laser has. Also, the Poisson distribution of the photon number in a coherent state is the same as we have found with the master equation of the previous chapter. This suggests that the state of a laser maybe is a coherent state.

However logical it may sound, a laser cannot be in a precise coherent state. A coherent state is a specific excited state of an oscillator – a closed conservative system. An operating laser is an open driven system with a finite coherence time, absorbing and emitting energy. It thus cannot be adequately described by a simple wave function. Most similarities between the coherent state and the coherent laser come from the fact that the lasing mode contains many photons and, at least in first approximation, can be regarded as a classical field configuration. And, as we know, a coherent state just expresses this classical state in terms of quantum mechanics. A better statement therefore is that the coherent state is useful to understand the quantum properties of laser.

Roy J. Glauber (b. 1925)

Won the Nobel Prize in 2005 for "his contribution to the quantum theory of optical coherence."

Roy Glauber was born in New York City. His father was a traveling salesman, and as soon as his child was old enough to travel, the Glaubers left New York for a nomadic existence in the mid-west. When Glauber reached the age of six and had to go to school, the family returned to New York to settle. After starting high school in 1937, his developing interest in science took the form of a fascination for astronomy. He spent almost a year building a working reflecting telescope, without a noteworthy budget. After graduating from the Bronx High School of Science, he won a scholarship for Harvard where he started to study physics in 1941.

After the US got involved in the war, many of Glauber's classmates were drawn into the army. He himself was too young, but when he turned 18 in 1943, he immediately offered to do war work. Several weeks later he was interviewed by a mysterious man who invited him to participate in what turned out to be the Manhattan project. Glauber thus went to Los Alamos, where he worked, as one of the youngest participants, for two years, mainly on finding the critical mass for the nuclear bomb.

Back at Harvard, Glauber got his Bachelor's degree in 1946 and in 1949 his Ph.D. degree under the supervision of Julian Schwinger. After graduating, he spent some years at Princeton, half a year with Pauli in Zurich, and one year at Caltech to replace Feynman who went to Brazil. In 1952, he returned to Harvard where he has stayed ever since. He worked on many different topics. During the late 1950s, he got fascinated, as many, by the newly developed laser. The quantum structure of its output remained a mystery for several years, but in the early 1960s Glauber managed to answer most open questions with the insight that the state of the photons in a laser beam in fact can be understood in terms of a coherent state.

10.4.1 Optical coherence time

To illustrate this, let us provide an estimate of the optical coherence time of a laser using a coherent state description. We start with the laser switched off, that is, there is no input of any energy. We still approximate the field in the lasing mode by a coherent state with a parameter λ (we leave out the trivial time-dependent phase of λ). If there are no losses nor pumping, the parameter should remain constant in time, $d\lambda/dt = 0$. Losses, photons escaping through the front mirror, cause relaxation of the parameter toward 0. As in the previous chapter, we describe this relaxation rate with the quality factor Q,

$$\frac{d\lambda}{dt} = \frac{d\lambda}{d\bar{n}}\frac{d\bar{n}}{dt} = -\frac{\omega}{2Q}\lambda. \tag{10.39}$$

However, the losses are in fact a stochastic process of emission of radiation quanta. To model the quantum fluctuations of emission, we add some random noise $\zeta(t)$ to the equation,[1]

$$\frac{d\lambda}{dt} = -\frac{\omega}{2Q}\lambda + \zeta(t), \quad \text{with} \quad \langle\zeta(t)\zeta(t')\rangle = S\delta(t - t'). \tag{10.40}$$

The δ correlation function of $\zeta(t)$ indicates a white noise spectrum. This noise causes fluctuations of λ proportional to the noise intensity S, explicitly $\langle|\lambda|^2\rangle = SQ/\omega$.

Control question. The differential equation (10.40) can be explicitly integrated to give $\lambda(t)$ for any noise source $\zeta(t)$. With this information in hand, can you prove the above relation for $\langle|\lambda|^2\rangle$?

Since the noise was introduced into the equation to mimic the quantum fluctuations, S should be chosen in such a way that the fluctuations of λ equal the vacuum fluctuations of the corresponding operators, $\langle|\lambda|^2\rangle = \frac{1}{2}$ (remember the diameter of the circle that represented the coherent state in Fig. 10.4),

$$\langle|\lambda|^2\rangle = S\frac{Q}{\omega} = \frac{1}{2}, \quad \text{so} \quad S = \frac{\omega}{2Q}. \tag{10.41}$$

The best way to understand (10.40) is to picture λ as two coordinates of a "particle" moving in a viscous medium with a velocity proportional to the force acting on the particle (the two coordinates being the real and imaginary part of λ). Such a particle experiences a central force trying to relax λ toward zero (this force is $\propto \lambda$) and a random force causing λ to fluctuate (this force is $\propto \zeta(t)$). It is important to recall that in absence of the central force, the particle exhibits Brownian motion with a diffusion coefficient $D = \frac{1}{2}S = \frac{1}{4}\omega/Q$.

Let us now switch the laser on! In this case, the pumping of energy into the laser pushes the particle away from the center. We note that equilibrium between pumping and losses is achieved along a whole circle with radius $|\lambda_0| = \sqrt{\bar{n}}$ (see Fig. 10.6). The balance of gain and loss does not, however, fix the phase of λ, and initially it may take any value.[2] If there were no noise, the phase would retain this value forever. The noise causes diffusion of the phase. We estimate the typical time for which the phase drift becomes of the order of 1 from the diffusion law, $|\lambda|^2 \simeq Dt$. This time is the optical coherence time sought,

$$\tau_{\text{oc}} = \frac{\bar{n}}{D} = \frac{4\bar{n}Q}{\omega}. \tag{10.42}$$

Besides this optical coherence time, one can define a quantum coherence time for which the quantum state is retained. The quantum state becomes different if the change of λ is of the order of 1, so this time is $\simeq Q/\omega$. In the previous chapter, we estimated $Q/\omega \simeq$

---

[1] As we present it, this approach looks heuristic if not controversial. In fact, it not only gives the correct results: it can be thoroughly justified using a path integral formulation of quantum mechanics. Such a justification, however, is far beyond the framework of this book.

[2] This resembles spontaneous symmetry breaking as studied in Chapters 4–6.

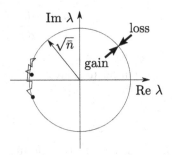

Fig. 10.6   Phase diffusion in lasers. Equilibrium between the gains and losses in the lasing mode fixes the modulus $|\lambda|$. The noise $\zeta(t)$ makes $\lambda(t)$ a fluctuating quantity. Fluctuations in the modulus result in $\langle|\lambda|^2\rangle = \frac{1}{2}$. The same fluctuations cause the phase arg λ to drift.

3×10^{-6} s and $r \simeq 10^{16}$ Hz. This gives $\bar{n} \simeq 3 \times 10^{10}$, and the typical coherence time is then $\tau_{oc} \simeq 3 \times 10^5$ s, roughly 100 hours.

Unfortunately, coherence times of modern lasers do not reach this limit: in practice, τ_{oc} is determined by the fact that the lasing frequency depends on macroscopic parameters of the laser and these parameters fluctuate in time.

> **Control question.** The traffic passing nearby causes vibrations that affect the distance between semi-transparent mirrors defining the lasing mode. How does this affect the optical phase of the emitted radiation?

10.4.2 Maxwell–Bloch equations

In Section 9.7, we treated the laser in an incoherent way. We condensed our description to describe all the physics in terms of probabilities and master equations. Only the transition rates appearing in these master equations were of a quantum nature, while the rest of the formalism only involved probabilities: those for an atom to be in the ground or excited state, and those to have n photons in the lasing mode. In a quantum language, we could say that we only kept the diagonal elements of the density matrix of our system, and disregarded the off-diagonal elements describing the quantum coherence between the different states. In the present chapter, we have found that the state of photons in a lasing mode can be approximated as a coherent state, a coherent superposition of the states with definite number of photons. Similarly, as we will see, the atoms in the laser can also be in a superposition of excited and ground states. Let us bring this into the picture.

However, we do not want to deal with the complex fully-quantum problem of a system of atoms in the field inside the laser. Rather, we introduce the following approximation. We treat the oscillating dipole moment of all atoms, which is in principle a quantum operator, as a classical variable. This is justified by the large number of atoms in the cavity. Similarly, the field in the lasing mode is approximated by the corresponding classical variable. In this form, we end up with two systems of equations: "Maxwell" equations for the field excited by the total dipole moment of atoms, and so-called "Bloch" equations describing the dipole moment of a single atom induced by the laser field. The self-consistent solution of these two systems then describes the lasing effect. We will see the differences and similarities with the previous approach.

Let us formulate a simplified model that permits a quick derivation of the equations. We assume a single lasing mode with its polarization in the x-direction. This allows us to take only three states of the Ne atom into account: the excited s-state, the p_x-state of intermediate energy, and the ground state $|g\rangle$. The energy difference between the excited state and the state $|p\rangle$, denoted by $\hbar\omega_{sp}$, is assumed to be close to the resonance frequency of the mode ω_0. This can provide coherence between the states $|s\rangle$ and $|p\rangle$, resulting in off-diagonal elements of the density matrix. All other off-diagonal elements can be disregarded. The atomic density matrix thus reads

$$\hat{\rho} = \begin{pmatrix} p_s & \rho_{ps} & 0 \\ \rho_{sp} & p_p & 0 \\ 0 & 0 & p_g \end{pmatrix}, \qquad (10.43)$$

where the indices s, p, and g refer to the excited, intermediate, and ground states, respectively. The operator of the dipole moment has an off-diagonal matrix element between the states $|s\rangle$ and $|p\rangle$, and reads

$$\hat{d} = er_{sp}\{|s\rangle\langle p| + |s\rangle\langle p|\} \qquad (10.44)$$

r_{sp} being the matrix element of the coordinate discussed in Chapter 9. The expectation value of this dipole moment reads therefore $\bar{d} = er_{sp}\left(\rho_{ps} + \rho_{ps}^*\right)$.

The interaction between the electric field of the mode and a single atom situated at the position $\mathbf{r}_0$ in the cavity is given by $\hat{H}_{int} = -\hat{\mathbf{E}}(\mathbf{r}_0) \cdot \hat{\mathbf{d}}$. The field operator in terms of the mode CAPs reads (see Chapter 8)

$$\hat{\mathbf{E}}(\mathbf{r}) = i\sqrt{\frac{\hbar\omega_0}{2\varepsilon_0}}\left(\hat{b}e^{-i\omega_0 t} - \hat{b}^\dagger e^{i\omega_0 t}\right)\mathbf{e}_m\Phi(\mathbf{r}), \qquad (10.45)$$

where the polarization vector $\mathbf{e}_m$ points in the x-direction, and the function $\Phi(\mathbf{r})$ normalized as $\int d\mathbf{r}\,\Phi^2(\mathbf{r}) = 1$ describes the details of the electric field distribution inside the cavity. For model purposes we can safely set $\Phi(\mathbf{r})$ to the same value for all atoms in the cavity, $\Phi(\mathbf{r}) = \mathcal{V}^{-1/2}$ for a cavity with volume $\mathcal{V}$. With this, the interaction term reads

$$\hat{H}_{int} = i\hbar g\left(\hat{b}^\dagger e^{i\omega_0 t} - \hat{b}e^{-i\omega_0 t}\right)\{|s\rangle\langle p| + |s\rangle\langle p|\}, \qquad (10.46)$$

with the coupling constant $g \equiv er_{sp}(\omega_0/2\hbar\varepsilon_0\mathcal{V})^{1/2}$. The coupling constant can be estimated as $g \simeq \omega_{at}(r_{at}^3/\mathcal{V})^{1/2}$, yielding $g \simeq 1$ kHz for $\mathcal{V} \simeq 1$ cm^3.

The Maxwell–Bloch equations for common systems allow for an efficient separation of time scales: the typical times for the ignition and damping of the lasing process are much longer than $(\omega_0)^{-1}$. To incorporate this, let us introduce the so-called rotating wave approximation. We make a time-dependent transformation $|s\rangle \to |s\rangle e^{i\omega_0 t}$, which results in a so-called *rotating frame*. In this rotating frame the effective Hamiltonian describing the splitting of the states $|s\rangle$ and $|p\rangle$ becomes

$$\hat{H}_{sp} = \hbar\omega_{sp}|s\rangle\langle s| \quad \Rightarrow \quad \hat{H}_{sp} = \hbar\nu_{sp}|s\rangle\langle s|, \qquad (10.47)$$

with $\nu_{sp} = \omega_{sp} - \omega_0$ the mismatch of the atomic and mode frequencies.

Control question. Can you derive this rotating frame Hamiltonian explicitly?

Applying the same transformation to $\hat{H}_{\text{int}}$ yields terms of two sorts: terms without an explicit time-dependence and terms oscillating with $e^{\pm 2i\omega_0 t}$. The latter highly oscillating terms are then disregarded, which is called the *rotating wave approximation*. We then find that the interaction part of the Hamiltonian (the so-called Jaynes–Cummings Hamiltonian) reads

$$\hat{H}_{\text{int}} = i\hbar g\left(\hat{b}^\dagger |p\rangle\langle s| - \hat{b}|s\rangle\langle p|\right). \tag{10.48}$$

From this point, we move in two directions as promised. Let us first elaborate on the "Maxwell part" of our model. We replace the operators $|s\rangle\langle p|$ and $|p\rangle\langle s|$ in the above equation by their average values ρ_{sp} and ρ_{ps}. The resulting Hamiltonian is of the same type as (10.8) where ρ_{ps} now plays the role of the external excitation. In the resulting time-evolution equation for the coherent state parameter λ we include the dissipation caused by the losses through the front mirror, as we did in (10.39). The source term $-g\rho_{ps}$ has to be multiplied by the number of atoms N in the cavity, and this, the Maxwell part reads

$$\dot{\lambda} + \frac{\omega_0}{2Q}\lambda = -gN\rho_{ps}. \tag{10.49}$$

Let us now consider the "Bloch part." To obtain a Hamiltonian for an atom, we treat the field as if it were in a coherent state and replace $\hat{b}, \hat{b}^\dagger \to \lambda, \lambda^*$ (thereby assuming a large number of photons in the mode). Combining this with the energy splitting described by $\hat{H}_{sp}$, we obtain

$$\hat{H}_B = \hbar v_{sp}|s\rangle\langle s| + i\hbar g\left(\lambda^* |p\rangle\langle s| - \lambda|s\rangle\langle p|\right). \tag{10.50}$$

The time-derivative of the density matrix is contributed to by two sorts of term. Some terms arise from the Hamiltonian, according to time-evolution equation $\dot{\hat{\rho}} = -\frac{i}{\hbar}[\hat{H}_B, \hat{\rho}]$, while others represent the (incoherent) transition rates. As in Section 9.7, we include transitions from $|g\rangle$ to $|s\rangle$ with rate Γ_r, and from $|s\rangle$ to $|p\rangle$ with rate Γ. For completeness, let us add the rate Γ_p of the transition from $|p\rangle$ to $|g\rangle$. These rates appear in the time-derivatives of the diagonal elements in the same way as in the master equations (the diagonal elements of $\hat{\rho}$ indeed simply describe occupation probabilities). The time-derivatives of the off-diagonal elements acquire a negative contribution proportional to half of the sum of the decay rates of the two corresponding states. With this, we write the four Bloch equations as follows

$$\dot{p}_g = -\Gamma_r p_g + \Gamma_p p_p, \tag{10.51}$$

$$\dot{p}_p = -\Gamma_p p_p + \Gamma p_s - g(\lambda^* \rho_{ps} + \lambda\rho_{ps}^*), \tag{10.52}$$

$$\dot{p}_s = \Gamma_r p_g - \Gamma p_s + g(\lambda^* \rho_{ps} + \lambda\rho_{ps}^*), \tag{10.53}$$

$$\dot{\rho}_{ps} = (iv_{sp} - \tfrac{1}{2}[\Gamma + \Gamma_p])\rho_{ps} + \lambda g(p_p - p_s). \tag{10.54}$$

Together with (10.49), this forms the set of Maxwell–Bloch equations sought.

They do not look like (9.93) and (9.94) describing the laser in the master equation approach, and were derived using concepts not invoked in Chapter 9. Nevertheless, there is a strict correspondence. To see this quickly, let us neglect the mismatch v_{sp}, assuming $|v_{sp}| \ll \Gamma, \Gamma_p$, and the time-dependence in (10.54). This permits us to express the off-diagonal element ρ_{sp} in terms of the diagonal ones,

$$\rho_{ps} = \frac{2g\lambda}{\Gamma + \Gamma_p}(p_p - p_s). \tag{10.55}$$

This we can then substitute into (10.51–10.53). The result is in fact a system of master equations! To get them into a usable form, we note that $|\lambda|^2 = \bar{n}$, and write

$$\dot{p}_g = -\Gamma_r p_g + \Gamma_p p_p, \tag{10.56}$$

$$\dot{p}_p = -\Gamma_p p_p + \Gamma p_s + \gamma \bar{n}(p_s - p_p), \tag{10.57}$$

$$\dot{p}_s = \Gamma_r p_g - \Gamma p_s + \gamma \bar{n}(p_p - p_s), \tag{10.58}$$

$$\text{with} \quad \gamma \equiv \frac{4g^2}{\Gamma + \Gamma_p}. \tag{10.59}$$

Recalling the reasoning of Section 9.7, we understand that γ is the spontaneous decay rate to the mode: the Maxwell–Bloch equations in fact provide a microscopic expression for this rate!

The enhanced rate $\gamma \bar{n}$ works in both directions: it pumps energy into the mode in the course of $s \to p$ transitions, and it extracts this energy in the course of $p \to s$ transitions. The latter is not good for lasing. To prevent this from happening, we should make sure that the p state is efficiently emptied, $p_p \approx 0$. This is guaranteed if the p state decays quickly to the ground state, that is, $\Gamma_p \gg \Gamma \gg \Gamma_r$. With this, we have $p_p \approx 0$ and $p_g \approx 1$, and we rewrite (10.58) as

$$\dot{p}_s = \Gamma_r - \Gamma p_s - \gamma \bar{n} p_s. \tag{10.60}$$

After multiplying with the number of atoms N, this indeed coincides with (9.93), since $N_{ex} = N p_s$ and $r = N \Gamma_r$.

Let us now return to the Maxwell equation (10.49) and use it to compute the time-derivative of the total number of photons, $\dot{\bar{n}} = \dot{\lambda}^* \lambda + \dot{\lambda} \lambda^*$. Substitution of the off-diagonal matrix element yields

$$\frac{d\bar{n}}{dt} = N_{ex} \gamma \bar{n} - \frac{\omega}{Q} \bar{n}, \tag{10.61}$$

which, in the limit $\bar{n} \gg 1$, indeed reproduces (9.94) obtained from a simpler reasoning.

We have thus established the equivalence of the two approaches, and also noted that the Maxwell–Bloch equations are more refined. For instance, with the Maxwell–Bloch equations one can find that the actual frequency of the lasing may differ slightly from both ω_0 and ω_{sp} (see Exercise 1): a problem which cannot even be formulated in the framework of master equations.

10.5 Coherent states of matter

We have seen that coherent states of radiation best approximate the classical states of the fields. Let us now think about matter. Along the lines of the usual dichotomy between radiation and matter, a classical state of particles corresponds to a fixed number of these particles. By no means can it be a coherent state.

However, nature is flexible enough to provide us with coherent states of particles. Such states can be macroscopic, that is, in a sense classical. We have already seen some states with no well-defined number of particles: the ground states of condensed matter systems with broken gauge symmetry. Examples were the Bose condensate of helium atoms in a superfluid and the condensate of Cooper pairs in a superconductor, see Chapters 5 and 6. Bose condensates are commonly referred to as quantum fluids, so one could say that the coherent state is a classical state of a quantum fluid. We have indeed seen that these states are characterized by a phase and are in fact superpositions of basis states with different numbers of particles. Since they are macroscopic, they allow us to bring quantum mechanics to a macroscopic level. Let us now undertake a more extensive discussion of the subject.

We start the discussion with revisiting the phase–number uncertainty relation (10.34). Let us derive it in a more rigorous manner for a boson condensate and obtain the *commutation relation* between the operators of phase and particle number. We know that the ground state of a condensate is characterized by large expectation values for the annihilation and creation operators $\hat{\Psi}_0$ and $\hat{\Psi}_0^\dagger$ for particles in the lowest energy level $\mathbf{k} = 0$. They obey the standard commutation relation

$$\hat{\Psi}_0 \hat{\Psi}_0^\dagger - \hat{\Psi}_0^\dagger \hat{\Psi}_0 = 1. \tag{10.62}$$

There is a macroscopically large average number of particles in the condensate, N_{back}. We are, however, interested in a smaller number of *extra* particles in the condensate and associate an operator $\hat{n}$ with this quantity, $\hat{n} \ll N_{\text{back}}$. Since the total number of particles $N_{\text{back}} + \hat{n} = \hat{\Psi}_0^\dagger \hat{\Psi}_0$, the operators can be presented in the following form,

$$\hat{\Psi}_0^\dagger = \sqrt{N_{\text{back}} + \hat{n}}\, e^{i\hat{\phi}} \quad \text{and} \quad \hat{\Psi}_0 = e^{-i\hat{\phi}}\sqrt{N_{\text{back}} + \hat{n}}, \tag{10.63}$$

which defines the operator of phase $\hat{\phi}$.

Control question. Can you explain the opposite order of factors in the above expressions?

Substituting this into the commutation relation (10.62), we obtain

$$e^{-i\hat{\phi}}(N_{\text{back}} + \hat{n})e^{i\hat{\phi}} - (N_{\text{back}} + \hat{n}) = 1, \quad \text{so} \quad [\hat{n}, e^{i\hat{\phi}}] = e^{i\hat{\phi}}. \tag{10.64}$$

Let us consider a wave function in ϕ-representation, $\psi(\phi)$. We see that the last relation is satisfied if we associate

$$\hat{n} \rightarrow -i\frac{\partial}{\partial\phi}. \tag{10.65}$$

Indeed, for any $\psi(\phi)$

$$-i\frac{\partial}{\partial\phi}\left\{e^{i\phi}\psi(\phi)\right\} = e^{i\phi}\psi(\phi) - ie^{i\phi}\frac{\partial}{\partial\phi}\psi(\phi). \qquad (10.66)$$

If we now commute $\hat{n}$ and $\hat{\psi}$, we obtain the canonical commutation relation

$$[\hat{n}, \hat{\phi}] = -i. \qquad (10.67)$$

We stress here the similarity with the single-particle coordinate–momentum commutation relation

$$[\hat{p}, \hat{x}] = -i\hbar. \qquad (10.68)$$

Apart from factors, operators of momentum and number of particles are mathematically equivalent. The same holds for operators of phase and coordinate. So, the phase-number uncertainty relation naturally follows from the Heisenberg relation for uncertainties of momentum and coordinate.

The main difference between coordinate and phase is that states which differ in phase by 2π are identical. This implies periodic boundary conditions in phase space, $\psi(\phi) = \psi(\phi+2\pi)$. The eigenfunctions of $\hat{n}$ satisfying these boundary conditions have only integer eigenvalues n,

$$\hat{n}\psi_n(\phi) = n\psi_n(\phi), \quad \text{so that} \quad \psi_n(\phi) = \frac{1}{\sqrt{2\pi}}e^{in\phi}. \qquad (10.69)$$

This is more than natural since the number of particles must be an integer. We took n to be the number of *extra* particles in the condensate, so it can be positive as well as negative.

So what is the wave function of the condensate? There are two extreme cases. If the condensate is completely isolated from all other particles of the same sort (a droplet of helium hanging in a vacuum, or a metallic superconducting island in an insulating environment) it surely contains a fixed number (n_0) of particles. Its wave function in n-representation is just $\psi(n) = \delta_{nn_0}$ and has a constant modulus in ϕ-representation, $\psi(\phi) = \exp(in_0\phi)/\sqrt{2\pi}$, that is, the phase is completely indefinite. In contrast to this, a bulk condensate connected to a reservoir of particles has a certain phase $\psi(\phi) = \delta(\phi - \phi_0)$ and a completely indefinite number of particles. The modulus of the wave function in n-representation does not depend on n, $\psi(n) \propto \exp(-i\phi_0 n)$.

These two extreme wave functions have infinite uncertainty in phase and particle number respectively. In all realistic cases, infinite variations of these quantities imply an infinite cost of energy so that a realistic wave function would compromise the definiteness of either quantity for the sake of lower energy. Below we consider an example system where such compromise actually occurs.

10.5.1 Cooper pair box

A Cooper pair box is a superconducting island of sub-micrometer size, which is connected to superconducting bulk material via a tunnel barrier – a *Josephson junction* – as depicted in Fig. 10.7. Although the island consists of billions of individual atoms and electrons,

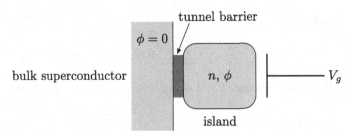

Fig. 10.7 Cooper pair box: a superconducting island is connected to a bulk superconductor via a tunnel barrier. The state of the island can be characterized by n, the number of (extra) Cooper pairs on it, or the phase ϕ relative to the bulk superconductor. The equilibrium number of Cooper pairs n_0 on the island can be tuned with a nearby electrostatic gate.

it is plausible to regard it as a quantum system with a single degree of freedom, either the superconducting phase of the island ϕ or the number of extra Cooper pairs n on the island. To operate the box, an extra gate electrode, not connected to the island, is placed nearby. The voltage applied to this electrode, V_g, shifts the electrostatic potential of the island with respect to the bulk electrode, and changes the energetically most favorable number of Cooper pairs n_0 on the island. Since the voltage is continuous, the parameter n_0 is continuous as well. We assume that the bulk superconductor is in a state with a well-defined phase. This is the reference phase that we conveniently set to zero.

The superconducting phase difference between the island and the bulk produces an electric supercurrent – a Josephson current – through the junction. Therefore, variance of the phase ϕ produces current fluctuations. This costs energy: kinetic energy of electrons involved in the current. Also, extra Cooper pairs on the island add electric charge to the island. This produces electric fields in the tunnel barrier and in the space surrounding the island and costs energy thereby. Thus, variance of n also costs energy: charging or potential energy. We see that neither variance in n nor ϕ can be infinite, and the actual wave function of the island can be neither of the two extreme wave functions considered above.

The actual wave function is determined from the Hamiltonian where both energies are present

$$\hat{H} = E_C(\hat{n} - n_0)^2 - E_J \cos \hat{\phi}. \tag{10.70}$$

The first term presents the charging energy and is proportional to the square of charge, as expected. The second term, the Josephson energy, exhibits a harmonic dependence on the phase. The origin of this dependence is the fact that Cooper pairs tunnel through the barrier one-by-one. To see this, we notice that the operator $e^{i\hat{\phi}}$ raises the number of Cooper pairs in the island by one,

$$e^{i\hat{\phi}}|n\rangle = |n+1\rangle, \tag{10.71}$$

while the inverse operator $e^{-i\hat{\phi}}$ decreases it by one,

$$e^{-i\hat{\phi}}|n\rangle = |n-1\rangle. \tag{10.72}$$

This property is proven from the explicit form of $\psi_n(\phi)$ given by (10.69). The Josephson energy in n-representation therefore reads

$$\frac{E_J}{2}\left(e^{i\hat{\phi}} + e^{-i\hat{\phi}}\right) = \frac{E_J}{2}\sum_n \Big\{|n\rangle\langle n+1| + |n\rangle\langle n-1|\Big\}, \tag{10.73}$$

and describes the transfer of single Cooper pairs between the island and the bulk electrode.

The eigenstates and energies can be determined from the Schrödinger equation and can be written either in phase-representation

$$E\psi(\phi) = \hat{H}\psi(\phi) = \left\{E_C\left(-i\frac{\partial}{\partial\phi} - n_0\right)^2 - E_J\cos\phi\right\}\psi(\phi), \tag{10.74}$$

or in n-representation

$$E\psi(n) = \hat{H}\psi(n) = E_C(n-n_0)^2\psi(n) - \frac{E_J}{2}\Big\{\psi(n-1) + \psi(n+1)\Big\}. \tag{10.75}$$

The equations are easy to analyze in the two complementary cases $E_C \gg E_J$ and $E_J \gg E_C$. In the first case, the charging energy dominates. The eigenstates are almost pure n-states, having a fixed charge, their energies are given by $E(n) = E_C(n-n_0)^2$. Tuning the parameter n_0, one can achieve a situation where the energies of two states with different n are the same – a level crossing. For instance, $E(0) = E(1)$ at $n_0 = \frac{1}{2}$. Close to such a level crossing, the small Josephson energy becomes important. It provides a coherent mixing of the states $|0\rangle$ and $|1\rangle$, so exactly at the crossing point the eigenfunctions $|\pm\rangle$ eventually become superpositions of the two basis states,

$$|\pm\rangle = \frac{1}{\sqrt{2}}\{|0\rangle \pm |1\rangle\}. \tag{10.76}$$

A Cooper pair box in this regime can be used as a macroscopic realization of a quantum two-level system – a qubit.

In the opposite limit, the Josephson energy E_J dominates. As far as low-lying levels ($E < 2E_J$) are concerned, they minimize their energy by concentrating near $\phi = 0$ where the Josephson energy reaches a minimum. In this case, we can expand $\cos\phi \approx 1 - \frac{1}{2}\phi^2$ and forget about the constant term. The Hamiltonian becomes

$$\hat{H} = E_C(\hat{n} - n_0)^2 + \frac{E_J}{2}\hat{\phi}^2. \tag{10.77}$$

Since the operators $\hat{n}$ and $\hat{\phi}$ satisfy the canonical commutation relation, they can be regarded as generalized momentum and coordinate. Therefore, the above Hamiltonian is that of an oscillator with its parabolic potential well centered around $n = n_0$.

Control question. Can you quantize this oscillator, that is, find a linear combination of $\hat{\phi}$ and $\hat{n}$ that represents the annihilation operator?

We know that this oscillator Hamiltonian gives rise to equidistant energy levels separated by $\hbar\omega_0 \equiv \sqrt{2E_C E_J}$. The uncertainty of the phase in the ground state is

$$\Delta\phi \simeq \left(\frac{E_C}{E_J}\right)^{1/4} \ll 1, \tag{10.78}$$

Fig. 10.8 There exists an analogy between a Cooper pair box and the simplest electric oscillator, the LC-circuit.

the phase is well-defined. The uncertainty of the number of Cooper pairs

$$\Delta n \simeq \left(\frac{E_J}{E_C}\right)^{1/4} \gg 1, \tag{10.79}$$

is large, satisfying (10.34). Higher-lying levels ($E > 2E_J$) are less sensitive to the Josephson term. At sufficiently high energy, they resemble the n-states.

    It is quite spectacular that the Cooper pair box in this limit is in fact an *electric* oscillator, such as the one considered in the beginning of the chapter (Fig. 10.8). Since the Cooper pair charge is $2e$, one can associate a capacitance $C = (2e)^2/2E_C$ with the charging energy involved. A Josephson junction at sufficiently small phase differences can be regarded as an inductor with an inductance $L = \pi^2\hbar^2/e^2 E_J$. This makes the Cooper pair such a wonderful device: it is able to exhibit simultaneously coherent states of matter and coherent states of radiation!

Table 10.1 Summary: Coherent states

Excitation of an oscillator with a "classical" pulse, LC-circuit as example

Hamiltonian: $\hat{H} = \frac{1}{2}C\hat{V}^2 + \frac{1}{2}L\hat{I}^2 - C\hat{V}V_{\text{ex}}(t) = \hbar\omega_0\hat{b}^\dagger\hat{b} - \sqrt{\hbar}v(t)(\hat{b}^\dagger + \hat{b})$, $v(t) \equiv V_{\text{ex}}(t)\sqrt{\frac{1}{2}C\omega_0}$

dynamics: from $\hat{H}$ we find $\hat{b}(t) = \hat{b}e^{-i\omega_0 t} + \frac{1}{\sqrt{\hbar}}d(t)$, with $d(t)$ solution of $\dot{d} = -i\omega_0 d + iv(t)$

for any function f: $\langle\psi(t)|f[\hat{b}]|\psi(t)\rangle = \langle 0|f[\hat{b}(t)]|0\rangle = \langle 0|f[\hat{b}e^{-i\omega_0 t} + \frac{1}{\sqrt{\hbar}}d(t)]|0\rangle = f[\frac{1}{\sqrt{\hbar}}d(t)]$

this can only hold if: $\hat{b}|\psi(t)\rangle = \lambda|\psi(t)\rangle$, with $\lambda = d(t)/\sqrt{\hbar}$ $\rightarrow$ *coherent state*

Properties of the coherent state $\hat{b}|\psi\rangle = \lambda|\psi\rangle$

explicitly: $|\psi\rangle = e^{-\frac{1}{2}|\lambda|^2}\sum_n \frac{\lambda^n}{\sqrt{n!}}|n\rangle$,   with $|n\rangle$ number state

statistics: mean $\bar{n} = \langle\psi|\hat{n}|\psi\rangle = |\lambda|^2$ and variance $\sqrt{\langle(n-\bar{n})^2\rangle} = \sqrt{\bar{n}}$

correspondence: coherent states provide best correspondence to classical states,
   they minimize the uncertainty relation $\rightarrow$ $\Delta n\,\Delta\phi \approx \frac{1}{2}$
   overlap between two coherent states $\langle\psi_{\lambda'}|\psi_\lambda\rangle = \exp\{-|\lambda'|^2/2 + \lambda'^*\lambda - |\lambda|^2/2\}$

Example: Laser, we approximate its state by a coherent state

dynamics: $\dfrac{d\lambda}{dt} = -\dfrac{\omega}{2Q}\lambda + \zeta(t)$,   losses and white noise with $\langle\zeta(t)\zeta(t')\rangle = S\delta(t-t')$

fluctuations: due to noise $\langle|\lambda|^2\rangle = SQ/\omega$ $\rightarrow$ in coherent state $\langle|\lambda|^2\rangle = 1/2$, so $S = \omega/2Q$

switch on: now equilibrium is $|\lambda| = \sqrt{\bar{n}}$, phase drifts due to noise (cf. Brownian motion)

coherence time: phase drifts ~ 1 when $\sqrt{\frac{1}{2}St} \sim |\lambda|$ $\rightarrow$ $\tau_{\text{oc}} = 4\bar{n}Q/\omega$

Example: Laser, Maxwell–Bloch equations

goal: find coupled set of time-evolution equations for configuration of lasing mode
        and density matrix of radiating atoms including states $|s\rangle$, $|p\rangle$, and $|g\rangle$

idea: (i) oscillating dipole moment of atoms $\rightarrow$ classical variable $er_{sp}\rho_{sp}$
        (ii) field of lasing mode $\rightarrow$ coherent state with parameter λ

Maxwell equation: $\dot{\lambda} + \frac{1}{2}(\omega_0/Q)\lambda = -gN\rho_{sp}$ with $g = er_{sp}(\omega_0/2\hbar\varepsilon_0\mathcal{V})^{1/2}$

Bloch equations: $\dot{p}_g = -\Gamma_r p_g + \Gamma_p p_p$, $\dot{p}_p = -\Gamma_p p_p + \Gamma_r p_s - g(\lambda^*\rho_{ps} + \lambda\rho_{ps}^*)$,
         $\dot{p}_s = -(\dot{p}_g + \dot{p}_p)$, $\dot{\rho}_{ps} = (iv_{sp} - \frac{1}{2}[\Gamma + \Gamma_p])\rho_{ps} + \lambda g(p_p - p_s)$
         with Γs being decay rates and $v_{sp} = \omega_{sp} - \omega_0$

result: for $|v_{sp}| \ll \Gamma, \Gamma_p$ and $\Gamma_p \gg \Gamma \gg \Gamma_r$ we recover the result from Chapter 9,
        but with more details, e.g. $\gamma = 4g^2/(\Gamma + \Gamma_p)$

Example: Coherent states of matter, Cooper pair box

phase and number operators are canonically conjugated, $[\hat{n}, \hat{\phi}] = -i$

Cooper pair box: superconducting island tunnel coupled to bulk superconductor

Hamiltonian: $\hat{H} = E_{\text{C}}(\hat{n} - n_0)^2 - E_{\text{J}}\cos\hat{\phi}$,   island with $\hat{n}$ Cooper pairs and phase $\hat{\phi}$

number representation: $\hat{H}|\psi\rangle = \left\{E_{\text{C}}(\hat{n} - n_0)^2 - \frac{1}{2}E_{\text{J}}\sum_n (|n\rangle\langle n+1| + |n\rangle\langle n-1|)\right\}|\psi\rangle$

phase representation: $\hat{H}\psi(\phi) = \left\{E_{\text{C}}\left(-i\frac{\partial}{\partial\phi} - n_0\right)^2 - E_{\text{J}}\cos\phi\right\}\psi(\phi)$

limit $E_{\text{C}} \gg E_{\text{J}}$:   eigenstates roughly $|n\rangle$ with $E_n = E_{\text{C}}(n - n_0)^2$, mixed by tunnel coupling

limit $E_{\text{J}} \gg E_{\text{C}}$:   for low energies $\hat{H} \approx E_{\text{C}}(\hat{n} - n_0)^2 + \frac{1}{2}E_{\text{J}}\hat{\phi}^2$ $\rightarrow$ harmonic oscillator
        the ground state is a coherent state

Exercises

1. *Mode pulling* (solution included). As mentioned, the actual frequency of lasing may deviate from the frequency of the lasing mode. Let us evaluate this frequency from the Maxwell–Bloch equations.

   a. We seek for solutions of the system of Maxwell–Bloch equations in the form $\lambda, \rho_{ps} \propto \exp\{-i\nu t\}$ and p_s, p_p, and p_g being constant in time. Here, ν is in fact the mismatch between the lasing frequency and the mode frequency. Derive an equation that determines ν and $(p_s - p_p)$ and does not explicitly contain $\bar{n}$.

   b. Assume $\nu, \nu_{sp} \ll (\Gamma + \Gamma_p)$ and find a simple expression for ν *not depending* on $(p_s - p_p)$.

2. *Cooper pair box.* In Section 10.5.1 we introduced the Cooper pair box: a small superconducting island, connected to a superconducting reservoir via a tunnel barrier. We showed how to write the Hamiltonian for the box both in phase-representation and in charge-representation. Here, we use the charge-representation,

$$\hat{H} = E_{\mathrm{C}}(\hat{n} - n_0)^2 + \frac{E_{\mathrm{J}}}{2} \sum_n \left\{ |n\rangle\langle n+1| + |n\rangle\langle n-1| \right\},$$

   where E_{C} is the charging energy, E_{J} the Josephson energy, and n_0 the "ideal" number of Cooper pairs on the island. This number n_0 is continuous and can be tuned by an electrostatic gate close to the island. The operator $\hat{n}$, with eigenstates $|n\rangle$, is the number operator for the number of Cooper pairs on the island.

   a. We assume that $E_{\mathrm{C}} \gg E_{\mathrm{J}}$, and treat the second term in the Hamiltonian as a small perturbation. What are the eigenstates of the unperturbed Hamiltonian $\hat{H}_0 = E_{\mathrm{C}}(\hat{n} - n_0)^2$? Sketch the spectrum of this Hamiltonian as a function of n_0.

   b. A small but finite Josephson energy E_{J} provides a coupling between the different charge states $|n\rangle$. We focus on the lowest lying crossing at $n_0 = 0$, where the two charge states involved are, $|-1\rangle$, and $|1\rangle$. These two charge states are, however, not directly coupled to each other, but only indirectly via $|0\rangle$. We thus have to consider all three states $|-1\rangle$, $|0\rangle$, and $|1\rangle$. Focusing on this 3×3 subspace, calculate the first-order corrections to the wave vectors $|-1\rangle$ and $|1\rangle$.

   c. Write down an effective Hamiltonian $\hat{H}_{\mathrm{eff}}$ for the 2×2 space of $|-1\rangle$ and $|1\rangle$, which takes into account the indirect coupling via $|0\rangle$.

   d. Due to the coupling via $|0\rangle$, the crossing of the two states becomes an anticrossing. What are the two eigenstates of $\hat{H}_{\mathrm{eff}}$ exactly at the crossing $n_0 = 0$? What are their eigenenergies?

3. *Coherent state in a harmonic oscillator.* In Section 10.3 we have given the wave function for a coherent particle state with expectation values $\langle \hat{p} \rangle = p_0$ and $\langle \hat{x} \rangle = x_0$,

$$\psi_0(x) = \frac{1}{\sqrt{\sigma\sqrt{2\pi}}} \exp\left\{ -\frac{(x - x_0)^2}{4\sigma^2} + i\frac{p_0 x}{\hbar} \right\}.$$

a. We assume that the particle has mass $m = 1$ and we place it in a parabolic potential $\frac{1}{2}\omega^2 x^2$, thus constructing a harmonic oscillator. Determine the σ for which the state given above is an eigenstate of the bosonic annihilation operator

$$\hat{a} = \frac{\omega\hat{x} + i\hat{p}}{\sqrt{2\hbar\omega}},$$

which annihilates excitations in the harmonic oscillator. What is the eigenvalue λ of $\hat{a}$?

b. We now assume that the coherent state given above is the state of the particle at time $t = 0$, so $\psi(x, 0) = \psi_0(x)$. From the Heisenberg equation of motion it follows that

$$e^{\frac{i}{\hbar}\hat{H}t}\hat{a}e^{-\frac{i}{\hbar}\hat{H}t} = \hat{a}(t) = \hat{a}e^{-i\omega t},$$

where $\hat{H} = \frac{1}{2}(\hat{p}^2 + \omega^2\hat{x}^2)$ is the Hamiltonian of the oscillator. Use this relation to show that the particle state at later times $\psi(x, t)$ will always be a coherent state.

c. At (b) you found that $\psi(x, t)$ is also a coherent state, but with a time-dependent coefficient $\lambda(t)$. For coherent particle states as considered in this exercise, we know that

$$\lambda = \frac{\omega\langle\hat{x}\rangle + i\langle\hat{p}\rangle}{\sqrt{2\hbar\omega}}.$$

Use this relation to find the time-dependent expectation values $\langle\hat{x}(t)\rangle$ and $\langle\hat{p}(t)\rangle$.

4. *Minimum uncertainty in a coherent state.* In this exercise we show how one can arrive at coherent states when looking for general states with minimum uncertainty. First, we refresh the derivation of the generalized uncertainty relation.

a. We want to know the variance of certain variables. For instance, for the operator $\hat{A}$ we are interested in

$$(\Delta A)^2 = \langle\psi|(\hat{A} - \langle\hat{A}\rangle)^2|\psi\rangle = \langle A|A\rangle,$$

using the notation $|A\rangle \equiv (\hat{A} - \langle\hat{A}\rangle)|\psi\rangle$. Use, *inter alia*, the Schwartz relation $|\langle a|b\rangle|^2 \leq \langle a|a\rangle\langle b|b\rangle$ to show that

$$(\Delta A)^2(\Delta B)^2 \geq \left(\frac{1}{2i}\langle[\hat{A}, \hat{B}]\rangle\right)^2.$$

b. In the derivation of the above uncertainty relation, we used two inequality relations. Minimum uncertainty would correspond with two *equalities* instead of inequalities. What relation should exist between $|A\rangle$ and $|B\rangle$ in order to have minimum uncertainty?

c. Let us now focus on the best-known canonically conjugated variables, position $\hat{x}$ and momentum $\hat{p}$. Give the condition found at (b) for minimum uncertainty in the form of an eigenvalue equation for a combination of $\hat{x}$ and $\hat{p}$. What does this say about the coherent state of a harmonic oscillator defined by

$$\hat{a}|\psi\rangle = \lambda|\psi\rangle?$$

d. Verify that the Gaussian wave packet given by

$$\psi(x) = C \exp\left\{\frac{i}{\hbar}p_0 x\right\} \exp\left\{\frac{(x - x_0)^2}{2\hbar a}\right\},$$

as given in (10.32) (with $a < 0$ of course), is indeed a general solution of the eigenvalue equation found at (c).

5. *Displacement and squeezing operators.* In quantum optics, where coherent states play an important role, two commonly encountered operators are the so-called *displacement* and *squeezing* operators, defined as

$$\hat{D}(\alpha) = e^{\alpha \hat{a}^\dagger - \alpha^* \hat{a}},$$

$$\hat{S}(z) = e^{\frac{1}{2}(z^* \hat{a}^2 - z \hat{a}^{\dagger 2})}, \quad \text{with} \quad z = re^{i\theta}.$$

The displacement operator acting on a coherent state with parameter λ brings it to the coherent state with $\lambda + \alpha$, i.e. it displaces the coherent state in its phase space over the distance α.

a. Let the operator $\hat{D}(\alpha)$ act on the vacuum (a coherent state with $\lambda = 0$), and show explicitly that this creates a coherent state with $\lambda = \alpha$.

b. Show that

$$\hat{D}^{-1}(\alpha)\hat{a}\hat{D}(\alpha) = \hat{a} + \alpha.$$

Suppose $|\psi\rangle$ is a coherent state with $\hat{a}|\psi\rangle = \lambda|\psi\rangle$. Use the above relation to prove that

$$\hat{a}\hat{D}(\alpha)|\psi\rangle = (\lambda + \alpha)|\psi\rangle.$$

c. Let us now focus on the harmonic oscillator again, the Hamiltonian being $\hat{H} = \frac{1}{2}(\hat{p}^2 + \omega^2 \hat{x}^2)$. Show that translation operator $\hat{T}_\mathbf{a}$ found in Section 1.7, which shifts the wave function $\hat{T}_\mathbf{a}\psi(\mathbf{r}) = \psi(\mathbf{r} + \mathbf{a})$, is indeed a displacement operator in the sense of this exercise. The displacement operator $\hat{D}(\alpha)$ with α real thus shifts the coherent state in space over a distance α.

d. What is the effect of $\hat{D}(\alpha)$ on a coherent state in the oscillator characterized by x_0 and p_0 (see (10.32)) if α is purely imaginary?

e. Suppose we have a coherent state with $\hat{a}|\psi\rangle = \lambda|\psi\rangle$. We know that the dimensionless conjugated variables $\hat{P} = \frac{i}{\sqrt{2}}(\hat{a}^\dagger - \hat{a})$ and $\hat{Q} = \frac{1}{\sqrt{2}}(\hat{a}^\dagger + \hat{a})$ have minimum uncertainty in this state, $(\Delta P)^2 = (\Delta Q)^2 = \frac{1}{2}$. Now let the squeezing operator act on this state, $|\psi_r\rangle = \hat{S}(r)|\psi\rangle$, where r is assumed real. Calculate the uncertainties in $\hat{P}$ and $\hat{Q}$ for the new state $|\psi_r\rangle$.

6. *A complicated representation of the unity matrix.* Show that the unity matrix can be represented as an integral over coherent states as

$$\hat{1} = \int \frac{d\lambda \, d\lambda^*}{\pi} |\psi_\lambda\rangle\langle\psi_\lambda|.$$

7. *A path to path integrals.* Path integral methods form a powerful tool set in quantum mechanics as well as in other stochastic theories. While we do not outline these methods in this book, we in fact come very close to them in our discussion of coherent states. Let

us show a path which would eventually lead to path integrals. To this end, we demonstrate that the evolution operator $\hat{U}(t_2, t_1)$ corresponding to the oscillator Hamiltonian $\hat{H} = \hbar\omega\hat{b}^\dagger\hat{b}$ can be represented as an integral over the *paths*: smooth functions $\lambda(t)$ defined in the interval $t_2 > t > t_1$,

$$\hat{U}(t_2, t_1) = \int \mathcal{D}\{\lambda(t)\}e^{-i\mathcal{S}(\{\lambda(t)/\hbar\})},$$

where

$$\mathcal{D}\{\lambda(t)\} = \prod_t \left(\frac{d\lambda(t)\,d\lambda^*(t)}{\pi} \right),$$

$$\mathcal{S} = \int_{t_1}^{t_2} dt \left\{ \frac{i\hbar}{2} \left(\dot{\lambda}^*(t)\lambda(t) - \lambda^*(t)\dot{\lambda}(t) \right) + \hbar\omega\lambda^*(t)\lambda(t) \right\}.$$

Taken as it is, the formula is good for nothing: we know very well the simple expression for $\hat{U}$ in the basis of number states. Yet the formula can be easily generalized to include many modes and more complex Hamiltonians and becomes a powerful tool thereby. Let us prove the formula.

a. Subdivide the interval into time intervals of length dt, the time moments corresponding to the ends of the intervals being t_i and $t_{i+i} = t_i + dt$. The evolution operator is a product of evolution operators for short intervals. Insert unity matrices between the shorter evolution operators and implement the representation of the unity matrices in terms of coherent states, such as found in Exercise 6. Give the result in terms of the product of the contributions coming from each interval.

b. Let us concentrate on a single interval, assuming $dt \to 0$. Assume $\lambda(t)$ is a smooth function and expand the contribution of an interval up to terms linear in dt. Represent the answer in terms of an exponent making use of $\exp\{A\,dt\} \approx 1 + A\,dt$, which is surely valid in the limit $dt \to 0$.

c. Collect the terms coming from all small intervals to prove the formula.

Solutions

1. *Mode pulling.*
   a. We substitute $\lambda(t) = \lambda \exp\{-i\nu t\}$ and $\rho_{ps}(t) = \rho_{ps} \exp\{-i\nu t\}$ into (10.49) and (10.54) to arrive at

$$0 = \lambda \left(-i\nu + \frac{\omega_0}{2Q} \right) + gN\rho_{ps},$$

$$0 = \left(i\nu + i\nu_{sp} - \frac{\Gamma + \Gamma_p}{2} \right) \rho_{ps} + \lambda g(p_p - p_s).$$

We solve ρ_{ps} from the second equation, substitute it into the first one, and divide the result by λ (thereby assuming that $\lambda \neq 0$). The resulting complex equation

$$-i\nu + \frac{\omega_0}{2Q} = \frac{g^2 N}{i(\nu + \nu_{sp}) - \frac{\Gamma + \Gamma_p}{2}} \left(p_p - p_s\right),$$

can be regarded as two real equations for two real parameters, $(p_s - p_p)$ and ν.

b. We take the real part of this equation disregarding ν and ν_{sp}. This yields

$$(p_s - p_p) = \frac{\omega_0(\Gamma + \Gamma_p)}{4g^2 NQ}.$$

Then we take the imaginary part linear in ν and ν_{sp} under the approximations made. This brings us to

$$\nu = -\nu_{sp}\frac{\beta}{1 + \beta} \quad \text{with} \quad \beta \equiv \frac{\omega_0}{Q(\Gamma + \Gamma_p)}.$$

The lasing frequency is close to ω_0 if $\beta \ll 1$ and close to ω_{sp} in the opposite limit.

PART IV

DISSIPATIVE QUANTUM MECHANICS

11

Dissipative quantum mechanics

Everyday experience tells us that any process, motion, or oscillation will stop after some time if we do not keep it going, by providing energy from some external source. The reason for this is the fact that energy is dissipated: no system is perfectly isolated from its *environment*, and any energy accumulated in the system is eventually dissipated into the environment. Table 11.1 summarizes the content of this chapter.

11.1 Classical damped oscillator

Let us first see how energy dissipation is treated in classical mechanics, by considering a generic model of a classical *damped oscillator* (Fig. 11.1). We take a single degree of freedom, x, and associate it with the coordinate of a particle. This particle has a mass M, is confined in a parabolic potential $U(x) = \frac{1}{2}ax^2$, and may be subjected to an external force F_{ext}. Dissipation of energy in the oscillator is due to *friction*, which is described by an extra force F_f, a *friction force* acting on the particle. The equations of motion for the particle then read

$$\dot{p} = -ax + F_f + F_{ext} \quad \text{and} \quad \dot{x} = \frac{p}{M}, \tag{11.1}$$

with p being the momentum of the particle.

The simplest assumption one can make about the friction force F_f is that it is proportional to the velocity of the particle,

$$F_f = -\gamma \dot{x}, \tag{11.2}$$

where γ is the damping coefficient. This simplest expression implies an *instant* relation between velocity and force. However, generally one expects some retardation: the force at the moment t depends not only on the velocity at the same time $\dot{x}(t)$, but also on the velocity of the particle in the past. In the time domain, the more general relation thus reads

$$F_f(t) = -\int_0^\infty d\tau\, \gamma(\tau)\, \dot{x}(t - \tau). \tag{11.3}$$

This relation is in fact simpler to express in terms of Fourier components of the force and coordinate. Using the definition of the Fourier transform of a function $f(t)$ of time,

$$f(\omega) = \int_{-\infty}^\infty dt\, f(t) e^{i\omega t}, \tag{11.4}$$

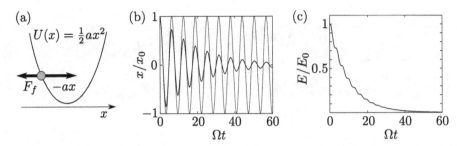

Fig. 11.1 (a) Forces acting on a particle in a damped oscillator. (b) Damped oscillations (thick line, using $\Omega = \sqrt{a/M}$ and $\tau_f = 10/\Omega$) and undamped oscillations (thin line). (c) Energy versus time for the damped oscillator.

we can write (11.3) as $F_f(\omega) = i\omega\gamma(\omega)x(\omega)$. When including retardation effects, the damping coefficient thus becomes frequency dependent.

Let us now solve the equations of motion (11.1) in the absence of the external force F_{ext}. As an ansatz, we assume that $x(t)$ has the form of a complex exponent, $x(t) = x_0 \exp(-i\Omega t)$. If we insert this into (11.1) and take for simplicity an instant damping coefficient, we obtain a quadratic equation for Ω,

$$0 = M\Omega^2 - a + i\Omega\gamma. \tag{11.5}$$

This equation has two roots, $\Omega_\pm = -i/2\tau_r \pm \Omega_0$, where we defined $\tau_r \equiv M/\gamma$ and $\Omega_0 \equiv \sqrt{a/M - (\gamma/2M)^2}$. If the friction is sufficiently small, that is, when $\gamma < 2\sqrt{aM}$, then Ω_0 is real and the two roots have the same imaginary part and opposite real parts. In that case, the solution with the initial conditions $x = x_0$ and $p = 0$ at $t = 0$ reads

$$x(t) = x_0 \left(\cos(\Omega_0 t) + \frac{\sin(\Omega_0 t)}{2\Omega_0 \tau_r} \right) \exp\left(-\frac{t}{2\tau_r}\right), \tag{11.6}$$

and describes damped oscillations with frequency Ω_0 and relaxation time τ_r, as illustrated in Fig. 11.1(b).

Control question. What does the solution look like when $\gamma > 2\sqrt{aM}$?

Importantly, this classical description does not say anything about the nature of the friction force. It can be used to describe damped oscillations of a particle in different environments, such as gas or water. It does not matter whether the friction arises from the particle colliding with gas molecules, or interacting with a moving liquid, all these details are incorporated into the friction coefficient $\gamma(\omega)$.

11.1.1 Dynamical susceptibility

Dynamical susceptibility is a fancy name for a simple thing: it is the linear response of a system to a weak periodic external driving. For our oscillator, the response is the coordinate x of the particle and the driving is the external force. In Fourier components, the dynamical susceptibility $\chi(\omega)$ is thus defined by

$$x(\omega) = \chi(\omega)F_{\text{ext}}(\omega). \tag{11.7}$$

Using the Fourier method, we can easily solve the equations of motion (11.1), now including the driving force, and find

$$\chi(\omega) = \frac{1}{a - M\omega^2 - i\omega\gamma(\omega)}. \tag{11.8}$$

The dynamical susceptibility $\chi(\omega)$ thus found yields the response of the damped oscillator (in terms of its coordinate x) to periodic driving with frequency ω. Generally this susceptibility is a complex quantity. Its modulo $|\chi(\omega)|$ gives the proportionality between the amplitude of the resulting oscillations in x and the amplitude of the driving force. If the susceptibility is complex, its argument $\arg[\chi(\omega)]$ introduces a phase shift between the driving and the response.

Control question. The imaginary part of $\chi(\omega)$ given above is an odd function of frequency while the real part is even. Can you argue that these properties are universal and should hold for *any* linear response?

So what is the use of this dynamical susceptibility? As we see in the next sections, this function $\chi(\omega)$ contains all relevant information to describe the dissipative dynamics of the system. The advantage of incorporating this information into $\chi(\omega)$ becomes enormous when the damped system is more complicated than the simple damped oscillator considered above, which can easily be the case. Think for instance of many coupled oscillators that all have their own damping coefficients, which is a typical setup for the case of propagation of light in dissipative media. Provided the equations that describe such a complicated system remain linear, one can still solve them to determine the dynamical susceptibility. All complications can be incorporated into the form of the function $\chi(\omega)$. Also, having a larger number of variables (for instance, more coordinates) is no problem. For a set $\{x_j\}$ of linear variables one can define a matrix dynamical susceptibility from the set of relations

$$x_i(\omega) = \sum_j \chi_{ij}(\omega)F_j(\omega), \tag{11.9}$$

F_j being the corresponding external forces.

The dynamical susceptibility of any system obeys an important relation that comes solely from the principle of *causality*: the response of the system can only depend on the external force in the past, not in the future. In terms of Fourier transforms, this implies that $\chi(\omega)$ has to be an analytical function of complex ω in the *upper* half of the complex plane (where Im $\omega > 0$). The properties of analytical functions then imply the celebrated Kramers–Kronig relation between the real and imaginary part of the dynamical susceptibility,

$$\mathrm{Re}\,\chi(\omega) = \int \frac{dz}{\pi} \frac{\mathrm{Im}\,\chi(z)}{z - \omega}. \tag{11.10}$$

This relation is widely applied in optics, where it relates the refractive index of a medium (the real part of its dielectric susceptibility, that is, the response of the dipole density on an external electric field) to the damping in the medium (the imaginary part).

Control question. Expression (11.8) diverges for some complex ω. Can you find the positions of these pole singularities? Explain how to reconcile the presence of these singularities with the analyticity mentioned?

The concept of dynamical susceptibility is a pillar of our formulation of dissipative quantum mechanics.

11.1.2 Damped electric oscillator

To stress the universality of the damped oscillator equations, let us consider oscillations in an electric circuit which consists of a capacitor of capacitance C, an inductance L, a resistor with resistance R, and a voltage source $V_{\text{ext}}(t)$, all connected in series as shown in Fig. 11.2(a). Note that this is the same circuit that we have already discussed in Chapters 7, 8, and 10. In Chapters 7 and 8, we only had two elements (L and C), which together behaved as a "perfect" oscillator. In Chapter 10 we added a voltage source to excite the system, and we showed how a short-pulse driving can create a coherent state. Now we go one step further, and also add a resistance R to the circuit which introduces dissipation into the system.

Let us take as the variable characterizing the state of the circuit the charge $q(t)$ in the capacitor. The equations of the circuit dynamics then read

$$-L\dot{I} = \frac{q}{C} + RI - V_{\text{ext}} \quad \text{and} \quad \dot{q} = I, \tag{11.11}$$

the same equations as (10.3) but with an extra term RI. As in (10.3), the first equation is obtained by summing up the voltage drops over all elements of the circuit, and by virtue of the Kirchhoff rule this sum must be zero. The second equation expresses charge conservation: the change of charge q in the capacitor should equal the current I in the circuit. By virtue of the same conservation, this current is the same in all elements of the circuit. The term RI, accounting for the presence of the resistor, introduces dissipation in the circuit. When $R \to 0$, the oscillator is undamped and we know from previous chapters that it oscillates with frequency $\Omega = 1/\sqrt{LC}$. We note that the substitution

$$q \to x, \quad \text{and} \quad p \to IL,$$
$$\text{with} \quad L = M, \quad \gamma = R, \quad 1/C = a, \quad \text{and} \quad V_{\text{ext}} = F_{\text{ext}}, \tag{11.12}$$

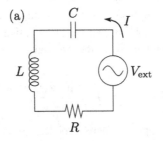
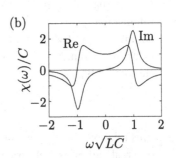

Fig. 11.2 (a) Electric damped oscillator. (b) Real and imaginary part of $\chi(\omega)$. The dissipation is relatively small, $R\sqrt{C/L} = 0.4$.

brings the equations exactly to the form (11.1), and they are thus equivalent. The resistor can in principle be replaced by a more complicated combination of circuit elements, which effectively leads to a frequency dependence of the resistance $R = R(\omega)$. Making the resistance frequency dependent is thus analogous to allowing for retardation effects in the damping of the mechanical oscillator we discussed before (having a more general damping function $\gamma(t)$ instead of the instantaneous coefficient γ).

The electrical circuit is usually characterized by the admittance $G(\omega)$ (the frequency-dependent conductance), defined by $I(\omega) = G(\omega)V_{\text{ext}}(\omega)$. The dynamical susceptibility for the description we chose above is defined as the response of the charge, rather than the current, to the external voltage. Since $\dot{q} = I$, the susceptibility differs from the admittance only by a factor,

$$q(\omega) = \frac{I(\omega)}{-i\omega} = \frac{G(\omega)}{-i\omega}V_{\text{ext}}(\omega), \quad \text{so that} \quad \chi(\omega) = \frac{G(\omega)}{-i\omega}. \tag{11.13}$$

The real and imaginary parts of $\chi(\omega)$ are plotted in Fig. 11.2(b) for the case of small dissipation.

11.2 Quantum description

For any classical *undamped* oscillator one can readily provide a consistent quantum description, as we have shown in Chapter 8. The canonically conjugated variables p and x become operators obeying the standard commutation relation $[\hat{p}, \hat{x}] = -i\hbar$. The oscillator energy also becomes an operator, the quantum Hamiltonian

$$\hat{H} = \frac{\hat{p}^2}{2M} + \frac{a\hat{x}^2}{2}, \tag{11.14}$$

that determines the evolution of the wave function, which is a function of a single variable, for instance x. Unfortunately, if one tries to generalize this quantization procedure for *damped* oscillators, one soon runs into trouble, as we explain below.

11.2.1 Difficulties with the quantum description

There have been several attempts to provide quantization rules or an analogue of the Schrödinger equation for a damped oscillator in terms of a single variable. However, none turned out to be successful. The main difficulty is rooted in the basic principles of quantum mechanics: the state of a system is represented by a wave function, a linear superposition of several states is also a state, and the evolution of the state is given by a unitary matrix, so that the norm of the wave function stays the same. Thinking about quantum dynamics, these principles inevitably lead to many stationary states of conserved energy, and to time-reversible evolution of a wave function. Neither of these concepts is compatible with dissipation. Dissipation is obviously time-irreversible, and the energy of the system is not conserved and the only real stationary state is the lowest energy state. Due to these difficulties, dissipative quantum mechanics has been formulated in its modern form only relatively recently, about 60 years after the advent of "non-dissipative" quantum mechanics.

11.2.2 Solution: Many degrees of freedom

It is interesting that the way to circumvent the difficulties with the quantum description is suggested by an analysis of the situation in classical mechanics. In fact, Newton's equations of classical mechanics also conserve the energy of a closed system and are time-reversible, precisely similar to the Schrödinger equation in quantum mechanics. To understand how Newton's equations are consistent with reality, including dissipation and friction, we consider an example illustrating the microscopic origin of friction. Let us think about the motion of a big massive particle in a gas. It experiences a friction force, and this friction force originates from collisions of the particle with gas molecules. These collisions conserve the momentum and energy of the *total* system (particle and gas), but transfer energy and momentum from the particle to the gas molecules. The dynamics of the whole system are Newtonian and time-reversible. The concept of dissipation and the associated breaking of time-reversibility exist in fact only in the minds of us, the observers. We are not willing nor able to follow the zillions of system variables, the coordinates and velocities of all the molecules in the gas. We wish to follow the massive particle only, and to be able to predict its dynamics from its velocity and coordinates. Dissipation and the problems with time-reversibility only arise due to us neglecting the degrees of freedom of all gas molecules.

This suggests that the correct *quantum* description of a dissipative system with a single degree of freedom must formally include (infinitely) many degrees of freedom to dissipate energy to. These extra degrees of freedom are commonly referred to as a *bath* or the environment. The analogy with classical mechanics also suggests that the details of the bath should not be important for the actual dynamics. For instance, in the example of the particle in the gas, all details about the gas are just incorporated into a single number, the damping coefficient γ which characterizes the friction force as in (11.2). This thus sets a task with contradictory goals. On the one hand we need infinitely many degrees of freedom to include dissipation, but on the other hand we want to get rid of these degrees of freedom as soon as possible.

11.2.3 Boson bath

A conventional, and the simplest, model of a bath capable of dissipating energy is the *boson bath*, consisting of an infinite set of non-interacting bosons. Why is this so? Colloquially, because everything is bosons. Anytime we talk about a system of which the degrees of freedom do not deviate much from their equilibrium positions, its classical dynamics can be described with linear equations. As we know, any set of linear equations can be approximated with an (infinite) set of oscillator equations. Such a set of oscillator equations can be diagonalized in terms of eigenmodes, yielding a set of uncoupled oscillator equations. Finally, the oscillator corresponding to each eigenmode can be quantized as described in Chapter 8, resulting in a single bosonic level. Quantizing the whole set of oscillators thus yields an infinite set of bosonic levels. Therefore, any environment is to linear approximation equivalent to a boson bath.

The Hamiltonian of an oscillator coupled to a boson bath reads

$$\hat{H} = \hat{H}_{\text{osc}} + \hat{H}_{\text{env}} + \hat{H}_{\text{coup}} - \hat{x}F_{\text{ext}}, \quad \text{where} \tag{11.15}$$

$$\hat{H}_{\text{osc}} = \frac{\hat{p}^2}{2M} + \frac{a\hat{x}^2}{2},$$

$$\hat{H}_{\text{env}} = \sum_k \hbar\omega_k \hat{b}_k^\dagger \hat{b}_k,$$

$$\hat{H}_{\text{coup}} = -\hat{x}\hat{F}, \quad \text{with} \quad \hat{F} \equiv \sum_k \hbar c_k(\hat{b}_k^\dagger + \hat{b}_k).$$

The first term is just the Hamiltonian of the undamped oscillator (11.14) and the second term describes the eigenmodes of the bath with eigenfrequencies ω_k. The third term is crucial since it couples the coordinate $\hat{x}$ of the oscillator to the bath bosons. The second and third terms together result in quite some parameters in the model: infinite sets of coupling coefficients c_k and frequencies ω_k. To understand how to deal with these, we have to consider the equations of motion for the system.

11.2.4 Quantum equations of motion

We would like to derive the equations of motion for the operators $\hat{x}$ and $\hat{p}$, and for the boson operators. We thus adopt the Heisenberg picture of quantum mechanics, where the evolution of an arbitrary operator $\hat{A}$ is determined by the Hamiltonian by means of the Heisenberg equation

$$\frac{d\hat{A}}{dt} = \frac{i}{\hbar}[\hat{H}, \hat{A}]. \tag{11.16}$$

We know the commutation relations for all operators appearing in (11.15),

$$[\hat{b}_k, \hat{b}_{k'}^\dagger] = \delta_{kk'} \quad \text{and} \quad [\hat{x}, \hat{p}] = i\hbar. \tag{11.17}$$

Computing commutators with the Hamiltonian $\hat{H}$ for $\hat{x}$ and $\hat{p}$, we obtain

$$\dot{\hat{p}} = -a\hat{x} + \hat{F} + F_{\text{ext}} \quad \text{and} \quad \dot{\hat{x}} = \frac{\hat{p}}{M}, \tag{11.18}$$

this being very similar to the classical equations (11.1) if we associate the operator $\hat{F}$ with the friction force.

Logically, the next step is to express this force $\hat{F}$ in terms of $\dot{\hat{x}}$. To do so, we write the equations for the time-evolution of the boson operators,

$$\dot{\hat{b}}_k = -i\omega_k \hat{b}_k + ic_k \hat{x}, \tag{11.19}$$

and concentrate on the terms proportional to $\hat{x}$. We express those in terms of their Fourier components,

$$\hat{b}_k(\omega) = \frac{c_k}{\omega_k - \omega - i0}\hat{x}(\omega). \tag{11.20}$$

Here we have implemented an important trick: we added an infinitesimally small *positive* imaginary part $i0$ to the frequency. This ensures that the expression has good causal properties, so that b_k at a given moment depends only on $\hat{x}(t)$ in the past, not on future values of $\hat{x}(t)$. This "negligibly small" imaginary part eventually breaks the time-reversal symmetry of the quantum equations. To see how it works, let us substitute the results into the equation for $\hat{F}$,

$$\hat{F}(\omega) = \sum_k c_k \left\{ \hat{b}(\omega) + \hat{b}^\dagger(\omega) \right\} = \sum_k c_k^2 \left(\frac{1}{\omega_k - \omega - i0} + \frac{1}{\omega_k + \omega + i0} \right) \hat{x}(\omega). \tag{11.21}$$

This equation thus presents a quantum version of (11.2): it shows how to express the analogue of the damping coefficient $\gamma(\omega)$ in terms of the eigenfrequencies ω_k of the bath modes and the corresponding coupling coefficients c_k.

Let us investigate the proportionality between $\hat{x}$ and $\hat{F}$ in more detail. We first concentrate on its *imaginary* part. Using the Cauchy formula from the theory of complex analysis,

$$\mathrm{Im}\left\{ \frac{1}{\omega - \omega_0 - i0} \right\} = \pi\delta(\omega - \omega_0), \tag{11.22}$$

this imaginary part reads

$$\mathrm{Im}\left\{ \sum_k c_k^2 \left(\frac{1}{\omega_k - \omega - i0} + \frac{1}{\omega_k + \omega + i0} \right) \right\} = \pi \sum_k c_k^2 \left\{ \delta(\omega - \omega_k) - \delta(-\omega - \omega_k) \right\}$$
$$= J(\omega) - J(-\omega), \tag{11.23}$$

where we have introduced the function

$$J(\omega) \equiv \sum_k \pi c_k^2 \delta(\omega - \omega_k). \tag{11.24}$$

For any set of discrete eigenfrequencies, this function consists of a series of δ-peaks. This is not what we want, since we would like the relation between the force and coordinate to be a smooth function of frequency. To achieve this, we assume that the eigenfrequencies are so close together that they form a continuous spectrum. This is a crucial assumption, it enables the bath to do what we expected it to do, that is, to provide dissipation and break time-reversibility. Finally, we require that

$$J(\omega) = \omega \, \mathrm{Re}\{\gamma(\omega)\}\Theta(\omega), \tag{11.25}$$

where the last factor is due to the fact that $J(\omega) \neq 0$ only for positive frequencies, as can be seen from (11.24). This particular choice of $J(\omega)$ in fact causes the above results to be consistent with the classical relation $F(\omega) = i\omega\gamma(\omega)x(\omega)$.

Control question. Do you see how the requirement (11.25) fits into the classical relation $F(\omega) = i\omega\gamma(\omega)x(\omega)$?

Hint. Remember that for any response function $\gamma(-\omega) = \gamma^*(\omega)$.

The above discussion also shows that the details of the bath – the coupling constants c_k and the frequencies ω_k – enter the dissipative dynamics only via the form of the function $J(\omega)$. One can choose the coefficients arbitrarily provided that the choice reproduces in the continuous limit the same $J(\omega)$.

The coefficient in (11.21) also has a *real* part,

$$\text{Re}\left\{\sum_k c_k^2\left(\frac{1}{\omega_k - \omega - i0} + \frac{1}{\omega_k + \omega + i0}\right)\right\} = \int \frac{dz}{\pi} J(z)\frac{2z}{z^2 - \omega^2}$$

$$= \omega\,\text{Im}\,\gamma(\omega) + \int \frac{dz}{\pi}\frac{J(z)}{z}. \tag{11.26}$$

Control question. Can you prove this equation?

The first term in the last line also appears in the classical equations (11.1)–(11.3), and the second term provides a certain modification of these equations. In fact, it looks as if the potential energy coefficient a has changed into

$$a_{\text{new}} = a - 2\int \frac{dz}{\pi}\frac{J(z)}{z}. \tag{11.27}$$

This change provides perhaps the simplest example of what is called *renormalization* in quantum physics. The way to look at it is to say: a_{new} is the observable coefficient of the potential energy, whereas a is just a parameter in the Hamiltonian. Usually, a is called the "bare" and a_{new} the "dressed" coefficient of the potential energy. The bare coefficient could be observed, that is, extracted from the motion of the particle, only if the coupling to the bath is switched off. This is of course usually not possible.

In fact, this renormalization allows one to handle divergent changes of a which sometimes arise from the coupling to a particular bath. In such cases, one just says that the bare coefficient a was already divergent as well, so that both divergences (the one in a and the one in the renormalization) compensate each other resulting in a finite observable a_{new}. We discuss these issues in more detail in Chapter 13.

11.2.5 Diagonalization

The origin of this renormalization can be traced to the fact that in the Hamiltonian (11.15) the original undamped oscillator and the bath are separated. If one thinks about it, one notices that this is at least not democratic. The original oscillator which we are considering is in fact a bosonic mode as well, very much like the modes of the bath. Why would we not treat all bosons on an equal footing? To do so, let us substitute into the Hamiltonian (11.15) $\hat{x}$ and $\hat{p}$ in their bosonic form (see Chapter 8)

$$\hat{x} = \sqrt{\frac{\hbar}{2}}(aM)^{-1/4}(\hat{b}_0^\dagger + \hat{b}_0) \quad \text{and} \quad \hat{p} = i\sqrt{\frac{\hbar}{2}}(aM)^{1/4}(\hat{b}_0^\dagger - \hat{b}_0). \tag{11.28}$$

The resulting total Hamiltonian $\hat{H}$ is quadratic in both $\hat{b}_k$ and $\hat{b}_0$. One can diagonalize this quadratic Hamiltonian with a linear unitary transformation which introduces a new equivalent set of bosons $\hat{b}_m$. These new bosonic operators are linear combinations of $\hat{b}_0$ and $\hat{b}_k$, and generally have a new set of eigenfrequencies. The resulting Hamiltonian takes the form

$$\hat{H} = \sum_m \hbar\omega_m \hat{b}_m^\dagger \hat{b}_m - \hat{x}F_{\text{ext}}, \tag{11.29}$$

where the operator $\hat{x}$ should be expressed in terms of new boson modes $\hat{b}_m$,

$$\hat{x} = \hbar \sum_m (C_m \hat{b}_m^\dagger + C_m^* \hat{b}_m). \tag{11.30}$$

The coefficients C_m (the spectral weights) could be obtained in the course of the diagonalization procedure from the coupling coefficients c_k in (11.15). However, we do not perform this diagonalization explicitly. Rather, we obtain the C_m in a different way. Let us derive the quantum equations of motion directly from the Hamiltonian (11.29),

$$\dot{\hat{b}}_m = -i\omega_m \hat{b}_m + iC_m F_{\text{ext}}. \tag{11.31}$$

Solving these equations with the Fourier method, we obtain

$$\hat{b}_m(\omega) = \frac{C_m}{\omega_m - \omega - i0} F_{\text{ext}}(\omega), \tag{11.32}$$

and summing over all modes we find a relation between $\hat{x}(\omega)$ and F_{ext},

$$\hat{x}(\omega) = \hbar \sum_m |C_m|^2 \left(\frac{1}{\omega_m - \omega - i0} + \frac{1}{\omega + \omega_m + i0} \right) F_{\text{ext}}(\omega). \tag{11.33}$$

Let us now recall the definition of dynamical susceptibility

$$x(\omega) = \chi(\omega)F_{\text{ext}}(\omega), \tag{11.34}$$

and to comply with this definition, we require

$$\text{Im}\,\chi(\omega) = \pi\hbar \sum_m |C_m|^2 \delta(\omega - \omega_m), \tag{11.35}$$

for $\omega > 0$. This defines the spectral weights C_m in terms of the dynamical susceptibility and finalizes our formulation of dissipative quantum mechanics.

Actually, we have obtained two formulations, expressed by (11.15) and (11.29) correspondingly. The first formulation relies on the separation of the system and the environment. Its essential element is the relation between the system coordinate and the back-action of the environment on this coordinate, the "friction" force coming from the environment. The second formulation is more democratic. All bosons are treated equally, and the essential element is the dynamical susceptibility of the system coordinate. The two formulations are equivalent, and which to use is a matter of taste and convenience.

11.3 Time-dependent fluctuations

To see how the scheme works, we apply it to evaluate the time-dependent fluctuations of the coordinate x. For a classical quantity, fluctuations are fully characterized by the time-dependent correlator $\langle x(t_1)x(t_2)\rangle$. In a stationary situation, the correlator only depends on the time difference of the arguments, $S(t) = \langle x(t_1)x(t_1 + t)\rangle = \langle x(0)x(t)\rangle$. Since classical variables commute, the correlator does not change upon exchange of the time arguments, and therefore $S(t) = S(-t)$. The *noise* spectrum of the quantity is given by the Fourier transform of $S(t)$,

$$S(\omega) = \int dt\, S(t)e^{i\omega t},\tag{11.36}$$

and is real and even in frequency.

Control question. Can you prove these properties?

Quantum time-dependent fluctuations are obtained by replacing the classical variables by the corresponding operators. It has to be noted that operators $\hat{x}(0)$ and $\hat{x}(t)$ generally do not commute. Thus the quantum "noise" defined as

$$S(\omega) = \int dt\langle \hat{x}(0)\hat{x}(t)\rangle e^{i\omega t},\tag{11.37}$$

is not even in frequency.

To see the physical meaning of the quantum noise, let us compute this noise. We make use of the "democratic formulation" of dissipative quantum mechanics, and substitute $\hat{x}$ in the form

$$\hat{x}(t) = \hbar \sum_m \{C_m \hat{b}_m^\dagger(t) + C_m^* \hat{b}_m(t)\},\tag{11.38}$$

where the time-dependent boson operators are as usual given by

$$\hat{b}_m^\dagger(t) = e^{i\omega_m t}\hat{b}_m^\dagger \quad \text{and} \quad \hat{b}_m(t) = e^{-i\omega_m t}\hat{b}_m.\tag{11.39}$$

Let us average over a number state of the boson bath where $\langle \hat{b}_m^\dagger \hat{b}_m\rangle = N_m$. The quantum noise is contributed to by each mode and reads

$$S(\omega) = 2\pi\hbar^2 \sum_m |C_m|^2 \{(1 + N_m)\delta(\omega + \omega_m) + N_m\delta(\omega - \omega_m)\}.\tag{11.40}$$

We see that at positive frequencies the noise is proportional to the number of bosons in each mode, while at negative frequencies it is proportional to $(1 + N_m)$ and persists in the absence of any bosons. A practical way to observe quantum noise is to connect the variable x to a detector that can either absorb or emit quanta at frequency ω. The signal of absorption by the detector is then proportional to number of bosons in the bath, that is, to $S(\omega)$ at positive ω. See Exercise 3 for an implementation of such detectors.

For a general state of the bath, we cannot proceed further than this. However, it is almost always plausible to assume that the bath is in a state of thermodynamic equilibrium. Then the average over a certain quantum state can be replaced by a statistical average. This is achieved by setting all N_m to their average values given by Bose–Einstein statistics,

$$N_m \rightarrow n_B(\omega_m) \equiv \frac{1}{e^{\hbar\omega_m/k_B T} - 1}. \tag{11.41}$$

Since in this case N_m only depends on the corresponding eigenfrequency, we can replace the summation over m by an integration over eigenfrequencies, and replace $n_B(\omega_m)$ by $n_B(\omega)$. Looking again at the definition of the dynamical susceptibility in terms of the coefficients C_m (11.35), we see that we can write

$$S(\omega) = 2\hbar \text{Im}[\chi(\omega)] n_B(\omega). \tag{11.42}$$

Control question. Can you derive this from (11.40) and (11.35)?
Hint. You will again need that $\chi(-\omega) = \chi^*(\omega)$.

We have thus managed to express the quantum noise in the coordinate x in terms of the dynamical susceptibility and the bath temperature.

11.3.1 Fluctuation–dissipation theorem

One can also define a symmetrized quantum noise $S_s(\omega) \equiv \frac{1}{2}\{S(\omega) + S(-\omega)\}$, which can be immediately related to the even-in-frequency classical noise. After a little bit of simple algebra, we derive the relation between the symmetrized noise and imaginary part of susceptibility,

$$S_s(\omega) = \hbar \text{Im}[\chi(\omega)] \coth\left(\frac{\hbar\omega}{2k_B T}\right). \tag{11.43}$$

This relation is commonly called the fluctuation–dissipation theorem (stated by Harry Nyquist in 1928 and proven by Herbert Callen and Theodore Welton in 1951). The theorem eventually holds for any system and variable, not only for the damped oscillator we considered. The reason for this is that fluctuations are in fact small deviations from equilibrium. Due to this smallness, the dynamics of fluctuations of any system can be linearized and therefore can be presented as dissipative dynamics of oscillators.

There are two interesting important limits of the fluctuation–dissipation theorem. For frequencies that are low in comparison with the temperature, $\hbar\omega \ll k_B T$, the noise

$$S_s(\omega) = \frac{2k_B T}{\omega} \text{Im}[\chi(\omega)], \tag{11.44}$$

does not depend on $\hbar$, meaning that these fluctuations are classical. In the opposite limit $\hbar\omega \gg k_B T$, one deals essentially with quantum fluctuations. Those correspond to the *vacuum* fluctuations we studied in Chapter 8, since at low temperatures the bath is in its ground (vacuum) state. The time-dependent symmetrized noise is then given by

$$S_s(\omega) = \hbar |\text{Im}\,\chi(\omega)|. \tag{11.45}$$

A non-symmetrized version of this formula, $S(\omega) = 2\hbar\Theta(-\omega)|\text{Im}\,\chi(\omega)|$, is even more elucidating: the noise is absent at positive frequencies and an external detector detects no emission from the system. Indeed, the system is in its ground state and emission is forbidden by energy conservation.

11.3.2 Kubo formula

Let us now also consider the *anti*symmetric superposition of quantum noises, $S_a(\omega) \equiv \frac{1}{2}\{S(\omega) - S(-\omega)\}$. One can see from (11.40) that this superposition is immediately related to the imaginary part of the dynamical susceptibility,

$$S_a(\omega) = \pi \hbar^2 \sum_m |C_m|^2 \{\delta(\omega + \omega_m) - \delta(\omega - \omega_m)\} = -\hbar \mathrm{Im}\, \chi(\omega). \tag{11.46}$$

The validity of this relation does not depend on N_m and consequently does not rely on the assumption of thermodynamic equilibrium. Let us see if we can obtain some relation for the whole function $\chi(\omega)$ rather than only for its imaginary part. Using the Kramers–Kronig relation and the Cauchy formula, we obtain

$$\int \frac{dz}{\pi} \frac{\mathrm{Im}\, \chi(z)}{z - \omega - i0} = \int \frac{dz}{\pi} \frac{\mathrm{Im}\, \chi(z)}{z - \omega} + i\mathrm{Im}\, \chi(\omega) = \chi(\omega). \tag{11.47}$$

We can thus write the full function $\chi(\omega)$ as an integral over its imaginary part, as done at the leftmost side of this equation. If we express $\mathrm{Im}\, \chi(\omega) = -\frac{1}{\hbar} S_a(\omega)$ in terms of the noise spectrum, and subsequently express $S(\omega)$ in terms of operator averages as in (11.37), then this brings us to

$$\chi(\omega) = -\frac{i}{\hbar} \int_{-\infty}^0 dt \langle [\hat{x}(t), \hat{x}(0)] \rangle e^{-i\omega t}. \tag{11.48}$$

Control question. Can you follow all steps of the derivation?

Equation 11.48 is commonly called the Kubo formula (derived by Ryogo Kubo in 1957) and relates the dynamical susceptibility to the operator average of the commutator. Its intended use is to compute the dynamical susceptibility for a quantum system with known dynamics. One can also revert this: if the susceptibility is known from measurements or a model, one can evaluate the quantum correlator in the right-hand side of the equation.

The Kubo formula as given above can be derived in many different ways: let us present here another insightful way to arrive at the same result. As we know, the susceptibility is defined as the response of the system in one of its coordinates to an external perturbation along the same (or a different) coordinate. Let us see if we can calculate this response explicitly using a simple first-order perturbation theory.

We describe the state of the system by a density matrix $\hat{\rho}$, which in equilibrium reads $\hat{\rho}_{\mathrm{eq}}$. The goal is now to find the linear response in $\hat{\rho}$ to a small external perturbation and from that find the expected change in the coordinate. We focus for simplicity on a situation where only one single coordinate is relevant so that the perturbation can be represented by the Hamiltonian

$$\hat{H}' = f(t)\hat{x}, \tag{11.49}$$

where $f(t)$ parameterizes the strength of the (time-dependent) perturbation. The correction to the density matrix linear in $f(t)$ due to this perturbation is given by (see Section 1.10)

$$\delta\hat{\rho}(t)^{(1)} = -\frac{i}{\hbar}\int^{t} dt'\,[\hat{H}'(t'),\hat{\rho}_{\text{eq}}].\tag{11.50}$$

The linear response in the coordinate x can then simply be calculated as

$$\delta\langle\hat{x}(t)\rangle = \text{Tr}\{\hat{x}(t)\delta\hat{\rho}(t)^{(1)}\} = -\frac{i}{\hbar}\int^{t} dt'\,\text{Tr}\{\hat{x}(t)[\hat{H}'(t'),\hat{\rho}_{\text{eq}}]\}$$
$$= -\frac{i}{\hbar}\int^{t} dt'\,\langle[\hat{x}(t),\hat{x}(t')]\rangle f(t'),\tag{11.51}$$

where the average in the last line is defined as $\langle\ldots\rangle = \text{Tr}\{\ldots\hat{\rho}_{\text{eq}}\}$. If we now assume that the correlations between $\hat{x}(t)$ and $\hat{x}(t')$ only depend on the time difference $t - t'$, we can write the response as

$$\delta\langle\hat{x}(t)\rangle = \int^{t} dt'\,\chi(t-t')f(t'),\tag{11.52}$$

where the function $\chi(t)$ is exactly the susceptibility we were looking for. In the frequency domain, the susceptibility reads

$$\chi(\omega) = -\frac{i}{\hbar}\int_{-\infty}^{0} dt\,\langle[\hat{x}(t),\hat{x}(0)]\rangle e^{-i\omega t},\tag{11.53}$$

this coincides with (11.48).

11.4 Heisenberg uncertainty relation

The standard quantum mechanics of a single degree of freedom predicts uncertainty relations for conjugated variables. The most famous uncertainty relation, for coordinate x and momentum p, reads

$$\langle\hat{x}^2\rangle\langle\hat{p}^2\rangle \geq \frac{\hbar^2}{4}.\tag{11.54}$$

Equality is eventually achieved for the ground state of an undamped oscillator, as we have seen in Chapter 10. Let us now look at the uncertainties, or fluctuations, in x and p for a damped oscillator. The fluctuations of x can be directly related to the quantum noise

$$\langle\hat{x}^2\rangle = S(t=0) = \int\frac{d\omega}{2\pi}S(\omega) = \hbar\int\frac{d\omega}{2\pi}\text{Im}[\chi(\omega)]\coth\left(\frac{\hbar\omega}{2k_BT}\right).\tag{11.55}$$

In the limit of zero temperature, at which the oscillator will be in its ground state, this reduces to

$$\langle\hat{x}^2\rangle = \hbar\int_{0}^{\infty}\frac{d\omega}{\pi}\text{Im}[\chi(\omega)].\tag{11.56}$$

As to the momentum, by virtue of the equations of motion $\hat{p} = M\dot{\hat{x}}$ we have

$$\langle\hat{p}^2\rangle = \int\frac{d\omega}{2\pi}M^2\omega^2\langle\hat{x}_\omega^2\rangle = \hbar\int_{0}^{\infty}\frac{d\omega}{\pi}M^2\omega^2\text{Im}[\chi(\omega)].\tag{11.57}$$

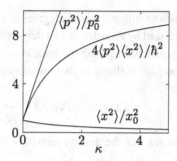

Fig. 11.3   Uncertainties in x and p, and their product versus the dissipation parameter $\kappa = \gamma/2\sqrt{aM}$.

We use the dynamical susceptibility given by (11.8), and then we find

$$\frac{\langle \hat{x}^2 \rangle}{x_0^2} = \frac{\ln\left(\frac{\kappa + \sqrt{\kappa^2 - 1}}{\kappa - \sqrt{\kappa^2 - 1}}\right)}{\pi\sqrt{\kappa^2 - 1}}, \tag{11.58}$$

where we have introduced dimensionless damping $\kappa = \gamma/2\sqrt{aM}$ and the uncertainty of the undamped oscillator $x_0^2 = \hbar/\sqrt{aM}$.

We see that $\langle x^2 \rangle > \langle x^2 \rangle_{\gamma=0}$ and $\langle x^2 \rangle \to 0$ if $\kappa \to \infty$. This means that the damping localizes the oscillator, decreasing the uncertainty in x. We understand that at the same time the kinetic energy of the oscillator must increase, so we expect the uncertainty in p to increase as well. There is a problem with the integral in (11.57). It diverges at large frequencies if we assume a frequency-independent damping γ. In reality, very fast fluctuations are not damped since the bath cannot react fast enough. We account for this by introducing a cut-off frequency ω_D, so that $\gamma(\omega) \approx 0$ for frequencies larger than ω_D, and we assume that ω_D is larger than any other frequency scale involved. Then the integral converges and

$$\frac{\langle \hat{p}^2 \rangle}{p_0^2} = (1 - 2\kappa^2)\frac{\langle \hat{x}^2 \rangle}{x_0^2} + \frac{4\kappa}{\pi}\ln\frac{\omega_D}{\omega_0}, \tag{11.59}$$

$p_0 = \hbar/2x_0$ being the momentum uncertainty of the undamped oscillator. We see that the divergence is only logarithmic, which is not so strong. For any practical frequency scale this implies that the term with the logarithm is not very large. The results for the uncertainties and their product are plotted versus κ in Fig. 11.3 assuming $\ln(\omega_D/\omega_0) = 7$.

11.4.1 Density matrix of a damped oscillator

The ground state of an undamped oscillator is characterized by a wave function. In the coordinate representation, it is an exponent of a quadratic form,

$$\Psi_0(x) = \frac{1}{\sqrt{\sqrt{2\pi}x_0}}\exp\left(-\frac{x^2}{4x_0^2}\right). \tag{11.60}$$

In the presence of dissipation, no x-dependent wave function can characterize the oscillator. The wave function is a property of the whole system, that is, of the oscillator plus the bath, and thus depends on infinitely many degrees of freedom. However, the state of the oscillator can be characterized by an x-dependent density matrix $\rho(x, x')$, which can be obtained from the full density matrix by taking the trace over all degrees of freedom except x.

For an undamped oscillator in its ground state, the density matrix reads

$$\hat{\rho}(x, x') = \Psi_0^*(x)\Psi_0(x') = \frac{1}{\sqrt{2\pi}x_0} \exp\left\{-\frac{x^2 + x'^2}{4x_0^2}\right\}. \tag{11.61}$$

Let us now *assume* that the density matrix of the damped oscillator can be expressed in a similar form,

$$\hat{\rho}(x, x') = C \exp(Ax^2 + Ax'^2 + Bxx'), \tag{11.62}$$

where A, B, and C are yet unknown coefficients. The proof that this is indeed the correct form for the density matrix is straightforward but boring, and is given as an exercise. To determine the unknown coefficients, we calculate $\langle \hat{x}^2 \rangle$ and $\langle \hat{p}^2 \rangle$ using (11.62), and then express A, B, and C in terms of the uncertainties calculated. We obtain

$$\hat{\rho}(x, x') = \frac{1}{\sqrt{2\pi \langle \hat{x}^2 \rangle}} \exp\left\{-\frac{(x + x')^2}{8\langle \hat{x}^2 \rangle} - \frac{\langle \hat{p}^2 \rangle}{2\hbar^2}(x - x')^2\right\}. \tag{11.63}$$

Let us make explicitly clear here that this density matrix does *not* correspond to a wave function, that is, it does not describe a pure state. The simplest "purity test" we know (see Section 1.10), is to compute $S_2 \equiv \text{Tr}[\hat{\rho}^2]$. For a pure state, $\hat{\rho}^2 = \hat{\rho}$ and $S_2 = \text{Tr}[\hat{\rho}] = 1$. Generally, $0 < S_2 < 1$, and $1/S_2$ gives an estimate of the number of pure states mixed together in a given density matrix. We can explicitly perform this purity test on the density matrix (11.63) of the damped oscillator, yielding

$$S_2 = \int dx\, dx'\, \hat{\rho}(x, x')\hat{\rho}(x', x) = \frac{\hbar}{2\sqrt{\langle \hat{p}^2 \rangle \langle \hat{x}^2 \rangle}}. \tag{11.64}$$

The purity is thus expressed in terms of the previous results for the uncertainties. We conclude that even a small dissipation makes the state "impure." Since any system is subject to dissipation, this implies that pure states hardly occur in Nature. Rather, the description of a system with a few degrees of freedom in terms of a wave function is an abstraction, very much like a completely isolated system in classical physics is an abstraction. On the other hand, the purity is still of the order of 1, even for large damping. This shows that the system retains quantum properties even if the dissipation is relatively strong.

To conclude, we have learned how to quantize any linear dissipative system: we only need to know its dynamical susceptibility. These results are straightforward to extend to many degrees of freedom by considering the dynamical susceptibility matrix of the corresponding variables. The real world implies dissipation, and dissipative quantum mechanics implies infinitely many degrees of freedom – those of the bath. The description of a system with several degrees of freedom in terms of a wave function is in fact an abstraction.

Table 11.1 Summary: Dissipative quantum mechanics

Damped oscillator: Classical description

equations of motion: $\dot{p} = -ax + F_f + F_{\text{ext}}$ and $\dot{x} = p/M$

dissipation: introduced by friction force, $F_f(t) = -\int_0^\infty d\tau\, \gamma(\tau)\dot{x}(t-\tau)$,

        with $\gamma(t)$ the damping coefficient

the dissipative dynamics of the oscillator can be characterized by the dynamical susceptibility

dynamical susceptibility: $\chi(\omega) = \dfrac{x(\omega)}{F_{\text{ext}}(\omega)}$, response in coordinate to external force

      for the oscillator: $\chi(\omega) = \dfrac{1}{a - M\omega^2 - i\omega\gamma(\omega)}$

Damped oscillator: Quantum description I

model: $\hat{H} = \dfrac{\hat{p}^2}{2M} + \dfrac{a\hat{x}^2}{2} + \sum_k \hbar\omega_k \hat{b}_k^\dagger \hat{b}_k - \hat{x}\sum_k \hbar c_k(\hat{b}_k^\dagger + \hat{b}_k) - \hat{x}F_{\text{ext}},$ boson bath

problem: to determine what ω_k and c_k to use

solution: relate ω_k and c_k to damping coefficient

     consistent picture when $\sum_k \pi c_k^2 \delta(\omega - \omega_k) = \omega\,\text{Re}\{\gamma(\omega)\}\Theta(\omega)$

Damped oscillator: Quantum description II, more "democratic"

describe oscillator ($\hat{p}$ and $\hat{x}$) also in terms of bosons $\rightarrow$ diagonalize resulting Hamiltonian

model: $\hat{H} = \sum_m \hbar\omega_m \hat{b}_m^\dagger \hat{b}_m - \hat{x}F_{\text{ext}},$ with $\hat{x} = \hbar\sum_m (C_m \hat{b}_m^\dagger + C_m^* \hat{b}_m)$

problem: to determine what ω_m and C_m to use

solution: relate ω_m and C_m to dynamical susceptibility

     consistent picture when $\text{Im}\,\chi(\omega) = \pi\hbar \sum_m |C_m|^2 \delta(\omega - \omega_m)$

Fluctuations

quantum noise: $S(\omega) = \int dt \langle \hat{x}(0)\hat{x}(t)\rangle e^{i\omega t}$

with description II: $S(\omega) = 2\hbar\text{Im}[\chi(\omega)]n_B(\omega)$, if bath is in thermal equilibrium

FDT: symmetrized noise $S_s(\omega) = \frac{1}{2}\{S(\omega) + S(-\omega)\} = \hbar\text{Im}[\chi(\omega)]\coth\left(\dfrac{\hbar\omega}{2k_B T}\right)$

Kubo formula: antisymmetrized noise $S_a(\omega) = \frac{1}{2}\{S(\omega) - S(-\omega)\} = -\hbar\text{Im}\,\chi(\omega)$

     with Kramers–Kronig relation follows $\chi(\omega) = -\dfrac{i}{\hbar}\int_{-\infty}^0 dt \langle[\hat{x}(t), \hat{x}(0)]\rangle e^{-i\omega t}$

Uncertainty relations for the damped oscillator

one finds at $T = 0$: $\langle \hat{x}^2 \rangle = \dfrac{\hbar}{\pi}\int_0^\infty d\omega\, \text{Im}[\chi(\omega)]$ and $\langle \hat{p}^2 \rangle = \dfrac{\hbar}{\pi}\int_0^\infty d\omega\, M^2\omega^2 \text{Im}[\chi(\omega)]$

assuming constant γ: $\langle \hat{x}^2 \rangle \rightarrow 0$ and $\langle \hat{p}^2 \rangle \rightarrow \infty$ when $\gamma \rightarrow \infty$, particle localizes when damped

density matrix: $\hat{\rho}(x, x') = \dfrac{1}{\sqrt{2\pi\langle \hat{x}^2 \rangle}} \exp\left\{-\dfrac{(x+x')^2}{8\langle \hat{x}^2 \rangle} - \dfrac{\langle \hat{p}^2 \rangle}{2\hbar^2}(x - x')^2\right\},$

     with "purity" $\text{Tr}[\hat{\rho}(x, x')^2] = \dfrac{\hbar}{2\sqrt{\langle \hat{p}^2 \rangle \langle \hat{x}^2 \rangle}}$

Exercises

1. *Alien atom in a one-dimensional chain* (solution included). Consider a one-dimensional lattice such as analyzed in Chapter 7. There is one alien atom in this chain at the discrete position "0." Let us describe its dissipative dynamics. To speed up calculations, we assume that the mass of this atom and its interactions with its nearest neighbors are the same as for all other atoms of the chain. We develop two alternative descriptions of the dynamics, differing in choice of coordinates. For the first description, we just use the physical coordinate of the atom $x \equiv x_0$. For the second description, we use the coordinate relative to two nearest neighbors, $X \equiv x_0 - \frac{1}{2}(x_{-1} + x_1)$.

    a. Compute the dynamical susceptibilities for variables x and X by solving the classical equations of motion for the atoms. Express the answer in terms of an integral over q, the phonon wave vector.

    b. Give expressions for the imaginary parts of the susceptibilities.

    c. Represent the operators $\hat{x}$ and $\hat{X}$ in terms of phonon CAPs. Describe the correspondence with the dynamical susceptibility.

    d. Find the friction force for both coordinates in the limit of low frequencies.

    e. Find the uncertainties in both coordinates x and X, as well as the uncertainties in the corresponding momenta. Quantify the purity.

2. *End of an RC-line.* An *RC*-line is a distributed linear electric circuit where the voltage $V(x)$ depends only on one coordinate x. The system is characterized by a resistance $\bar{R}$ and capacitance $\bar{C}$ per unit length, so that the charge density in the line is given by $q(x) = \bar{C}V(x)$ and the current is $I(x) = -(dV/dx)/\bar{R}$.

    a. From charge conservation, derive the evolution equation for $V(x)$. Demonstrate the formal equivalence with the diffusion equation. What is the diffusion coefficient?

    We consider a semi-infinite *RC*-line with a capacitor C_0 connected to its end. We are interested in the (quantum) fluctuations of the voltage $V(0)$ at the end of the wire and the current I_c in the capacitor.

    b. Compute, using the Fourier method, the voltage response to the current injected into the end of the wire. Use the evolution equations derived.

    c. Find the dynamical susceptibility $\chi(\omega)$ for the variable $V(0)$. Find the time scale t_r associated with the discharge of the capacitor C_0.

    d. Attempt to compute the quantum uncertainty of $V(0)$ through the capacitor and demonstrate it diverges at high frequencies.

    e. Compute the fluctuation of voltage at high temperatures $k_B T \gg \hbar/t_r$. The fluctuation is finite. Does this imply that the temperature reduces quantum uncertainty?

3. *Atom as a quantum noise detector.*

    a. Consider two quantum states, $|A\rangle$ and $|B\rangle$, that are weakly coupled by the variable x which is a part of a boson bath described by the Hamiltonian (11.29). The coupling Hamiltonian

$$\hat{H}_{\text{int}} = \gamma |A\rangle \langle B| \hat{x} + \text{H.c.},$$

causes transitions between the states. The corresponding transition rates can be evaluated with Fermi's golden rule. Show that the transition rate from $|A\rangle$ to $|B\rangle$ is proportional to the quantum noise of quantity x at the frequency $(E_B - E_A)/\hbar$.

b. Consider two states of an atom, $|s\rangle$ and $|p_z\rangle$. The quantum noise of which physical quantity can be detected by measuring the transition rates between these states? What plays the role of the coupling coefficient γ?

c. Compute the dynamical susceptibility of the electric field in vacuum, determine the quantum noise and demonstrate that the atomic transition rates computed in Chapter 9 can be reproduced in this way.

4. *Dissipation through Fermi's golden rule.* Let us turn to the model given by the Hamiltonian (11.15). We treat the coupling term $\hat{H}_{\text{coup}}$ as a perturbation which causes transitions between the states $|n\rangle$ of the original undamped oscillator.

a. Show that transitions are only possible between $|n\rangle$ and $|n-1\rangle$. Compute the corresponding rates assuming a frequency-independent friction coefficient.

b. Compute the energy dissipation as a function of energy from the classical equations of motion in the limit of small dissipation.

c. Find and explain the correspondence between the "classical" and "quantum" results.

5. *Quantum Brownian motion.* Consider a particle moving in a (three-dimensional) dissipative medium. It is subject to a friction force but, in distinction from the damped oscillator considered, it is not confined by any potential. The equation of motion thus reads

$$M\ddot{\mathbf{r}} = \mathbf{F}_f = -\gamma\dot{\mathbf{r}}.$$

Suppose we fix the particle at some initial position at $t = 0$. The probability of finding the particle at a distance d from its initial position is Gaussian, $P(d) \propto \exp\{-d^2/2\sigma(t)\}$. We compute and analyze $\sigma(t)$.

a. Relate $\sigma(t)$ to the (quantum) noise of any of the three projections of $\mathbf{r}$.

b. Find $\chi(\omega)$ for this variable, evaluate the quantum noise and represent $\sigma(t)$ in the form of an integral over ω.

c. An important time scale in the problem is $t_r = M/\gamma$. We will assume a small but non-vanishing temperature $k_B T \ll \hbar/t_r$. Evaluate $\sigma(t)$ at $t \ll t_r$.

d. Evaluate $\sigma(t)$ for intermediate time-scales $t_r \ll t \ll \hbar/k_B T$.

e. Evaluate $\sigma(t)$ in the long-time limit. Why did we mention Brownian motion?

6. *Density matrix of the damped oscillator.* We promised to prove that the density matrix $\hat{\rho}(x, x')$ of the variable x is an exponent of quadratic form in x and x' in the ground state of the bath. To do this, we take the bath model of (11.29).

a. Express the CAPs $\hat{b}_m^\dagger$ and $\hat{b}_m$ in terms of the corresponding coordinates $\hat{x}_m$ and momenta, and express $\hat{x}$ as a weighted sum over $\hat{x}_m$.

b. Find the ground state wave function of the bath in the coordinate representation. Write down the corresponding density matrix $\hat{\rho}(\{x_m\}, \{x'_m\})$.

c. Taking the partial trace corresponds to Gaussian integration over x_m. Perform this integration to find the $\hat{\rho}(x, x')$ sought.

7. *Entanglement entropy.* Evaluate the entanglement entropy $S = \mathrm{Tr}[\hat{\rho} \ln \hat{\rho}]$ for the density matrix given by (11.63). Proceed as follows:

    a. Note that a product of any two matrices of the form (11.62) is also a matrix of this form. Derive a composition law for such multiplication.

    b. Apply this law to compute the nth power of the density matrix (11.63) and find $S_n = \mathrm{Tr}[\rho^n]$.

    c. Obtain S by taking the formal limit $S = -\lim_{n \to 1} \frac{dS_n}{dn}$.

Solutions

1. *Alien atom in a one-dimensional chain.*

    a. We add a time- and atom-dependent force to the classical equation (7.3) and solve it by the Fourier method. For the coordinate x, the force is applied to a single atom, its Fourier component does not depend on q and the susceptibility reads

$$\chi_x(\omega) = \frac{1}{m} \int_{-\pi}^{\pi} \frac{dq}{2\pi} \frac{1}{\Omega^2 \sin^2(q/2) - (\omega + i0)^2},$$

$\Omega \equiv 2\sqrt{K/m}$ being the maximum phonon frequency. For the coordinate X, the conjugated force also persists at two neighboring atoms, $F_{-1} = F_1 = -F/2$ and $F_0 = F$. Its Fourier component equals $2F \sin^2(q/2)$ and the susceptibility reads

$$\chi_X(\omega) = \frac{1}{m} \int_{-\pi}^{\pi} \frac{dq}{2\pi} \frac{4 \sin^4(q/2)}{\Omega^2 \sin^2(q/2) - (\omega + i0)^2}.$$

In both cases, we have added an infinitesimally small imaginary part to ω to account for proper causality of $\chi(\omega)$.

    b. This "small" addition becomes crucial when computing the imaginary parts with the aid of the Cauchy formula,

$$\mathrm{Im}[\chi_x(\omega)] = \frac{1}{m\Omega\omega} \frac{1}{\sqrt{1 - (\omega/\Omega)^2}},$$

$$\mathrm{Im}[\chi_X(\omega)] = \frac{4\omega^3}{m\Omega^5} \frac{1}{\sqrt{1 - (\omega/\Omega)^2}}.$$

This holds at $|\omega| < \Omega$ and the imaginary parts are zero otherwise.

    c. Adjusting the formulas of Chapter 8 for the field operators of the chain, and substituting the definitions of x and X, we obtain

$$\hat{x} = \sum_q \frac{\sqrt{\hbar}}{2\sqrt{Km}|\sin(q/2)|} \left(\hat{a}_q + \hat{a}_q^\dagger \right),$$

$$\hat{X} = \sum_q \frac{\sqrt{\hbar}|\sin(q/2)|^3}{\sqrt{Km}} \left(\hat{a}_q + \hat{a}_q^\dagger \right).$$

The coefficients can be related to the spectral weights C_m introduced in (11.30), where m numbers the discrete phonon modes. The relation between C_m and the imaginary part of the susceptibility is set by (11.35).

d. Let us consider the variable x first. To start, we need to express the force from the other atoms $\tilde{F}$ acting on the alien atom in terms of the susceptibility. We use the equation of motion $m\ddot{x} = F + \tilde{F}$, which is conveniently rewritten as $x_\omega = \chi_0(\omega)(F_\omega + \tilde{F}_\omega)$. Since $F_\omega = x_\omega/\chi(\omega)$, this gives

$$\tilde{F}_\omega = \left(\frac{1}{\chi_0} - \frac{1}{\chi}\right) x_\omega.$$

The friction force, being dissipative, is proportional to the imaginary part of the coefficient, $\mathrm{Im}[\chi(\omega)^{-1}]$. At low frequencies, the susceptibility is purely imaginary and the friction force is given by

$$F_f = i\omega m\Omega x_\omega, \quad \text{so that} \quad F_f(t) = -m\Omega\dot{x}.$$

As common to assume, the friction force is proportional to the velocity of the alien atom.

For the variable X, the calculation is the same except that χ_0 is evaluated with the effective mass $m^* = \frac{2}{3}m$ corresponding to this variable (to make sure that this is the effective mass, one computes $\chi(\omega)$ at large frequencies $\omega \gg \Omega$ when the forces from the neighbouring atoms can be disregarded).

In the limit of low frequencies, the susceptibility has a real part $\mathrm{Re}[\chi_X(\omega)] = 2/m\Omega^2 \gg \mathrm{Im}[\chi_X(\omega)]$. Then it follows for the friction force

$$F_f = i\mathrm{Im}[\chi_X(\omega)]/\mathrm{Re}[\chi_X(\omega)]^2 \, X_\omega = \frac{im\omega^3}{\Omega} X_\omega.$$

Thus the friction force is proportional to the third derivative of x with respect to time.

e. The uncertainties are evaluated with (11.56) and (11.57). The integral for $\langle \hat{x}^2 \rangle$ diverges logarithmically at low frequencies. Let us cut this divergence at $1/\tau$, where τ is the time of phonon propagation through the whole chain, and we take the integral with logarithmic accuracy,

$$\langle \hat{x}^2 \rangle = \frac{\hbar}{\pi m\Omega} \ln(2\Omega\tau).$$

The momentum uncertainty does not diverge, $\langle \hat{p}^2 \rangle = \hbar m\Omega/\pi$.

The purity evaluated with (11.64) thus reads $S_2 = \pi/2\sqrt{\ln(2\Omega\tau)}$. This indicates (infinitely) many states involved in the quantum dynamics of x.

As to the alternative variable X, the conjugated momentum is $P = m^*\dot{X}$, both uncertainties converge and

$$\langle \hat{X}^2 \rangle = \frac{8}{3\pi}\frac{\hbar}{\Omega m}, \quad \langle \hat{P}^2 \rangle = \frac{128}{135\pi}\hbar m\Omega, \quad \text{and} \quad S_2 = \frac{9\sqrt{5}\pi}{64} \approx 0.988.$$

The quantum dynamics of X therefore correspond to those of an almost pure state!

12 Transitions and dissipation

In this second chapter on dissipative quantum mechanics, we focus on the role of dissipation in transitions between different quantum states, as summarized in Table 12.2. We consider a generic system: a qubit embedded in a dissipative environment. We introduce the spin–boson model to understand the role the environment plays during transitions between the two levels. Elaborating further on this model, we find a general way to classify all possible environments. We look at a specific environment: the electromagnetic vacuum studied in previous chapters.

12.1 Complicating the damped oscillator: Towards a qubit

In the previous chapter, we considered in detail a simplest quantum dissipative system, the damped oscillator. Its generic realization is a particle which is subject to a friction force and is confined in a parabolic potential $U(x) = \frac{1}{2}ax^2$. We can easily make the system more interesting and complicated. The model considered in the previous chapter can be generalized to any potential $U(x)$, so that the corresponding classical equations of motion become

$$\dot{p} = -\frac{\partial U(x)}{\partial x} + F_{\text{f}} + F_{\text{ext}} \quad \text{and} \quad \dot{x} = \frac{p}{M}. \tag{12.1}$$

The corresponding quantum Hamiltonian consists of three parts, cf. (11.15),

$$\hat{H} = \hat{H}_{\text{osc}} + \hat{H}_{\text{env}} + \hat{H}_{\text{coup}}, \quad \text{where} \quad \hat{H}_{\text{osc}} = \frac{\hat{p}^2}{2M} + U(\hat{x}), \tag{12.2}$$

and with $\hat{H}_{\text{env}}$ and $\hat{H}_{\text{coup}}$ being the same as for the damped oscillator. Since the resulting equations of motion, both the quantum and the classical, are not linear anymore, the problem becomes more complicated. On the classical level, the complication is minor: one needs to solve two differential equations to find the time-evolution of the system, a task easily delegated to a computer. On the quantum level, the complication is overwhelming, since the non-linearity forbids us to reduce the problem to a set of non-interacting bosonic modes. One needs to work with infinitely many degrees of freedom that cannot be separated from each other. No computer can help with a numerically exact solution. The only reliable method would be to do perturbation theory in terms of a weak non-linearity.

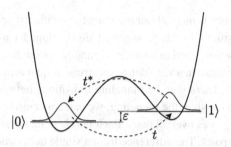

A potential with two minima of almost the same depth. A classical damped particle will freeze in either the left or right minimum. A quantum particle can persist in any superposition of the localized states $|0\rangle$ and $|1\rangle$.

A relevant question arising is whether these complications are worth addressing at all. If the potential $U(x)$ has only a single minimum, we can readily guess the result in qualitative terms: the classical particle at zero temperature freezes in the minimum, finite temperature and/or quantum mechanical effects spread the particle around this minimum. Any potential minimum can be approximated with a parabolic well for particles with energies not too far above the minimum. To leading order, the dynamics in a single minimum are thus always qualitatively similar to those of the damped oscillator considered.

The search for a more challenging and interesting problem leads us to a potential $U(x)$ with *two* minima of (almost) the same depth, like the potential pictured in Fig. 12.1. In this case, the qualitative difference with an oscillator is apparent already at the classical level. A classical system with such a potential is *bistable*: it may freeze in either minimum, provided there is some friction present. Once frozen in a given minimum, it will remain there forever. The system does not have to relax always to the same unique state, and can in fact thus be used as a binary memory cell. In contrast to a classical particle, a quantum particle does not have to sit in either one of the potential minima, it may be delocalized. This results from coherent tunneling through the potential barrier separating the minima. If we call the states with the particle localized in a certain minimum $|0\rangle$ and $|1\rangle$, then the stationary states could be the superpositions $\frac{1}{\sqrt{2}}\{|0\rangle \pm |1\rangle\}$ where the particle is equally distributed between both minima. Such superposition states are frequently used to illustrate the basic principles of quantum mechanics.

In this chapter we learn how a particle *tunnels* between two states in the presence of friction. Although the complete solution of this problem is still not presented, its qualitative understanding was achieved about 25 years ago and was a major development of quantum mechanics. A very concise formulation of this understanding is as follows. The system is *either* quantum (potentially delocalized between the minima) *or* classical (bistable memory cell) depending on the properties of the environment. It appears that the same holds for more complicated situations, for instance for a particle in a periodic potential. It also holds qualitatively for any number of quantum states near the minima. But let us here truncate the problem, and keep things as simple as possible. From all possible quantum states of a system with a potential $U(x)$, we concentrate on two states localized in two respective local minima and consider the effect of dissipation on the *transitions* between these two states. In other words, we treat a qubit in the presence of dissipation.

The story is more about the transitions rather than the qubit. As a complementary setup for transitions in the presence of dissipation, let us briefly turn to the damped electric oscillator described in Chapter 11, and focus on the electrons in the plates of the capacitor. They are there in a very dense spectrum of quantum levels, that is continuous for practical purposes. There may be tunneling transitions between the levels on the left and the right plate. To consider the tunneling, we can concentrate on each pair of levels and sum up the tunneling rates over all pairs. Each pair is in fact a qubit and can be treated separately from all other pairs. The difference from a single qubit situation is that tunneling in the opposite direction (against the direction of the current) is unlikely to occur within the same pair of states, that is, in the same qubit.

12.1.1 Delocalization criterion

The ideal quantum two-state system, a qubit, has been extensively treated in the introductory chapter and is quickly refreshed now. The basis states of the qubit corresponding to the two minima are called $|0\rangle$ and $|1\rangle$. The energetics of the qubit are determined by two parameters: the energy splitting ε between the two minima and the tunneling amplitude t_q (for the purposes of this chapter we may disregard the phase of this amplitude; we assign the index q to it so as not to mix it up with time). The two eigenstates of the qubit, $|+\rangle$ and $|-\rangle$, are normalized linear superposition of the two basis states,

$$|+\rangle = \cos(\tfrac{\theta}{2})|1\rangle + \sin(\tfrac{\theta}{2})|0\rangle \quad \text{and} \quad |-\rangle = \cos(\tfrac{\theta}{2})|0\rangle - \sin(\tfrac{\theta}{2})|1\rangle, \tag{12.3}$$

with $\tan(\theta) \equiv 2t_q/\varepsilon$. These two eigenstates are separated by the energy $2\sqrt{(\varepsilon/2)^2 + t_q^2}$.

The ground state of the qubit is $|-\rangle$, whereas the state localized in the lower energy minimum is $|0\rangle$. Let us introduce a parameter to characterize the *delocalization* of the qubit: the probability p_{dec} of finding the qubit in the "wrong" higher energy minimum $|1\rangle$, provided that the qubit is in its ground state. From the expression for $|-\rangle$, we derive that

$$p_{\text{dec}} = |\langle 1|-\rangle|^2 = \sin^2(\tfrac{\theta}{2}), \quad \text{so that} \quad p_{\text{dec}} = \frac{t_q^2}{\varepsilon^2} + \dots \quad \text{for} \quad t_q \to 0. \tag{12.4}$$

We see that the delocalization probability is in fact small if the tunneling amplitude is weak in comparison with the energy splitting, $|t_q| \ll |\varepsilon|$, and the leading term given above can be obtained by a perturbation expansion in t_q. We note, however, that this leading order quickly increases and eventually diverges with decreasing ε. This signals the delocalization of the qubit. We use this signal to identify delocalization in the context of dissipative quantum mechanics.

12.2 Spin–boson model

Let us formulate a framework model for a qubit embedded in a dissipative environment. This model was first elaborated about 30 years ago and for historical reasons is called the *spin–boson* model. The word "spin" appears here since a qubit was commonly mapped onto the two states of a spin $-\frac{1}{2}$ particle. The word "boson" refers to the bosonic model of environment described in the previous chapter. The total Hamiltonian of the spin–boson model consists of the combination of those of the qubit, the environment and the coupling between the latter. The qubit Hamiltonian is a 2×2 matrix in the space spanned by $\{|1\rangle, |0\rangle\}$,

$$\hat{H}_q = \frac{\varepsilon}{2}\hat{\sigma}_z + t_q\hat{\sigma}_x. \tag{12.5}$$

As usual, the environment consists of non-interacting bosonic modes,

$$\hat{H}_{\text{env}} = \sum_m \hbar\omega_m \hat{b}_m^\dagger \hat{b}_m. \tag{12.6}$$

As to the coupling Hamiltonian, it is assumed that only one spin component, $\hat{\sigma}_z$, is coupled to the environment. The strength of this coupling is controlled by a parameter q_0

$$\hat{H}_{\text{coup}} = -q_0\hat{\sigma}_z\hat{F} \equiv -q_0\hat{\sigma}_z\hbar \sum_m (C_m\hat{b}_m^\dagger + C_m^*\hat{b}_m). \tag{12.7}$$

We should make sense out of the spectral weight coefficients C_m that enter the definition of the operator $\hat{F}$. How should we do this? The best way is to replace q_0 by an auxiliary continuous variable q, which for the special case of the spin–boson model is restricted to the two quantized values $\pm q_0$. If our model stems from that of a particle in a two-minimum potential, we can interpret q simply as the coordinate of the particle. The operator $\hat{F}$ corresponds then to the friction force due to the environment which the particle may experience on its way between the minima (compare (12.7) with (11.15) describing the damped oscillator).

Let us now provide some action. Let us move the particle with our hands, so that this auxiliary variable changes in time following a fixed $q(t)$. The environment will react on this change of q, which results in a friction force $\hat{F}$ on the particle. As in Chapter 11, we introduce a dynamical susceptibility $\chi(\omega)$ that relates the action and the reaction,

$$F(\omega) = \chi(\omega)q(\omega). \tag{12.8}$$

As shown in the previous chapter, this fixes the spectral weights such that

$$\text{Im}\,\chi(\omega) = \pi\hbar \sum_m |C_m|^2\delta(\omega - \omega_m), \tag{12.9}$$

for positive frequencies ω. To avoid misunderstanding, we stress that the dynamical susceptibility always defines the reaction to an action. The "force" is now the reaction, while the "coordinate" is the action. In the previous chapter, it was the other way around.

To provide an alternative illustration of this approach, let us turn to the example of tunneling between two capacitor plates – tunneling in an electromagnetic environment. In this case, the continuous variable q is associated with the charge transferred between the plates in course of motion of a single electron between the plates. Since we transfer a single electron with elementary charge e, we know that $2q_0 = e$. The "force" conjugated to this variable is nothing but the voltage drop $\hat{V}$ between the plates, and the coupling Hamiltonian reads $-\frac{1}{2}e\hat{V}\hat{\sigma}_z$. Let us imagine that we connect an external current source to the capacitor so that we can provide any desired charge transfer $q(t) = \int dt' I(t')$ between the plates. The voltage drop $V(t)$ is related to this $q(t)$ by the total impedance $Z(\omega)$ of the circuit,

$$V(\omega) = Z(\omega)I(\omega) = -i\omega Z(\omega)q(\omega). \qquad (12.10)$$

This again fixes the spectral weights to

$$\omega \operatorname{Re} Z(\omega) = -\pi\hbar \sum_m |C_m|^2 \,\delta(\omega - \omega_m). \qquad (12.11)$$

Control question. The spin–boson model just formulated is rather restrictive, including coupling of the environment to z-projection of qubit "spin" only. Explain how to generalize the model?

12.3 Shifted oscillators

The spin–boson model formulated above cannot be solved exactly for general parameters t_q and ε. However, we can obtain very instructive solutions if we set $t_q = 0$. We present these solutions, and then use them to proceed further with a perturbation expansion in terms of t_q.

If $t_q = 0$, no tunneling may occur and the system is fixed in one of the two classical states, 1 or 0. The Hamiltonian for either of the states depends on boson variables only and reads

$$\hat{H}_{1,0} = \sum_m \left\{ \hbar\omega_m \hat{b}_m^\dagger \hat{b}_m \mp q_0\hbar(C_m\hat{b}_m^\dagger + C_m^*\hat{b}_m) \right\}, \qquad (12.12)$$

where $+(-)$ refers to the state 0(1). If this Hamiltonian had not had the term linear in q_0, we would have known very well how to handle it. The stationary wave functions would have been those with a fixed number of bosons n_m in each mode, denoted by $|\{n_m\}\rangle$, and the corresponding energies would be $\sum_m \hbar\omega_m n_m$. We are going to use a trick that removes the linear term from the Hamiltonian.

An inspiration to find this trick, one can get from a simple analogy with a single oscillator. Let us try to "spoil" the parabolic potential of the oscillator with an extra term linear in x,

$$U(x) = \frac{ax^2}{2} + \lambda x. \qquad (12.13)$$

The resulting potential in fact stays parabolic but with a *shifted* coordinate x,

$$U(x) = \frac{ax^2}{2} + \lambda x = \frac{a}{2}\left(x - \frac{\lambda}{a}\right)^2 - \left(\frac{\lambda}{a}\right)^2. \qquad (12.14)$$

Thus inspired, let us extend this concept and shift the boson operators by adding a constant to each. The shift is opposite for the states 1 and 0, so the new shifted operators differ for these two states,

$$\hat{b}_m^{(1,0)} = \hat{b}_m \pm \frac{1}{2}\lambda_m. \qquad (12.15)$$

For each state, we express the old operators in terms of new ones and substitute them into
(12.12). We check that the resulting Hamiltonian contains no linear terms provided

$$\lambda_m = \frac{2q_0 C_m}{\omega_m}, \tag{12.16}$$

and reads

$$\hat{H}_{1,0} = \sum_m \hbar\omega_m \hat{b}_m^{(1,0)\dagger} \hat{b}_m^{(1,0)} + \text{const.} \tag{12.17}$$

It is important to recognize that these shifted operators satisfy the same boson commutation
relations as the original operators. We thus know how to deal with both the $\hat{H}_1$ and $\hat{H}_0$ given
above. The eigenstates of $\hat{H}_1$ are $|\{n_m\}_1\rangle$ and those of $\hat{H}_0$ are $|\{n_m\}_0\rangle$, both corresponding to
a set of bosonic occupation numbers $\{n_m\}$. These sets of states, however, are not the same
for 0 and 1 since the annihilation and creation operators have been shifted. Let us find the
relation between the (bosonic) vacua in the two states, $|0_1\rangle$ and $|0_0\rangle$. By definition of the
vacuum,

$$\hat{b}_m^{(0)}|0_0\rangle = 0, \tag{12.18}$$

one can extract no boson from it. We can now see how the bosonic vacuum of one of the
states, say 0, looks in terms of the annihilation operators associated with the *other* state 1,

$$\hat{b}_m^{(1)}|0_0\rangle = \lambda_m|0_0\rangle. \tag{12.19}$$

But wait a second, we have seen this equation before, in Chapter 10 (see (10.17))! We then
found that a coherent state $|\psi\rangle$ for a certain bosonic mode is defined as the eigenstate of
the annihilation operator of this mode,

$$\hat{b}|\psi\rangle = \lambda|\psi\rangle. \tag{12.20}$$

As we know, this coherent state can be written as a superposition of states with a certain
number of bosons n in the mode,

$$|\psi(\lambda)\rangle = e^{-\frac{1}{2}|\lambda|^2} \sum_n \frac{\lambda^n}{\sqrt{n!}}|n\rangle, \tag{12.21}$$

the average number of bosons being $\bar{N} = |\lambda|^2$. Taking all modes into account we come to
an important understanding: one of the vacua is a coherent state with respect to the other,
and the shifts λ_m determine the parameters of this coherent state. The resulting vacuum of
the one state is thus in the other state a direct product of coherent states in each mode,

$$|0_0\rangle = \prod_m |\psi_m(\lambda_m)\rangle. \tag{12.22}$$

This can be expanded as a sum over the basis vectors of the Fock space, those being labeled
by $\{n_m\}$,

$$|0_0\rangle = \sum_{\{n_m\}} \left(\prod_m e^{-\frac{1}{2}|\lambda_m|^2} \frac{(\lambda_m)^{n_m}}{\sqrt{n_m!}} \right) |\{n_m\}_1\rangle. \tag{12.23}$$

12.4 Shake-up and *P(E)*

If we let the system relax while in the state 0, it finally gets to the ground state $|0_0\rangle$. Let us (again just with our hands) suddenly flip the value of q_0 from q_0 to $-q_0$. This is called a *shake-up*. If we manage to do this shake-up quickly enough, the wave function of the system does not change and is still $|0_0\rangle$. However, this is no longer the ground state. By the shake-up, we have brought the Hamiltonian of the system to that of the state 1, while the system is still in the ground state of $\hat{H}_0$. We should thus reference all states with respect to $\hat{H}_1$, and the wave function after the shake-up has become a superposition of the ground and excited states of $\hat{H}_1$ with the coefficients given by (12.23).

In simple classical terms, a shake-up supplies some energy to the environment, as illustrated in Fig. 12.2(a). Flipping q_0 to $-q_0$ is equivalent to shifting the coordinate of the environmental oscillators over a distance $|\lambda_m|$. A particle that initially is in its ground state (solid lines) does not change its position in the course of the shake-up but does acquire some energy and ends up in an excited state (dashed lines).

In the context of our "damped qubit" we can appreciate the concept of the shake-up as follows. Performing the shake-up corresponds to suddenly flipping the state of the qubit from 0 to 1 with our hands (not worrying about an eventual energy difference ε). We found that in the course of this flip, the state of the environment changes from a vacuum to the coherent state (12.23). The modulo squares of the coefficients of the components $|n_{m,1}\rangle$ in (12.23) thus give the probabilities of having supplied the energy $n_m \hbar \omega_m$ to the environment during the flip. If one of these possible energies exactly matches the energy difference ε of the qubit (and of course assuming now that there is a finite coupling t_q between $|0\rangle$ and $|1\rangle$), then there is a finite chance for the qubit to make a transition from its excited state to its ground state while transferring the energy ε to the environment.

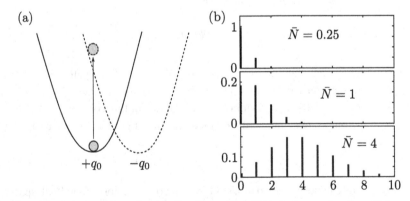

Fig. 12.2 (a) The effect of the shake-up on the environmental oscillators. The solid(dashed) curve gives the oscillator potential before(after) the shake-up. (b) For a single-mode environment, *P(E)* consists of a series of equidistant peaks corresponding to absorption of *n* quanta (12.25).

To start with, let us consider a single bosonic mode m and estimate the probability of supplying energy E to it in course of the shake-up. We start by writing the partial probabilities p_n to excite n bosons in the mode,

$$p_n = |\langle 0_0|n_{m,1}\rangle|^2 = e^{-\bar{N}_m}\frac{\bar{N}_m^n}{n!}, \qquad (12.24)$$

where the average number of bosons $\bar{N}_m = |\lambda_m|^2 = |2q_0 C_m/\omega_m|^2$ depends on the mode index m. We see that the probability distribution for n is Poissonian, the same as we have encountered when working with coherent states in Chapter 10. The total probability distribution $P_m(E)$ in terms of energy consists for this single mode of a series of equidistant δ-peaks, each of the peaks corresponding to absorption of an integer number of quanta during the shake up,

$$P_m(E) = \sum_n e^{-\bar{N}}\frac{\bar{N}^n}{n!}\delta(E - n\hbar\omega_m). \qquad (12.25)$$

Very complex things in dissipative quantum mechanics are easier to understand if we now inspect the above simple expression for a single-mode $P_m(E)$ in two different limits (see also the plots in Fig. 12.2b). If $\bar{N} \ll 1$, the most probable thing to happen is to excite no bosons at all. The shake-up thus would most likely occur without any dissipation. If, however, there are bosons excited, the probability is the largest to excite precisely one boson. Exciting a larger number of bosons quickly becomes very unlikely with increasing this number. The opposite limit is $\bar{N} \gg 1$. In this case, the function $P_m(E)$ is centered around $\bar{N}$ with a spread $\simeq \sqrt{\bar{N}}$, and it is highly improbable to excite more bosons. Importantly, the probability of exciting none or few bosons is also strongly suppressed.

As we know, an environment in general does not consist of one single mode, but usually rather infinitely many modes that together form a continuous spectrum. The total function $P(E)$ is thus contributed to by all modes and all possible sets of boson numbers $\{n_m\}$ in the modes. Summing them all up gives

$$P(E) = \sum_{\{n_m\}} e^{-\sum_m \bar{N}_m}\left(\prod_m \frac{(\bar{N}_m)^{n_m}}{n_m!}\right)\delta\left(E - \sum_m n_m\hbar\omega_m\right). \qquad (12.26)$$

This cumbersome expression may be reduced to a much more comprehensive form, which we do in the following sections. However, before doing this, we take a short break to understand a concept of fundamental significance.

12.5 Orthogonality catastrophe

Let us first concentrate on a simple quantity: the probability of exciting strictly no bosons during the shake-up and thus having a "quiet" shake-up, without disturbing the environment and without any dissipation. Including the qubit in the picture again, this corresponds to the "willingness" of the environment to allow for elastic transitions, that is, transitions between $|0\rangle$ and $|1\rangle$ when $\varepsilon = 0$. The probability of a quiet shake-up is the product of the probabilities of emitting no bosons in each mode, and therefore reads

$$P(0) = e^{-\sum_m \bar{N}_m}. \tag{12.27}$$

For any finite number of boson modes this probability stays *finite*. However, this may change if we go to the continuous limit. The integral corresponding to the sum over modes in (12.27) could diverge, resulting in a strictly zero probability of a quiet shake-up. This situation is called the *orthogonality catastrophe* and was first described by P. W. Anderson in 1967. The point concerning orthogonality is that $P(0)$ is just the square of the overlap between the two vacua $\langle 0_1 | 0_0 \rangle$. If $P(0) = 0$, the overlap vanishes and the two vacua must be orthogonal. Remember, orthogonality of two quantum states implies that these states in a way belong to two different worlds: they neither know nor care about each other, do not mix and no (elastic) tunneling is possible between them. This would strictly forbid elastic transitions in any system embedded in such an environment, which does not sound very sensible.

We can use the relation (12.9) to express the probability $P(0)$ in terms of a physical quantity, the frequency-dependent dynamical susceptibility $\chi(\omega)$,

$$\sum_m \bar{N}_m = \sum_m \left| \frac{2q_0 C_m}{\omega_m} \right|^2 = \int_0^\infty d\omega \sum_m \delta(\omega - \omega_m) |C_m|^2 \frac{4q_0^2}{\omega^2} \\
= \int_0^\infty d\omega \frac{4q_0^2}{\pi \hbar \omega^2} \operatorname{Im} \chi(\omega). \tag{12.28}$$

This integral could diverge at either the upper or the lower limit. The upper ("ultraviolet") divergence is usually irrelevant since it can be removed by introducing an upper cut-off frequency as we have several times in previous chapters. The divergence at low frequencies, however, is generally impossible to cut. We see that the integral diverges and the orthogonality catastrophe takes place only if $\operatorname{Im} \chi(\omega)$ approaches zero at $\omega \to 0$ slower than $|\omega|$.

12.6 Workout of $P(E)$

Let us now bring the expression (12.25) derived for the function $P(E)$ to a more convenient and instructive form. The idea is to introduce an extra integration to write the delta function in (12.26) as an integral over an exponent,[1]

$$\delta(E - \hbar\omega n) = \int \frac{dt}{2\pi} \exp\{i(E - \hbar\omega n)t\}. \tag{12.29}$$

---

[1] To avoid excessive use of $\hbar$, the variable t here, although it looks like time, has the dimension of inverse energy.

Philip Warren Anderson (b. 1923)
Shared the Nobel Prize in 1977 with Sir Nevill Mott and John H. van Vleck for "their fundamental theoretical investigations of the electronic structure of magnetic and disordered systems."

A statistical citation analysis claimed that Philip Anderson is the most creative scientist in the world, which in fact may be a rare case of statistics telling the truth. In the field of quantum condensed matter theory, Anderson has an unbelievably long record: an explanation of the Josephson effect, the Anderson orthogonality catastrophe, Anderson localization, the Anderson model for a ferromagnetic impurity in a metal, a novel approach to the Kondo model, theories of spin glasses, high-temperature superconductivity, and many other works. In fact, even economics has not evaded his grasp.

Fellow scientists call his works "prophetic" since his ideas are not only original, but their comprehension requires thorough attention at the level of single words and strongly developed interpretation skills. Yet he is a very successful author. The title of his famous popular article on many-body physics, "More is different," has achieved a status of mantra.

Anderson had grown up in Urbana-Champaign in a professor's family. However, unlike many other US Nobel Prize winners, he had never been affiliated with this university. As a war-time undergraduate, he developed antennas for Naval Research. This left him with a life-long admiration for electric equipment and engineers. For most of his career, from 1949 till 1984, he was employed by Bell Laboratories. He went to Princeton only when business and fundamental research finally stopped getting along. His public stands are few but memorable: he fiercely opposed President Ronald Reagan's *Star Wars* project.

The resulting exponent can be straightforwardly summed up over n_m,

$$
\begin{aligned}
P(E) &= \sum_{\{n_m\}} \delta\left(E - \sum_m n_m \hbar\omega_m\right) e^{-\sum_m \bar{N}_m} \prod_m \frac{(\bar{N}_m)^{n_m}}{n_m!} \\
&= \int \frac{dt}{2\pi} \sum_{\{n_m\}} e^{i(E - \sum_m n_m \hbar\omega_m)t} \, e^{-\sum_m \bar{N}_m} \prod_m \frac{(\bar{N}_m)^{n_m}}{n_m!} \\
&= \int \frac{dt}{2\pi} e^{iEt} \prod_m \sum_{n_m} \frac{(\bar{N}_m)^{n_m}}{n_m!} e^{-in_m \hbar\omega_m t - \bar{N}_m} \\
&= \int \frac{dt}{2\pi} e^{iEt} \prod_m e^{\bar{N}_m(e^{-i\hbar\omega_m t}-1)} = \int \frac{dt}{2\pi} e^{iEt} e^{\sum_m \bar{N}_m(e^{-i\hbar\omega_m t}-1)} \\
&= \int \frac{dt}{2\pi} e^{iEt} e^{J(t)-J(0)},
\end{aligned}
\qquad (12.30)
$$

where in the last line we define[2]

$$J(t) \equiv \sum_m e^{-i\omega_m t} \bar{N}_m = \int_0^\infty d\omega\, e^{-i\omega t} \frac{4q_0^2}{\pi \hbar \omega^2} \mathrm{Im}\, \chi(\omega). \tag{12.31}$$

Finally, let us try to express the whole function $P(E)$ in terms of the leading order Fourier component $J(t)$. We start by defining the function $P_1(E)$,

$$P_1(E) = \frac{4q_0^2}{\pi E^2} \mathrm{Im}\, \chi(E/\hbar), \tag{12.32}$$

which physically corresponds to the effective density of states characterizing the environment. In the limit of small q_0, that is, weak coupling between the qubit and the environment, the shake-up can be treated by perturbation theory in q_0. The leading perturbation corrections come from processes involving the excitation of a single boson only in the course of the shake-up. In this case, $P_1(E)dE$ is the (small) probability of emitting this boson in the energy interval $(E, E + dE)$. The perturbation theory is applicable provided $P_1(E) \ll E^{-1}$ and fails otherwise. We conditionally refer to $P_1(E)$ as the single-boson probability regardless of the applicability of perturbation theory. We note that $P(E)$ is a probability distribution normalized in the usual way, $\int_0^\infty dE P(E) = 1$, while it follows from the definition of $J(t)$ that

$$\int_0^\infty dE\, P_1(E) = \sum_m \bar{N}_m \equiv \bar{N}, \tag{12.33}$$

with $\bar{N}$ being the average number of bosons excited. As we know, this number is related to the probability of a quiet shake-up, $P(0) = \exp\{-\bar{N}\}$ and diverges under conditions of the orthogonality catastrophe.

In terms of $P_1(E)$, the function $P(E)$ finally reads

$$P(E) = \int \frac{dt}{2\pi\hbar} e^{-\frac{i}{\hbar}Et} \exp\left\{ \int_0^\infty dE'\, P_1(E')\left[e^{\frac{i}{\hbar}E't} - 1 \right] \right\}. \tag{12.34}$$

In the limit of small q_0, one can expand this expression in powers of q_0, that is, in powers of $P_1(E)$. A term of nth power in $P_1(E)$ is a contribution of a process involving excitation of n bosons. The contributions of the first three orders read

$$P(E) = \delta(E) + \left[-\delta(E)\bar{N} + P_1(E) \right]$$
$$+ \frac{1}{2}\left[\tfrac{1}{2}\delta(E)\bar{N}^2 - 2\bar{N}P_1(E) + \int dE_1 P_1(E_1)P_1(E - E_1) \right], \tag{12.35}$$

the orders being grouped by brackets. The zeroth order gives a δ-peak corresponding to the absence of dissipation. The first order encompasses: (i) a correction to the magnitude of this δ-peak and therefore to $P(0)$, and (ii) the probability of the excitation of a single boson with energy E. The second order gives the next order correction to $P(0)$, a correction to the single-boson probability, and a contribution coming from excitations of two bosons with total energy E.

Control question. Can you reproduce (12.35) starting from (12.34)?

---

[2] Here we restore the dimension of t to time.

12.7 Transition rates and delocalization

So far we have considered the process of the shake-up and figured out the probability distribution $P(E)$ to transfer the energy E to the environment in course of a shake-up. This shake-up situation was, however, a result of *extrinsic* action, since we started by shaking the system with our hands. It is important to recognize that the very same $P(E)$ is also useful to understand the *intrinsic* properties of our system, such as spontaneous transitions between the qubit states and delocalization of the qubit ground state.

Let us do the transitions first. We concentrate on two states, $|0\rangle$ and $|1\rangle$, separated by the energy ε and we consider now a small but finite tunneling matrix element t_q. We start with our qubit in the excited state and expect it to transit to the ground state. The corresponding transition rate is evaluated with Fermi's golden rule, generally written

$$\Gamma_i = \frac{2\pi}{\hbar} \sum_f |\langle f|\hat{H}'|i\rangle|^2 \delta(E_i - E_f). \tag{12.36}$$

In our case, the matrix element $\langle f|\hat{H}'|i\rangle$ is proportional to t_q, the different final states are those of the environment, and the energy to be dissipated to the environment is ε. The matrix element is also proportional to the overlap between the initial and final states of the Hamiltonians $\hat{H}_{0,1}$, similarly to our computation of the function $P(E)$. No wonder that the transition rate reads

$$\Gamma(\varepsilon) = \frac{2\pi t_q^2}{\hbar} P(\varepsilon). \tag{12.37}$$

The transition rate Γ thus measures the probability of giving energy ε to the environment in the course of the transition. If one-boson processes dominate, this probability is the one-boson probability $P_1(\varepsilon)$,

$$\Gamma(\varepsilon) = \frac{2\pi t_q^2}{\hbar} P_1(\varepsilon). \tag{12.38}$$

In this expression, the δ-peak at zero energy does not contribute to the transition rate since the transition is accompanied by a finite energy transfer.

The delocalization probability is evaluated along similar lines and can also be expressed in terms of $P(E)$. In first order in t_q, the ground state $|0_0\rangle$ acquires a correction that includes all possible states $|\{n_m\}_1\rangle$ in the qubit state 1,

$$|g\rangle = |0_0\rangle + \sum_{\{n_m\}} \psi_{\{n_m\}} |\{n_m\}_0\rangle, \quad \text{where} \quad \psi_{\{n_m\}} = \frac{t_q \langle 0_0|\{n_m\}_1\rangle}{\varepsilon + \sum_m \hbar n_m \omega_m}. \tag{12.39}$$

The delocalization probability in the leading order in t_q is the sum of the partial probabilities of being in any of these states, that is $p_{\text{dec}} = \sum_{\{n_m\}} |\psi_{\{n_m\}}|^2$. Let us note that

$$P(E) = \sum_{\{n_m\}} |\langle 0_0|\{n_m\}_1\rangle|^2 \delta \left(E - \sum_m \hbar n_m \omega_m \right), \tag{12.40}$$

and with this, we can express p_{dec} in terms of $P(E)$ as an integral over energies,

$$p_{\text{dec}} = t_q^2 \int_0^\infty dE \frac{P(E)}{(\varepsilon + E)^2}. \tag{12.41}$$

If there is no dissipation, $P(E) = \delta(E)$ and the expression reduces to the result for an ideal qubit, $p_{\text{dec}} = t_q^2/\varepsilon^2$.

We now proceed by using the above results to provide a broad overview of all possible situations encountered in the context of the spin–boson model: we present a classification of all possible environments.

12.8 Classification of environments

There are two main types of $P(E)$, the difference between them is visualized in Fig. 12.3. For the first type of $P(E)$ (Fig. 12.3(a)), the orthogonality catastrophe takes place so that $P(0) = 0$ and $P(E)$ is small at low energies. Thereby, $P(E)$ bears a similarity with the single-mode case for $\bar{N} \gg 1$. The function $P(E)$ has a peak at large energies, this peak becoming relatively narrow in the classical limit $\bar{N} \gg 1$.

For the second type of $P(E)$ (Fig. 12.3(b)), no orthogonality catastrophe is present. The probability $P(0)$ of a quiet shake-up remains finite, and in the function $P(E)$ this is manifested as a characteristic δ-peak at zero energy. A tail following the peak is mainly due to single-boson emissions, $P(E) \approx P_1(E)$, the contributions involving more bosons are increasingly suppressed. One can see the similarity with the single-mode $P(E)$ at $\bar{N} \ll 1$.

The borderline between these two types is rather unusual. At this borderline, the function $P(E)$ does not have a δ-peak at zero, so the orthogonality catastrophe takes place. However, $P(E)$ is more intensive at lower energies in comparison with the first type: it assumes a power-law dependence. The exponent of the power law may be negative, in this case $P(E)$ even increases with decreasing energy.

Control question. What is the most negative exponent possible for $P(E)$?

(a) 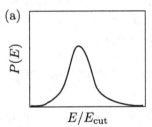 (b)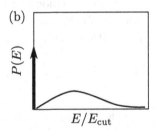

Fig. 12.3   Two types of $P(E)$. (a) Orthogonality catastrophe, $P(E) = 0$ at $E = 0$. (b) No catastrophe, a δ-peak at $E = 0$. One notices the similarity of the shape of $P(E)$ for these two types with the $P(E)$ for a single-mode environment for $\bar{N} \gg 1$ and $\bar{N} \ll 1$, respectively (cf. the plots in Fig. 12.2).

Sir Anthony James Leggett (b. 1938)
Shared the Nobel Prize in 2003 with Alexei Abrikosov and Vitaly Ginzburg for "pioneering contributions to the theory of superconductors and superfluids."

Anthony Leggett entered the University of Oxford in 1955 with the intention of gaining the famous Greats Degree: that is in classics. He wrote: "I certainly do feel the philosophy component of the degree, at least, has helped to shape the way at which I look at the world and in particular at the problems of physics."

He thanks the Soviet *Sputnik* project for his career change: the re-evaluation of the Western education system that followed made available fellowships for humanity students to switch to natural sciences. Why the switch? "I wanted the possibility of being wrong without being stupid – of being wrong, if you like, for interesting and non-trivial reasons". Dirk ter Haar was brave enough to supervise this unusual student with no formal background in physics, and postdoctoral positions in Urbana-Champaign, Kyoto, and Harvard shaped Leggett as a scientist. Soon after leaving the University of Sussex, Leggett published a series of spectacular works on the theory of superfluid ^{3}He and ^{4}He that eventually brought him the Nobel Prize.

Since the beginning of the 1980s Leggett has been back in snowy Urbana-Champaign, a place he once swore never to come back to. His scientific interests have shifted now to the foundations of quantum mechanics, conforming to his philosophical background. He has been particularly interested in the possibility of using special condensed matter systems, such as Josephson devices, to test the validity of the extrapolation of the quantum formalism to the macroscopic level. This interest has led to a considerable amount of pioneering technical work on the application of quantum mechanics to collective variables and in particular on ways of incorporating dissipation into the calculations. He put forward the Caldeira–Leggett model of bosonic environments and proposed the classification of dissipative environments outlined in this chapter.

With this notion, we are ready to outline the classification of all possible environments, proposed by Sir Anthony Leggett in 1981. The starting point of the classification is the energy dependence of the *single-boson* probability at small energies. By virtue of (12.32) it is related to the frequency dependence of the dynamical susceptibility at low frequencies. To avoid possible problems related to convergence of the integrals at high energies, let us introduce a typical cut-off energy E_{cut} such that $P_1(E)$ vanishes at $E \gg E_{cut}$. We assume a power-law dependence of $P_1(E)$ at $E \to 0$,

$$P_1(E) \propto E^s \exp\{-E/E_{cut}\}, \tag{12.42}$$

with an exponent s.

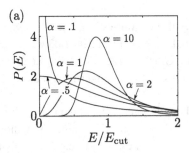
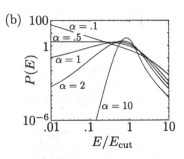

Fig. 12.4 The borderline $s = -1$: an Ohmic environment. (a) The function $P(E)$ is plotted for various values of α. (b) The same plots in log-log coordinates, revealing the power-law dependence at low energies.

If $s < -1$, the environment is called *subohmic*. The integral of $P_1(E)$ over energies, and thus $\bar{N}$ (see (12.33)), diverges at small energies resulting in an orthogonality catastrophe. The function $P(E)$ in this case is thus of the first type shown in Fig. 12.3. If $s > -1$, the environment is *superohmic* and no orthogonality catastrophe takes place, $P(E)$ being thus of the second type. A special case is the so-called *Ohmic* environment, which corresponds to the borderline $s = -1$ between the two types (Fig. 12.4). We distinguish this special case because of its practical importance: the Ohmic case corresponds to a simple instantaneous friction model ($\gamma(\omega) = \gamma$) for a particle in a potential, and to a frequency-independent impedance $Z(\omega) = Z$ of an environment for electrons tunneling in an electric circuit. In terms of the dynamical susceptibility, $s = -1$ corresponds to Im $\chi(\omega) \propto \omega$. An interesting point concerning the Ohmic environment is that the corresponding $P(E)$ assumes a power-law with an exponent that depends on the coupling strength to the environment, that is, on the intensity of single-boson processes,

$$P_1(E) \propto \frac{2\alpha}{E}, \quad \text{so} \quad P(E) \propto E^{2\alpha-1} \quad \text{for} \quad s = -1. \tag{12.43}$$

Control question. Can you derive the latter relation from (12.34)?

For example, for electrons tunneling in an electric circuit characterized by an impedance Z, one finds $\alpha = e^2 Z / 2\pi\hbar$. Let us in the following sections briefly describe the specifics of these three different classes of environment.

12.8.1 Subohmic

A subohmic environment suppresses the tunneling transitions between qubit states $|0\rangle$ and $|1\rangle$ in the most efficient way. One can say that a qubit embedded in a subohmic environment loses its quantum properties and behaves very much like a classical binary memory cell. The transition rate vanishes at zero energy and is exponentially small at low energies. To derive this from (12.34), we note that the integral over E' converges at low energies resulting in a growing function of t,

$$\int_0^\infty dE' \, P_1(E') \left[e^{-\frac{i}{\hbar}E't} - 1 \right] \propto t^{-s-1}, \tag{12.44}$$

the exponent $-s - 1$ being positive. The integral over E is approximated with the saddle-point method, yielding

$$\Gamma(\varepsilon) \propto P(\varepsilon) \propto \exp\{-\varepsilon^{-\gamma}\}, \quad \text{with} \quad \gamma \equiv \frac{1-s}{2+s} > 0. \qquad (12.45)$$

Control question. Do you think that subohmic environments with $s < -2$ can exist? Give the arguments.

It looks as if the environment provides an additional tunnel barrier separating the two states. This barrier becomes impenetrable at $\varepsilon = 0$ and gives an exponentially small tunneling probability at non-zero energies.

Control question. A realization of a subohmic environment for electron tunneling is a long RC line, corresponding to $s = -3/2$. At sufficiently low voltage bias, the typical energy differences for electrons tunneling are of the order of the temperature. What can you say about the temperature dependence of the tunneling conductance?

Extra support to picture this as a classical memory cell is given by the delocalization probability. The integral in (12.41) is contributed to by energies $E \sim E_{\text{cut}}$. Not only the result of integration does not diverge at small ε, but we can completely neglect the energy splitting dependence provided $\varepsilon \ll E_{\text{cut}}$. The resulting delocalization probability is small, $p_{\text{dec}} \simeq t_q^2/E_{\text{cut}}^2$, and, in sharp contrast with a dissipationless qubit, stays small whatever we do with ε.

12.8.2 Ohmic

For an Ohmic environment, $P(E)$ is not exponentially small and may even increase with decreasing E. However, transitions with a given energy loss generally involve many bosons. We can see this from the fact that $P(E)$ is suppressed in comparison with $P_1(E)$,

$$\frac{P_1(E)}{P(E)} \propto E^{-2\alpha} \gg 1, \qquad (12.46)$$

as follows from (12.43). The latter ratio gives an estimate for the number of bosons involved in dissipating E.

Let us inspect the transition rate as a function of the energy splitting ε,

$$\Gamma(\varepsilon) = \frac{2\pi t_q^2}{\hbar} \left(\frac{\varepsilon}{E_{\text{cut}}} \right)^{2\alpha - 1}. \qquad (12.47)$$

Transitions between the two states introduce an energy uncertainty $\simeq \hbar\Gamma$. Let us compare this uncertainty with the energy itself. A slow decay rate produces a "good" state, where the energy uncertainty is much smaller than the splitting between the levels, $\hbar\Gamma(\varepsilon)/\varepsilon \ll 1$. In other words, a "good" excited state lives long enough to prove that it is a state with an energy distinct from that of the ground state. A "bad" state can hardly be used as an excited state of a quantum system, for instance, in the context of quantum information processing: it decays before we are able to use it.

As usual, let us concentrate on small energy differences. We see that $\alpha > 1$ implies a good state, where the uncertainty decreases when we let $\varepsilon \to 0$. As to $\alpha < 1$, the ratio of uncertainty and energy now increases with decreasing energy, and at sufficiently low energy the state inevitably becomes bad. This signals that something special happens at a coupling strength corresponding to $\alpha = 1$. What precisely?

The answer to this question was given by the German physicist Albert Schmid in 1983. He proved that the qubit states are localized at $\alpha > 1$ and delocalized otherwise. This "localization transition" at $\alpha = 1$ is called the Schmid transition. Let us note that delocalization corresponds to bad states, while good states are localized thus missing an important quantum property. This seems wrong, since a qubit without dissipation is a good state, meaning that it has an (infinitely) slow decay rate, but on the other hand it also exhibits delocalization. To comprehend the situation, let us again look at the integral (12.41), which for the case of an Ohmic environment can be rewritten as

$$p_{\text{dec}} = t_q^2 \int_0^\infty dE \frac{E^{2\alpha-1}}{(\varepsilon + E)^2} e^{-E/E_{\text{cut}}}. \tag{12.48}$$

If $\alpha < 1$, the integral converges at energies $\simeq \varepsilon$ and the delocalization probability

$$p_{\text{dec}} \simeq \frac{t_q^2}{\varepsilon^2} \left(\frac{\varepsilon}{E_{\text{cut}}} \right)^{2\alpha} \tag{12.49}$$

increases with decreasing ε, reaching values $\simeq 1$ at sufficiently small ε. Otherwise, the integral converges at $\varepsilon \simeq E_{\text{cut}}$. In this case, as in the case of the subohmic environment, we can disregard the energy dependence and demonstrate that the delocalization probability remains small everywhere, $p_{\text{dec}} = t_q^2/E_{\text{cut}}^2$.

12.8.3 Superohmic

For a superohmic environment ($s > -1$), the transition rate is dominated by single-boson processes, $P(\varepsilon) \approx P_1(\varepsilon)$ at $\varepsilon \neq 0$. The transition rate is a power law at low energies

$$\Gamma(\varepsilon) = \frac{2\pi t_q^2}{\hbar} P_1(\varepsilon) \propto \varepsilon^s. \tag{12.50}$$

Repeating the reasoning for the Ohmic case, we conclude that the excited state in the superohmic regime is bad provided $-1 < s < 1$, and good when $s > 1$.

In contrast to the Ohmic regime, this does not imply any localization transition at $s = 1$. In fact, the qubit states are always delocalized in a superohmic environment. The origin of the delocalization is the δ-peak of $P(E)$ at zero energy, which is absent for the Ohmic environment. This peak provides the possibility to tunnel without dissipation, a feature which was absent in the previously considered environments. The integral (12.41) for the delocalization probability is mainly contributed to by this δ-peak. Let us single out this contribution to the integral, rewriting it as

$$p_{\text{dec}} = P(0)\frac{t_q^2}{\varepsilon^2} + \int_\eta^\infty t_q^2 dE \frac{P_1(E)}{(\varepsilon + E)^2}. \tag{12.51}$$

Table 12.1 Overview of the properties of the different classes of environment.

Environment	Delocalization	Good state	δ-peak
Superohmic, $s > 1$	Yes	Yes	Yes
Superohmic, $-1 < s < 1$	Yes	No	Yes
Ohmic, $s = -1, \alpha < 1$	Yes	No	No
Ohmic, $s = -1, \alpha > 1$	No	Yes	No
Subohmic $-2 < s < -1$	No	Yes	No

The first term looks similar to the delocalization probability of a dissipationless qubit, signaling the delocalization at small energies. The second term gives for $s > 1$ a small ε-independent contribution similar to that in subohmic case. For $s < 1$, this contribution is $\propto \varepsilon^{s-1}$. In both cases, the first term dominates at small energies.

The results obtained for all environments are summarized in Table 12.1. From the highest to the lowest row, the qubit changes its behavior from standard quantum to that of a classical binary memory cell.

12.9 Vacuum as an environment

The spin–boson model as formulated above looks rather specific, since the interaction with the environment is chosen such as not to cause transitions between the qubit states. Transitions between the states could be caused by variables of the environment coupling to $\hat{\sigma}_x$ or $\hat{\sigma}_y$ in the qubit basis. In fact, in a way it looks more natural to work in the basis where the qubit Hamiltonian is diagonal while the transitions are caused by dissipation. For such a model, we would have

$$\hat{H}_q = \frac{E}{2}\hat{\sigma}_z \quad \text{and} \quad \hat{H}_{\text{coup}} = -q_0\hat{\sigma}_x\hat{F}. \tag{12.52}$$

However, this model also fits in the spin–boson model presented above. To show this explicitly, we exchange $\hat{\sigma}_z \leftrightarrow \hat{\sigma}_x$, and with this we reproduce the spin–boson model with $\varepsilon = 0$ and $t_q = E/2$. Under these circumstances, we cannot make use of the limit $t_q \ll \varepsilon$ as we did above. However, all qualitative results, including the classification of the environment, do not depend on this limit and are thus valid for $\varepsilon = 0$ as well. The only point is that we need to modify expression (12.37) for the transition rate. The energy change is now E rather than ε, and $t_q = E/2$. Assuming that single-boson processes dominate, we obtain

$$\Gamma(E) = \frac{\pi E^2}{2\hbar}P_1(E). \tag{12.53}$$

This expression thus provides a way to determine the function $P_1(E)$ from a known or measured energy dependence of the transition rate.

Let us make here a connection with the material of Chapter 9, the part on emission and absorption of irradiation by atoms. We have learned that the electromagnetic waves in vacuum are bosons. Therefore, the vacuum by itself can be seen as a dissipative environment that induces transitions between the atomic states. Let us investigate what kind of environment the electromagnetic vacuum is. We found in Section 9.1 an expression for the transition rate between two atomic states A and B,

$$\Gamma(\hbar\omega) = \frac{4\alpha\omega^3}{3c^2}|\mathbf{r}_{BA}|^2, \tag{12.54}$$

with $\mathbf{r}_{AB}$ being the non-diagonal matrix element of the dipole moment between the states, and $\alpha \approx 1/137$ the fine structure constant. This is a single-boson transition, so it makes sense to compare it with expression (12.53). Thereby we can obtain the function $P_1(E)$ for the vacuum as felt by the atomic states,

$$P_1(E) = \frac{8\alpha}{3\pi}\left(\frac{|\mathbf{r}_{BA}|}{c\hbar}\right)^2 E. \tag{12.55}$$

We can thus conclude that the atoms see the vacuum as a superohmic environment with $s = 1$. This ensures that the atomic states are "good" quantum states, not subject to the orthogonality catastrophe, and suitable for quantum manipulation.

However, the transition rate is small because the atoms are fairly small in comparison with the photon wavelength at energy E, so that the dipole moment of the transition is small. Larger values of the dipole moment obviously increase the coupling to the environment, the dissipation, and thus the transition rate. We can think of antennas, those are specially made to emit electromagnetic radiation in the most efficient way. The dipole moment in antennas is only limited by the photon wavelength c/ω. If we replace the atomic dipole moment with this larger dipole moment,

$$|\mathbf{r}_{BA}|_{\max} \sim \frac{c}{\omega} \quad \Rightarrow \quad P_1(E) \sim \frac{\alpha}{E}, \tag{12.56}$$

we understand that the effective environment becomes Ohmic, $s = -1$. The anomalous power is small, of the order of $\alpha \approx 1/137$. We know that having an Ohmic environment leads in principle to the orthogonality catastrophe. This is, however, hardly relevant for atoms since they feel a superohmic environment, but it may become relevant for charges moving at larger distances. However, the orthogonality catastrophe is eventually a minor problem since the fine structure constant is very small, $\alpha \ll 1$. If this had not been the case, we would live in a very different world. We come back to this problem at the end of Chapter 13.

Table 12.2 Summary: Transitions and dissipation

Qubit in a dissipative environment, originates from a modification of the damped oscillator

$U(x)$ is now *double* well potential → two minima with localized states $|0\rangle$ and $|1\rangle$

classically: particle freezes in either 0 or 1

quantum mechanically: stationary state of particle can be superposition $\alpha|0\rangle + \beta|1\rangle$

delocalization probability: chance to find ground state qubit in higher energy minimum

$$\text{if}\quad \hat{H} = \tfrac{1}{2}\varepsilon\hat{\sigma}_z + t_q\hat{\sigma}_x, \quad \text{then}\quad p_{\text{dec}} \approx (t_q/\varepsilon)^2 \quad \text{for}\quad t_q \to 0$$

Spin–boson model, couple a qubit to boson bath

model: $$\hat{H} = \frac{\varepsilon}{2}\hat{\sigma}_z + t_q\hat{\sigma}_x + \sum_m \hbar\omega_m \hat{b}_m^\dagger \hat{b}_m - q_0\hat{\sigma}_z\hbar\sum_m (C_m\hat{b}_m^\dagger + C_m^*\hat{b}_m),$$

with $\text{Im}\,\chi(\omega) = \pi\hbar\sum_m |C_m|^2\,\delta(\omega - \omega_m)$ (see Chapter 11)

problem: cannot be solved for general t_q and ε

Spin–boson model, limit of $t_q \to 0$, i.e. no tunneling

particle freezes in either $|0\rangle$ or $|1\rangle$, correspondingly $\hat{H}_{1,0} = \sum_m \left\{ \hbar\omega_m\hat{b}_m^\dagger\hat{b}_m \mp q_0\hbar(C_m\hat{b}_m^\dagger + C_m^*\hat{b}_m) \right\}$

terms linear in $\hat{b}_m^{(\dagger)}$:   shift operators $\hat{b}_m^{(1,0)} = \hat{b}_m \pm \dfrac{q_0 C_m}{\omega_m}$, then $\hat{H}_{1,0} = \sum_m \hbar\omega_m\hat{b}_m^{(1,0)\dagger}\hat{b}_m^{(1,0)}$

consequence: the boson vacuum in one qubit state is a coherent state in the other qubit state

---

Shake-up and $P(E)$

shake-up: start in $|0_0\rangle$ and suddenly flip $q_0 \to -q_0$

$P(E)$:   probability distribution to transfer energy E to the bath during the shake-up

$$P(E) = \sum_{\{n_m\}} e^{-\sum_m \bar{N}_m} \left(\prod_m \frac{(\bar{N}_m)^{n_m}}{n_m!} \right) \delta\left(E - \sum_m n_m\hbar\omega_m\right), \quad \text{with}\quad \bar{N} = \left|\frac{2q_0 C_m}{\omega_m}\right|^2$$

$$= \int \frac{dt}{2\pi} e^{-\frac{i}{\hbar}Et} \exp\left\{ \int_0^\infty dE'\, P_1(E')\left[e^{-\frac{i}{\hbar}E't} - 1\right] \right\}, \quad \text{with}\quad P_1(E) = \frac{4q_0^2}{\pi E^2}\,\text{Im}\,\chi(E/\hbar)$$

interpretations:

$P(0) = \exp\{-\sum_m \bar{N}_m\}$, probability of quiet shake-up (elastic transition from $|0\rangle$ to $|1\rangle$))

$P(0) = 0$,   if $\text{Im}\,\chi(\omega) \to 0$ at $\omega \to 0$ slower than $|\omega|$ →   *orthogonality catastrophe* (OC)

$P_1(E)$ is single-boson contribution to $P(E)$: effective density of states of the bath

---

Applications of $P(E)$, we now add a finite tunnel coupling t_q

transitions: $\Gamma_{1\to 0}(\varepsilon) = \dfrac{2\pi t_q^2}{\hbar} P(\varepsilon)$,   delocalization: $p_{\text{dec}} = t_q^2 \displaystyle\int_0^\infty dE\, \frac{P(E)}{(\varepsilon + E)^2}$

Classification of environments, we assume $P_1(E) \propto E^s \exp\{-E/E_{\text{cut}}\}$ →   distinguish in s

subohmic, $-2 < s < -1$:   OC, transitions suppressed as $\Gamma(\varepsilon) \propto \exp\{-\varepsilon^{-\gamma}\}$ with $\gamma \equiv \frac{1-s}{2+s} > 0$,

tiny delocalization $p_{\text{dec}} \simeq t_q^2/E_{\text{cut}}^2$ →   almost classical behavior

Ohmic, $s = -1$: $P_1(E) \propto 2\alpha/E$, which implies $P(E) \propto E^{2\alpha-1}$ →   distinguish in α

all α: OC, transitions $\Gamma(\varepsilon) = \frac{2\pi}{\hbar}t_q^2(\varepsilon/E_{\text{cut}})^{2\alpha-1}$

$\alpha > 1$:   "good" state ($\hbar\Gamma(\varepsilon)/\varepsilon \ll 1$), tiny delocalization $p_{\text{dec}} \simeq t_q^2/E_{\text{cut}}^2$

$\alpha < 1$:   "bad" state, $p_{\text{dec}} \simeq (t_q/\varepsilon)^2(\varepsilon/E_{\text{cut}})^{2\alpha}$ is large at small enough ε

superohmic, $s > -1$:   no OC, transitions $\Gamma(\varepsilon) \propto \varepsilon^s$, "good" state only when $s > 1$,

$p_{\text{dec}} \simeq (t_q/\varepsilon)^2 P(0)$ at small energies

Exercises

1. *Electron tunneling in an electromagnetic environment* (solution included). There is a tunnel junction between two metallic leads L and R. The voltage V applied to the junction results in a difference of chemical potentials $\mu_L - \mu_R = eV$ between the leads. An elementary tunneling process is a transfer of a single electron from a filled level i in the left lead to an empty level j in the right lead. We assume a vanishing temperature and a Fermi distribution of the electrons in both leads. In this case, the tunneling processes can only happen if $E_i < \mu_L$ and $E_j > \mu_R$. If we denote the initial and final state of the process $|0\rangle$ and $|1\rangle$ respectively, the Hamiltonian is similar to that of the spin–boson model,

$$E_i |0\rangle\langle 0| + E_j |1\rangle\langle 1| + t_{ij}\Big\{|0\rangle\langle 1| + |1\rangle\langle 0|\Big\} + e\hat{V}\Big\{|0\rangle\langle 0| - |1\rangle\langle 1|\Big\},$$

where the electromagnetic environment is represented by an operator of fluctuating voltage $\hat{V}$. The corresponding response function is related to the impedance of the environment, $\chi(\omega) = i\omega Z(\omega)$. We need to compute the tunneling current, which is proportional to the number of electrons that tunnel per unit of time.

   a. Compute the tunneling current, disregarding the interaction with the environment and assuming that the quantity

$$A(\varepsilon, \varepsilon') \equiv \sum_{i,j} \delta(\varepsilon - E_i)\delta(\varepsilon' - E_j)|t_{ij}|^2,$$

   does not depend on ε and ε' in the energy interval of interest. Express the tunneling conductance in terms of A.

   b. Let us bring the environment into the picture. Explain how to define $P(E)$ in this case and how to express the tunneling current in terms of $P(E)$ and the tunneling conductance.

   c. Assume a single-mode environment corresponding to $\operatorname{Re} Z(\omega) = Z_0\omega_0\delta(\omega - \omega_0)$. Compute $P(E)$ for this case.

   d. Compute the differential conductance dI/dV for the single-mode case.

2. *Finite temperatures and $P(E)$.* If the temperature of the environment is finite, the qubit can switch spontaneously between its two states. The switching is characterized by two rates, $\Gamma_{0\to 1}$ and $\Gamma_{1\to 0}$. A straightforward but long calculation not given here shows that the rates can be presented in the form

$$\Gamma_{0\to 1} = \frac{t_q^2}{2\pi\hbar}P(\varepsilon, T) \quad \text{and} \quad \Gamma_{1\to 0} = \frac{t_q^2}{2\pi\hbar}P(-\varepsilon, T),$$

where $P(E, T)$ generalizes the relation (12.30) to the case of finite temperature. The concrete expression for $P(E)$ reads

$$P(E, T) = \int \frac{dt}{2\pi} e^{-\frac{i}{\hbar}Et} \exp\left\{\int_{-\infty}^{\infty} dE'\, P_1(E', T)\Big[e^{\frac{i}{\hbar}E't} - 1\Big]\right\}.$$

where

$$P_1(E,T) = \frac{4q_0^2}{\pi E^2} \mathrm{Im}[\chi(E/\hbar)] \frac{1}{1 - \exp(-E/T)}.$$

a. Describe similarities and differences of the above relations and those in the chapter that are valid in the limit of vanishing temperature.

b. Find the general relation between $P(E,T)$ and $P(-E,T)$ *without* using the expression derived under point (a).

c. Concentrate on the limit of high temperatures, such that the integrals in the expressions containing $\mathrm{Re}\,\chi(\omega)$ converge at frequencies much smaller than $k_B T/\hbar$. Derive the expression for $P(E,T)$ in this limit.

d. Give an interpretation of the expression derived under (c) in terms of the thermal fluctuation of the qubit energy splitting.

3. *Transitions in a molecule.* A diatomic molecule can be found in two electron states A and B. The energies of these two states are functions of the interatomic distance a. The equilibrium length of the molecule is different in these two states, $a_B \neq a_A$. We assume that we can approximate the dependence of the energy on a with two parabolas of the same curvature (see Fig. 12.5(a)), $E_{A,B}(a) = -\frac{1}{2}k(a - a_{A,B})^2 + E_{A,B}$, which permits us to map the setup on the spin–boson model. Tunneling between the states is weak, $|t| \ll |E_A - E_B|$. The dynamics of the interatomic distance are those of a weakly damped oscillator with frequency ω_0 and quality factor $Q \gg 1$. Our goal is to find the transition rates assuming $|E_A - E_B| \gg k_B T \gg \omega_0$.

a. Recognize the spin–boson model. Write down the required dynamical susceptibility. Compute the quantum (a_q^2) and thermal (a_T^2) fluctuations of a. How do these compare to $|a_A - a_B|$?

b. Compute the function $P(E)$ and the transition rates using the results of Exercise 2.

c. Find the energy E_c at which two parabolas cross. Interpret the result in terms of the energy differences $E_c - E_A$ and $E_c - E_B$.

4. *Tunneling of an alien atom.* The equilibrium position of an alien atom (black) in a square two-dimensional lattice (white atoms) is in the center of a square (see Fig. 12.5b). The atom can tunnel between equivalent equilibrium positions, as indicated with the arrow. This tunneling is affected by emission or absorption of lattice phonons. The interaction between the alien atom and a lattice atom is given by $U(r)$, r being the distance between

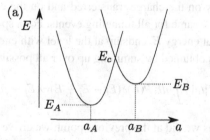
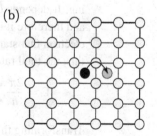

Fig. 12.5 (a) Exercise 3. Energies of molecular states A and B versus interatomic distance a. (b) Exercise 4. Tunneling of an alien atom (black) between equilibrium positions in a two-dimensional square lattice.

the atoms, $U(r) \propto r^{-\beta}$ at $r \to \infty$. We are interested in low-energy phonons only, and for those the wave vector $\mathbf{q}$ by far exceeds the lattice spacing.

a. Compute the coupling term between a phonon mode with a given $\mathbf{q}$ and the discrete position of the alien atom using first-order perturbation theory in U.

b. From this, compute the single-phonon function $P_1(E)$ concentrating on its energy dependence in the limit of small energies. Give a classification of the possible environments depending on the value of β. Is there a possibility to localize the atom?

c. Extend the qualitative analysis to the cases of one- and three-dimensional lattices.

Solutions

1. *Electron tunneling in an electromagnetic environment.*

   a. Without interactions, electron energy is conserved in the course of tunneling. The overall rate of tunneling electron transitions from all levels i is given by Fermi's golden rule

   $$\Gamma = \frac{2\pi}{\hbar} \sum_{i,j} |t_{ij}|^2 \delta(E_i - E_j)\Theta(E_j - \mu_R)\Theta(\mu_L - E_i).$$

   Let us introduce a continuous energy variable ε and use the following identity

   $$\delta(E_i - E_j) = \int d\varepsilon\, \delta(\varepsilon - E_j)\delta(\varepsilon - E_i).$$

   With this,

   $$\Gamma = \frac{2\pi}{\hbar} \int d\varepsilon \sum_{i,j} |t_{ij}|^2 \delta(\varepsilon - E_j)\delta(\varepsilon - E_i)\Theta(\varepsilon - \mu_R)\Theta(\mu_L - \varepsilon)$$

   $$= \frac{2\pi}{\hbar} \int_{\mu_L}^{\mu_R} d\varepsilon\, A(\varepsilon, \varepsilon) = \frac{2\pi}{\hbar} A(\mu_R - \mu_L) = \frac{2\pi e}{\hbar} AV.$$

   Since $I = e\Gamma$ and $I = G_T V$, the tunneling conductance is expressed as $G_T = 2\pi A(e^2/\hbar)$.

   b. $P(E)$ is the probability of emitting the energy E into the environment when tunneling. It depends only on the charge transferred and on the environment properties, and therefore it is the same for all tunneling events. For a specific energy loss E the electron that starts at energy E_i ends up at the level with energy $E_j = E_i - E$. With this, the total rate is obtained by summing up over all possible losses E,

   $$\Gamma = \frac{2\pi}{\hbar} \sum_{i,j} |t_{ij}|^2 \int dE\, P(E)\delta(E_j - E_i + E)\Theta(E_j - \mu_R)\Theta(\mu_L - E_i).$$

   Transforming this as we did at the previous point, we arrive at

   $$I(V) = \frac{G_T}{e} \int_0^{eV} d\varepsilon\, (eV - \varepsilon)P(\varepsilon).$$

To specify $P(E)$, we note that the force q in our case is equivalent to charge, the charge transferred is e so that $q_0 = e/2$. Substituting this into (12.32), and recalling that $\chi(\omega) = i\omega Z(\omega)$, we obtain

$$P_1(E) = \frac{e^2}{\pi\hbar} \frac{\text{Re}[Z(E/\hbar)]}{E}.$$

After this, $P(E)$ is obtained from (12.34).

c. This is given by (12.25) with $\bar{N} = \int dE P(E) = Z_0 e^2/\pi\hbar$.

d. A quick way to arrive at the answer is to note that from the general relation for the tunneling current derived above it follows that

$$\frac{\partial^2 I}{\partial V^2} = e G_T P(eV).$$

The second derivative for the case under consideration is thus given by a set of δ-functions at $eV = n\hbar\omega_0$. Integration yields the differential conductance that is a step function with jumps at $eV = n\hbar\omega_0$,

$$\frac{\partial I}{\partial V} = G_T \sum_{n=0}^{m} e^{-\bar{N}} \frac{(\bar{N})^n}{n!} \quad \text{for} \quad \hbar\omega_0 m < eV < \hbar\omega_0(m+1).$$

PART V

RELATIVISTIC QUANTUM MECHANICS

Relativistic quantum mechanics

This chapter presents a short introduction to relativistic quantum mechanics that conforms to the style and methods of this book. In this context, relativity is a special kind of symmetry. It turns out that it is a fundamental symmetry of our world, and any understanding of elementary particles and fields should be necessarily relativistic – even if the particles do not move with velocities close to the speed of light. The symmetry constraints imposed by relativity are so overwhelming that one is able to construct, or to put it better, guess the correct theories just by using the criteria of beauty and simplicity. This method is certainly of value although not that frequently used nowadays. Examples of the method are worth seeing, and we provide some.

We start this chapter with a brief overview of the basics of relativity, refresh your knowledge of Lorentz transformations, and present the framework of Minkowski spacetime. We apply these concepts first to relativistic classical mechanics and see how Lorentz covariance allows us to see correspondences between quantities that are absent in non-relativistic mechanics, making the basic equations way more elegant. The Schrödinger equation is not Lorentz covariant, and the search for its relativistic form leads us to the Dirac equation for the electron. We find its solutions for the case of free electrons, plane waves, and understand how it predicted the existence of positrons. This naturally brings us to the second quantization of the Dirac equation. We combine the Dirac and Maxwell equations and derive a Hamiltonian model of quantum electrodynamics that encompasses the interaction between electrons and photons. The scale of the interaction is given by a small dimensionless parameter which we have encountered before: the fine structure constant $\alpha \approx 1/137 \ll 1$. One thus expects perturbation theory to work well. However, the perturbation series suffer from ultraviolet divergences of fundamental significance. To show the ways to handle those, we provide a short introduction to renormalization. The chapter is summarized in Table 13.1.

13.1 Principles of the theory of relativity

The laws of classical mechanics are the same in any reference frame which does not move or moves with a constant velocity. Indeed, since Newton's laws relate forces to *acceleration* (and not velocity) and since forces in classical mechanics are functions of the relative instantaneous positions of objects, it should make no difference whether we describe the motion of classical objects in a stationary reference frame or in a frame moving with a constant velocity.

Let us rephrase the above statement in more scientific terms. We consider two different reference frames, K and K', which move with respect to each other with a constant velocity $\mathbf{v}$. We can express the coordinates $\mathbf{r}'$ in K' in terms of the coordinates $\mathbf{r}$ in K as

$$\mathbf{r}' = \mathbf{r} - \mathbf{v}t + \mathbf{r}_0 \quad \text{and} \quad t' = t + t_0, \tag{13.1}$$

with $\mathbf{r}_0$ being a constant vector, and t_0 being a constant time shift. This is known as a *Galilean transformation* of time and coordinates. From (13.1) we see that accelerations, $\ddot{\mathbf{r}}' = \ddot{\mathbf{r}}$ are not affected by transformation. The distance between two particles 1 and 2 remains the same as well, $|\mathbf{r}'_1 - \mathbf{r}'_2| = |\mathbf{r}_1 - \mathbf{r}_2|$. We therefore say that Newton's laws of classical mechanics are *covariant*[1] under Galilean transformations: the principle of relativity of classical mechanics.

Several problems with this principle of relativity became apparent in the nineteenth century. First of all, the force between two moving charges turned out to be a function of their velocities, and it therefore *changes* under a Galilean transformation. Second, the experiments by Michelson and Morley (1887) demonstrated that the propagation speed of light is in fact the *same* for observers moving with *different* velocities. If all laws of physics were covariant under Galilean transformations, then *all* velocities, including the speed of light, should vary depending on the relative velocity of the observer. Clearly, this is not compatible with the experiments mentioned. To express these problems in more theoretical terms, we note that Maxwell's laws of electrodynamics are not covariant under Galilean transformations.

This, however, did not stop Einstein in 1905 from formulating the basic postulate of his special theory of relativity: all laws of Nature should obey one single principle of relativity, that is, they should be the same in any reference frame which is stationary or moving with a constant velocity. If one applies this postulate to Maxwell's equations, one sees that this can only hold if the speed of light in vacuum c (in our SI notation $c = 1/\sqrt{\mu_0 \varepsilon_0}$) is a universal constant, not depending on the choice of reference frame. It is clear that the Galilean transformation does *not* give the right relation between time and coordinates of reference frames moving with different velocities. One needs something else to conform to Einstein's postulate.

13.1.1 Lorentz transformation

Under the assumptions that space and time are homogeneous and isotropic, the transformation sought is linear (so that motion with a constant velocity transforms to motion with a constant velocity), and it is not too difficult to find the correct relation. The only price one has to pay is to accept that not only the differences of the coordinates, but also the time intervals in the two reference frames can be different. Let us consider two reference frames K and K' moving along the x-axis, so that the other spatial coordinates can be chosen to be

---

[1] The term *covariant* means that the *form* of the laws does not change. Classical mechanics is not *invariant* under Galilean transformations: the actual motion of the objects (their velocity, momentum, kinetic energy, etc.) does change under a transformation.

the same, $y = y'$ and $z = z'$. The most general linear transformation in this case is given by a 2×2 matrix $\hat{A}$,

$$\begin{pmatrix} x' \\ ct' \end{pmatrix} = \hat{A} \begin{pmatrix} x \\ ct \end{pmatrix} + \begin{pmatrix} x_0 \\ ct_0 \end{pmatrix}. \tag{13.2}$$

Here we multiply t by the speed of light c to make sure that all elements of $\hat{A}$ have the same dimensionality. Let us rewrite this relation for *intervals* of x and t so that the constants x_0, t_0 drop out of the equations,

$$\begin{pmatrix} \Delta x' \\ c\Delta t' \end{pmatrix} = \hat{A} \begin{pmatrix} \Delta x \\ c\Delta t \end{pmatrix}. \tag{13.3}$$

The two-vectors $(1, 1)^T$ and $(1, -1)^T$ correspond to motion with velocity $\pm c$: for these two vectors $\Delta x = \pm c \Delta t$. Since the speed of light should be the same in all reference frames, these vectors should be eigenvectors of $\hat{A}$. An arbitrary matrix of this form can be written with two real parameters β, γ as

$$\hat{A} = \gamma \begin{pmatrix} 1 & -\beta \\ -\beta & 1 \end{pmatrix}. \tag{13.4}$$

Applying the matrix to a vector $(0, 1)^T$ describing no motion, we obtain $\gamma(-\beta, 1)^T$ so that the parameter β gives the velocity of the reference frame, $\beta = v/c$. The product of the matrices $\hat{A}(\beta)\hat{A}(-\beta)$ describes the transformation from K to a moving frame K' and back to K, and therefore must be an identity matrix: this fixes $\gamma = \sqrt{1 - \beta^2}$. Finally, the transformation thus reads

$$\begin{pmatrix} \Delta x' \\ c\Delta t' \end{pmatrix} = \gamma \begin{pmatrix} 1 & -\beta \\ -\beta & 1 \end{pmatrix} \begin{pmatrix} \Delta x \\ c\Delta t \end{pmatrix}, \tag{13.5}$$

which can be rewritten in an explicit form,

$$x' = \gamma(x - vt) \quad \text{and} \quad t' = \gamma \left(t - \frac{v}{c^2} x \right). \tag{13.6}$$

Such a transformation is called a *Lorentz transformation*. We note that when K and K' do not move too fast with respect to each other, that is, when $v \ll c$, the Lorentz transformation reduces to the Galilean transformation.

Since we dropped the intuitively appealing assumption that $t = t'$, the Lorentz transformation probably leads to counterintuitive predictions for observations made in different reference frames. This is indeed the case, and we mention here the most famous consequences of the special theory of relativity.

First of all, events which happen simultaneously in one reference frame are generally not simultaneous in another frame. This can be directly seen from the Lorentz transformation (13.6). Suppose we have in reference frame K two events happening at the same time $t = 0$, one at coordinate $(x_1, 0, 0)$ and one at $(x_2, 0, 0)$. We see that, according to the Lorentz transformation, the times associated with the two events in K' are *different*: $t_1' = -(\gamma v/c^2)x_1$ and $t_2' = -(\gamma v/c^2)x_2$, so there is a finite time difference $|t_1' - t_2'| = (\gamma v/c^2)|x_1 - x_2|$. This means that if you synchronize two clocks and keep them at fixed positions separated in space, then the two clocks seem to run out of pace to any observer moving with a finite velocity relatively to the clocks. It is thus impossible to decide in an absolute sense

Hendrik Antoon Lorentz (1853–1928)
Shared the Nobel Prize in 1902 with Pieter Zeeman "in recognition of the extraordinary service they rendered by their researches into the influence of magnetism upon radiation phenomena."

Hendrik Lorentz was born in the east of the Netherlands as the son of a wealthy nursery owner. In 1870 he entered the University of Leiden, where he obtained within a year a Bachelor's degrees in mathematics and physics, and in 1875 a doctor's degree in physics. He refused an offer in 1877 from the University of Utrecht to become professor in mathematics, hoping for a teaching position in Leiden. One year later he was indeed invited to take a newly created chair in Leiden, which was one of the first professor positions in Europe entirely focused on *theoretical* physics.

In the 1870s, Lorentz was one of the first who fully understood the significance of Maxwell's work. In his doctoral thesis and during his first years as a professor in Leiden, he mainly worked on the theory of electromagnetism. He answered several open questions in classical electrodynamics, and presented one of the first theories of the electron. This explained the splitting of spectral lines in a magnetic field observed by Zeeman in 1896, for which they shared the Nobel Prize.

To explain the apparent immobility of the "ether" (the medium in which electromagnetic waves were supposed to exist), Lorentz introduced in 1904 the Lorentz transformation and the concept of Lorentz contraction, ideas which formed the basis on which Einstein built his theory of relativity. Many regard Lorentz as the greatest physicist of the turn of the century. He paved the way for most of the new theories of the 20th century, including quantum theory and relativity. Until his death in 1928, he was internationally regarded as the Nestor of theoretical physics.

whether two events separated in space happened simultaneously or not. This consequence of the theory of relativity is called the relativity of simultaneity.

Further, the apparent size of objects is different for observers in different reference frames. Suppose that in the frame K' there is lying a rigid rod along the x'-axis with its end points at the coordinates x'_1 and x'_2. If an observer in the same reference frame wants to measure the length of this rod, he does so by looking at the two ends at the same time and deciding how far apart they are. Since in K' the two end points always lie at x'_1 and x'_2, its length is simply $l' = |x'_2 - x'_1|$. If now an observer in the other frame K wants to measure the length of this rod, he also must determine the coordinates of the end points of the rod, but now at the same time in *his* reference frame. Let us for simplicity assume that he does the measurement at $t = 0$. We see from (13.6) that then the simple relation $x' = \gamma x$ follows, so that the length observed in K is $l = l'/\gamma$. Since $\gamma \geq 1$ the rod seems always *smaller* to the observer moving relatively to the rod. This effect is called Lorentz contraction.

An effect related to Lorentz contraction is called time dilation: the duration of a time interval between two events depends on the velocity of the observer with respect to the point where the events take place. This too can be easily seen from (13.6). If the two events occur at the origin $(0, 0, 0)$ of the frame K, one at t_1 and one at t_2, then the times associated with the events for an observer in the moving frame K' are $t'_1 = \gamma t_1$ and $t'_2 = \gamma t_2$. The time interval between the two events thus seems *longer* to the observer moving relatively to the place where the events take place.

All these effects are indeed rather counterintuitive, and all stem from assuming that not only coordinates but also time can be relative. Since this revolutionary ingredient was able to resolve the apparent contradictions between experiments and the traditional notion of (Newtonian) relativity, Einstein's theory of relativity rapidly gained acceptance under theoretical physicists. Starting in the 1930s the weird consequences of the theory, such as the effect of time dilation, have even been confirmed explicitly by many experiments, and now it is generally accepted that the Lorentz transformation gives the correct relation between time and coordinates of different reference frames.

13.1.2 Minkowski spacetime

In 1907, two years after the publication of Einstein's first papers on the theory of relativity, Hermann Minkowski formulated perhaps the biggest discovery in the history of natural sciences. He understood that the best way to consider space and time is not as separate or separable things, but to regard them together as a four-dimensional space, a continuum of points. Each point – an event – is characterized by a four-dimensional vector, three of its components being space coordinates and one being proportional to time, $(\mathbf{r}, ct)$. All components of a vector should have the same dimension (of length). To achieve this, the time-component has been rescaled by the speed of light c, so that "one meter of time" corresponds to the time it takes a photon to travel one meter.

A pointlike object moving through our "real" three-dimensional space is represented in four-dimensional spacetime by its *worldline*. In Fig. 13.1(a) we show a possible worldline

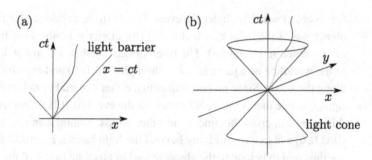

(a)   (b)

Fig. 13.1   Illustration of Minkowski spacetime. (a) A possible worldline connecting the event $(0, 0, 0, 0)$ with other events. The dotted line marks the light barrier, that is, the boundary of the region reachable by a worldline. (b) The corresponding light cone in higher-dimensional spacetime.

Hermann Minkowski (1864–1909)

Hermann Minkowski was born in Kaunas in Lithuania, then part of the Russian empire. When Minkowski was eight years old, his German parents decided to move back to Germany, and the family settled in the town of Königsberg, present-day Kaliningrad in Russia.

In 1880, Minkowski enrolled in the University of Königsberg to study mathematics. One year later, the French Académie des Sciences announced that the 1883 Grand Prix would be awarded for a general answer to the question "In how many ways can a given integer be written as a sum of five squared integers?" Minkowski, only eighteen years old, delivered a solution and shared the prize with Henry Smith, a renowned Irish mathematician who had produced an answer independently. Minkowski continued working in this direction, and wrote his doctoral thesis on the same topic, on which he graduated in 1885.

In the following years, Minkowski moved to Bonn, back to Königsberg, to Zurich (where he was one of Einstein's teachers), and in 1902 finally to Göttingen. His fellow student in Königsberg and close friend David Hilbert was Head of the Department of Mathematics in Göttingen, and had created a professorial position especially for Minkowski.

Under the influence of Hilbert, Minkowski became interested in mathematical physics. He learned about the new developments in the theory of electromagnetism, and about the work of Lorentz and Einstein. Thus inspired, he developed his theory of four-dimensional spacetime and derived a four-dimensional treatment of electromagnetism. His ideas were to become key ingredients of the further development of the theory of relativity. Minkowski, however, did not live long enough to witness this: he died suddenly of appendicitis in 1909, at the young age of 44.

for an object which includes the event $(0, 0, 0, 0)$. In a colloquial way we would say that the object was at the point $\mathbf{r} = 0$ at $t = 0$ (for clarity we only show the x- and t-components of spacetime in the figure). The tilted dotted line has a slope of 1 and would correspond to the worldline of a particle going through $(0, 0, 0, 0)$ and moving with the speed of light along the x-axis. Since no corporal entity, either material or radiative, can travel faster than light, this line marks the *light barrier* for the event $(0, 0, 0, 0)$: no worldline which includes this event can cross this line, or in other words, nothing can participate both in the event $(0, 0, 0, 0)$ and in an event lying beyond the light barrier. Events lying within this boundary are thus said to belong to the absolute past or absolute future of the event $(0, 0, 0, 0)$. When including more spatial dimensions, the light barrier becomes a (hyper)cone which is called the *light cone* of an event, as illustrated in Fig. 13.1(b).

13.1.3 The Minkowski metric

Now that we have decided to combine space and time into a four-dimensional spacetime, it is a good idea to set up a notation which treats the four components of spacetime on a more equal footing. We thus introduce a four-dimensional vector $\mathbf{x} \equiv (x^0, x^1, x^2, x^3)$ where

$$x^0 = ct, \quad x^1 = x, \quad x^2 = y, \quad \text{and} \quad x^3 = z. \tag{13.7}$$

We use Greek indices to denote the four components of the vector, x^μ, where $\mu = 0 \ldots 4$. In this notation, the Lorentz transformation, which is linear, can be represented by a 4×4 matrix Λ and follows from

$$(x')^\mu = \Lambda^\mu_\nu x^\nu, \tag{13.8}$$

where summation over repeated indices is implied. The elements of the matrix Λ are given by (13.6), so we find

$$\Lambda = \begin{pmatrix} \gamma & -\gamma v/c & 0 & 0 \\ -\gamma v/c & \gamma & 0 & 0 \\ 0 & 0 & 1 & 0 \\ 0 & 0 & 0 & 1 \end{pmatrix}. \tag{13.9}$$

The Lorentz transformation in fact can be seen as a kind of rotation in four-dimensional spacetime.

Control question. Suppose we define the angle φ by $\tanh \varphi = v/c$. Can you show that the above transformation is identical to a rotation in the (x^0, x^1)-plane over the imaginary angle $i\varphi$?

Let us now consider two points in four-dimensional spacetime, the events $\mathbf{x}_a$ and $\mathbf{x}_b$. If we were to think again in non-relativistic terms, we would guess that both the distance in space $\sqrt{(x^1_a - x^1_b)^2 + (x^2_a - x^2_b)^2 + (x^3_a - x^3_b)^2}$ and the time interval $x^0_a - x^0_b$ do not depend on the coordinate system in which we measure. However, we know better by now. Since space and time get mixed up during transformations from one coordinate system to another, it is impossible to separate spacetime and define two invariant metrics. Minkowski realized this, and was able to find the "correct" metric which must contain all four components of the vector x^μ. The resulting Minkowski metric

$$(ds)^2 = (x^0_a - x^0_b)^2 - (x^1_a - x^1_b)^2 - (x^2_a - x^2_b)^2 - (x^3_a - x^3_b)^2, \tag{13.10}$$

is indeed invariant under Lorentz transformations. It thus defines a kind of distance between two events in spacetime, such as $dr = |\mathbf{x}_a - \mathbf{x}_b|$ and $dt = |t_a - t_b|$ do in our common-day space and time. As the distance dr is invariant under rotations of three-dimensional space, the metric ds stays the same under the "Lorentz rotation" (13.9) in four-dimensional spacetime. When $(ds)^2 < 0$ the two events are separated by the light barrier, meaning that nothing can participate in both events.

Yet here is the problem: the "distance" of the Minkowski metrics is not always positive thus not satisfying the mathematical notion of distance. The squared "length" (or inner product) of a four-vector defined by this metric cannot be simply written as $x^\mu x^\mu$ due to the relative minus signs in (13.10): otherwise it would be strictly positive. The most common notation is to define a *contravariant* and a *covariant* form of the four-vector. The vector with the components x^μ as defined above is called contravariant, and its covariant counterpart is given by

$$x_0 = x^0, \quad x_1 = -x^1, \quad x_2 = -x^2, \quad \text{and} \quad x_3 = -x^3, \tag{13.11}$$

and is distinguished by having its indices at the bottom. With these two definitions the inner product of a four-vector reads concisely $x_\mu x^\mu$.

13.1.4 Four-vectors

The relativity principle states that all physics is the same in all reference frames moving with a constant velocity. All physical quantities must therefore be Lorentz covariant, and their values in two different frames are related by the corresponding Lorentz transformation. The four-vector **x** corresponding to the coordinates of spacetime is just one example of a Lorentz covariant physical quantity. In general, quantities can be scalars, four-vectors, or four-dimensional tensors of any higher rank. A tensor of rank n has n indices, and the transformation to another reference frame includes n Lorentz matrices, each acting on one index. For instance, a tensor R of rank 3 is transformed as

$$(R')^{\alpha\beta\gamma} = \Lambda^\alpha_\delta \Lambda^\beta_\epsilon \Lambda^\gamma_\zeta R^{\delta\epsilon\zeta}. \tag{13.12}$$

Many physical quantities (such as momentum, current density, etc.) are in a non-relativistic context represented by three-dimensional vectors. A three-vector, however, is not a Lorentz-covariant quantity, and it must thus be complemented with one more component to form a four-vector. Let us illustrate this for the case of the momentum vector.

A naive guess to construct the momentum four-vector would be to define a four-velocity like $u^\mu \equiv dx^\mu/dt$ and then define $p^\mu = mu^\mu$. But if we have a closer look at the elements of u^μ, for instance $u^2 = dx/dt = c(dx^2/dx^0)$, we see that both dx and dt are in fact components of the coordinate four-vector. Therefore the Lorentz transformation of the element $c(dx^2/dx^0)$ *cannot* be simply given by Λ, and the vector u^μ is thus not Lorentz covariant.

The way out is to define velocity in terms of the time experienced by the object moving with this velocity. The effect of time dilatation predicts that a time interval $d\tau$ from the point of view of the moving object corresponds to a longer time interval $dt = \gamma d\tau$ for a stationary observer. To the observer, the object moves with the velocity

$$v = \sqrt{\left(\frac{dx}{dt}\right)^2 + \left(\frac{dy}{dt}\right)^2 + \left(\frac{dz}{dt}\right)^2}, \tag{13.13}$$

where x, y, z, and t are the space and time coordinates in the frame of the observer. An infinitesimally small time interval $d\tau$ in the frame of the moving object reads in terms of the coordinates of the observer

$$dτ = dt\sqrt{1 - \frac{v^2}{c^2}} = \frac{1}{c}\sqrt{(ds)^2}.$$ (13.14)

We see that the time interval as experienced by the moving object is invariant! We can thus define a Lorentz covariant four-velocity as

$$u^μ = \frac{dx^μ}{dτ} = γ(c, v_x, v_y, v_z),$$ (13.15)

of which the inner product $u_μ u^μ = c^2$ is invariant, as it should be.

The four-momentum follows then simply as $p^μ = mu^μ$, where m is the mass of the object in its own reference frame, its so-called rest mass.[2] We thus have

$$p^μ = γ(mc, mv_x, mv_y, mv_z),$$ (13.16)

its inner product $p_μ p^μ = m^2 c^2$ being a scalar and therefore the same in all reference frames. A relevant question to ask is what is the *meaning* of the "extra" first component of the momentum four-vector, p^0. In the coordinate four-vector it was time, but here it reads $γmc$, which has at first sight no obvious interpretation. The answer can be found by calculating the time-derivative of p^0 from the point of view of the stationary observer. We find

$$mc\frac{dγ}{dt} = \frac{γm}{c}\mathbf{v}\cdot\frac{d\mathbf{v}}{dt} = \frac{1}{c}\mathbf{v}\cdot\frac{d\mathbf{p}}{dt}.$$ (13.17)

If we then use the same definitions of force and work as we know from non-relativistic mechanics, we see that this relation can be rewritten in terms of the energy E of the object,

$$\frac{dp^0}{dt} = \frac{1}{c}\frac{dE}{dt}.$$ (13.18)

We can thus identify $p^0 = E/c$ (in fact formally only up to a constant, but we can safely set it to zero), and the momentum four-vector thus reads

$$p^μ = \left(\frac{E}{c}, γmv_x, γmv_y, γmv_z\right),$$ (13.19)

where

$$E = γmc^2.$$ (13.20)

For an object at rest this reduces to $E = mc^2$, Einstein's famous equation describing the mass–energy equivalence. For small velocities we can expand the energy as $E ≈ mc^2 + \frac{1}{2}mv^2$, indeed yielding the non-relativistic kinetic energy, added to the rest energy mc^2. Using the invariance of the inner product of the momentum four-vector, $p_μ p^μ = m^2 c^2$, we can also derive the relation between relativistic energy and momentum,

$$E^2 = m^2 c^4 + p^2 c^2.$$ (13.21)

---

[2] This rest mass is an invariant. If we were to take the mass as it appears to a stationary observer (thus moving relatively to the object) we would run into the same trouble as we did when we used dt instead of $dτ$ for the velocity.

Let us mention a few other four-vectors representing physical quantities. The gradient vector simply becomes a four-vector including the time derivative,

$$\frac{\partial}{\partial x^\mu} = \left(\frac{\partial}{\partial x^0}, \frac{\partial}{\partial x^1}, \frac{\partial}{\partial x^2}, \frac{\partial}{\partial x^3} \right), \tag{13.22}$$

which gives the rules how to construct a Lorentz covariant differential equation. The inner product of the gradient operator with itself is the d'Alembertian

$$\Box = \frac{\partial}{\partial x_\mu} \frac{\partial}{\partial x^\mu} = \frac{1}{c^2} \frac{\partial^2}{\partial t^2} - \nabla^2, \tag{13.23}$$

which is an invariant Lorentz scalar by construction.

Also, the vector and scalar potentials of the electromagnetic field can be united into a single four-vector,

$$A^\mu = \left(\frac{\phi}{c}, \mathbf{A} \right). \tag{13.24}$$

The density of electric current is similarly united with the charge density ρ,

$$j^\mu = (c\rho, \mathbf{j}). \tag{13.25}$$

In these notations many physical laws become particularly elegant. For instance, the conservation of charge is expressed as

$$\frac{\partial j^\mu}{\partial x_\mu} = 0, \tag{13.26}$$

and the Maxwell equations can be united into a single equation,

$$\Box A^\mu = \mu_0 j^\mu, \tag{13.27}$$

as we promised in Chapter 7.

We will not rewrite here all non-relativistic physics in relativistic form. The above examples sufficiently illustrate the power of the principle of relativity. The mere condition of Lorentz covariance allows one to see links between quantities which are not related in non-relativistic physics and it makes many equations so simple that we almost seem invited to guess them.

13.2 Dirac equation

Let us now finally turn to quantum mechanics and try to guess a "relativistic Schrödinger equation" for the wave function of a single particle in vacuum. The Schrödinger equation in its common form

$$i\hbar \frac{\partial \psi}{\partial t} = \frac{\hat{\mathbf{p}}^2}{2m} \psi, \tag{13.28}$$

is not relativistic, since it approximates the true relativistic energy $E(\mathbf{p}) = \sqrt{m^2 c^4 + p^2 c^2}$ with the non-relativistic expression $\hat{\mathbf{p}}^2 / 2m$.

We could act pragmatically and replace $\hat{\mathbf{p}}^2/2m$ by the correct energy $E(\hat{\mathbf{p}})$,

$$i\hbar\frac{\partial \psi}{\partial t} = \sqrt{m^2c^4 + \hat{\mathbf{p}}^2c^2}\,\psi. \tag{13.29}$$

This equation, however, is not Lorentz covariant, which is enough to reject it. Indeed, the separate appearance of the operator $\partial/\partial t$ cannot fit in a relativistic description since time and coordinate should be treated on an equal footing: taking out one of the four coordinates and assigning to it a special role cannot lead to a covariant equation. Apart from this, it would also have been quite unpleasant if (13.29) were the correct equation. The problem is that in the coordinate representation the suggested Hamiltonian operator would have to be expressed as an integral kernel $\hat{H}(\mathbf{r} - \mathbf{r}')$ so that the equation reads

$$i\hbar\frac{\partial \psi}{\partial t} = \int d\mathbf{r}'\hat{H}(\mathbf{r} - \mathbf{r}')\psi(\mathbf{r}'). \tag{13.30}$$

This is a *non-local* equation: the evolution of the wave function at the point $\mathbf{r}$ actually depends on the values of the wave function in all other points. This is not only ugly: it is in contradiction with the relativity principle since this implies an instant information exchange between the distinct points $\mathbf{r}$ and $\mathbf{r}'$.

Let us proceed in a covariant way and assume that the wave function ψ is a scalar and therefore Lorentz invariant. If we take two more Lorentz scalars as ingredients, the d'Alembertian (13.23) and the inner product m^2c^2 of the momentum four-vector, we can construct the covariant equation

$$\left\{\Box + \frac{m^2c^2}{\hbar^2}\right\}\psi = 0 \quad \text{or} \quad \left\{\frac{\partial}{\partial x_\mu}\frac{\partial}{\partial x^\mu} + \frac{m^2c^2}{\hbar^2}\right\}\psi = 0, \tag{13.31}$$

which is exactly what you would get if you were to take the relativistic energy–momentum relation $E^2 = m^2c^4 + p^2c^2$ and make the usual replacements $E \to i\hbar\partial_t$ and $\mathbf{p} \to -i\hbar\nabla$ and assume that these operators act on a wave function.

Equation (13.31) is the so-called Klein–Gordon equation and was the first relativistic quantum evolution equation proposed. It is able to produce solutions in the form of plane waves, $\psi(\mathbf{r}, t) \propto \exp\{\frac{i}{\hbar}\mathbf{p} \cdot \mathbf{r}\}$ having the correct relativistic energy $E(\mathbf{p}) = \sqrt{m^2c^4 + p^2c^2}$. At first sight the Klein–Gordon equation thus seems to be a reasonable guess.

However, for any given $\mathbf{p}$ there always exists a second solution of the Klein–Gordon equation, having the negative energy $E(\mathbf{p}) = -\sqrt{m^2c^4 + p^2c^2}$. At the time, no physical explanation of negative energies was known, and it looked like the equation has twice the number of solutions it should have, which was a serious problem to many physicists. But in fact, the existence of negative energy solutions is not catastrophic. Also the Dirac equation turns out to have such solutions, and below we explain how to deal with them.

There exists another fundamental yet more subtle problem with the Klein–Gordon equation. In non-relativistic theory, the square of the wave function $|\psi|^2$ can be interpreted as the probability density ρ for finding the particle somewhere in space. This probability density is then related to the probability current $\mathbf{j} = -\frac{i\hbar}{2m}(\psi^*\nabla\psi - \psi\nabla\psi^*)$ by the continuity equation $\nabla \cdot \mathbf{j} = \partial_t\rho$. In relativistic theory this continuity equation reads $\partial j^\mu/\partial x_\mu = 0$, where probability density and current must be united in a single four-vector. To make it covariant this four-vector must read $j^\mu = \frac{i\hbar}{2m}(\psi^* \partial^\mu \psi - \psi \partial^\mu \psi^*)$, where $\partial^\mu \equiv \partial/\partial x_\mu$. The first component of this four-vector gives the probability density,

$$\rho = \frac{i\hbar}{2mc^2}\left(\psi^* \frac{\partial}{\partial t}\psi - \psi \frac{\partial}{\partial t}\psi^*\right). \tag{13.32}$$

Since we have the correspondence $E \leftrightarrow i\hbar\partial_t$, we see that for a stationary state this results in $\rho = (E/mc^2)|\psi|^2$, which at a first glance seems reasonable: in the non-relativistic limit we have $E \approx mc^2$ and the expression reduces to $\rho = |\psi|^2$ as it should. But wait, before we showed that we can have solutions of the Klein–Gordon equation with positive *and* negative energies. We then said that the mere existence of negative energy solutions is not catastrophic, and we promised to explain later what they mean. If we, however, unite these negative energy solutions with the expression for ρ we found, we *do* have a serious problem: ρ can become negative, and this obviously does not make sense. In fact, the problem is more general and also persists for non-stationary states. It all boils down to the Klein–Gordon equation being a second order differential equation in t. Mathematically this means that when looking for solutions, we are free to choose *two* initial conditions: ψ and $\partial_t \psi$ at time $t = 0$. But since ρ is only a function of these two freely choosable parameters, there is nothing that forbids us choosing them such that the probability density can become negative.

Dirac wanted to solve this problem by finding an equation that has fewer solutions than the Klein–Gordon equation and can be written as a first order differential equation in t. Suppose one finds an operator $\hat{w}$ that represents the square root of the d'Alembertian, $\hat{w}^2 = \Box$. The operator of the Klein–Gordon equation can be presented as the product of two commuting factors,

$$\Box + \frac{m^2 c^2}{\hbar^2} = \left(\hat{w} + i\frac{mc}{\hbar}\right)\left(\hat{w} - i\frac{mc}{\hbar}\right). \tag{13.33}$$

Each of the equations

$$\left\{\hat{w} + i\frac{mc}{\hbar}\right\}\psi = 0 \quad \text{and} \quad \left\{\hat{w} - i\frac{mc}{\hbar}\right\}\psi = 0, \tag{13.34}$$

must have fewer solutions than the Klein–Gordon equation, while all solutions of either equation satisfy the Klein–Gordon equation.

The square root of the d'Alembertian can be extracted in a variety of ways. Dirac wanted to find the most elegant way. He suggested a local operator that contains only first derivatives of the four coordinates,

$$\hat{w} = \hat{\gamma}^\mu \frac{\partial}{\partial x_\mu}. \tag{13.35}$$

The $\hat{\gamma}^\mu$ here are four yet unknown operators, not depending on coordinate. Since $\hat{w}^2 = \Box$, the $\hat{\gamma}$s should satisfy the following relations,

$$\hat{\gamma}^n \hat{\gamma}^m = -\hat{\gamma}^m \hat{\gamma}^n \quad \text{for} \quad n \neq m, \tag{13.36}$$

and

$$\hat{\gamma}^0 \hat{\gamma}^0 = 1, \quad \hat{\gamma}^1 \hat{\gamma}^1 = -1, \quad \hat{\gamma}^2 \hat{\gamma}^2 = -1, \quad \text{and} \quad \hat{\gamma}^3 \hat{\gamma}^3 = -1. \tag{13.37}$$

Dirac immediately realized that these operators can be represented by *matrices*. He picked up a maths book and figured out that the simplest matrices fulfilling the above relations are 4×4 matrices. Since the matrices must work on the wave function ψ, the wave function should have four components as well, and such a four-component wave function is called a *bispinor*. The Dirac equation takes the form

$$\left\{ \hat{\gamma}^{\mu} \frac{\partial}{\partial x_{\mu}} + i\frac{mc}{\hbar} \right\} \psi = 0, \tag{13.38}$$

for a four-component ψ. We sometimes use Latin indices to label the bispinor components, $\psi \rightarrow \psi_{a}$, yet mostly skip this indexing and imply, for instance, $\hat{\gamma}^{\mu}\psi \rightarrow \gamma^{\mu}_{ab}\psi_{b}$. Let us try to find a concrete representation of the $\hat{\gamma}$s. Looking again at the condition (13.36), we recognize that it looks very much like the anti-commutation relations fulfilled by the 2×2 Pauli matrices describing the non-relativistic electron spin, $\hat{\boldsymbol{\sigma}} = (\hat{\sigma}_{1}, \hat{\sigma}_{2}, \hat{\sigma}_{3})$. Indeed, we know that

$$\hat{\sigma}_{n}\hat{\sigma}_{m} = -\hat{\sigma}_{m}\hat{\sigma}_{n} \quad \text{for} \quad n \neq m, \quad \text{and} \quad \hat{\sigma}_{n}^{2} = 1, \tag{13.39}$$

which suggests that we could use the Pauli matrices to build the $\hat{\gamma}$s. A problem with this is that there are only three Pauli matrices, and, moreover, they work in a two-dimensional, not four-dimensional, space. However, nothing forbids us constructing our four-dimensional space as the direct product space of two two-dimensional spaces, $S_{4D} = S_{2D}^{(1)} \otimes S_{2D}^{(2)}$. We can then start by assuming that we can write the last three $\hat{\gamma}$s as

$$\hat{\gamma}^{n} = i\hat{\sigma}_{n}^{(1)} \otimes \hat{X}^{(2)} \quad \text{for} \quad n = 1, 2, 3. \tag{13.40}$$

The unknown operator $\hat{X}^{(2)}$ acts only in $S_{2D}^{(2)}$ and thus commutes with all $\hat{\sigma}^{(1)}$s. If we further require that $[\hat{X}^{(2)}]^{2} = 1$, then we see that the $\hat{\gamma}^{1}$, $\hat{\gamma}^{2}$, and $\hat{\gamma}^{3}$ thus defined behave exactly as required. Taking now the first matrix $\hat{\gamma}^{0}$ into play, we see that requirements (13.36) and (13.37) can be written as

$$\hat{\gamma}^{0}[i\hat{\sigma}_{n}^{(1)} \otimes \hat{X}^{(2)}] = -[i\hat{\sigma}_{n}^{(1)} \otimes \hat{X}^{(2)}]\hat{\gamma}^{0} \quad \text{and} \quad \hat{\gamma}^{0}\hat{\gamma}^{0} = 1. \tag{13.41}$$

These relations can easily be satisfied if one assumes that $\hat{\gamma}^{0}$ acts solely in the space $S_{2D}^{(2)}$, so that $\hat{\gamma}^{0} = \hat{Y}^{(2)} \otimes 1^{(1)}$. The requirement (13.41) then reduces to

$$\hat{Y}^{(2)}\hat{X}^{(2)} = -\hat{X}^{(2)}\hat{Y}^{(2)} \quad \text{and} \quad \hat{Y}^{(2)}\hat{Y}^{(2)} = 1, \tag{13.42}$$

which can be satisfied by choosing for $\hat{Y}^{(2)}$ and $\hat{X}^{(2)}$ two different Pauli matrices in the space $S_{2D}^{(2)}$. A common choice is $\hat{Y}^{(2)} = \hat{\sigma}_{3}^{(2)}$ and $\hat{X}^{(2)} = \hat{\sigma}_{2}^{(2)}$. Now we can write explicit expressions for the $\hat{\gamma}$s (boldface denotes a three-dimensional vector)

$$\hat{\gamma}^{0} = \begin{pmatrix} 1 & 0 \\ 0 & -1 \end{pmatrix} \quad \text{and} \quad \hat{\boldsymbol{\gamma}} = \begin{pmatrix} 0 & \hat{\boldsymbol{\sigma}} \\ -\hat{\boldsymbol{\sigma}} & 0 \end{pmatrix}, \tag{13.43}$$

where the explicitly written 2×2 matrices act in the space $S_{2D}^{(2)}$ and each element of the matrices is in fact a matrix in $S_{2D}^{(1)}$. The Dirac equation (13.38) thus becomes

$$\begin{pmatrix} -i\partial_{t} & -ic\hat{\boldsymbol{\sigma}} \cdot \nabla \\ ic\hat{\boldsymbol{\sigma}} \cdot \nabla & i\partial_{t} \end{pmatrix} \begin{pmatrix} \psi_{A} \\ \psi_{B} \end{pmatrix} = -\frac{mc^{2}}{\hbar} \begin{pmatrix} \psi_{A} \\ \psi_{B} \end{pmatrix}, \tag{13.44}$$

where the Dirac bispinor is represented by two usual spinors, $\psi \rightarrow (\psi_{A}, \psi_{B})^{T}$.

One of the advantages of the Dirac equation is that it can be written in a familiar Hamiltonian form. Although this form is not Lorentz covariant, it brings us back to the solid ground of common quantum mechanics and thus facilitates an interpretation of the Dirac equation. We obtain the Hamiltonian form by multiplying the covariant form (13.38) with $\hbar \hat{\gamma}^0$,

$$i\hbar \frac{\partial \psi}{\partial t} = \left\{ c\hat{\boldsymbol{\alpha}} \cdot \hat{\mathbf{p}} + \hat{\beta} mc^2 \right\} \psi. \tag{13.45}$$

This is just an equation of the Schrödinger type with a Hamiltonian that has a 4×4 matrix structure. The explicit form of the matrices entering this Hamiltonian is

$$\hat{\boldsymbol{\alpha}} = \hat{\gamma}^0 \hat{\boldsymbol{\gamma}} = \begin{pmatrix} 0 & \hat{\boldsymbol{\sigma}} \\ \hat{\boldsymbol{\sigma}} & 0 \end{pmatrix} \quad \text{and} \quad \hat{\beta} = \hat{\gamma}^0 = \begin{pmatrix} 1 & 0 \\ 0 & -1 \end{pmatrix}. \tag{13.46}$$

Although the Dirac matrices $\hat{\gamma}^\mu$ bear the same index as the gradient operator, they do *not* form a four-vector. The true four-vector is that of the probability current and can be found as follows. Defining the so-called adjoint bispinor $\bar{\psi} \equiv \psi^\dagger \hat{\gamma}^0$ and making use of the relation $(\hat{\gamma}^\mu)^\dagger \hat{\gamma}^0 = \hat{\gamma}^0 \hat{\gamma}^\mu$, we can write the adjoint Dirac equation

$$\frac{\partial}{\partial x_\mu} \bar{\psi} \hat{\gamma}^\mu - i \frac{mc}{\hbar} \bar{\psi} = 0. \tag{13.47}$$

Control question. Do you see how to derive this adjoint equation?

When we multiply (13.47) with ψ from the right, the original Dirac equation (13.38) with $\bar{\psi}$ from the left, and add the two resulting equations, we find

$$\frac{\partial}{\partial x_\mu} \bar{\psi} \hat{\gamma}^\mu \psi = 0, \tag{13.48}$$

thus yielding a continuity equation for the object $\bar{\psi} \hat{\gamma}^\mu \psi$. We can therefore relate this object to the four-vector of probability current as

$$j^\mu = c \, \bar{\psi} \hat{\gamma}^\mu \psi, \tag{13.49}$$

resulting in a probability density $\rho = j^0/c = \bar{\psi} \hat{\gamma}^0 \psi = \sum_a \psi_a^\dagger \psi_a$. The Dirac equation thus indeed leads to a definition of probability density which is always positive definite: a big improvement in comparison with the Klein–Gordon equation!

13.2.1 Solutions of the Dirac equation

Let us search for a solution of the Dirac equation in the form of plane waves,

$$\psi(x^\mu) = \frac{1}{\sqrt{\mathcal{V}}} \psi \exp\{-\tfrac{i}{\hbar} p_\mu x^\mu\}, \tag{13.50}$$

ψ on the right hand side being a coordinate-independent bispinor. Substituting this into (13.38) yields

$$\left\{ -\hat{\gamma}^\mu p_\mu + mc \right\} \psi = 0, \tag{13.51}$$

which is nothing but a system of four algebraic equations for four unknowns. We rewrite it using the explicit form of the Dirac matrices $\hat{\gamma}^\mu$ and the covariant energy–momentum vector p_μ. The equations become

$$
\begin{pmatrix} -\dfrac{E}{c} + mc & \boldsymbol{\sigma} \cdot \mathbf{p} \\ -\boldsymbol{\sigma} \cdot \mathbf{p} & \dfrac{E}{c} + mc \end{pmatrix} \begin{pmatrix} \psi_A \\ \psi_B \end{pmatrix} = 0. \tag{13.52}
$$

We note that this equation still has solutions at both positive and negative energies, $E = \pm\sqrt{m^2 c^4 + p^2 c^2}$. As we reported in the previous section, Dirac's trick does *not* save us from having negative energies.

Let our goal be to find solutions that have a well-defined non-relativistic limit, that is, when $\mathbf{p} \to 0$. In this limit we have $E \approx \pm(mc^2 + \frac{1}{2}mv^2)$. For the positive energy solution, we see that in this limit three coefficients in (13.52) are small, of the order of p, whereas one coefficient is $E/c + mc \approx 2mc$. The second equation in (13.52) then tells us that ψ_B will be very small compared to ψ_A, which invites us to express ψ_B in terms of a small factor times ψ_A. Similarly, for negative energies we express ψ_A in terms of ψ_B, so that

$$
\begin{aligned}
\psi_B &= \frac{\boldsymbol{\sigma} \cdot \mathbf{p}}{E/c + mc}\psi_A \quad \text{for} \quad E > 0, \\
\psi_A &= \frac{-\boldsymbol{\sigma} \cdot \mathbf{p}}{-E/c + mc}\psi_B \quad \text{for} \quad E < 0.
\end{aligned} \tag{13.53}
$$

The spinors ψ_A (at positive energies) and ψ_B (at negative energies) have a direct correspondence with the non-relativistic spin $\frac{1}{2}$. The Dirac equation is therefore suitable to describe spin-$\frac{1}{2}$ particles, for instance electrons. The spinors can be chosen to be the eigenfunctions of the spin operator along a certain axis. For instance, we can pick the z-axis and define the eigenfunctions with $s_z = \pm\frac{1}{2}$ to correspond to the spinors

$$
\psi_+ = \begin{pmatrix} 1 \\ 0 \end{pmatrix} \quad \text{and} \quad \psi_- = \begin{pmatrix} 0 \\ 1 \end{pmatrix}. \tag{13.54}
$$

With this, the four independent normalized solutions of the Dirac equation read

$$
\psi_{\pm,+} = \sqrt{\frac{|E| + mc^2}{2|E|}} \begin{pmatrix} \psi_\pm \\ \frac{c\mathbf{p}\cdot\boldsymbol{\sigma}}{|E|+mc^2}\psi_\pm \end{pmatrix}, \tag{13.55}
$$

$$
\psi_{\pm,-} = \sqrt{\frac{|E| + mc^2}{2|E|}} \begin{pmatrix} \frac{-c\mathbf{p}\cdot\boldsymbol{\sigma}}{|E|+mc^2}\psi_\pm \\ \psi_\pm \end{pmatrix}, \tag{13.56}
$$

the second index labeling positive (negative) energy. At $\mathbf{p} = 0$, these four solutions correspond to spin $s_z = \pm\frac{1}{2}$ and energy $E = \pm mc^2$, and form a standard basis in a four-dimensional space,

$$
\begin{aligned}
\psi_{+,+} &= (1,0,0,0)^T, \quad \psi_{-,+} = (0,1,0,0)^T, \\
\psi_{+,-} &= (0,0,1,0)^T, \quad \psi_{-,-} = (0,0,0,1)^T.
\end{aligned} \tag{13.57}
$$

In fact, there is a more convenient choice for the axis of the spin projection than just an arbitrary one. The three-momentum of the particle points in a particular direction and thus provides a natural quantization axis for spin. Let us parameterize the energy–momentum as

$$E = \pm mc^2 \cosh \chi \quad \text{and} \quad \mathbf{p} = mc\,\mathbf{n} \sinh \chi, \tag{13.58}$$

where $\mathbf{n}$ is a three-vector of unit length.

Control question. Can you relate the angle χ to the γ of Section 13.1?

The two possible projections of spin on the vector $\mathbf{n}$ correspond to

$$\mathbf{n} \cdot \boldsymbol{\sigma} = \pm 1, \tag{13.59}$$

which in fact reminds us of the two possible circular polarizations of light. The two solutions for positive energy,

$$\frac{1}{\sqrt{\cosh \chi}} \begin{pmatrix} \cosh \frac{\chi}{2} \\ 0 \\ \sinh \frac{\chi}{2} \\ 0 \end{pmatrix} \quad \text{and} \quad \frac{1}{\sqrt{\cosh \chi}} \begin{pmatrix} 0 \\ \cosh \frac{\chi}{2} \\ 0 \\ -\sinh \frac{\chi}{2} \end{pmatrix}, \tag{13.60}$$

correspond to the projections "+" and "−" respectively. The two corresponding solutions for negative energies are

$$\frac{1}{\sqrt{\cosh \chi}} \begin{pmatrix} -\sinh \frac{\chi}{2} \\ 0 \\ \cosh \frac{\chi}{2} \\ 0 \end{pmatrix} \quad \text{and} \quad \frac{1}{\sqrt{\cosh \chi}} \begin{pmatrix} 0 \\ \sinh \frac{\chi}{2} \\ 0 \\ \cosh \frac{\chi}{2} \end{pmatrix}. \tag{13.61}$$

We conclude with a remark concerning the covariance of the elements of the Dirac equation. A bispinor is a four-component Lorentz covariant quantity, but distinct from a four-vector. This means that its components are transformed from one reference frame to another by 4×4 matrices B_{ab}, so that $\psi'_a = B_{ab}\psi_b$, but these matrices are distinct from the matrix Λ which usually represents the Lorentz transformation.

Control question. Do you see how to find the matrices B_{ab} from the above solutions of the Dirac equation?

Still the biggest problem for Dirac was the interpretation of the negative energies. To emphasize the importance of the problem, let us consider energy relaxation of an electron by emission of photons.[3] We expect any excited electron to loose energy until it comes to rest in its lowest energy state. However, if negative energies are present in the spectrum, the electron could lower its energy further and further. First, it would emit a photon with an energy exceeding the gap $2mc^2$ in the spectrum and then it would accelerate further emitting more and more photons. An electron would be an inexhaustible energy source!

Dirac overcame this problem by assuming that all states with negative energies are already completely filled. Since electrons are fermions, the Pauli exclusion principle tells

---

[3] As we have seen in Chapter 9, this process is forbidden in vacuum by virtue of momentum conservation. We thus assume some slowly-moving protons present, so that the electron can emit photons while scattering off the protons, see Section 9.6

us that no electron from a state with positive energy can enter the filled states in the "cellar" of negative energies. Thereby Dirac has put forward a revolutionary concept of the vacuum. Instead of being an empty place, it is actually full of particles!

But if you start thinking about it, you could soon get confused. First of all, an infinitely deep Dirac sea completely filled with electrons must carry an infinite charge! Why would we not notice the presence of this immense charge? With a little bit of creativity one can come up with an easy answer. Other elementary fermionic particles (such as protons) carry a positive charge, and if there are both an infinite number of electrons and protons in the Dirac sea, there is no reason why the charges of all electrons and protons could not cancel, leading to a total charge of zero in the vacuum. Apart from the charge, we also have a problem with the gravitation produced by the mass of all the particles. This is in fact similar to the problem with the zero-point energy of the bosons we encountered in Chapter 8. The most sensible thing to do is to assume that there is some cut-off energy above which (and below which) the Dirac equation does not hold anymore. What governs the physics beyond this energy scale is not known, but hopefully it resolves the conceptual difficulties with the Dirac sea. The best strategy is pragmatic: we define the completely filled Dirac sea to have zero energy and charge and only consider changes in this situation, that is, excitations of the vacuum.

13.2.2 Second quantization

Now we are ready to do the second quantization for the Dirac equation. For each discrete value of electron momentum, we introduce four electron-creation operators corresponding to the four solutions (13.55) and (13.56) of the Dirac equation. We label these operators with the projection of the electron's spin ($s_z = \frac{1}{2}\sigma$, where $\sigma = \pm 1$) and with the sign of the energy of the electron ($\epsilon = \pm 1$). In general, the electronic creation operator thus reads $\hat{a}^\dagger_{\sigma,\epsilon,\mathbf{k}}$. The field operator of the Dirac field can be expressed as

$$\hat{\psi}(\mathbf{r}, t) = \sum_{\mathbf{k},\sigma,\epsilon} \psi_{\sigma,\epsilon}(\mathbf{k})\hat{a}_{\sigma,\epsilon,\mathbf{k}} \exp\left\{-\frac{i}{\hbar}E(\mathbf{k})t + i\mathbf{k} \cdot \mathbf{r}\right\}. \tag{13.62}$$

To obtain the Dirac vacuum $|v\rangle$, we fill all states with negative energy,

$$|v\rangle = \prod_{\mathbf{k}} \hat{a}^\dagger_{+,-,\mathbf{k}} \hat{a}^\dagger_{-,-,\mathbf{k}}|0\rangle. \tag{13.63}$$

But we see that with this definition it is no longer the case that applying an electronic *annihilation* operator to the vacuum, $\hat{a}|v\rangle$, always produces zero. Indeed, since all negative energy states are filled, it is perfectly fine to apply an annihilation operator with $\epsilon = -1$. So if we were to do this, what is actually happening? In fact, we would obtain a new state – a hole – where, in comparison with the vacuum, the charge is decreased by one electron charge e, the spin is decreased by $\frac{1}{2}\sigma$, the momentum is decreased by $\hbar\mathbf{k}$, and the energy is "decreased by a negative value," that is, increased. This we can readily interpret as the creation of a *positron* with momentum $-\hbar\mathbf{k}$, spin $-\frac{1}{2}\sigma$, positive charge $-e$, and positive energy. We can thus redefine the original CAPs introducing "proper" electron creation operators $\hat{c}^\dagger_{\sigma,\mathbf{k}} = \hat{a}^\dagger_{\sigma,+,\mathbf{k}}$ (acting at positive energies only) and positron creation operators

$$\hat{b}^{\dagger}_{+,\mathbf{k}} = -\hat{a}_{-,-,(-\mathbf{k})} \quad \text{and} \quad \hat{b}^{\dagger}_{-,\mathbf{k}} = \hat{a}_{+,-,(-\mathbf{k})}, \tag{13.64}$$

acting at negative energies. These newly introduced operators satisfy the usual fermionic commutation relations.

The Hamiltonian reads in terms of these operators

$$\hat{H} = \sum_{\mathbf{k},\sigma} \sqrt{m^2c^4 + \hbar^2 k^2 c^2} \left\{ \hat{c}^{\dagger}_{\sigma,\mathbf{k}} \hat{c}_{\sigma,\mathbf{k}} + \hat{b}^{\dagger}_{\sigma,\mathbf{k}} \hat{b}_{\sigma,\mathbf{k}} \right\}, \tag{13.65}$$

and the Dirac vacuum satisfies the standard definition of the vacuum,

$$\hat{c}_{\sigma,\mathbf{k}}|v\rangle = \hat{b}_{\sigma,\mathbf{k}}|v\rangle = 0, \tag{13.66}$$

for any σ and $\mathbf{k}$.

This is in fact how Dirac predicted the existence of positrons – positively charged particles with the same mass as the electron.[4] This revolutionary hypothesis of Dirac was confirmed in 1933 with the experimental discovery of the positron. Since then it has become generally accepted that almost every particle has a counterpart: an antiparticle of opposite charge. Exceptions are particles like photons that do not bear any charge. In this case, particles and antiparticles are identical. The symmetry between particles and antiparticles raises the question why we see so many particles around us and so few antiparticles. An anti-world with positrons instead of electrons and anti-protons and anti-neutrons instead of protons and neutrons has the same chance to exist as our world. The present situation is believed to originate from a symmetry breaking at the first milliseconds of the history of the Universe.

It is clear that particles and antiparticles cannot be together for a long time. They meet and annihilate each other, emitting energy in the form of photons. However, due to the constraint of momentum conservation in such a process, an electron and a positron can annihilate only while emitting two or more photons. The inverse process is prohibited as well: a single photon that has in principle enough energy to excite an electron–positron pair does not do so in the vacuum.

There is an interesting analogy between the Dirac vacuum and the ground state of some condensed matter systems like semiconductors. In most semiconductors, two sorts of carrier are present: electrons and holes. A typical spectrum of a semiconductor consists of an upper energy band (conduction band) and a lower energy band (valence band) separated by a gap. The states in the valence band are completely filled and the excitations in this band are in fact positively charged holes. The differences with the Dirac vacuum are the energy scales (it costs about 10^6 eV to excite an electron–hole pair in vacuum and about 1 eV in a typical semiconductor), but also concern symmetry. The holes in semiconductors are never precisely identical to the electrons in the conduction band. Another interesting analogy to "Dirac physics" in a condensed matter system is provided by graphene, a graphite sheet

---

[4] The strength of a scientist is not the ability to avoid errors but the ability to recover them quickly. Originally Dirac suggested that the positively charged hole in the vacuum is in fact a proton, because that was the only positively charged elementary particle known at the time. Weyl then pointed out that a hole in the vacuum must have the same mass as the corresponding particle. Since the electron mass and proton mass differ by a factor of roughly 1800, "missing" electrons cannot be identified with "extra" protons. Dirac immediately acknowledged this and adjusted his hypothesis, predicting the existence of the positron.

consisting of a single atomic layer. Its low-energy spectrum consists of two cones symmetric in energy space, the states in the lower cone being completely filled. This spectrum is a close match to the spectrum of (two-dimensional) massless Dirac fermions, see Exercise 4 at the end of this chapter.

We can re-express the operator of a bispinor Dirac field in terms of the operators and the solutions (13.55) and (13.56)

$$\hat{\psi}(\mathbf{r}) = \frac{1}{\sqrt{\mathcal{V}}} \sum_{\mathbf{k},\alpha} \left(\psi_\alpha^{(e)}(\mathbf{k})\hat{c}_{\alpha,\mathbf{k}} \exp\{i\mathbf{k}\cdot\mathbf{r}\} + \psi_\alpha^{(p)}(\mathbf{k})\hat{b}_{\alpha,\mathbf{k}}^\dagger \exp\{-i\mathbf{k}\cdot\mathbf{r}\} \right), \qquad (13.67)$$

with $\psi_\alpha^{(e)}(\mathbf{k}) = \psi_{\alpha,+}(\mathbf{k})$ and $\psi_\alpha^{(p)}(\mathbf{k}) = i(\sigma_y)_{\alpha\beta}\psi_{\beta,-}(-\mathbf{k})$ to make it conform to our definition of the positron CAPs. The indices α and β are ± 1 and indicate spin, and $\psi_{\pm,\pm}(\mathbf{k})$ is a four-component bispinor.

Control question. Do you see how the definitions of $\psi_\alpha^{(e)}(\mathbf{k})$ and $\psi_\beta^{(p)}(\mathbf{k})$ follow?

Let us introduce the compact notations

$$\mathbf{u}(\mathbf{k}) = \frac{\hbar c \mathbf{k}}{\sqrt{2E(\mathbf{k})[mc^2 + E(\mathbf{k})]}} \quad \text{and} \quad w(\mathbf{k}) = \sqrt{1 - \mathbf{u}(\mathbf{k})^2}, \qquad (13.68)$$

where all $E(\mathbf{k})$ are taken positive. With this,

$$\psi_\alpha^{(e)} = \begin{pmatrix} w\psi_\alpha \\ (\mathbf{u}\cdot\boldsymbol{\sigma})\psi_\alpha \end{pmatrix} \quad \text{and} \quad \psi_\alpha^{(p)} = \begin{pmatrix} i(\mathbf{u}\cdot\boldsymbol{\sigma})\sigma_y\psi_\alpha \\ iw\sigma_y\psi_\alpha \end{pmatrix}. \qquad (13.69)$$

Let us emphasize here for the sake of clarity that the functions $\psi_\alpha^{(e)}$ and $\psi_\alpha^{(p)}$ are four-component bispinors while the function ψ_α is a two-component spinor. In both cases the index α labels spin, and can take the two values ± 1.

As we know, the fermion field is not a physical quantity by itself: physical quantities are at least bilinear in terms of the fields. The most important quantities are the fermion density and current. Let us express the corresponding operators in terms of CAPs, we will need these relations when dealing with interactions. We substitute (13.67) into (13.49),

$$\hat{\rho}(\mathbf{r}) = \frac{1}{\mathcal{V}} \sum_{\mathbf{k},\mathbf{q},\alpha,\beta} \left(R_{\alpha\beta}^{ee}(\mathbf{k}-\mathbf{q},\mathbf{k})\hat{c}_{\mathbf{k}-\mathbf{q},\alpha}^\dagger \hat{c}_{\mathbf{k},\beta} + R_{\alpha\beta}^{pp}(\mathbf{k}-\mathbf{q},\mathbf{k})\hat{b}_{\mathbf{k}-\mathbf{q},\alpha}^\dagger \hat{b}_{\mathbf{k},\beta} \right.$$
$$\left. + R_{\alpha\beta}^{ep}(\mathbf{k}-\tfrac{1}{2}\mathbf{q}, -\mathbf{k}-\tfrac{1}{2}\mathbf{q})\hat{c}_{\mathbf{k}-\mathbf{q}/2,\alpha}^\dagger \hat{b}_{-\mathbf{k}-\mathbf{q}/2,\beta}^\dagger \right. \qquad (13.70)$$
$$\left. + R_{\alpha\beta}^{pe}(\mathbf{k}+\tfrac{1}{2}\mathbf{q}, -\mathbf{k}+\tfrac{1}{2}\mathbf{q})\hat{b}_{\mathbf{k}+\mathbf{q}/2,\alpha} \hat{c}_{-\mathbf{k}+\mathbf{q}/2,\beta} \right) \exp\{i\mathbf{q}\cdot\mathbf{r}\},$$

$$\hat{\mathbf{j}}(\mathbf{r}) = \frac{c}{\mathcal{V}} \sum_{\mathbf{k},\mathbf{q},\alpha,\beta} \left(\mathbf{J}_{\alpha\beta}^{ee}(\mathbf{k}-\mathbf{q},\mathbf{k})\hat{c}_{\mathbf{k}-\mathbf{q},\alpha}^\dagger \hat{c}_{\mathbf{k},\beta} + \mathbf{J}_{\alpha\beta}^{pp}(\mathbf{k}-\mathbf{q},\mathbf{k})\hat{b}_{\mathbf{k}-\mathbf{q},\alpha}^\dagger \hat{b}_{\mathbf{k},\beta} \right.$$
$$\left. + \mathbf{J}_{\alpha\beta}^{ep}(\mathbf{k}-\tfrac{1}{2}\mathbf{q}, -\mathbf{k}-\tfrac{1}{2}\mathbf{q})\hat{c}_{\mathbf{k}-\mathbf{q}/2,\alpha}^\dagger \hat{b}_{-\mathbf{k}-\mathbf{q}/2,\beta}^\dagger \right. \qquad (13.71)$$
$$\left. + \mathbf{J}_{\alpha\beta}^{pe}(\mathbf{k}+\tfrac{1}{2}\mathbf{q}, -\mathbf{k}+\tfrac{1}{2}\mathbf{q})\hat{b}_{\mathbf{k}+\mathbf{q}/2,\alpha} \hat{c}_{-\mathbf{k}+\mathbf{q}/2,\beta} \right) \exp\{i\mathbf{q}\cdot\mathbf{r}\},$$

the coefficients R and $\mathbf{J}$ being given by

$$R^{ee}(\mathbf{k}_1, \mathbf{k}_2) = -R^{pp}(\mathbf{k}_1, \mathbf{k}_2) = w_1 w_2 + \mathbf{u}_1 \cdot \mathbf{u}_2 + i\boldsymbol{\sigma} \cdot (\mathbf{u}_1 \times \mathbf{u}_2), \tag{13.72}$$

$$R^{ep}(\mathbf{k}_1, \mathbf{k}_2) = [R^{pe}(\mathbf{k}_2, \mathbf{k}_1)]^\dagger = (w_1 \mathbf{u}_2 + w_2 \mathbf{u}_1) \cdot \boldsymbol{\sigma} i\sigma_y, \tag{13.73}$$

$$\mathbf{J}^{ee}(\mathbf{k}_1, \mathbf{k}_2) = -\mathbf{J}^{pp}(\mathbf{k}_1, \mathbf{k}_2) = w_1 \mathbf{u}_2 + w_2 \mathbf{u}_1 + i\boldsymbol{\sigma} \times (\mathbf{u}_2 w_1 - \mathbf{u}_1 w_2), \tag{13.74}$$

$$\mathbf{J}^{ep}(\mathbf{k}_1, \mathbf{k}_2) = [\mathbf{J}^{pe}(\mathbf{k}_2, \mathbf{k}_1)]^\dagger = w_1 w_2 \boldsymbol{\sigma} i\sigma_y + (\mathbf{u}_1 \cdot \boldsymbol{\sigma})\boldsymbol{\sigma}(\mathbf{u}_2 \cdot \boldsymbol{\sigma}) i\sigma_y, \tag{13.75}$$

where we use the notation $\mathbf{u}_a \equiv \mathbf{u}(\mathbf{k}_a)$ and $w_a \equiv w(\mathbf{k}_a)$. The coefficients with ee and pp are diagonal in spin space, i.e. they come with a 2×2 unity matrix. Note that here $\hat{\mathbf{j}}$ represents the particle current density: to obtain the (charge) current density one has to include a factor e.

In the non-relativistic limit of $k_{1,2} \ll mc/\hbar$ we recover the familiar spin-independent expressions

$$R^{ee} = 1 \quad \text{and} \quad \mathbf{J}^{ee}(\mathbf{k}_1, \mathbf{k}_2) = \frac{\hbar(\mathbf{k}_1 + \mathbf{k}_2)}{2mc}. \tag{13.76}$$

For positrons, they are of opposite sign, as expected for particles of opposite charge. At $\mathbf{k}_1 = \mathbf{k}_2$, that is, for contributions to slowly varying densities and currents, we have $R^{ee} = 1$ and $\mathbf{J}^{ee} = \mathbf{v}(\mathbf{k})/c$, in accordance with the assumptions made in Chapter 9. A new element typical for relativistic physics is the *spin dependence* of density and current at $\mathbf{k}_1 \neq \mathbf{k}_2$. This is a manifestation of spin–orbit coupling, see Exercise 3 at the end of this chapter.

Another new element is that in general the density and current cannot just be simply separated into electron and positron contributions: there are interference terms coming with R^{ep}, $\mathbf{J}^{ep}$, etc. These terms manifest the fact that electrons and positrons are not independent particles, but arising from the same Dirac field. The interference terms can create and annihilate electron–positron pairs. Terms proportional to $i\sigma_y$ create a pair in a spin-singlet state, while those coming with $i\sigma_y \boldsymbol{\sigma}$ create pairs in spin-triplet states.

13.2.3 Interaction with the electromagnetic field

Electrons possess electric charge and should therefore interact with the electromagnetic field. We can derive this interaction with the same gauge trick we used in Chapter 9. The relativistic Dirac equation

$$\left\{ \hat{\gamma}^\mu \frac{\partial}{\partial x_\mu} + i\frac{mc}{\hbar} \right\} \psi = 0, \tag{13.77}$$

is gauge-invariant. This means that if we transform the bispinor

$$\psi'(x^\mu) \rightarrow \psi(x^\mu) \exp\{i\Phi(x^\mu)\}, \tag{13.78}$$

and simultaneously transform the equation into

$$\gamma^\mu \left\{ \frac{\partial}{\partial x_\mu} - i\frac{\partial}{\partial x_\mu}\Phi \right\} \psi' + i\frac{mc}{\hbar}\psi' = 0, \tag{13.79}$$

it describes precisely the same physics in the sense that all observable quantities remain the same. One can see this, for instance, from the fact that the four-dimensional current density remains invariant under this transformation,

$$\bar{\psi}'\gamma^\mu\psi' \to \bar{\psi}\gamma^\mu\psi. \tag{13.80}$$

As discussed in Chapter 9, gauge invariance is not a void abstraction. The fundamental interactions are believed to originate from gauge invariance, so (13.79) should suffice to fix the form of interaction between radiation and relativistic electrons. This is achieved by the substitution

$$\frac{\partial}{\partial x_\mu}\Phi \to \frac{e}{\hbar}A_\mu, \tag{13.81}$$

which corresponds to the use of "long" derivatives in the presence of an electromagnetic field,

$$\frac{\partial}{\partial x_\mu} \to \frac{\partial}{\partial x_\mu} - i\frac{e}{\hbar}A_\mu. \tag{13.82}$$

So, the Dirac electron in the presence of an electromagnetic field is described by the following Lorentz covariant equation,

$$\gamma^\mu\left\{\frac{\partial}{\partial x_\mu} - i\frac{e}{\hbar}A_\mu\right\}\psi + i\frac{mc}{\hbar}\psi = 0. \tag{13.83}$$

At the moment, this equation is for a single-electron wave function in the presence of a classical electromagnetic field. This equation has many applications by itself. It describes, for instance, the relativistic hydrogen atom and spin–orbit interaction in solids. Besides, the equation is ready for a second-quantization procedure.

13.3 Quantum electrodynamics

To perform the quantization, one uses the analogy between the equation for a wave function and the equation for a corresponding fermionic field operator. So the same equation is valid for bispinor field operators $\hat{\psi}(x^\mu)$. The classical field A_μ is also replaced by the bosonic field operator, extensively studied in Chapters 7–9. This then defines the reaction of the electrons to the electromagnetic field. But, as we know, the electrons also affect the field. To take this into account one quantizes the Maxwell equation (13.27) and uses expression (13.49) for the four-vector current density. The resulting equations define a quantum field theory – *quantum electrodynamics*. Being non-linear, the set of equations cannot be solved exactly.

There are several ways to implement this quantization. A consistently Lorentz-invariant way is to use quantum Lagrangians rather than Hamiltonians: indeed, any Hamiltonian distinguishes time from the other spacetime coordinates. One would also impose the Lorentz-invariant Lorenz gauge $\partial_\mu A^\mu = 0$. This would be the most adequate treatment, resulting in elegant short forms and formulas. However, it requires learning specific techniques not applied in general quantum mechanics and not discussed in this book.

Instead, we do it in a simpler way. We stick to the Hamiltonian description and deal with the electromagnetic field in the Coulomb gauge. We use the electron and photon operators in the same form as in the previous chapters. The main difference is that we bring positrons into the picture.

13.3.1 Hamiltonian

The full Hamiltonian consists of terms describing the free fermions and photons, the Coulomb repulsion between the fermions, and their interaction with the radiation,

$$\hat{H} = \hat{H}_{\rm F} + \hat{H}_{\rm C} + \hat{H}_{\rm int}. \tag{13.84}$$

The free particle Hamiltonian assumes its usual form,

$$\hat{H}_{\rm F} = \sum_{\sigma,\mathbf{k}} E(\mathbf{k}) \left\{ \hat{c}^\dagger_{\sigma,\mathbf{k}} \hat{c}_{\sigma,\mathbf{k}} + \hat{b}^\dagger_{\sigma,\mathbf{k}} \hat{b}_{\sigma,\mathbf{k}} \right\} + \hbar\omega(\mathbf{k}) \hat{a}^\dagger_{\mathbf{k},\sigma} \hat{a}_{\mathbf{k},\sigma}, \tag{13.85}$$

where $\hat{a}^{(\dagger)}$ are photon CAPs and σ labels both the projections of the fermion spin and the photon polarization. We write the Coulomb term in the Fourier representation,

$$\hat{H}_{\rm C} = 2\pi\alpha\hbar c \mathcal{V} \sum_{\mathbf{q}} \frac{1}{q^2} \hat{\rho}_{-\mathbf{q}} \hat{\rho}_{\mathbf{q}}, \tag{13.86}$$

where α is the fine structure constant $\alpha = e^2/4\pi\varepsilon_0\hbar c$. The Fourier components of the density operator $\hat{\rho}(\mathbf{r}) = \sum_a \hat{\psi}^\dagger_a(\mathbf{r})\hat{\psi}_a(\mathbf{r})$ are given by (13.70). The interaction with the photons is described by the familiar term $\hat{H}_{\rm int} = -e \int d\mathbf{r}\, \hat{\mathbf{A}}(\mathbf{r}) \cdot \hat{\mathbf{j}}(\mathbf{r})$. We thus simply substitute the operators in terms of the corresponding CAPs to arrive at

$$\hat{H}_{\rm int} = -\sqrt{2\pi\alpha}\,\frac{c\hbar}{\sqrt{\mathcal{V}}} \sum_{\mathbf{k},\mathbf{q}} \sum_{\alpha,\beta,\gamma} \frac{1}{\sqrt{q}} \Big\{ \hat{a}^\dagger_{\mathbf{q},\gamma} \Big(K^{ee}_{\alpha\beta\gamma}(\mathbf{k}-\mathbf{q},\mathbf{k})\hat{c}^\dagger_{\mathbf{k}-\mathbf{q},\alpha}\hat{c}_{\mathbf{k},\beta}$$

$$+ K^{pp}_{\alpha\beta\gamma}(\mathbf{k}-\mathbf{q},\mathbf{k})\hat{b}^\dagger_{\mathbf{k}-\mathbf{q},\alpha}\hat{b}_{\mathbf{k},\beta} + K^{ep}_{\alpha\beta\gamma}(\mathbf{k}-\tfrac{1}{2}\mathbf{q},-\mathbf{k}-\tfrac{1}{2}\mathbf{q})\hat{c}^\dagger_{\mathbf{k}-\mathbf{q}/2,\alpha}\hat{b}^\dagger_{-\mathbf{k}-\mathbf{q}/2,\beta}$$

$$+ K^{pe}_{\alpha\beta\gamma}(\mathbf{k}+\tfrac{1}{2}\mathbf{q},-\mathbf{k}+\tfrac{1}{2}\mathbf{q})\hat{b}_{\mathbf{k}+\mathbf{q}/2,\alpha}\hat{c}_{-\mathbf{k}+\mathbf{q}/2,\beta} \Big) + \text{H.c.} \Big\}, \tag{13.87}$$

where $K^{ee}_{\alpha\beta\gamma}(\mathbf{k}_1,\mathbf{k}_2) = \mathbf{e}_\gamma \cdot \mathbf{J}^{ee}_{\alpha\beta}(\mathbf{k}_1,\mathbf{k}_2)$ and similar relations hold for K^{ep}, K^{pe}, and K^{pp}.

So, what is this Hamiltonian good for? It turns out that relativistic quantum electrodynamics predicts a set of new experimentally observable effects at sufficiently high energies $\simeq mc^2$ that are absent in usual electrodynamics. For instance, electrons and positrons can annihilate while producing light. Also, virtual emission of electron–positron pairs results in non-linear electric effects present in the vacuum, like the scattering of a photon on a phonon. In general one can say that the number of fermionic particles is not conserved in course of a scattering event: new electron–positron pairs could be created. The above theory also predicts relativistic corrections to the more common scattering processes. For instance, theoretical predictions for scattering cross-sections can be obtained with the help of the Hamiltonian (13.84).

13.3.2 Perturbation theory and divergences

We see that for both interaction terms – Coulomb and radiative – the interaction enters the Hamiltonian with the small dimensionless parameter α. This suggests that also in the relativistic world the interactions can be efficiently described by perturbation theory.

Indeed, perturbation theory has been successfully used to evaluate cross-sections of unusual scattering processes as mentioned above, as well as relativistic corrections to the usual scattering processes. However, we noted already at the first steps of the development of quantum electrodynamics that the perturbation series suffer from serious problems: *divergences* contributed to by virtual states of high energy. It took several decades to overcome these problems and to understand how and why the divergences cancel for observable quantities. At the end of the day, the *problems* with the calculations appeared to be more interesting and important than the *results* of the calculations. The understanding how to handle the divergences of the perturbation series has changed the understanding of what a physical theory is and contributed to the concept of *renormalization*, the most important concept that has emerged in physics since the middle of the twentieth century.

To see the divergences, it is sufficient to look at the interaction corrections to very simple quantities. To start with, let us evaluate such correction to the energy of an electron with wave vector $\mathbf{k}$ and spin σ. Without interactions, the energy is $E(\mathbf{k}) = \sqrt{(mc^2)^2 + (\hbar ck)^2}$. It is important for further calculation to recognize that the particle energy is in fact an excitation energy: it is the difference of the energies of the excited state $\hat{c}^\dagger_{\mathbf{k},\sigma}|v\rangle$ and of the vacuum $|v\rangle$. We thus need to compute the corrections to the energies of both states and take the difference.

Let us consider first the effect of the Coulomb term. In this case, we deal with the first-order correction $\simeq \alpha$ that reads

$$[\delta E(\mathbf{k})]_C = \langle v|\hat{c}_{\mathbf{k},\sigma}\hat{H}_C\hat{c}^\dagger_{\mathbf{k},\sigma}|v\rangle - \langle v|\hat{H}_C|v\rangle. \tag{13.88}$$

The Hamiltonian $\hat{H}_C$ (see (13.86)) contains two density operators $\hat{\rho}_{\pm\mathbf{q}}$. We see that the only terms in $\hat{\rho}$ giving a non-zero result when acting on the Dirac vacuum are those $\propto \hat{c}^\dagger\hat{b}^\dagger$ with all possible wave vectors in the density operator on the right. After the action, the vacuum is in one of the virtual states with an electron–positron pair. Most of the terms creating electron–positron pairs however commute with $\hat{c}^\dagger_{\mathbf{k},\sigma}$ and therefore provide the same contribution to the energies of the vacuum and the excited state. They thus cancel in (13.88) and do not contribute to the change of the particle energy. There are two exceptions: (i) terms in $\hat{H}_C$ which create and annihilate an electron–positron pair with the electron in the state $\mathbf{k}, \sigma$ contribute to the second term in (13.88) but not to the first term, and (ii) terms with $\hat{c}^\dagger_{\mathbf{k}-\mathbf{q},\sigma'}\hat{c}_{\mathbf{k},\sigma}$ in the right density operator and $\hat{c}^\dagger_{\mathbf{k},\sigma}\hat{c}_{\mathbf{k}-\mathbf{q},\sigma'}$ in the left one contribute to the first term in (13.88) but not to the second. The states thus involved in the correction are schematically depicted in Fig. 13.2a, and we see that the correction of interest is the difference of the two contributions

$$[\delta E(\mathbf{k})]_C = \alpha \int \frac{d\mathbf{q}}{(2\pi)^3} \frac{2\pi\hbar c}{q^2} \sum_{\sigma'} \left\{ R^{ee}_{\sigma\sigma'}(\mathbf{k}, \mathbf{k}-\mathbf{q}) R^{ee}_{\sigma'\sigma}(\mathbf{k}-\mathbf{q}, \mathbf{k}) \right.$$
$$\left. - R^{ep}_{\sigma\sigma'}(\mathbf{k}, \mathbf{k}-\mathbf{q}) R^{pe}_{\sigma'\sigma}(\mathbf{k}-\mathbf{q}, \mathbf{k}) \right\}. \tag{13.89}$$

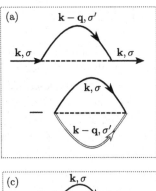
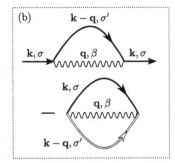
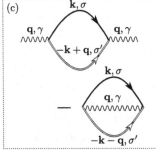
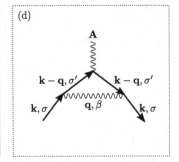

Fig. 13.2 Quantum states involved in divergent corrections to particle energies and interaction constants. The black arrows represent electron states, the white arrows positron states, and the wiggly lines photon states. (a) The Coulomb correction to the electron energy. (b) The radiation correction to the electron energy. (c) Correction to the photon energy. (d) Vertex correction.

Let us investigate the convergence of the integral at $q \gg k$. A rough estimation would come from a dimensional analysis. The coefficients R are dimensionless and tend to a constant limit at $q \to \infty$. The integral is then estimated as $[\delta E]_C \simeq \alpha \hbar c \int dq$, and we see that it diverges linearly at the upper limit! Fortunately, the situation is not so bad since the terms with R^{ee} and R^{ep} cancel each other at $q \to \infty$. Therefore, the existence of antiparticles helps us to fight the divergences. Substituting explicit expressions for R^{ee}, R^{ep}, and R^{pe} yields

$$[\delta E(\mathbf{k})]_C = 2\pi\alpha\hbar c \int \frac{d\mathbf{q}}{(2\pi)^3 q^2} \left(\frac{m^2 c^4}{E(\mathbf{k})E(\mathbf{k} - \mathbf{q})} \right), \tag{13.90}$$

the term in brackets emerging from the cancellation of electron–electron and electron–positron terms. We note that at large q the energy goes to the limit $E(\mathbf{k} - \mathbf{q}) \to \hbar c q$. Therefore, the integral over q still diverges logarithmically at large q. Let us assume an upper cut-off $q_{\rm up}$ in momentum space and set $E(\mathbf{k} - \mathbf{q}) \to \hbar c q$, which is valid at $q \gg q_{\rm low} \simeq mc/\hbar$, and we take the integral over q with the logarithmic accuracy cutting it at $q_{\rm up}$ at the upper limit and at $q_{\rm low}$ at the lower limit,

$$[\delta E(\mathbf{k})]_C = 2\pi\alpha \frac{m^2 c^4}{E(\mathbf{k})} \int \frac{d\mathbf{q}}{(2\pi)^3 q^3} = \alpha \frac{m^2 c^4}{\pi E(\mathbf{k})} \mathcal{L},$$

$$\text{where} \quad \mathcal{L} = \int_{q_{\rm low}}^{q_{\rm up}} \frac{dq}{q} = \ln \frac{q_{\rm up}}{q_{\rm low}}. \tag{13.91}$$

A compact way to represent this correction is to ascribe the change of the electron's energy to a change of its rest mass. Indeed, from $E = \sqrt{(mc^2)^2 + (c\hbar k)^2}$ we see that $\delta E = (\delta m)mc^4/E$, and the energy shift is equivalent to a relative change of the rest mass

$$\left(\frac{\delta m}{m}\right)_C = \frac{\alpha}{\pi}\mathcal{L}. \tag{13.92}$$

We see that the Coulomb term leads to a positive interaction-induced correction to the electron (and, by symmetry, to the positron) mass that eventually logarithmically diverges at large wave vectors. We postpone the discussion of this rather surprising fact until we have found all divergent quantities in quantum electrodynamics.

There is also a radiation correction to $E(\mathbf{k})$. It results from a second order perturbation in $\hat{H}_{\text{int}}$, and therefore is also proportional to α. The correction to the energy of the one-particle state involves a virtual state composed of a photon with wave vector $\mathbf{q}$ and polarization β and and an electron with wave vector $\mathbf{k} - \mathbf{q}$ and polarization σ'. The corresponding "missing" contribution for the correction to the vacuum energy involves the virtual state composed of an electron with $\mathbf{k}, \sigma$, a photon with $\mathbf{q}, \beta$, and a positron with $-\mathbf{k} - \mathbf{q}, \sigma'$. (Fig. 13.2(b)). We thus arrive at

$$[\delta E(\mathbf{k})]_R = -2\pi\alpha \sum_{\beta,\sigma'} \int \frac{d\mathbf{q}(\hbar c)^2}{(2\pi)^3 q} \left(\frac{|K_{\sigma\sigma'\beta}^{ee}(\mathbf{k}, \mathbf{k}-\mathbf{q})|^2}{E(\mathbf{k}-\mathbf{q}) + \hbar cq - E(\mathbf{k})} - \frac{|K_{\sigma\sigma'\beta}^{ep}(\mathbf{k}, -\mathbf{k}-\mathbf{q})|^2}{E(\mathbf{k}-\mathbf{q}) + \hbar cq + E(\mathbf{k})} \right). \tag{13.93}$$

Similar to the Coulomb correction, a dimensional estimation leads to a linear divergence. Electron–electron and electron–positron terms cancel each other at large q, lowering the degree of divergence again to a logarithmic one. The calculation yields

$$[\delta E(\mathbf{k})]_R = \frac{5\alpha}{6\pi} \frac{m^2 c^4}{E(\mathbf{k})}\mathcal{L} - \frac{2\alpha}{3\pi} E(\mathbf{k})\mathcal{L}. \tag{13.94}$$

We know how to deal with the first term in this expression: it gives a change of the electron mass. The second term is more intriguing since it gives a correction even at large k, where the spectrum does not depend on the mass. One can regard this correction as a change of the speed of light. Indeed, for small changes

$$\delta E = \frac{\delta c}{c}\left(E + \frac{m^2 c^4}{E}\right), \tag{13.95}$$

and with this we obtain

$$\frac{\delta m}{m} = \left(\frac{\delta m}{m}\right)_C + \left(\frac{\delta m}{m}\right)_R = \frac{5\alpha}{2\pi}\mathcal{L} \quad \text{and} \quad \frac{\delta c}{c} = -\frac{2\alpha}{3\pi}\mathcal{L}. \tag{13.96}$$

We can check if the change computed is indeed a change of the speed of the light: we can actually compute the corresponding correction to the photon spectrum. The correction arises in second order in $\hat{H}_{\text{int}}$. Let us compute this correction to the energy of the state with a photon having $\mathbf{q}, \gamma$. The correction involves a virtual state with an electron–positron pair, the electron and positron being in the states $\mathbf{k}, \sigma$ and $-\mathbf{k} + \mathbf{q}, \sigma'$. The "missed" virtual state in the correction to the vacuum energy consists of a photon with $\mathbf{q}, \gamma$, an electron and positron with $\mathbf{k}, \sigma$ and $-\mathbf{k} - \mathbf{q}, \sigma'$, respectively. The total correction is schematically depicted in Fig. 13.2(c), and reads explicitly

$$\delta\omega(\mathbf{q}) = \frac{2\pi\alpha}{q} \int \frac{d\mathbf{k}\hbar c^2}{(2\pi)^3} \left(\frac{|K_{\sigma\sigma'\gamma}^{ep}(\mathbf{k}, -\mathbf{k} + \mathbf{q})|}{E(-\mathbf{k} + \mathbf{q}) + E(\mathbf{k}) - \hbar cq} - \frac{|K_{\sigma\sigma'\gamma}^{ep}(\mathbf{k}, -\mathbf{k} - \mathbf{q})|}{E(-\mathbf{k} - \mathbf{q}) + E(\mathbf{k}) + \hbar cq} \right).$$
(13.97)

Singling out the logarithmically divergent part, we indeed confirm that the correction to the speed of light is given by (13.96).

Another logarithmic divergence emerges from the corrections to the matrix elements, so-called vertex corrections. Let us consider an external field with a uniform vector potential $\mathbf{A}$. It enters the Hamiltonian with the corresponding current and gives rise to the matrix element between two electron states $|\mathbf{k}, \sigma\rangle = \hat{c}_{\mathbf{k},\sigma}^\dagger |v\rangle$. As we know from Chapter 9, the element is proportional to the electron velocity and charge,

$$M = -e\mathbf{A} \cdot \mathbf{v}(\mathbf{k}).$$
(13.98)

With the interaction, the original electron states acquire a small admixture of the states with a photon and an electron with another wave vector $\mathbf{k} - \mathbf{q}$,

$$|\mathbf{k}, \sigma\rangle = \hat{c}_{\mathbf{k},\sigma}^\dagger |v\rangle + \sum_{\mathbf{q},\sigma',\gamma} \beta_{q,\sigma',\gamma} \hat{a}_{\mathbf{q},\gamma}^\dagger \hat{c}_{\mathbf{k}-\mathbf{q},\sigma'}^\dagger |v\rangle,$$
(13.99)

where the admixture coefficients β_n of the state $|n\rangle$ into the state $|m\rangle$ are given by first-order perturbation theory, $\beta_n = (\hat{H}_{\text{int}})_{mn}/(E_m - E_n)$. These admixtures change the matrix element M. This correction is pictured in Fig. 13.2(d), and reads

$$(\delta M)_1 = -e \sum_{q,\sigma',\gamma} |\beta_{q,\sigma',\gamma}|^2 \mathbf{A} \cdot \mathbf{v}(\mathbf{k} - \mathbf{q})$$

$$= -e2\pi\alpha(\hbar c)^2 \int \frac{d\mathbf{q}}{(2\pi)^3 q} \mathbf{A} \cdot \mathbf{v}(\mathbf{k} - \mathbf{q}) \frac{|K_{\sigma\sigma'\gamma}^{ee}(\mathbf{k}, \mathbf{k} - \mathbf{q})|^2}{(E(\mathbf{k} - \mathbf{q}) - E(\mathbf{k}) + c\hbar q)^2}.$$
(13.100)

Singling out the logarithmically divergent term at large q gives (see Exercise 5)

$$(\delta M)_1 = M\frac{\alpha}{12\pi}\mathcal{L}.$$
(13.101)

It is interesting to note that the integral also diverges at small q, that is, at $q \ll k$. Such divergences are called *infrared*, since they are associated with small wave vectors and long wave lengths, while the divergences at large wave vectors are called *ultraviolet*. This divergence is also a logarithmic one. The calculation (Exercise 5) gives

$$(\delta M)_{1,\text{infrared}} = M\frac{\alpha}{\pi} \left\{ 1 + \frac{1}{2}\left(\frac{c}{v} - \frac{v}{c}\right) \ln\left(\frac{c - v}{v + c}\right) \right\} \mathcal{L}_{\text{infrared}},$$
(13.102)

where we cut the integral over q at q_{low} and $q_{\text{up}} \simeq k$ and introduce $\mathcal{L}_{\text{infrared}} \equiv \ln(q_{\text{up}}/q_{\text{low}})$.

Now let us note that the $(\delta M)_1$ we concentrated on is not the only correction to the matrix element. The wave function actually acquires a second-order correction,

$$|\mathbf{k}, \sigma\rangle \to \left\{ 1 - \frac{1}{2}\sum_{q,\sigma',\gamma} |\beta_{q,\sigma',\gamma}|^2 \right\} |\mathbf{k}, \sigma\rangle,$$
(13.103)

which leads to another correction $(\delta M)_2$ to the matrix element. As a matter of fact, the first-order and second-order corrections can be collected to a nice expression. One can argue about this using gauge invariance: in fact, a time- and space-independent vector potential corresponds to a shift of the wave vector, and the diagonal matrix element considered is the derivative of energy $E(\mathbf{k})$ with respect to this shift. We reckon that $M = -e\mathbf{A} \cdot \mathbf{v}$, not dependent on the interaction. Therefore, the sum of the corrections is proportional to the correction to the velocity,

$$(\delta M)_1 + (\delta M)_2 = -e\mathbf{A} \cdot \delta\mathbf{v}. \tag{13.104}$$

We come to the important conclusion that the electron charge is not affected by the interaction, $\delta e = 0$. The interaction is controlled by so-called dimensionless charge α. We remember that, $\alpha = e^2/4\pi\varepsilon_0\hbar c$. Since the speed of light is changing, α itself is modified as well,

$$\frac{\delta\alpha}{\alpha} = -\frac{\delta c}{c} = \frac{2\alpha}{3\pi}\mathcal{L}. \tag{13.105}$$

If the vector potential is not uniform, the vertex interaction correction is less trivial. Choosing a vector potential in the form corresponding to a uniform magnetic field, and computing the matrix element gives an addition to the electron energy proportional to the field. At zero velocity, this is proportional to the spin and is in fact a Zeeman energy $\propto \mathbf{B} \cdot \hat{\sigma}$. The interaction correction to the matrix element is thus a change of the actual magnetic moment μ_e of the electron in comparison with $2\mu_B$. The correction does not contain logarithmic divergences and is given by

$$\frac{\mu_e}{2\mu_B} = 1 + \frac{\alpha}{2\pi}, \tag{13.106}$$

up to first order in α. This change, called the anomalous magnetic moment, can be readily measured and conforms to the theoretical predictions.

To summarize the results of this section, we attempted to apply quantum perturbation theory in α to quantum electrodynamics. We have found corrections $\propto \alpha$ to the electron mass, the speed of light, and the interaction strength α itself. The corrections are formally infinite. They diverge either at large (ultraviolet) or small (infrared) values of the wave vectors.

13.4 Renormalization

This presents a problem: one does not expect such behavior from a good theory. To put it more precisely, the calculation done reveals a set of problems of different significance and importance. Let us recognize the problems and filter out those that are easy to deal with.

First of all, we note that there would be no problem with ultraviolet divergences if the Hamiltonian in question described excitations in a solid, rather than fundamental excitations in the Universe. For a solid, we would have a natural cut-off above which the theory would not be valid any longer. This cut-off q_{up} can be set to, say, the atomic scale. With this, the corrections would not diverge. Moreover, they would hardly exceed the scale of

α since the factor $\mathcal{L}$ is only a logarithm of large numbers. However, in the context of this chapter, nothing suggests the existence of a natural short-distance cut-off below which the known physical laws cease to work. The ad hoc assumption of the existence of such a cut-off is a blow to the elegance and simplicity of the theory.

Infrared divergences would indicate trouble for a solid-state theory as well. However, in Chapter 12 we gave some clues how to handle these divergences. The point we made in Section 12.9 is that the vacuum can be regarded as an Ohmic environment for charged particles and leads to the orthogonality catastrophe at low energies (and, therefore, at small wave vectors). The orthogonality catastrophe taking place indicates that in this case multi-photon processes are important. Once those are taken into account – this would require summing up the contributions of all orders of perturbation theory in α – the divergences are gone.

This may inspire a hypothesis that the ultraviolet divergences can also be removed by summing up all orders of perturbation theory. If this were the case, it would indicate a non-analytical dependence of observable quantities on α. To give a hypothetical example, the interaction correction to the electron mass could in the end look like $\delta m/m = \alpha \ln(1/\alpha)$. This expression cannot be expanded in Taylor series in α, each term of perturbation theory would diverge. Physically, this would indicate the emergence of a short-distance scale $\simeq \alpha(\hbar/mc)$ below which the electromagnetic interactions become weaker. This is a fair hypothesis to check, and in fact it has been checked. The result is negative: summing up all orders of perturbation theory does not remove the ultraviolet divergences. Moreover, as we will see, electromagnetic interactions are in fact enhanced at short distance scale.

Another problem is that, from the point of view of relativistic symmetry, it seems very troublesome that we have found corrections to the speed of light, whether these corrections are divergent or finite. Indeed, relativistic symmetry is introduced at the spacetime level, long before any material fields come about in this spacetime. One can trace the origin of the correction to the fact that the perturbation theory is formulated at the Hamiltonian level, this explicitly breaking the Lorentz invariance. In particular, our upper cut-off has been set in a non-invariant way: we only cut wave vectors, and do not take any care about the frequencies. However, it remains unclear why the result of the calculation does depend on the way – invariant or non-invariant – we perform it. In principle, observable quantities should not depend on this.

The problem of the ultraviolet divergences persisted for decades. The solution – the principle of *renormalization* – was not accepted commonly and instantly, as happens with most advances in physics. Dirac doubted its validity till his last days, yearning for a more consistent theory. Feynman, despite his crucial role in the development of quantum electrodynamics, called renormalization a "hocus-pocus." However, the ideas and tools of renormalization have slowly spread among the next generation of physicists and nowadays they are indispensable for the analysis of any sufficiently complex physical model.

The key idea of renormalization has already been mentioned in Chapter 11, where we discussed the difference between "bare" and "dressed" quantities. Bare quantities are parameters entering the Hamiltonian, while dressed quantities can be observed in an experiment. Let us take the electron mass as an example. The quantity m we measure in experiments and find in the tables is a dressed mass, corresponding to our concrete world with an interaction constant $\alpha \approx 1/137$. The fact that the mass depends on the interaction suggests that the bare mass – the quantity m^* entering the free single-particle Hamiltonian $\hat{H}_F$ – differs from m. Up to first order in α, we have $m = m^* + \delta m$, where the interaction correction is given by (13.96).

We can now re-express the bare mass $m^* = m - \delta m$, substitute this into the Hamiltonian and expand in $\delta m \propto \alpha$. In comparison with the original Hamiltonian, we get an extra term

$$\delta \hat{H} = (\delta m)\frac{\partial \hat{H}_F}{\partial m} = \frac{\delta m}{m} \sum_{\mathbf{k},\sigma} \frac{m^2 c^4}{E(\mathbf{k})} \left(\hat{c}^\dagger_{\sigma,\mathbf{k}} \hat{c}_{\sigma,\mathbf{k}} + \hat{b}^\dagger_{\sigma,\mathbf{k}} \hat{b}_{\sigma,\mathbf{k}} \right). \tag{13.107}$$

This extra term is called a *counterterm*. It is proportional to α and as such should be taken into account while computing perturbative corrections of first order in α. This cancels the ultraviolet divergences in this order! The results of perturbation theory then do not depend on the cut-off wave vector q_{up} and are therefore well defined.

In our Hamiltonian approach, we need to add counterterms that cancel ultraviolet divergent corrections to the mass and speed of light. Since the coefficients R in the density operator and K in the current density operator depend on these quantities, the explicit form of the counterterms is rather cumbersome. In a Lorentz-invariant (Lagrangian) formulation, one needs three simple counterterms that correspond to the renormalization of the Dirac field, the photon field, and the electron mass. Quantum electrodynamics was shown to be *renormalizable*. This means that the finite set of three counterterms is sufficient to cancel the divergences in all the orders of perturbation theory. There are examples of *unrenormalizable* models. In that case, the number of counterterms required for the cancellation increases with the order of the perturbation series and is thus infinite. The current consensus is that these models are "bad," not describing any physics and have to be modified to achieve renormalizability.

Renormalization brings about the concept of *rescaling*: a sophisticated analogue of the method of dimensional analysis. Let us provide a simple illustration of this rescaling. We know now that quantum electrodynamics is a renormalizable theory so that all physical results do not depend on the upper cut-off. The upper cut-off, however, explicitly enters the counterterms and integrals one has to deal with when computing perturbation corrections. This makes us rich: we have a set of *equivalent* theories that differ in the parameter q_{up}, or, better to say, $\mathcal{L}$.

The theories differ not only in $\mathcal{L}$. Since the corrections to observables like m, c, and α depend on $\mathcal{L}$, these observables are also different for different equivalent theories. One can interpret this as a dependence of these observables on the scale. In relativistic theory, the spatial scale is related to the particle energies involved, $q_{\text{up}} \simeq E/(\hbar c)$. Let us assume that we are attempting to measure α using experiments involving particle scattering with different particle energies. We get a fine structure constant that depends on the scale, $\alpha(\mathcal{L})$. How can we evaluate this?

Let us use (13.105) and, instead of writing a perturbation series, we change the scale $\mathcal{L}$ in small steps, $\mathcal{L} \to \mathcal{L} + \delta \mathcal{L}$. In this way, we rewrite (13.105) as a differential equation

$$\frac{d\alpha}{d\mathcal{L}} = \frac{2}{3\pi}\alpha^2. \tag{13.108}$$

This equation has a solution

$$\alpha(\mathcal{L}) = \frac{\alpha_0}{1 - \alpha_0(3\pi/2)\mathcal{L}}, \tag{13.109}$$

where $\alpha_0 = \alpha(\mathcal{L} = 0)$. Let us interpret this result. The scale $\mathcal{L} = 0$ corresponds to low energies of the order of mc^2. We know from our experience that $\alpha_0 \approx 1/137$. What happens if we increase the energy? The effective strength of the interaction increases as well. The increase, which is quite slow in the beginning, accelerates so that $\alpha(\mathcal{L})$ reaches infinity at a certain scale $\mathcal{L}_c = 2/(3\pi\alpha_0)$ corresponding to $q_c \approx 4 \cdot 10^{124} mc/\hbar$. The momentum (or energy) where this happens is called a Landau pole. In fact, we cannot honestly say that α turns to infinity at this scale. Equation 13.105 was derived perturbatively, that is, under the assumption that $\alpha \ll 1$. It will not work if α grows to values $\simeq 1$. However, even the first-order perturbative calculation indicates that α grows and reaches values of the order of 1 at sufficiently short distances or high energies. Quantum electrodynamics, being a perturbative theory at the usual scales, becomes a non-perturbative theory at very short distances.

This illustration of renormalization concludes this book on Advanced Quantum Mechanics. We have been through a variety of topics, trying to keep the technical side as simple as possible. In this way, any person of a practical inclination can get a glimpse of various fields, while making calculations involving CAPs and quantum states. This tool set constitutes the backbone of quantum mechanics and will be useful for many future applications of this marvelous theory. Dear reader, best of luck with using this!

Table 13.1 Summary: Relativistic quantum mechanics

Principle of relativity, all physics is the same in a stationary or constant velocity references frame

Galilean transformation: $\mathbf{r}' = \mathbf{r} - \mathbf{v}t + \mathbf{r}_0$ and $t' = t + t_0$ relate different reference frames

                       treats $\mathbf{r}$ and t separately, turned out inconsistent with experiments

Lorentz transformation: $x' = \gamma(x - vt)$, $y' = y$, $z' = z$,   and $t' = \gamma[t - (v/c^2)x]$

                       with $\gamma = \sqrt{1 - (v/c)^2}$ is correct relation for $\mathbf{v} = v\mathbf{e}_x$

                       mixes $\mathbf{r}$ and t $\rightarrow$ concept of four-dimensional spacetime

all three-dimensional vectorial quantities are four-vectors in relativistic physics:

     spacetime $x^\mu = (ct, \mathbf{r})$, velocity $u^\mu = \gamma(c, \mathbf{v})$, momentum $p^\mu = (E/c, \gamma m\mathbf{v})$,

     vector and scalar potential $A^\mu = (\phi/c, \mathbf{A})$, current and charge density $j^\mu = (c\rho, \mathbf{j})$

these vectors are contravariant, their covariant counterparts read $x_\mu = (ct, -\mathbf{r})$, $u_\mu = \gamma(c, -\mathbf{v})$, etc.

Dirac equation, the search for a "relativistic Schrödinger equation"

Klein–Gordon equation: $\{\hbar^2 \Box + m^2 c^2\}\psi = 0$,   with $\Box = (1/c^2)\partial_t^2 - \nabla^2$

problem: Lorentz covariant but can lead to negative probability densities

Dirac equation: $\left\{\hat{\gamma}^\mu \dfrac{\partial}{\partial x_\mu} + i\dfrac{mc}{\hbar}\right\}\psi = 0$,   with $\hat{\gamma}^0 = \begin{pmatrix} 1 & 0 \\ 0 & -1 \end{pmatrix}$, $\hat{\boldsymbol{\gamma}} = \begin{pmatrix} 0 & \hat{\boldsymbol{\sigma}} \\ -\hat{\boldsymbol{\sigma}} & 0 \end{pmatrix}$

                       Lorentz covariant and always positive definite probability density

explicitly: $\begin{pmatrix} -i\partial_t & -ic\hat{\boldsymbol{\sigma}} \cdot \nabla \\ ic\hat{\boldsymbol{\sigma}} \cdot \nabla & i\partial_t \end{pmatrix}\psi = -\dfrac{mc^2}{\hbar}\psi$,   with bispinor $\psi = \begin{pmatrix} \psi_A \\ \psi_B \end{pmatrix}$

look for solutions: $\psi \propto \exp\{-\frac{i}{\hbar}p_\mu x^\mu\}$ $\rightarrow$ momentum states with $E = \pm\sqrt{m^2c^4 + p^2c^2}$

field operator: $\hat{\psi}(\mathbf{r}, t) = \sum_{\mathbf{k},\sigma,\epsilon} \psi_{\sigma,\epsilon}(\mathbf{k})\hat{a}_{\sigma,\epsilon,\mathbf{k}} \exp\left\{-\frac{i}{\hbar}E(\mathbf{k})t + i\mathbf{k}\cdot\mathbf{r}\right\}$, with $\sigma, \epsilon = \pm$

Dirac vacuum: all negative energy states are filled, $|v\rangle = \prod_{\mathbf{k}} \hat{a}^\dagger_{+,-,\mathbf{k}} \hat{a}^\dagger_{-,-,\mathbf{k}} |0\rangle$

                       a "hole" in the Dirac sea corresponds to an *antiparticle*

electron CAPs: $\hat{c}^\dagger_{+,\mathbf{k}} = \hat{a}^\dagger_{+,+,\mathbf{k}}$ and $\hat{c}^\dagger_{-,\mathbf{k}} = \hat{a}^\dagger_{-,+,\mathbf{k}}$

positron CAPs: $\hat{b}^\dagger_{+,\mathbf{k}} = -\hat{a}_{-,-,(-\mathbf{k})}$ and $\hat{b}^\dagger_{-,\mathbf{k}} = \hat{a}_{+,-,(-\mathbf{k})}$

in general, operators like $\hat{\rho}$ and $\hat{\mathbf{j}}$ now contain terms proportional to $\hat{c}^\dagger\hat{c}$, $\hat{b}^\dagger\hat{b}$, $\hat{c}^\dagger\hat{b}^\dagger$, and $\hat{b}\hat{c}$

Quantum electrodynamics, describes Dirac particles in the presence of an electromagnetic field

Hamiltonian: (free electrons, positrons, and photons) + (Coulomb interaction)

                      + (interaction between matter and radiation)

perturbation theory: small parameter in interaction terms is $\alpha = e^2/4\pi\varepsilon_0\hbar c \approx 1/137$

divergences: interaction corrections to particle energies diverge logarithmically

introduce cut-off: corrections can be interpreted as corrections to m, c, and α

                     $\dfrac{\delta m}{m} = \dfrac{5\alpha}{2\pi}\mathcal{L}$ and $\dfrac{\delta c}{c} = -\dfrac{\delta\alpha}{\alpha} = -\dfrac{2\alpha}{3\pi}\mathcal{L}$,   with $\mathcal{L} = \ln\dfrac{q_{\text{up}}}{q_{\text{low}}}$

Renormalization

"bare" quantities are parameters in the Hamiltonian, "dressed" quantities are observable

counterterms: extra terms in the Hamiltonian to get "correct" observable quantities

                in QED these counterterms *repair* the divergences $\rightarrow$ QED is *renormalizable*

Exercises

1. *Schrödinger–Pauli equation* (solution included). The Hamiltonian describing the effect of a magnetic field on the spin of an electron, $\hat{H} = -\mu_B \mathbf{B} \cdot \hat{\boldsymbol{\sigma}}$, was introduced in Chapter 1 by arguing that spin is a form of angular momentum, and that it therefore adds to the magnetic moment of the electron. We now show how this Hamiltonian follows from the Dirac equation.

   a. Let us assume for simplicity a time-independent vector potential $\mathbf{A}$. Write the stationary version of the Dirac equation in the presence of an electromagnetic field (13.83) in terms of $\mathbf{A}$ and φ, and as a coupled set of equations for the electron part ψ_A and positron part ψ_B of the bispinor, such as in (13.52). Eliminate the small ψ_B from the equations.

   b. We now use the fact that both the non-relativistic energy $E_{\mathrm{nr}} \equiv E - mc^2$ and $e\varphi$ are much smaller than mc^2. Rewrite the equation for ψ_A found above in terms of E_{nr} instead of E, which makes mc^2 the only large scale in the equation. Expand the fraction containing E_{nr} in the denominator to first order in $(v/c)^2$.

   c. Keep only the zeroth-order term of the expansion and show that to this order the equation for ψ_A reads

   $$\left\{ \frac{1}{2m}(\hat{\mathbf{p}} - e\mathbf{A})^2 - \frac{e\hbar}{2m}\hat{\boldsymbol{\sigma}} \cdot \mathbf{B} + e\varphi \right\} \psi_A = E_{\mathrm{nr}}\psi_A,$$

   indeed the ordinary Schrödinger equation including the interaction of the spin of the electron with the magnetic field.

   Hint: $(\boldsymbol{\sigma} \cdot \mathbf{A})(\boldsymbol{\sigma} \cdot \mathbf{B}) = \mathbf{A} \cdot \mathbf{B} + i\boldsymbol{\sigma} \cdot (\mathbf{A} \times \mathbf{B})$.

2. *Contraction or dilation?* A first-year Klingon cadet studies Lorentz transformations. He inspects (13.6) and finds that the length intervals in two reference systems are related by $L' = \sqrt{1 - (v/c)^2}L$. This brings him to the conclusion that space ships appear bigger while moving. He elaborates an example where a Klingon *B'rel*-class star ship of length 160 m appears as long as 1000 m, moving with a certain velocity v along the enemy lines.

   a. The inexperienced cadet made a common mistake based on our non-relativistic intuition. Which one?

   b. Help him to derive the correct formula for the apparent size change.

   c. What is the correct answer for the apparent size of a *B'rel* ship under the conditions of the example.

3. *Spin–orbit interaction.* In Exercise 1 we started expanding the electronic part of the Dirac equation in orders of $(v/c)^2$. If one includes the next order correction, one finds, inter alia, a term describing the spin–orbit coupling of the electrons. Similarly to the Zeeman term, this coupling was originally added as a quasi-phenomenological term to the electronic Hamiltonian. We now derive it from the Dirac equation.

   a. Write the decoupled Dirac equation for ψ_A, keeping the next term in the expansion of Exercise 1(b). Assume for simplicity that we have no magnetic field, $\mathbf{A} = 0$.

   There are several problems with this expression. One of them is that the resulting ψ_A (and ψ_B) can no longer satisfy the normalization condition $\int d\mathbf{r}|\boldsymbol{\psi}(\mathbf{r})|^2 = 1$. This can be seen as follows. We are interested in corrections to ψ_A of order $\sim (v/c)^2$. However, ψ_B, which up to now we forgot about, is of order $\sim (v/c)\psi_A$, see (13.53). The normalization condition for the whole bispinor thus becomes

$$1 = \int d\mathbf{r}\,[\psi_A^\dagger(\mathbf{r})\psi_A(\mathbf{r}) + \psi_B^\dagger(\mathbf{r})\psi_B(\mathbf{r})] \approx \int d\mathbf{r}\,\psi_A^\dagger(\mathbf{r})\left(1 + \frac{\hat{p}^2}{4m^2c^2}\right)\psi_A(\mathbf{r}),$$

which also forces corrections to ψ_A of the same order $(v/c)^2$.

b. We introduce a "rescaled" two-component spinor $\psi = \hat{A}\psi_A$ which is properly normalized, i.e. it satisfies $\int d\mathbf{r}|\psi(\mathbf{r})|^2 = 1$. Find $\hat{A}$.

c. Write the equation found at (a) for ψ. This equation still contains E_{nr} on both sides. In order to be able to write it in the form $\hat{H}\psi = E_{nr}\psi$, multiply it from the left by $\hat{A}^{-1}$. Keep only terms up to order $(v/c)^2$.

d. Show that

$$\hat{p}^2\varphi = \hbar^2(\boldsymbol{\nabla}\cdot\mathbf{E}) + 2i\hbar\mathbf{E}\cdot\hat{\mathbf{p}} + \varphi\hat{p}^2,$$

and that

$$(\boldsymbol{\sigma}\cdot\hat{\mathbf{p}})\varphi(\boldsymbol{\sigma}\cdot\hat{\mathbf{p}}) = -\hbar\boldsymbol{\sigma}\cdot(\mathbf{E}\times\hat{\mathbf{p}}) + i\hbar\mathbf{E}\cdot\hat{\mathbf{p}} + \varphi\hat{p}^2.$$

e. Use the relations derived at (d) to show that the equation found at (c) can be written as

$$\left\{\frac{\hat{p}^2}{2m} + e\varphi - \frac{\hat{p}^4}{8m^3c^2} - \frac{e\hbar\boldsymbol{\sigma}\cdot(\mathbf{E}\times\hat{\mathbf{p}})}{4m^2c^2} - \frac{e\hbar^2}{8m^2c^2}(\boldsymbol{\nabla}\cdot\mathbf{E})\right\}\psi = E_{nr}\psi, \quad (13.110)$$

the third term describing the spin–orbit coupling of the electron, and the fourth term, the *Darwin term*, accounting for a small energy shift proportional to the charge density $\boldsymbol{\nabla}\cdot\mathbf{E}$.

f. Explain the significance of the third term in (13.110).

4. *Graphene.* The Dirac equation is traditionally only relevant in high-energy physics, when relativistic effects become important. For condensed matter systems at low temperatures, its main merit is that one can derive from it small corrections to the Hamiltonian which describe spin–orbit interaction, hyperfine coupling, etc. Some ten years ago, however, the Dirac equation suddenly started playing an important role in condensed matter physics. In 2004, it was for the first time demonstrated experimentally possible to isolate a single monatomic layer of graphite (called graphene), to connect flakes of graphene to electrodes, and to perform transport measurements on the flakes. The band structure of graphene has the peculiar property that the electronic spectrum close to the Fermi level can be exactly described by a Dirac equation for massless particles.

Graphene consists of a hexagonal lattice of carbon atoms, see Fig. 13.3. Three of the four valence electrons of each carbon atom are used to form sp^2-bonds which hold the lattice together. The fourth electron occupies an (out-of-plane) p_z-orbital, and is not very strongly bound to the atom. These electrons can thus move through the lattice, hopping from one p_z-orbital to a neighboring one, and participate in electrical transport.

We denote the lattice vectors with $\mathbf{b}_1$ and $\mathbf{b}_2$ (see figure), and the set of all lattice sites is thus given by $\mathbf{R} = n_1\mathbf{b}_1 + n_2\mathbf{b}_2$, the ns being integers. There are two atoms per unit cell (indicated with gray and white in the figure) which are separated by $\mathbf{a}$. The available electronic states are thus $\psi_{p_z}(\mathbf{r} - \mathbf{R})$ and $\psi_{p_z}(\mathbf{r} - \mathbf{R} - \mathbf{a})$, where $\psi_{p_z}(\mathbf{r})$ is the

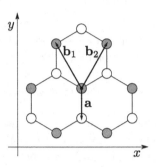

Fig. 13.3 Part of the hexagonal lattice of graphene. The basis vectors $\mathbf{b}_1$ and $\mathbf{b}_2$ are indicated, as well as the vector $\mathbf{a}$ connecting two nearest-neighbor atoms. The unit cell of the lattice contains two atoms: atoms belonging to the two different sublattices are colored white and gray.

wave function of a p_z-state. We denote these states respectively by $|\mathbf{R}\rangle$ and $|\mathbf{R} + \mathbf{a}\rangle$, and assume the resulting electronic basis $\{|\mathbf{R}\rangle, |\mathbf{R} + \mathbf{a}\rangle\}$ to be orthonormal.

We have a periodic lattice with a two-atom basis, so we can write the electronic states on the two sublattices in terms of Fourier components,

$$|\mathbf{k}(1)\rangle = \frac{1}{\sqrt{N}} \sum_{\mathbf{R}} e^{i\mathbf{k}\cdot\mathbf{R}}|\mathbf{R}\rangle \quad \text{and} \quad |\mathbf{k}(2)\rangle = \frac{1}{\sqrt{N}} \sum_{\mathbf{R}} e^{i\mathbf{k}\cdot\mathbf{R}}|\mathbf{R} + \mathbf{a}\rangle.$$

a. Hopping between neighboring p_z-states is described by

$$\hat{H} = t \sum_{\mathbf{R}} \left\{ |\mathbf{R}\rangle\langle\mathbf{R} + \mathbf{a}| + |\mathbf{R}\rangle\langle\mathbf{R} + \mathbf{b}_1 + \mathbf{a}| + |\mathbf{R}\rangle\langle\mathbf{R} + \mathbf{b}_2 + \mathbf{a}| + \text{H.c.} \right\}, \quad (13.111)$$

where for simplicity we assume t to be real. Write this Hamiltonian in the basis $\{|\mathbf{k}(1)\rangle, |\mathbf{k}(2)\rangle\}$.

b. Give the x- and y-components of the two lattice vectors $\mathbf{b}_{1,2}$ in terms of the lattice constant $b = |\mathbf{b}_1| = |\mathbf{b}_2|$.

c. Diagonalize the Hamiltonian (13.111) and show that the eigenenergies read

$$E(\mathbf{k}) = \pm t\sqrt{1 + 4\cos(\tfrac{1}{2}bk_x)\cos(\tfrac{1}{2}\sqrt{3}bk_y) + 4\cos^2(\tfrac{1}{2}bk_x)}. \quad (13.112)$$

We found two electronic bands, corresponding to the positive and negative energies of (13.112). In pure undoped graphene, each atom contributes one electron. This means that in equilibrium and at zero temperature, all electronic states of the lower band are filled, and all of the upper band are empty. In analogy to semiconductors, we can thus call the lower band the valence band and the upper band the conduction band.

d. Calculate the two allowed energies for an electronic state with wave vector $\mathbf{K} \equiv b^{-1}(\tfrac{2}{3}\pi, \tfrac{2}{\sqrt{3}}\pi)$. What does this imply for the band structure? Give the Fermi energy.

e. Electronic states with wave vectors close to $\mathbf{K}$ thus have energies close to the Fermi energy and therefore describe low-energy excitations. Let us investigate these states. We consider a state with wave vector $\mathbf{K} + \mathbf{k}$, where $|\mathbf{k}| \ll |\mathbf{K}|$ is assumed. Write the Hamiltonian for this state in the basis $\{|\mathbf{k}(1)\rangle, |\mathbf{k}(2)\rangle\}$ as you did in (a). Expand the result to linear order in $|\mathbf{k}|$, and show that it can be written as

$$\hat{H} = v_F \boldsymbol{\sigma} \cdot \hat{\mathbf{p}},$$

a Dirac equation without the term mc^2, that is, for massless particles. Give the Fermi velocity v_F in terms of t and b.

5. *Ultraviolet and infrared vertex divergences.* Compute the divergences of the vertex corrections given by (13.100) at large and small q.

    a. Express $\sum_{\sigma',\gamma} |K^{ee}_{\sigma\sigma'\gamma}(\mathbf{k}_1, \mathbf{k}_2)|^2$ in terms of $\mathbf{u}_{1,2}$ and $w_{1,2}$.

    b. Give the limit of the integrand at $q \to \infty$.

    c. Give the limit of the integrand at $q \to 0$.

    d. Perform the integration over the direction of vector $\mathbf{q}$ in both cases.

Solutions

1. *Schrödinger–Pauli equation.*

    a. The two coupled equations read

$$\boldsymbol{\sigma} \cdot (\hat{\mathbf{p}} - e\mathbf{A})\psi_B = \frac{1}{c}(E - e\varphi - mc^2)\psi_A,$$

$$-\boldsymbol{\sigma} \cdot (\hat{\mathbf{p}} - e\mathbf{A})\psi_A = -\frac{1}{c}(E - e\varphi + mc^2)\psi_B.$$

Eliminating ψ_B results in

$$\boldsymbol{\sigma} \cdot (\hat{\mathbf{p}} - e\mathbf{A}) \frac{c^2}{E - e\varphi + mc^2} \boldsymbol{\sigma} \cdot (\hat{\mathbf{p}} - e\mathbf{A})\psi_A = (E - e\varphi - mc^2)\psi_A.$$

    b. The energy E_{nr} and $e\varphi$ are both assumed to be of the order of $\sim mv^2$, and thus smaller than mc^2 by a factor $(v/c)^2$. If we expand in this small parameter, we find

$$\frac{c^2}{E - e\varphi + mc^2} = \frac{1}{2m} \frac{2mc^2}{2mc^2 + E_{nr} - e\varphi} \approx \frac{1}{2m}\left(1 - \frac{E_{nr} - e\varphi}{2mc^2}\right).$$

    c. To zeroth order the equation thus reads

$$\frac{1}{2m}\big[\boldsymbol{\sigma} \cdot (\hat{\mathbf{p}} - e\mathbf{A})\big]\big[\boldsymbol{\sigma} \cdot (\hat{\mathbf{p}} - e\mathbf{A})\big]\psi_A = (E_{nr} - e\varphi)\psi_A.$$

Applying the rule given in the hint yields

$$\left\{\frac{1}{2m}(\hat{\mathbf{p}} - e\mathbf{A})^2 + e\varphi + i\boldsymbol{\sigma} \cdot (-i\hbar\boldsymbol{\nabla} - e\mathbf{A}) \times (-i\hbar\boldsymbol{\nabla} - e\mathbf{A})\right\}\psi_A = E_{nr}\psi_A.$$

The cross product can then be simplified, using (i) $\mathbf{A} \times \mathbf{A} = 0$ and (ii) $\boldsymbol{\nabla} \times \mathbf{A} = (\boldsymbol{\nabla} \times \mathbf{A}) - \mathbf{A} \times \boldsymbol{\nabla}$, where the differentiation operator inside the parentheses works only on the vector potential whereas all others work on everything to the right of it. Substituting $\boldsymbol{\nabla} \times \mathbf{A} = \mathbf{B}$ in the resulting expression yields the Schrödinger–Pauli equation.

Index

Printed in the United States
By Bookmasters